Emerging Topics
in Computer Vision

IMSC Press Multimedia Series

ANDREW TESCHER, Series Editor, *Compression Science Corporation*

Advisory Editors

LEONARDO CHIARIGLIONE, *CSELT*
TARIQ S. DURRANI, *University of Strathclyde*
JEFF GRALNICK, *E-splosion Consulting, LLC*
CHRYSOSTOMOS L. "MAX" NIKIAS, *University of Southern California*
ADAM C. POWELL III, *The Freedom Forum*

▶ Emerging Topics in Computer Vision
   Edited by Gérard Medioni and Sing Bing Kang

▶ Streaming Media Server Design
   *Ali E. Dashti, Seon Ho Kim, Cyrus Shahabi,
   and Roger Zimmermann*

▶ Touch in Virtual Environments:
   Haptics and the Design of Interactive Systems
   *Edited by Margaret L. McLaughlin,
   João P. Hespanha, and Gaurav S. Sukhatme*

▶ Text to Speech Synthesis: New Paradigms and Advances
   Edited by Shrikanth Narayanan and Abeer Alwan

▶ The MPEG-4 Book
   Edited by Fernando Pereira and Touradj Ebrahimi

▶ iTV Handbook: Technologies and Standards
   Edward M. Schwalb

▶ Quality of Service for Internet Multimedia
   Jitae Shin, Daniel C. Lee, and C.-C. Jay Kuo

▶ Multimedia Fundamentals:
   Media Coding and Content Processing
   Ralf Steinmetz and Klara Nahrstedt

▶ Intelligent Systems for Video Analysis and Access Over
   the Internet
   Wensheng Zhou and C.-C. Jay Kuo

The Integrated Media Systems Center (IMSC), a National Science Foundation Engineering Research Center in the University of Southern California's School of Engineering, is a preeminent multimedia and Internet research center. IMSC seeks to develop integrated media systems that dramatically transform the way we work, communicate, learn, teach, and entertain. In an integrated media system, advanced media technologies combine, deliver, and transform information in the form of images, video, audio, animation, graphics, text, and haptics (touch-related technologies). IMSC Press, in partnership with Prentice Hall, publishes cutting-edge research on multimedia and Internet topics. IMSC Press is part of IMSC's educational outreach program.

Emerging Topics
in Computer Vision

Gérard Medioni
Sing Bing Kang

PRENTICE
HALL
PTR

PRENTICE HALL
PROFESSIONAL TECHNICAL REFERENCE
UPPER SADDLE RIVER, NJ 07458
WWW.PHPTR.COM

Library of Congress Cataloging-in-Publication Data

Emerging Topics in Computer Vision / edited by Sing Bing Kang and Gérard Medioni
   p. cm.
   Includes bibliographical references and index.
   ISBN 0-13-101366-1 (alk. paper)
   1. Computer vision. I. Kang, Sing Bing. II. Medioni, Gérard.

TA1634.E52 2004
006.3'7--dc22

2004044631

©2005 Pearson Education, Inc.
Publishing as Prentice Hall Professional Technical Reference
Upper Saddle River, New Jersey 07458

Prentice Hall PTR offers excellent discounts on this book when ordered in quantity for bulk purchases or special sales. For more information, please contact:
   U.S. Corporate and Government Sales, 1-800-382-3419,
   corpsales@pearsontechgroup.com.
   For sales outside of the U.S., please contact:
   International Sales, 1-317-581-3793, international@pearsontechgroup.com.

Company and product names mentioned herein are trademarks or registered trademarks of their respective owners.

Printed in the United States of America
10 9 8 7 6 5 4 3 2 1
First Printing, July 2004

ISBN 0-13-101366-1

Pearson Education LTD.
Pearson Education Australia PTY, Limited
Pearson Education South Asia Pte. Ltd.
Pearson Education Asia Ltd.
Pearson Education Canada, Ltd.
Pearson Educación de Mexico, S.A. de C.V.
Pearson Education—Japan
Pearson Malaysia SDN BHD

Contents

PREFACE

One of the major changes instituted at the 2001 Conference on Computer Vision and Pattern Recognition (CVPR) in Kauai, Hawaii, was the replacement of the traditional tutorial sessions with a set of short courses. The topics of these short courses were carefully chosen to reflect the diversity in computer vision and represent very promising areas. The response was a very pleasant surprise, with sometimes more than 200 people attending a single short course. This overwhelming response was the inspiration for this book.

There are three parts in this book. The first part covers some of the more fundamental aspects of computer vision, the second describes a few interesting applications, and the third details specific approaches to facilitate programming for computer vision. This book is not intended to be a comprehensive coverage of computer vision; it can, however, be used as a complement to most computer vision textbooks.

We would like to thank all the contributors for all their hard work, and Bernard Goodwin for his support and enthusiasm for our book project.

Gérard Medioni, University of Southern California
Sing Bing Kang, Microsoft Research
June, 2004

CONTRIBUTORS

Gary Bradski
Intel Labs, Machine Learning Group
SC12-303
2200 Mission College Blvd.
Santa Clara, CA 95052-8119
USA
www.intel.com/research/mrl/research/opencv
www.intel.com/research/mrl/research/media-visual.htm

Paul Debevec
USC Institute for Creative Technologies
13274 Fiji Way, 5th Floor
Marina del Rey, CA 90292
USA
www.debevec.org/

Alexandre R. J. François
PHE-222 MC-0273
Institute for Robotics and Intelligent Systems
University of Southern California
Los Angeles, CA 90089-0273
USA
afrancoi@usc.edu
iris.usc.edu/~afrancoi

Theo Gevers
University of Amsterdam
Kruislaan 403
1098 SJ Amsterdam
The Netherlands
gevers@science.uva.nl
carol.science.uva.nl/~gevers/

Anders Heyden
Centre for Mathematical Sciences
Lund University
Box 118
SE-221 00 Lund
Sweden
heyden@maths.lth.se
www.maths.lth.se/matematiklth/personal/andersp/

Mathias Kölsch
Computer Science Department
University of California
Santa Barbara, CA 93106
USA
matz@cs.ucsb.edu
www.cs.ucsb.edu/~matz

Stan Z. Li
Microsoft Research Asia
5/F, Beijing Sigma Center
No. 49, Zhichun Road, Hai Dian District
Beijing, China 100080
szli@microsoft.com
www.research.microsoft.com/~szli

Juwei Lu
Bell Canada Multimedia Lab
University of Toronto
Bahen Centre for Information Technology
Room 4154, 40 St George Str.
Toronto, ON, M5S 3G4
Canada

juwei@dsp.utoronto.ca
www.dsp.utoronto.ca/∼juwei/

Gerard Medioni
SAL 300, MC-0781
Computer Science Department
University of Southern California
Los Angeles, CA 90089-0781
USA
medioni@iris.usc.edu
iris.usc.edu/home/iris/medioni/User.html

Peter Meer
Electrical and Computer Engineering Department
Rutgers University
94 Brett Road
Piscataway, NJ 08854-8058
USA
meer@caip.rutgers.edu
www.caip.rutgers.edu/∼meer

Philippos Mordohai
PHE 204, MC-0273
3737 Watt Way
Los Angeles, CA 90089-0273
USA
mordohai@usc.edu
iris.usc.edu/home/iris/mordohai/User.html

Marc Pollefeys
Department of Computer Science
University of North Carolina
Sitterson Hall, CB#3175
Chapel Hill, NC 27599-3175
USA
marc@cs.unc.edu
www.cs.unc.edu/∼marc/

Doug Roble
Digital Domain
300 Rose Av

Venice, CA 90291
USA
www.d2.com

Arnold Smeulders
ISIS Group, University of Amsterdam
Kruislaan 403
1098SJ Amsterdam
The Netherlands
smeulder@science.uva.nl
www.science.uva.nl/isis/

Matthew Turk
Computer Science Department
University of California
Santa Barbara, CA 93106
USA
mturk@cs.ucsb.edu
www.cs.ucsb.edu/~mturk

Zhengyou Zhang
Microsoft Corporation
One Microsoft Way
Redmond, WA 98052
USA
zhang@microsoft.com
www.research.microsoft.com/~zhang/

Chapter 1

INTRODUCTION

The topics in this book were handpicked to showcase the developments we consider to be exciting and promising in computer vision. They are a mix of more well-known and traditional topics (such as camera calibration, multiview geometry, and face detection) and newer ones (such as vision for special effects and tensor voting framework). All have the common denominator of either demonstrated longevity or potential for endurance in computer vision.

The book is designed to accompany computer vision textbooks. Each chapter is self-contained and is written by well-known authorities in the area. The book is organized into three parts, covering various fundamentals, applications, and programming aspects of computer vision.

Part I, "Fundamentals in Computer Vision," consists of four chapters. Two of the chapters deal with the more conventional but still popular areas: camera calibration and multiview geometry. They deal with the most fundamental operations associated with vision. The chapter on robust estimation techniques will be very useful for researchers and practitioners of computer vision alike. There is also a chapter on a more recently developed tool (the tensor voting framework) that can be customized for a variety of problems.

Part II, "Applications in Computer Vision," covers two more recent applications (image-based lighting and vision for visual effects) and three in more conventional areas (image search engines, face detection and recognition, and perceptual interfaces).

One of the more overlooked areas in computer vision is the programming aspect of computer vision. While generic commercial packages can be used, there exist popular libraries or packages that are specifically geared for computer vision. Part III, "Programming for Computer Vision," describes two different approaches to facilitate programming for computer vision.

PART I
FUNDAMENTALS IN
COMPUTER VISION

It is fitting that we start with some of the more fundamental concepts in computer vision. The range of topics covered in Part I is wide: camera calibration, structure from motion, dense stereo, 3D modeling, robust techniques for model fitting, and a more recently developed concept called tensor voting.

In Chapter 2, Zhang reviews the different techniques for calibrating a camera. More specifically, he describes calibration techniques that use 3D reference objects, 2D planes, and 1D lines, as well as self-calibration techniques.

One of the more popular (and difficult) areas in computer vision is stereo. Heyden and Pollefeys describe how camera motion and scene structure can be reliably extracted from image sequences in Chapter 3. Once this is accomplished, dense depth distributions can be extracted for 3D surface reconstruction and image-based rendering applications.

A basic task in computer vision is hypothesizing models (e.g., 2D shapes) and using input data (typically image data) to corroborate and fit the models. In practice, however, robust techniques for model fitting must be used to handle input noise. In Chapter 4, Meer describes various robust regression techniques such as M-estimators, RANSAC, and Hough transform. He also covers the mean shift algorithm for the location estimation problem.

The claim by Medioni and his colleagues that computer vision problems can be addressed within a Gestalt framework is the basis of their work on tensor voting. In Chapter 5, Medioni and Mordohai provide an introduction to the concept of tensor voting, which is a form of binning according to proximity to ideal primitives such as edges and points. They show how this scheme can be applied to a variety of applications, such as curve and surface extraction from noisy 2D and 3D points (respectively), stereo matching, and motion-based grouping.

Chapter 2

CAMERA CALIBRATION

Zhengyou Zhang

Camera calibration is a necessary step in 3D computer vision in order to extract metric information from 2D images. It has been studied extensively in computer vision and photogrammetry, and even recently, new techniques have been proposed. In this chapter, we review the techniques proposed in the literature including those using 3D apparatus (two or three planes orthogonal to each other, or a plane undergoing a pure translation, etc.), 2D objects (planar patterns undergoing unknown motions), 1D objects (wand with dots), and unknown scene points in the environment (self-calibration). The focus is on presenting these techniques within a consistent framework.

2.1 Introduction

Much work in camera calibration has been done in the photogrammetry community (see [3, 6] to cite a few) and more recently in computer vision ([12, 11, 33, 10, 37, 35, 22, 9] to cite a few). According to the dimension of the calibration objects, we can classify those techniques roughly into four categories.

3D reference object-based calibration. Camera calibration is performed by observing a calibration object whose geometry in 3D space is known with very good precision. Calibration can be done very efficiently [8]. The calibration object usually consists of two or three planes orthogonal to each other. Sometimes, a plane undergoing a precisely known translation is also used [33], which equivalently provides 3D reference points. This approach requires an expensive calibration apparatus and an elaborate setup.

2D plane-based calibration. Techniques in this category require observation of a planar pattern shown at a few different orientations [42, 31]. Different from Tsai's technique [33], the knowledge of the plane motion is not necessary. Because almost anyone can make such a calibration pattern by himself or herself, the setup is easier for camera calibration.

1D line-based calibration. Calibration objects used in this category are composed of a set of collinear points [44]. As will be shown, a camera can be calibrated by observing a moving line around a fixed point, such as a string of balls hanging from the ceiling.

Self-calibration. Techniques in this category do not use any calibration object and can be considered a 0D approach because only image point correspondences are required. Just by moving a camera in a static scene, the rigidity of the scene provides, in general, two constraints [22, 21] on the cameras' internal parameters from one camera displacement by using image information alone. Therefore, if images are taken by the same camera with fixed internal parameters, correspondences between three images are sufficient to recover both the internal and external parameters, which allow us to reconstruct 3D structure up to a similarity [20, 17]. Although no calibration objects are necessary, a large number of parameters need to be estimated, resulting in a much harder mathematical problem.

Other techniques exist: vanishing points for orthogonal directions [4, 19] and calibration from pure rotation [16, 30].

No single calibration technique is the best for all conditions. It depends on the situation a user needs to deal with. Following are a few recommendations:

– Calibration with apparatus versus self-calibration. Whenever possible, if we can precalibrate a camera, we should do it with a calibration apparatus. Self-calibration cannot usually achieve an accuracy comparable to that of precalibration, because self-calibration needs to estimate a large number of parameters, resulting in a much harder mathematical problem. When precalibration is impossible (e.g., scene reconstruction from an old movie), self-calibration is the only choice.

– Partial versus full self-calibration. Partial self-calibration refers to the case where only a subset of camera intrinsic parameters are to be calibrated. Along the same line as the previous recommendation, whenever possible, partial self-calibration is preferred because the number of parameters to be estimated is smaller. Take an example of 3D reconstruction with a camera with variable focal length. It is preferable to precalibrate the pixel aspect ratio and the pixel skewness.

– Calibration with 3D versus 2D apparatus. Highest accuracy can usually be obtained by using a 3D apparatus, so it should be used when accuracy is indispensable and when it is affordable to make and use a 3D apparatus. According to feedback from computer vision researchers and practitioners around the world in the last couple of years, calibration with a 2D apparatus seems to be the best choice in most situations because of its ease of use and good accuracy.

– Calibration with 1D apparatus. This technique is relatively new, and it is hard for the moment to predict how popular it will be. It, however, should be useful especially for calibration of a camera network. To calibrate the relative geometry between multiple cameras as well as their intrinsic parameters, it is necessary for all involved cameras to simultaneously observe a number of points. It is hardly possible to achieve this with 3D or 2D calibration apparatus[1] if one camera is mounted in the front of a room while another in the back. This is not a problem for 1D objects. We can, for example, use a string of balls hanging from the ceiling.

This chapter is organized as follows. Section 2.2 describes the camera model and introduces the concept of the absolute conic, which is important for camera calibration. Section 2.3 presents the calibration techniques using a 3D apparatus. Section 2.4 describes a calibration technique by observing a freely moving planar pattern (2D object). Its extension for stereo calibration is also addressed. Section 2.5 describes a relatively new technique that uses a set of collinear points (1D object). Section 2.6 briefly introduces the self-calibration approach and provides references for further reading. Section 2.7 concludes the chapter with a discussion on recent work in this area.

2.2   Notation and Problem Statement

We start with the notation used in this chapter.

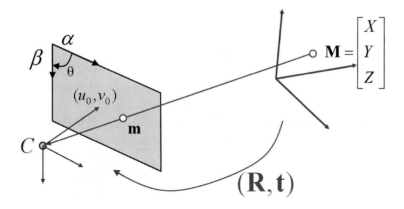

Figure 2.1. Pinhole camera model.

2.2.1   Pinhole camera model

A 2D point is denoted by $\mathbf{m} = [u, v]^T$. A 3D point is denoted by $\mathtt{M} = [X, Y, Z]^T$. We use $\widetilde{\mathbf{x}}$ to denote the augmented vector by adding 1 as the last element: $\widetilde{\mathbf{m}} = [u, v, 1]^T$ and $\widetilde{\mathtt{M}} = [X, Y, Z, 1]^T$. A camera is modeled by the usual pinhole (see Figure 2.1): The image of a 3D point $\mathtt{M}$, denoted by $\mathbf{m}$, is formed by an optical ray from $\mathtt{M}$ passing through the optical center C and intersecting the image plane. The three points $\mathtt{M}$, $\mathbf{m}$, and C are collinear. In Figure 2.1, for illustration purpose, the image plane is positioned between the scene point and the optical center, which is mathematically equivalent to the physical setup under which the image plane is in the other side with respect to the optical center. The relationship between the 3D point $\mathtt{M}$ and its image projection $\mathbf{m}$ is given by

$$s\widetilde{\mathbf{m}} = \underbrace{\mathbf{A}\begin{bmatrix}\mathbf{R} & \mathbf{t}\end{bmatrix}}_{\mathbf{P}}\widetilde{\mathtt{M}} \equiv \mathbf{P}\widetilde{\mathtt{M}}, \qquad (2.1)$$

$$\text{with } \mathbf{A} = \begin{bmatrix} \alpha & \gamma & u_0 \\ 0 & \beta & v_0 \\ 0 & 0 & 1 \end{bmatrix} \qquad (2.2)$$

$$\text{and} \quad \mathbf{P} = \mathbf{A}\begin{bmatrix}\mathbf{R} & \mathbf{t}\end{bmatrix}, \qquad (2.3)$$

where s is an arbitrary scale factor, $(\mathbf{R}, \mathbf{t})$, called the extrinsic parameters, is the rotation and translation that relates the world coordinate system to

---

[1]An exception is when those apparatus are made transparent; then the cost is much higher.

the camera coordinate system, and $\mathbf{A}$ is called the camera intrinsic matrix, with (u_0, v_0) the coordinates of the principal point, α and β the scale factors in image u and v axes, and γ the parameter describing the skew of the two image axes. The 3×4 matrix $\mathbf{P}$ is called the camera projection matrix, which mixes both intrinsic and extrinsic parameters. In Figure 2.1, the angle between the two image axes is denoted by θ, and we have $\gamma = \alpha \cot \theta$. If the pixels are rectangular, then $\theta = 90°$ and $\gamma = 0$.

The task of camera calibration is to determine the parameters of the transformation between an object in 3D space and the 2D image observed by the camera from visual information (images). The transformation includes:

- Extrinsic parameters (sometimes called external parameters): orientation (rotation) and location (translation) of the camera $(\mathbf{R}, \mathbf{t})$;

- Intrinsic parameters (sometimes called internal parameters): characteristics of the camera $(\alpha, \beta, \gamma, u_0, v_0)$.

The rotation matrix, although consisting of nine elements, has only three degrees of freedom. The translation vector $\mathbf{t}$ obviously has 3 parameters. Therefore, there are six extrinsic parameters and five intrinsic parameters, totaling 11 parameters.

We use the abbreviation $\mathbf{A}^{-T}$ for $(\mathbf{A}^{-1})^T$ or $(\mathbf{A}^T)^{-1}$.

2.2.2   Absolute conic

Now let us introduce the concept of the absolute conic. For more details, refer to [7, 15].

A point $\mathbf{x}$ in 3D space has projective coordinates $\widetilde{\mathbf{x}} = [x_1, x_2, x_3, x_4]^T$. The equation of the plane at infinity, Π_∞, is $x_4 = 0$. The absolute conic Ω is defined by a set of points satisfying the equation

$$
\begin{aligned}
x_1^2 + x_2^2 + x_3^2 &= 0 \\
x_4 &= 0 .
\end{aligned}
\tag{2.4}
$$

Let $\mathbf{x}_\infty = [x_1, x_2, x_3]^T$ be a point on the absolute conic (see Figure 2.2). By definition, we have $\mathbf{x}_\infty^T \mathbf{x}_\infty = 0$. We also have $\widetilde{\mathbf{x}}_\infty = [x_1, x_2, x_3, 0]^T$ and $\widetilde{\mathbf{x}}_\infty^T \widetilde{\mathbf{x}}_\infty = 0$. This can be interpreted as a conic of purely imaginary points on Π_∞. Indeed, let $x = x_1/x_3$ and $y = x_2/x_3$ be a point on the conic, then $x^2 + y^2 = -1$, which is an imaginary circle of radius $\sqrt{-1}$.

An important property of the absolute conic is its invariance to any rigid transformation. Let the rigid transformation be $\mathbf{H} = \begin{bmatrix} \mathbf{R} & \mathbf{t} \\ \mathbf{0} & 1 \end{bmatrix}$. Let $\mathbf{x}_\infty$ be

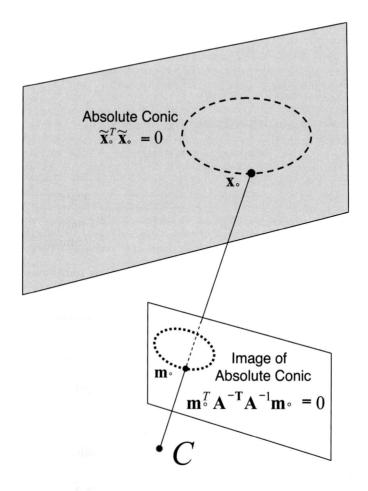

Figure 2.2. Absolute conic and its image.

a point on Ω. By definition, its projective coordinates: $\widetilde{\mathbf{x}}_\infty = \begin{bmatrix} \mathbf{x}_\infty \\ 0 \end{bmatrix}$ with $\mathbf{x}_\infty^T \mathbf{x}_\infty = 0$. The point after the rigid transformation is denoted by $\mathbf{x}'_\infty$, and

$$\widetilde{\mathbf{x}}'_\infty = \mathbf{H}\widetilde{\mathbf{x}}_\infty = \begin{bmatrix} \mathbf{R}\mathbf{x}_\infty \\ 0 \end{bmatrix}.$$

Thus, $\mathbf{x}'_\infty$ is also on the plane at infinity. Furthermore, $\mathbf{x}'_\infty$ is *on the same* Ω because

$$\mathbf{x}'^T_\infty \mathbf{x}'_\infty = (\mathbf{R}\mathbf{x}_\infty)^T (\mathbf{R}\mathbf{x}_\infty) = \mathbf{x}_\infty^T (\mathbf{R}^T \mathbf{R})\mathbf{x}_\infty = 0 \ .$$

The image of the absolute conic, denoted by ω, is also an imaginary conic and is determined only by the intrinsic parameters of the camera. This can be seen as follows. Consider the projection of a point $\mathbf{x}_\infty$ on Ω, denoted by $\mathbf{m}_\infty$, which is given by

$$\widetilde{\mathbf{m}}_\infty = s\mathbf{A}[\mathbf{R} \quad \mathbf{t}] \begin{bmatrix} \mathbf{x}_\infty \\ 0 \end{bmatrix} = s\mathbf{A}\mathbf{R}\mathbf{x}_\infty \ .$$

It follows that

$$\widetilde{\mathbf{m}}^T \mathbf{A}^{-T} \mathbf{A}^{-1} \widetilde{\mathbf{m}} = s^2 \mathbf{x}_\infty^T \mathbf{R}^T \mathbf{R} \mathbf{x}_\infty = s^2 \mathbf{x}_\infty^T \mathbf{x}_\infty = 0 \ .$$

Therefore, the image of the absolute conic is an imaginary conic and is defined by $\mathbf{A}^{-T}\mathbf{A}^{-1}$. It does not depend on the extrinsic parameters of the camera.

If we can determine the image of the absolute conic, then we can solve the camera's intrinsic parameters, and the calibration is solved.

We show several ways in this chapter to determine ω, the image of the absolute conic.

2.3   Camera Calibration with 3D Objects

The traditional way to calibrate a camera is to use a 3D reference object such as those shown in Figure 2.3. In Figure 2.3a, the calibration apparatus used at INRIA [8] is shown, which consists of two orthogonal planes; a checker pattern is printed on each. A 3D coordinate system is attached to this apparatus, and the coordinates of the checker corners are known very accurately in this coordinate system. A similar calibration apparatus is a cube with a checker pattern painted in each face, so three faces are generally visible to the camera. Figure 2.3b illustrates the device used in Tsai's technique [33], which uses only one plane with checker pattern, but the plane needs to be displaced at least once with known motion. This is equivalent to knowing the 3D coordinates of the checker corners.

A popular technique in this category consists of four steps [8]:

1. Detect the corners of the checker pattern in each image;

2. Estimate the camera projection matrix $\mathbf{P}$ using linear least squares;

3. Recover intrinsic and extrinsic parameters $\mathbf{A}$, $\mathbf{R}$, and $\mathbf{t}$ from $\mathbf{P}$;

4. Refine $\mathbf{A}$, $\mathbf{R}$, and $\mathbf{t}$ through a nonlinear optimization.

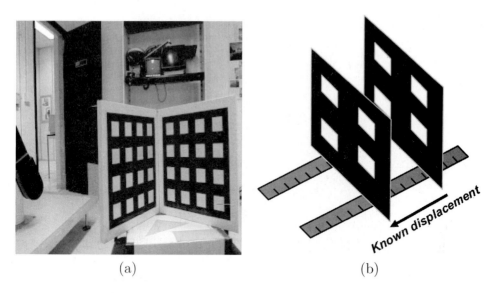

(a)   (b)

Figure 2.3. 3D apparatus for calibrating cameras.

It is also possible to first refine **P** through a nonlinear optimization, and then determine **A**, **R**, and **t** from the refined **P**.

It is worth noting that using corners is not the only possibility. We can avoid corner detection by working directly in the image. In [25], calibration is realized by maximizing the gradients around a set of control points that define the calibration object. Figure 2.4 illustrates the control points used in that work.

2.3.1   Feature extraction

If we use a generic corner detector, such as Harris corner detector, to detect the corners in the checker pattern image, the result is usually not good because the detector corners have poor accuracy (about one pixel). A better solution is to leverage the known pattern structure by first estimating a line for each side of the square and then computing the corners by intersecting the fitted lines. There are two common techniques to estimate the lines. The first is to first detect edges, and then fit a line to the edges on each side of the square. The second technique is to directly fit a line to each side of a square in the image such that the gradient on the line is maximized. One possibility is to represent the line by an elongated Gaussian and estimate the parameters of the elongated Gaussian by maximizing the total gradient covered by the Gaussian. We should note that if the lens distortion is not

Figure 2.4. Control points used in a gradient-based calibration technique.

severe, a better solution is to fit just one single line to all the collinear sides. This allows a much more accurate estimation of the position of the checker corners.

2.3.2   Linear estimation of the camera projection matrix

Once we extract the corner points in the image, we can easily establish their correspondences with the points in the 3D space because of knowledge of the patterns. Based on the projection equation (2.1), we can now estimate the camera parameters. However, the problem is quite nonlinear if we try to estimate directly $\mathbf{A}$, $\mathbf{R}$, and $\mathbf{t}$. If, on the other hand, we estimate the camera projection matrix $\mathbf{P}$, a linear solution is possible.

Given each 2D-3D correspondence $\mathbf{m}_i = (u_i, v_i) \leftrightarrow \mathsf{M}_i = (X_i, Y_i, Z_i)$, we can write down two equations based on (2.1):

$$\underbrace{\begin{bmatrix} X_i & Y_i & Z_i & 1 & 0 & 0 & 0 & 0 & u_i X_i & u_i Y_i & u_i Z_i & u_i \\ 0 & 0 & 0 & 0 & X_i & Y_i & Z_i & 1 & v_i X_i & v_i Y_i & v_i Z_i & v_i \end{bmatrix}}_{\mathbf{G}_i} \mathbf{p} = \mathbf{0},$$

where $\mathbf{p} = [p_{11}, p_{12}, \ldots, p_{34}]^T$ and $\mathbf{0} = [0, 0]^T$.

For n point matches, we can stack all equations together:

$$\mathbf{G}\mathbf{p} = \mathbf{0} \qquad \text{with } \mathbf{G} = [\mathbf{G}_1^T, \ldots, \mathbf{G}_n^T]^T \ .$$

Matrix $\mathbf{G}$ is a $2n \times 12$ matrix. The projection matrix can now be solved by

$$\min_{\mathbf{p}} \|\mathbf{G}\mathbf{p}\|^2 \quad \text{subject to } \|\mathbf{p}\| = 1.$$

The solution is the eigenvector of $\mathbf{G}^T\mathbf{G}$ associated with the smallest eigenvalue.

In the above, in order to avoid the trivial solution $\mathbf{p} = \mathbf{0}$ and considering that $\mathbf{p}$ is defined up to a scale factor, we have set $\|\mathbf{p}\| = 1$. Other normalizations are possible. In [1], $p_{34} = 1$, which, however, introduce a singularity when the correct value of p_{34} is close to zero. In [10], the constraint $p_{31}^2 + p_{32}^2 + p_{33}^2 = 1$ was used, which is singularity free.

Anyway, the above linear technique minimizes an algebraic distance and yields a *biased* estimation when data are noisy. We present an unbiased solution later.

2.3.3   Recover intrinsic and extrinsic parameters from P

Once the camera projection matrix $\mathbf{P}$ is known, we can uniquely recover the intrinsic and extrinsic parameters of the camera. Let us denote the first 3×3 submatrix of $\mathbf{P}$ by $\mathbf{B}$ and the last column of $\mathbf{P}$ by $\mathbf{b}$, that is, $\mathbf{P} \equiv [\mathbf{B} \quad \mathbf{b}]$. Since $\mathbf{P} = \mathbf{A}[\mathbf{R} \quad \mathbf{t}]$, we have

$$\mathbf{B} = \mathbf{A}\mathbf{R}, \tag{2.5}$$

$$\mathbf{b} = \mathbf{A}\mathbf{t}. \tag{2.6}$$

From (2.5), we have

$$\mathbf{K} \equiv \mathbf{B}\mathbf{B}^T = \mathbf{A}\mathbf{A}^T = \begin{bmatrix} \underbrace{\alpha^2 + \gamma^2 + u_0^2}_{k_u} & \underbrace{u_0\, v_0 + c\,\beta}_{k_c} & u_0 \\ \underbrace{u_0\, v_0 + c\,\alpha}_{k_c} & \underbrace{\alpha_v^2 + v_0^2}_{k_v} & v_0 \\ u_0 & v_0 & 1 \end{bmatrix}.$$

Because $\mathbf{P}$ is defined up to a scale factor, the last element of $\mathbf{K} = \mathbf{B}\mathbf{B}^T$ is usually not equal to 1, so we have to *normalize* it such that $\mathbf{K}_{33}$(the last element) =

1. After that, we immediately obtain

$$u_0 = \mathbf{K}_{13}, \tag{2.7}$$

$$v_0 = \mathbf{K}_{23}, \tag{2.8}$$

$$\beta = \sqrt{k_v - v_0^2}, \tag{2.9}$$

$$\gamma = \frac{k_c - u_0\,v_0}{\beta}, \tag{2.10}$$

$$\alpha = \sqrt{k_u - u_0^2 - \gamma^2}. \tag{2.11}$$

The solution is unambiguous because $\alpha > 0$ and $\beta > 0$.

Once the intrinsic parameters, or equivalently matrix $\mathbf{A}$, are known, the extrinsic parameters can be determined from (2.5) and (2.6) as

$$\mathbf{R} = \mathbf{A}^{-1}\mathbf{B}, \tag{2.12}$$

$$\mathbf{t} = \mathbf{A}^{-1}\mathbf{b}\,. \tag{2.13}$$

2.3.4   Refine calibration parameters
through a nonlinear optimization

The above solution is obtained through minimizing an algebraic distance, which is not physically meaningful. We can refine it through maximum likelihood inference.

We are given n 2D–3D correspondences $\mathbf{m}_i = (u_i, v_i) \leftrightarrow \mathtt{M}_i = (X_i, Y_i, Z_i)$. Assume that the image points are corrupted by independent and identically distributed noise. The maximum likelihood estimate can be obtained by minimizing the distances between the image points and their predicted positions:

$$\min_{\mathbf{P}} \sum_i \|\mathbf{m}_i - \boldsymbol{\phi}(\mathbf{P}, \mathtt{M}_i)\|^2, \tag{2.14}$$

where $\boldsymbol{\phi}(\mathbf{P}, \mathtt{M}_i)$ is the projection of $\mathtt{M}_i$ onto the image according to (2.1).

This is a nonlinear minimization problem, which can be solved with the Levenberg-Marquardt algorithm, as implemented in `Minpack` [23]. It requires an initial guess of $\mathbf{P}$, which can be obtained using the linear technique described earlier. Note that since $\mathbf{P}$ is defined up to a scale factor, we can set the element having the largest initial value as 1 during the minimization.

Alternatively, instead of estimating $\mathbf{P}$ as in (2.14), we can directly estimate the intrinsic and extrinsic parameters, $\mathbf{A}$, $\mathbf{R}$, and $\mathbf{t}$, using the same criterion. The rotation matrix can be parameterized with three variables such as Euler angles or scaled rotation vector.

2.3.5 Lens distortion

Up to this point, we use the pinhole model to describe a camera. It says that the point in 3D space, its corresponding point in image, and the camera's optical center are collinear. This linear projective equation is sometimes not sufficient, especially for low-end cameras (such as webcams) or wide-angle cameras; lens distortion has to be considered.

According to [33], there are four steps in camera projection, including lens distortion:

Step 1: *Rigid transformation* from world coordinate system (X_w, Y_w, Z_w) to camera one (X, Y, Z):

$$[X \quad Y \quad Z]^T = \mathbf{R} \, [X_w \quad Y_w \quad Z_w]^T + \mathbf{t}.$$

Step 2: *Perspective projection* from 3D camera coordinates (X, Y, Z) to *ideal* image coordinates (x, y) under pinhole camera model:

$$x = f\frac{X}{Z} \,, \qquad y = f\frac{Y}{Z},$$

where f is the effective focal length.

Step 3: *Lens distortion*[2]:

$$\breve{x} = x + \delta_x \,, \qquad \breve{y} = y + \delta_y,$$

where $(\breve{x}, \breve{y})$ are the *distorted* or *true* image coordinates, and (δ_x, δ_y) are distortions applied to (x, y).

Step 4: *Affine transformation* from real image coordinates $(\breve{x}, \breve{y})$ to *frame buffer* (pixel) image coordinates $(\breve{u}, \breve{v})$:

$$\breve{u} = d_x^{-1}\breve{x} + u_0 \,, \qquad \breve{v} = d_y^{-1}\breve{y} + v_0 \,,$$

where (u_0, v_0) are coordinates of the principal point, and d_x and d_y are distances between adjacent pixels in the horizontal and vertical directions respectively.

There are two types of distortions:

---

[2]The lens distortion described here is different from Tsai's treatment. Here, we go from ideal to real image coordinates, similar to [36].

Radial distortion: It is symmetric; ideal image points are distorted along radial directions from the distortion center. This is caused by imperfect lens shape.

Decentering distortion: This is usually caused by improper lens assembly; ideal image points are distorted in both radial and tangential directions.

Refer to [29, 3, 6, 37] for more details.

The distortion can be expressed as power series in radial distance $r = \sqrt{x^2 + y^2}$:

$$\delta_x = x(k_1 r^2 + k_2 r^4 + k_3 r^6 + \cdots) + [p_1(r^2 + 2x^2) + 2p_2 xy](1 + p_3 r^2 + \cdots) ,$$
$$\delta_y = y(k_1 r^2 + k_2 r^4 + k_3 r^6 + \cdots) + [2p_1 xy + p_2(r^2 + 2y^2)](1 + p_3 r^2 + \cdots) ,$$

where $k_1, k_2, \ldots$ are coefficients of radial distortion and $p_1, p_2, \ldots$ are coefficients of decentering distortion.

Based on the reports in the literature [3, 33, 36], it is likely that the distortion function is totally dominated by the radial components and especially dominated by the first term. It has also been found that any more elaborated modeling not only would not help (negligible when compared with sensor quantization), but also would cause numerical instability [33, 36].

Denote the ideal pixel image coordinates by $u = x/d_x$ and $v = y/d_y$. By combining Step 3 and Step 4 and using only the first two radial distortion terms, we obtain the following relationship between $(\breve{u}, \breve{v})$ and (u, v):

$$\breve{u} = u + (u - u_0)[k_1(x^2 + y^2) + k_2(x^2 + y^2)^2], \tag{2.15}$$
$$\breve{v} = v + (v - v_0)[k_1(x^2 + y^2) + k_2(x^2 + y^2)^2] . \tag{2.16}$$

Following the same reasoning as in (2.14), camera calibration including lens distortion can be performed by minimizing the distances between the image points and their predicted positions:

$$\min_{\mathbf{A}, \mathbf{R}, \mathbf{t}, k_1, k_2} \sum_i \|\mathbf{m}_i - \breve{\mathbf{m}}(\mathbf{A}, \mathbf{R}, \mathbf{t}, k_1, k_2, \mathbf{M}_i)\|^2, \tag{2.17}$$

where $\breve{\mathbf{m}}(\mathbf{A}, \mathbf{R}, \mathbf{t}, k_1, k_2, \mathbf{M}_i)$ is the projection of $\mathbf{M}_i$ onto the image according to (2.1), followed by distortion according to (2.15) and (2.16).

2.3.6 An example

Figure 2.5 displays an image of a 3D reference object taken by a camera to be calibrated at INRIA. Each square has four corners, and there are in total 128 points used for calibration.

Figure 2.5. Camera calibration with a 3D apparatus.

Without considering lens distortion, the estimated camera projection matrix is

$$\mathbf{P} = \begin{bmatrix} 7.025659e{-}01 & -2.861189e{-}02 & -5.377696e{-}01 & 6.241890e{+}01 \\ 2.077632e{-}01 & 1.265804e{+}00 & 1.591456e{-}01 & 1.075646e{+}01 \\ 4.634764e{-}04 & -5.282382e{-}05 & 4.255347e{-}04 & 1 \end{bmatrix}.$$

From $\mathbf{P}$, we can calculate the intrinsic parameters: $\alpha = 1380.12$, $\beta = 2032.57$, $\gamma \approx 0$, $u_0 = 246.52$, and $v_0 = 243.68$. So, the angle between the two image axes is $90°$, and the aspect ratio of the pixels is $\alpha/\beta = 0.679$. For the extrinsic parameters, the translation vector $\mathbf{t} = [-211.28, -106.06, 1583.75]^T$ (in mm); that is, the calibration object is about 1.5m away from the camera; the rotation axis is $[-0.08573, -0.99438, 0.0621]^T$ (i.e., almost vertical), and the rotation angle is $47.7°$.

Other notable works in this category include [27, 38, 36, 18].

2.4   Camera Calibration with 2D Objects:
Plane-Based Technique

In this section, we describe how a camera can be calibrated using a moving plane. We first examine the constraints on the camera's intrinsic parameters provided by observing a single plane.

2.4.1   Homography between the model plane and its image

Without loss of generality, we assume the model plane is on $Z = 0$ of the world coordinate system. Let's denote the i-th column of the rotation matrix $\mathbf{R}$ by $\mathbf{r}_i$. From (2.1), we have

$$s \begin{bmatrix} u \\ v \\ 1 \end{bmatrix} = \mathbf{A} \begin{bmatrix} \mathbf{r}_1 & \mathbf{r}_2 & \mathbf{r}_3 & \mathbf{t} \end{bmatrix} \begin{bmatrix} X \\ Y \\ 0 \\ 1 \end{bmatrix} = \mathbf{A} \begin{bmatrix} \mathbf{r}_1 & \mathbf{r}_2 & \mathbf{t} \end{bmatrix} \begin{bmatrix} X \\ Y \\ 1 \end{bmatrix} .$$

By abuse of notation, we still use M to denote a point on the model plane, but $\mathbf{M} = [X, Y]^T$, since Z is always equal to 0. In turn, $\widetilde{\mathbf{M}} = [X, Y, 1]^T$. Therefore, a model point M and its image $\mathbf{m}$ is related by a homography $\mathbf{H}$:

$$s\widetilde{\mathbf{m}} = \mathbf{H}\widetilde{\mathbf{M}} \qquad \text{with} \quad \mathbf{H} = \mathbf{A} \begin{bmatrix} \mathbf{r}_1 & \mathbf{r}_2 & \mathbf{t} \end{bmatrix} . \qquad (2.18)$$

As is clear, the 3×3 matrix $\mathbf{H}$ is defined up to a scale factor.

2.4.2   Constraints on the intrinsic parameters

Given an image of the model plane, a homography can be estimated (see the appendix in Section 2.8). Let's denote it by $\mathbf{H} = [\mathbf{h}_1 \quad \mathbf{h}_2 \quad \mathbf{h}_3]$. From (2.18), we have

$$[\mathbf{h}_1 \quad \mathbf{h}_2 \quad \mathbf{h}_3] = \lambda \mathbf{A} \begin{bmatrix} \mathbf{r}_1 & \mathbf{r}_2 & \mathbf{t} \end{bmatrix} ,$$

where λ is an arbitrary scalar. Using the knowledge that $\mathbf{r}_1$ and $\mathbf{r}_2$ are orthonormal, we have

$$\mathbf{h}_1^T \mathbf{A}^{-T} \mathbf{A}^{-1} \mathbf{h}_2 = 0, \qquad (2.19)$$

$$\mathbf{h}_1^T \mathbf{A}^{-T} \mathbf{A}^{-1} \mathbf{h}_1 = \mathbf{h}_2^T \mathbf{A}^{-T} \mathbf{A}^{-1} \mathbf{h}_2 . \qquad (2.20)$$

These are the two basic constraints on the intrinsic parameters, given one homography. Because a homography has eight degrees of freedom and there are six extrinsic parameters (three for rotation and three for translation), we can only obtain two constraints on the intrinsic parameters. Note that $\mathbf{A}^{-T}\mathbf{A}^{-1}$ actually describes the image of the absolute conic [20]. Section 2.4.3 provides a geometric interpretation.

2.4.3 Geometric interpretation

We are now relating (2.19) and (2.20) to the absolute conic [22, 20].

It is not difficult to verify that the model plane, under our convention, is described in the camera coordinate system by the following equation:

$$
\begin{bmatrix} \mathbf{r}_3 \\ \mathbf{r}_3^T \mathbf{t} \end{bmatrix}^T \begin{bmatrix} x \\ y \\ z \\ w \end{bmatrix} = 0 \; ,
$$

where $w = 0$ for points at infinity and $w = 1$ otherwise. This plane intersects the plane at infinity at a line, and we can easily see that $\begin{bmatrix} \mathbf{r}_1 \\ 0 \end{bmatrix}$ and $\begin{bmatrix} \mathbf{r}_2 \\ 0 \end{bmatrix}$ are two particular points on that line. Any point on it is a linear combination of these two points; that is,

$$
\mathbf{x}_\infty = a \begin{bmatrix} \mathbf{r}_1 \\ 0 \end{bmatrix} + b \begin{bmatrix} \mathbf{r}_2 \\ 0 \end{bmatrix} = \begin{bmatrix} a\mathbf{r}_1 + b\mathbf{r}_2 \\ 0 \end{bmatrix} \; .
$$

Now, let's compute the intersection of the above line with the absolute conic. By definition, the point $\mathbf{x}_\infty$, known as the *circular point* [26], satisfies $\mathbf{x}_\infty^T \mathbf{x}_\infty = 0$; that is, $(a\mathbf{r}_1 + b\mathbf{r}_2)^T(a\mathbf{r}_1 + b\mathbf{r}_2) = 0$, or $a^2 + b^2 = 0$. The solution is $b = \pm ai$, where $i^2 = -1$. That is, the two intersection points are

$$
\mathbf{x}_\infty = a \begin{bmatrix} \mathbf{r}_1 \pm i\mathbf{r}_2 \\ 0 \end{bmatrix} \; .
$$

The significance of this pair of complex conjugate points is that they are invariant to Euclidean transformations. Their projection in the image plane is given, up to a scale factor, by

$$
\widetilde{\mathbf{m}}_\infty = \mathbf{A}(\mathbf{r}_1 \pm i\mathbf{r}_2) = \mathbf{h}_1 \pm i\mathbf{h}_2 \; .
$$

Point $\widetilde{\mathbf{m}}_\infty$ is on the image of the absolute conic, described by $\mathbf{A}^{-T}\mathbf{A}^{-1}$ [20]. This gives

$$
(\mathbf{h}_1 \pm i\mathbf{h}_2)^T \mathbf{A}^{-T}\mathbf{A}^{-1}(\mathbf{h}_1 \pm i\mathbf{h}_2) = 0 \; .
$$

Requiring that both real and imaginary parts be zero yields (2.19) and (2.20).

2.4.4 Closed-form solution

We now provide the details on how to effectively solve the camera calibration problem. We start with an analytical solution. This initial estimation is followed by a nonlinear optimization technique based on the maximum likelihood criterion, to be described in Section 2.4.5.

Let

$$
\mathbf{B} = \mathbf{A}^{-T}\mathbf{A}^{-1} \equiv \begin{bmatrix} B_{11} & B_{12} & B_{13} \\ B_{12} & B_{22} & B_{23} \\ B_{13} & B_{23} & B_{33} \end{bmatrix},
\tag{2.21}
$$

$$
= \begin{bmatrix} \frac{1}{\alpha^2} & -\frac{\gamma}{\alpha^2\beta} & \frac{v_0\gamma - u_0\beta}{\alpha^2\beta} \\ -\frac{\gamma}{\alpha^2\beta} & \frac{\gamma^2}{\alpha^2\beta^2} + \frac{1}{\beta^2} & -\frac{\gamma(v_0\gamma - u_0\beta)}{\alpha^2\beta^2} - \frac{v_0}{\beta^2} \\ \frac{v_0\gamma - u_0\beta}{\alpha^2\beta} & -\frac{\gamma(v_0\gamma - u_0\beta)}{\alpha^2\beta^2} - \frac{v_0}{\beta^2} & \frac{(v_0\gamma - u_0\beta)^2}{\alpha^2\beta^2} + \frac{v_0^2}{\beta^2} + 1 \end{bmatrix}.
\tag{2.22}
$$

Note that $\mathbf{B}$ is symmetric, defined by a 6D vector

$$
\mathbf{b} = [B_{11}, B_{12}, B_{22}, B_{13}, B_{23}, B_{33}]^T .
\tag{2.23}
$$

Let the i-th column vector of $\mathbf{H}$ be $\mathbf{h}_i = [h_{i1}, h_{i2}, h_{i3}]^T$. Then, we have

$$
\mathbf{h}_i^T\mathbf{B}\mathbf{h}_j = \mathbf{v}_{ij}^T\mathbf{b},
\tag{2.24}
$$

with $\mathbf{v}_{ij} = [h_{i1}h_{j1}, h_{i1}h_{j2}+h_{i2}h_{j1}, h_{i2}h_{j2}, h_{i3}h_{j1}+h_{i1}h_{j3}, h_{i3}h_{j2}+h_{i2}h_{j3}, h_{i3}h_{j3}]^T$. Therefore, the two fundamental constraints (2.19) and (2.20), from a given homography, can be rewritten as two homogeneous equations in $\mathbf{b}$:

$$
\begin{bmatrix} \mathbf{v}_{12}^T \\ (\mathbf{v}_{11} - \mathbf{v}_{22})^T \end{bmatrix} \mathbf{b} = \mathbf{0} .
\tag{2.25}
$$

If n images of the model plane are observed, by stacking n such equations as (2.25), we have

$$
\mathbf{V}\mathbf{b} = \mathbf{0} ,
\tag{2.26}
$$

where $\mathbf{V}$ is a $2n \times 6$ matrix. If $n \geq 3$, we have, in general, a unique solution $\mathbf{b}$ defined up to a scale factor. If $n = 2$, we can impose the skewless constraint $\gamma = 0$, that is, $[0, 1, 0, 0, 0, 0]\mathbf{b} = 0$, which is added as an additional equation to (2.26). (If $n = 1$, we can solve only two camera intrinsic parameters, e.g., α and β, assuming u_0 and v_0 are known [e.g., at the image center] and $\gamma = 0$, and that is indeed what we did in [28] for head pose determination based on the fact that eyes and mouth are reasonably coplanar. In fact, Tsai [33] already mentions that focal length from one plane is possible, but incorrectly says that aspect ratio is not.) The solution to (2.26) is well known as the eigenvector of $\mathbf{V}^T\mathbf{V}$ associated with the smallest eigenvalue (equivalently, the right singular vector of $\mathbf{V}$ associated with the smallest singular value).

Once $\mathbf{b}$ is estimated, we can compute all camera intrinsic parameters as follows. The matrix $\mathbf{B}$, as described in Section 2.4.4, is estimated up to a scale factor, i.e.,, $\mathbf{B} = \lambda\mathbf{A}^{-T}\mathbf{A}$ with λ an arbitrary scale. Without difficulty, we can uniquely extract the intrinsic parameters from matrix $\mathbf{B}$.

$$v_0 = (B_{12}B_{13} - B_{11}B_{23})/(B_{11}B_{22} - B_{12}^2)$$
$$\lambda = B_{33} - [B_{13}^2 + v_0(B_{12}B_{13} - B_{11}B_{23})]/B_{11}$$
$$\alpha = \sqrt{\lambda/B_{11}}$$
$$\beta = \sqrt{\lambda B_{11}/(B_{11}B_{22} - B_{12}^2)}$$
$$\gamma = -B_{12}\alpha^2\beta/\lambda$$
$$u_0 = \gamma v_0/\alpha - B_{13}\alpha^2/\lambda\,.$$

Once $\mathbf{A}$ is known, the extrinsic parameters for each image are readily computed. From (2.18), we have

$$\mathbf{r}_1 = \lambda\mathbf{A}^{-1}\mathbf{h}_1\,, \quad \mathbf{r}_2 = \lambda\mathbf{A}^{-1}\mathbf{h}_2\,, \quad \mathbf{r}_3 = \mathbf{r}_1 \times \mathbf{r}_2\,, \quad \mathbf{t} = \lambda\mathbf{A}^{-1}\mathbf{h}_3$$

with $\lambda = 1/\|\mathbf{A}^{-1}\mathbf{h}_1\| = 1/\|\mathbf{A}^{-1}\mathbf{h}_2\|$. Of course, because of noise in data, the so-computed matrix $\mathbf{R} = [\mathbf{r}_1, \mathbf{r}_2, \mathbf{r}_3]$ does not generally satisfy the properties of a rotation matrix. The best rotation matrix can then be obtained through, for example, singular value decomposition [13, 41].

2.4.5   Maximum likelihood estimation

The above solution is obtained through minimizing an algebraic distance which is not physically meaningful. We can refine it through maximum likelihood inference.

We are given n images of a model plane, and there are m points on the model plane. Assume that the image points are corrupted by independent and identically distributed noise. The maximum likelihood estimate can be obtained by minimizing the following functional:

$$\sum_{i=1}^{n}\sum_{j=1}^{m}\|\mathbf{m}_{ij} - \widehat{\mathbf{m}}(\mathbf{A}, \mathbf{R}_i, \mathbf{t}_i, \mathtt{M}_j)\|^2\,, \tag{2.27}$$

where $\widehat{\mathbf{m}}(\mathbf{A}, \mathbf{R}_i, \mathbf{t}_i, \mathbf{M}_j)$ is the projection of point $\mathbf{M}_j$ in image i, according to equation (2.18). A rotation $\mathbf{R}$ is parameterized by a vector of three parameters, denoted by $\mathbf{r}$, which is parallel to the rotation axis and whose magnitude is equal to the rotation angle. $\mathbf{R}$ and $\mathbf{r}$ are related by the Rodrigues formula [8]. Minimizing (2.27) is a nonlinear minimization problem, which is solved with the Levenberg-Marquardt algorithm as implemented in `Minpack` [23]. It requires an initial guess of $\mathbf{A}, \{\mathbf{R}_i, \mathbf{t}_i | i = 1..n\}$, which can be obtained using the technique described in the previous section.

Desktop cameras usually have visible lens distortion, especially the radial components. We have included these while minimizing (2.27). See my technical report [41] for more details.

2.4.6   Dealing with radial distortion

Up to now, we have not considered lens distortion of a camera. However, a desktop camera usually exhibits significant lens distortion, especially radial distortion. Refer to Section 2.3.5 for distortion modeling. In this section, we consider the first two terms of radial distortion.

Estimating Radial Distortion by Alternation.   As the radial distortion is expected to be small, we would expect to estimate the other five intrinsic parameters, using the technique described in Section 2.4.5, reasonably well by simply ignoring distortion. One strategy is then to estimate k_1 and k_2 after having estimated the other parameters, which will give us the ideal pixel coordinates (u, v). Then, from (2.15) and (2.16), we have two equations for each point in each image:

$$\begin{bmatrix} (u-u_0)(x^2+y^2) & (u-u_0)(x^2+y^2)^2 \\ (v-v_0)(x^2+y^2) & (v-v_0)(x^2+y^2)^2 \end{bmatrix} \begin{bmatrix} k_1 \\ k_2 \end{bmatrix} = \begin{bmatrix} \breve{u}-u \\ \breve{v}-v \end{bmatrix} .$$

Given m points in n images, we can stack all equations together to obtain in total $2mn$ equations, or in matrix form as $\mathbf{D}\mathbf{k} = \mathbf{d}$, where $\mathbf{k} = [k_1, k_2]^T$. The linear least-squares solution is given by

$$\mathbf{k} = (\mathbf{D}^T\mathbf{D})^{-1}\mathbf{D}^T\mathbf{d} . \tag{2.28}$$

Once k_1 and k_2 are estimated, we can refine the estimate of the other parameters by solving (2.27) with $\widehat{\mathbf{m}}(\mathbf{A}, \mathbf{R}_i, \mathbf{t}_i, \mathbf{M}_j)$ replaced by (2.15) and (2.16). We can alternate these two procedures until convergence.

Complete Maximum Likelihood Estimation.   Experimentally, we found the convergence of the above alternation technique is slow. A natural extension to (2.27) is then to estimate the complete set of parameters by minimizing the following functional:

$$\sum_{i=1}^{n}\sum_{j=1}^{m}\|\mathbf{m}_{ij} - \breve{\mathbf{m}}(\mathbf{A}, k_1, k_2, \mathbf{R}_i, \mathbf{t}_i, \mathbf{M}_j)\|^2 , \qquad (2.29)$$

where $\breve{\mathbf{m}}(\mathbf{A}, k1, k2, \mathbf{R}_i, \mathbf{t}_i, \mathbf{M}_j)$ is the projection of point $\mathbf{M}_j$ in image i according to equation (2.18), followed by distortion according to (2.15) and (2.16). This is a nonlinear minimization problem, which is solved with the Levenberg-Marquardt algorithm as implemented in `Minpack` [23]. A rotation is again parameterized by a 3-vector $\mathbf{r}$, as in Section 2.4.5. An initial guess of $\mathbf{A}$ and $\{\mathbf{R}_i, \mathbf{t}_i | i = 1, ..., n\}$ can be obtained using the technique described in Section 2.4.4 or in Section 2.4.5. An initial guess of k_1 and k_2 can be obtained with the technique described in the last paragraph or simply by setting them to 0.

2.4.7   Summary

The recommended calibration procedure is as follows:

1. Print a pattern and attach it to a planar surface;

2. Take a few images of the model plane under different orientations by moving either the plane or the camera;

3. Detect the feature points in the images;

4. Estimate the five intrinsic parameters and all the extrinsic parameters using the closed-form solution described in Section 2.4.4;

5. Estimate the coefficients of the radial distortion by solving the linear least-squares (2.28);

6. Refine all parameters, including lens distortion parameters, by minimizing (2.29).

There is a degenerate configuration in my technique when planes are parallel to each other. See my technical report [41] for a more detailed description.

In summary, this technique requires the camera to observe a planar pattern from only a few different orientations. Although the minimum number of orientations is two if pixels are square, we recommend four or five different orientations for better quality. We can move either the camera or the planar pattern. The motion does not need to be known, but should not be a pure translation. When the number of orientations is only two, one should avoid positioning the planar pattern parallel to the image plane. The pattern could be anything, as long as we know the metric on the plane. For example, we can print a pattern with a laser printer and attach the paper to a reasonable planar surface such as a hard book cover. We can even use a book with known size because the four corners are enough to estimate the plane homographies.

2.4.8   Experimental results

The proposed algorithm has been tested on both computer-simulated data and real data. The closed-form solution involves finding a singular value decomposition of a small $2n \times 6$ matrix, where n is the number of images. The nonlinear refinement within the Levenberg-Marquardt algorithm takes three to five iterations to converge. We describe in this section one set of experiments with real data when the calibration pattern is at different distances from the camera. Refer to [41] for more experimental results with both computer-simulated and real data, and to `http://research.microsoft.com/~zhang/Calib/` for some experimental data and the software.

The example is shown in Figure 2.6. The camera to be calibrated is an off-the-shelf PULNiX CCD camera with 6 mm lens. The image resolution is 640×480. As can be seen in Figure 2.6, the model plane contains 9×9 squares with nine special dots used to identify automatically the correspondence between reference points on the model plane and square corners in images. It was printed on an A4 paper with a 600 DPI laser printer and attached to a cardboard.

Ten images of the plane were taken (six of them are shown in Figure 2.6). Five of them (Set A) were taken at close range, while the other five (Set B) were taken at a larger distance. We applied our calibration algorithm to

Table 2.1. Calibration results with the images shown in Figure 2.6

image set	α	β	ϑ	u_0	v_0	k_1	k_2
A	834.01	839.86	89.95°	305.51	240.09	−0.2235	0.3761
B	836.17	841.08	89.92°	301.76	241.51	−0.2676	1.3121
A+B	834.64	840.32	89.94°	304.77	240.59	−0.2214	0.3643

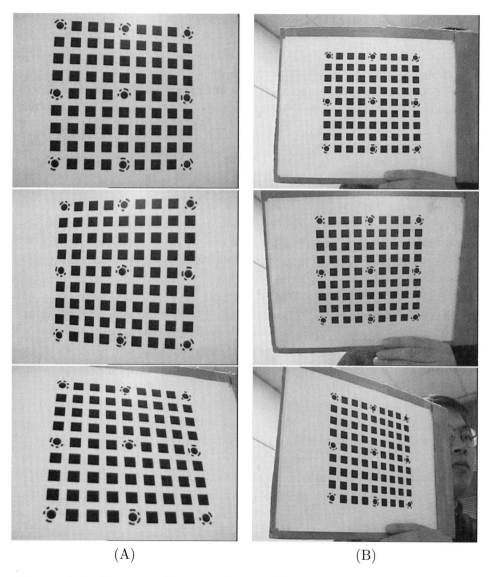

(A)   (B)

Figure 2.6. Two sets of images taken at different distances to the calibration pattern. Each set contains five images. On the left, three images from the set taken at a close distance are shown. On the right, three images from the set taken at a larger distance are shown.

Set A, Set B, and also to the whole set (Set A+B). The results are shown
in Table 2.1. For intuitive understanding, we show the estimated angle
between the image axes, ϑ, instead of the skew factor γ. We can see that
the angle ϑ is very close to 90°, as expected with almost all modern CCD
cameras. The camera parameters were estimated consistently for all three
sets of images, except the distortion parameters with Set B. The reason is
that the calibration pattern occupies only the central part of the image in Set
B, where lens distortion is not significant and therefore cannot be estimated
reliably.

2.4.9   Related work

Almost at the same time, Sturm and Maybank [31], independent from us,
developed the same technique. They assumed the pixels are square (i.e., $\gamma =$
0) and have studied the degenerate configurations for plane-based camera
calibration.

Gurdjos *et al.* [14] re-derived the plane-based calibration technique from
the center line constraint.

My original implementation (only the executable) is available at
`http://research.microsoft.com/~zhang/calib/`. Bouguet has reimplemented my technique in Matlab, which is available at
`http://www.vision.caltech.edu/bouguetj/calib_doc/`.

In many applications, such as stereo, multiple cameras need to be calibrated simultaneously in order to determine the relative geometry between
cameras. In 2000, I extended (not published) this plane-based technique
to stereo calibration for my stereo-based gaze-correction project [40, 39].
The formulation is similar to (2.29). Consider two cameras, and denote the
quantity related to the second camera by $'$. Let $(\mathbf{R}_s, \mathbf{t}_s)$ be the rigid trans-
formation between the two cameras such that $(\mathbf{R}', \mathbf{t}') = (\mathbf{R}, \mathbf{t}) \circ (\mathbf{R}_s, \mathbf{t}_s)$, or
more precisely, $\mathbf{R}' = \mathbf{R}\mathbf{R}_s$ and $\mathbf{t}' = \mathbf{R}\mathbf{t}_s + \mathbf{t}$. Stereo calibration is then to
solve $\mathbf{A}, \mathbf{A}', k_1, k_2, k_1', k_2', \{(\mathbf{R}_i, \mathbf{t}_i) | i = 1, \ldots, n\}$, and $(\mathbf{R}_s, \mathbf{t}_s)$ by minimizing
the following functional:

$$\sum_{i=1}^{n} \sum_{j=1}^{m} \left[\delta_{ij} \| \mathbf{m}_{ij} - \check{\mathbf{m}}(\mathbf{A}, k_1, k_2, \mathbf{R}_i, \mathbf{t}_i, \mathsf{M}_j) \|^2 + \right.$$

$$\left. \delta_{ij}' \| \mathbf{m}_{ij}' - \check{\mathbf{m}}(\mathbf{A}', k_1', k_2', \mathbf{R}_i', \mathbf{t}_i', \mathsf{M}_j) \|^2 \right] \qquad (2.30)$$

subject to

$$\mathbf{R}_i' = \mathbf{R}_i \mathbf{R}_s \quad \text{and} \quad \mathbf{t}_i' = \mathbf{R}_i \mathbf{t}_s + \mathbf{t}_i \ .$$

In the above formulation, $\delta_{ij} = 1$ if point j is visible in the first camera, and $\delta_{ij} = 0$ otherwise. Similarly, $\delta'_{ij} = 1$ if point j is visible in the second camera. This formulation thus does not require the same number of feature points to be visible over time or across cameras. Another advantage of this formulation is that the number of extrinsic parameters to be estimated has been reduced from $12n$ if the two cameras are calibrated independently to $6n + 6$. This is a reduction of 24 dimensions in parameter space if five planes are used.

Obviously, this is a nonlinear optimization problem. To obtain the initial guess, we first run single-camera calibration independently for each camera, and compute $\mathbf{R}_s$ through SVD from $\mathbf{R}'_i = \mathbf{R}_i \mathbf{R}_s$ $(i = 1, \ldots, n)$ and $\mathbf{t}_s$ through least-squares from $\mathbf{t}'_i = \mathbf{R}_i \mathbf{t}_s + \mathbf{t}_i$ $(i = 1, \ldots, n)$. Recently, a closed-form initialization technique through factorization of homography matrices is proposed in [34].

2.5   Solving Camera Calibration with 1D Objects

In this section, we describe in detail how to solve the camera calibration problem from a number of observations of a 1D object consisting of three collinear points moving around one of them [43, 44]. We consider only this minimal configuration, but it is straightforward to extend the result if a calibration object has four or more collinear points.

2.5.1   Setups with free-moving 1D calibration objects

We now examine possible setups with 1D objects for camera calibration. As already mentioned, we need to have several observations of the 1D objects. Without loss of generality, we choose the camera coordinate system to define the 1D objects; therefore, $\mathbf{R} = \mathbf{I}$ and $\mathbf{t} = \mathbf{0}$ in (2.1).

Two points with known distance.   This could be the two endpoints of a stick, and we take a number of images while waving freely the stick. Let A and B be the two 3D points, and $\mathbf{a}$ and $\mathbf{b}$ be the observed image points. Because the distance between A and B is known, we only need five parameters to define A and B. For example, we need three parameters to specify the coordinates of A in the camera coordinate system, and two parameters to define the orientation of the line AB. On the other hand, each image point provides two equations according to (2.1), giving in total four equations. Given N observations of the stick, we have five intrinsic parameters and $5N$ parameters for the point positions to estimate, that is, the total number of unknowns is

$5 + 5N$. However, we have only $4N$ equations. Camera calibration is thus impossible.

Three collinear points with known distances. By adding an additional point, say C, the number of unknowns for the point positions still remains the same, i.e., $5+5N$, because of known distances of C to A and B. For each observation, we have three image points, yielding in total $6N$ equations. Calibration seems to be plausible, but is in fact not. This is because the three image points for each observation must be collinear. Collinearity is preserved by perspective projection. We therefore have only five independent equations for each observation. The total number of independent equations, $5N$, is always smaller than the number of unknowns. Camera calibration is still impossible.

Four or more collinear points with known distances. As seen above, when the number of points increases from two to three, the number of independent equations (constraints) increases by one for each observation. If we have a fourth point, will we have in total $6N$ independent equations? If so, we would be able to solve the problem because the number of unknowns remains the same, i.e., $5 + 5N$, and we would have more than enough constraints if $N \geq 5$. The reality is that the addition of the fourth point or even more points does not increase the number of independent equations. It will always be $5N$ for any four or more collinear points. This is because the cross ratio is preserved under perspective projection. With known cross ratios and three collinear points, whether they are in space or in images, other points are determined exactly.

2.5.2   Setups with 1D calibration objects moving around a fixed point

From the previous discussion, calibration is impossible with a free-moving 1D calibration object, no matter how many points on the object. Now let us examine what happens if one point is fixed. In the sequel, without loss of generality, point A is the fixed point, and **a** is the corresponding image point. We need three parameters, which are unknown, to specify the coordinates of A in the camera coordinate system, while image point **a** provides two scalar equations according to (2.1).

Two points with known distance. They could be the endpoints of a stick, and we move the stick around the endpoint that is fixed. Let B be the free endpoint and **b**, its corresponding image point. For each observation, we need two parameters to define the orientation of the line AB and therefore the position of B because the distance between A and B is known. Given N observations of the stick, we have five intrinsic parameters, three parameters for A and $2N$ parameters for the free endpoint positions to estimate; that is, the total number of unknowns is $8 + 2N$. However, each observation of **b** provides two equations, so together with **a** we have in total only $2 + 2N$ equations. Camera calibration is thus impossible.

Three collinear points with known distances.   As explained in the last subsection, by adding an additional point, say C, the number of unknowns for the point positions still remains the same: $8 + 2N$. For each observation, **b** provides two equations, but **c** provides only one additional equation because of the collinearity of **a**, **b**, and **c**. Thus, the total number of equations is $2 + 3N$ for N observations. By counting the numbers, we see that if we have six or more observations, we should be able to solve camera calibration, and this is the case, as we shall show in the next section.

Four or more collinear points with known distances.   Again, the number of unknowns and the number of independent equations remain the same because of invariance of cross-ratios. This said, the more collinear points we have, the more accurate camera calibration will be in practice because data redundancy can combat the noise in image data.

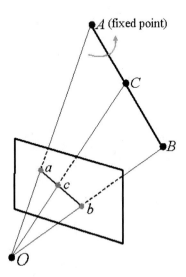

Figure 2.7. Illustration of 1D calibration objects.

2.5.3   Basic equations

Refer to Figure 2.7. Point A is the fixed point in space, and the stick AB moves around A. The length of the stick AB is known to be L:

$$\|B - A\| = L \; . \tag{2.31}$$

The position of point C is also known with respect to A and B, and therefore

$$C = \lambda_A A + \lambda_B B \; , \tag{2.32}$$

where λ_A and λ_B are known. If C is the midpoint of AB, then $\lambda_A = \lambda_B = 0.5$. Points $\mathbf{a}$, $\mathbf{b}$, and $\mathbf{c}$ on the image plane are projections of space points A, B, and C respectively.

Without loss of generality, we choose the camera coordinate system to define the 1D objects; therefore, $\mathbf{R} = \mathbf{I}$ and $\mathbf{t} = \mathbf{0}$ in (2.1). Let the unknown depths for A, B, and C be z_A, z_B, and z_C, respectively. According to (2.1), we have

$$A = z_A \mathbf{A}^{-1} \widetilde{\mathbf{a}}, \tag{2.33}$$

$$B = z_B \mathbf{A}^{-1} \widetilde{\mathbf{b}}, \tag{2.34}$$

$$C = z_C \mathbf{A}^{-1} \widetilde{\mathbf{c}} \; . \tag{2.35}$$

Substituting them into (2.32) yields

$$z_C \widetilde{\mathbf{c}} = z_A \lambda_A \widetilde{\mathbf{a}} + z_B \lambda_B \widetilde{\mathbf{b}} \tag{2.36}$$

after eliminating $\mathbf{A}^{-1}$ from both sides. By performing cross-product on both sides of the above equation with $\widetilde{\mathbf{c}}$, we have

$$z_A \lambda_A (\widetilde{\mathbf{a}} \times \widetilde{\mathbf{c}}) + z_B \lambda_B (\widetilde{\mathbf{b}} \times \widetilde{\mathbf{c}}) = \mathbf{0} \; .$$

In turn, we obtain

$$z_B = -z_A \frac{\lambda_A (\widetilde{\mathbf{a}} \times \widetilde{\mathbf{c}}) \cdot (\widetilde{\mathbf{b}} \times \widetilde{\mathbf{c}})}{\lambda_B (\widetilde{\mathbf{b}} \times \widetilde{\mathbf{c}}) \cdot (\widetilde{\mathbf{b}} \times \widetilde{\mathbf{c}})} \; . \tag{2.37}$$

From (2.31), we have

$$\|\mathbf{A}^{-1}(z_B \widetilde{\mathbf{b}} - z_A \widetilde{\mathbf{a}})\| = L \; .$$

Substituting z_B by (2.37) gives

$$z_A \| \mathbf{A}^{-1} \big(\widetilde{\mathbf{a}} + \frac{\lambda_A (\widetilde{\mathbf{a}} \times \widetilde{\mathbf{c}}) \cdot (\widetilde{\mathbf{b}} \times \widetilde{\mathbf{c}})}{\lambda_B (\widetilde{\mathbf{b}} \times \widetilde{\mathbf{c}}) \cdot (\widetilde{\mathbf{b}} \times \widetilde{\mathbf{c}})} \widetilde{\mathbf{b}} \big) \| = L \, .$$

This is equivalent to

$$z_A^2 \mathbf{h}^T \mathbf{A}^{-T} \mathbf{A}^{-1} \mathbf{h} = L^2 \qquad (2.38)$$

with

$$\mathbf{h} = \widetilde{\mathbf{a}} + \frac{\lambda_A (\widetilde{\mathbf{a}} \times \widetilde{\mathbf{c}}) \cdot (\widetilde{\mathbf{b}} \times \widetilde{\mathbf{c}})}{\lambda_B (\widetilde{\mathbf{b}} \times \widetilde{\mathbf{c}}) \cdot (\widetilde{\mathbf{b}} \times \widetilde{\mathbf{c}})} \widetilde{\mathbf{b}} \, . \qquad (2.39)$$

Equation (2.38) contains the unknown intrinsic parameters $\mathbf{A}$ and the unknown depth, z_A, of the fixed point A. It is the basic constraint for camera calibration with 1D objects. Vector $\mathbf{h}$, given by (2.39), can be computed from image points and known λ_A and λ_B. Since the total number of unknowns is six, we need at least six observations of the 1D object for calibration. Note that $\mathbf{A}^{-T} \mathbf{A}$ actually describes the image of the absolute conic [20].

2.5.4 Closed-form solution

Let

$$\mathbf{B} = \mathbf{A}^{-T} \mathbf{A}^{-1} \equiv \begin{bmatrix} B_{11} & B_{12} & B_{13} \\ B_{12} & B_{22} & B_{23} \\ B_{13} & B_{23} & B_{33} \end{bmatrix} \qquad (2.40)$$

$$= \begin{bmatrix} \frac{1}{\alpha^2} & -\frac{\gamma}{\alpha^2 \beta} & \frac{v_0 \gamma - u_0 \beta}{\alpha^2 \beta} \\ -\frac{\gamma}{\alpha^2 \beta} & \frac{\gamma^2}{\alpha^2 \beta^2} + \frac{1}{\beta^2} & -\frac{\gamma(v_0 \gamma - u_0 \beta)}{\alpha^2 \beta^2} - \frac{v_0}{\beta^2} \\ \frac{v_0 \gamma - u_0 \beta}{\alpha^2 \beta} & -\frac{\gamma(v_0 \gamma - u_0 \beta)}{\alpha^2 \beta^2} - \frac{v_0}{\beta^2} & \frac{(v_0 \gamma - u_0 \beta)^2}{\alpha^2 \beta^2} + \frac{v_0^2}{\beta^2} + 1 \end{bmatrix} \, . \qquad (2.41)$$

Note that $\mathbf{B}$ is symmetric and can be defined by a 6D vector

$$\mathbf{b} = [B_{11}, B_{12}, B_{22}, B_{13}, B_{23}, B_{33}]^T \, . \qquad (2.42)$$

Let $\mathbf{h} = [h_1, h_2, h_3]^T$, and $\mathbf{x} = z_A^2 \mathbf{b}$, then equation (2.38) becomes

$$\mathbf{v}^T \mathbf{x} = L^2 \qquad (2.43)$$

with

$$\mathbf{v} = [h_1^2, 2h_1 h_2, h_2^2, 2h_1 h_3, 2h_2 h_3, h_3^2]^T \, .$$

When N images of the 1D object are observed, by stacking n such equations as (2.43), we have

$$\mathbf{V}\mathbf{x} = L^2 \mathbf{1} , \qquad (2.44)$$

where $\mathbf{V} = [\mathbf{v}_1, \ldots, \mathbf{v}_N]^T$ and $\mathbf{1} = [1, \ldots, 1]^T$. The least-squares solution is then given by

$$\mathbf{x} = L^2 (\mathbf{V}^T \mathbf{V})^{-1} \mathbf{V}^T \mathbf{1} . \qquad (2.45)$$

Once $\mathbf{x}$ is estimated, we can compute all the unknowns based on $\mathbf{x} = z_A^2 \mathbf{b}$. Let $\mathbf{x} = [x_1, x_2, \ldots, x_6]^T$. Without difficulty, we can uniquely extract the intrinsic parameters and the depth z_A as

$$v_0 = (x_2 x_4 - x_1 x_5)/(x_1 x_3 - x_2^2)$$
$$z_A = \sqrt{x_6 - [x_4^2 + v_0(x_2 x_4 - x_1 x_5)]/x_1}$$
$$\alpha = \sqrt{z_A/x_1}$$
$$\beta = \sqrt{z_A x_1/(x_1 x_3 - x_2^2)}$$
$$\gamma = -x_2 \alpha^2 \beta/z_A$$
$$u_0 = \gamma v_0/\alpha - x_4 \alpha^2/z_A .$$

At this point, we can compute z_B according to (2.37), so points A and B can be computed from (2.33) and (2.34), while point C can be computed according to (2.32).

2.5.5 Nonlinear optimization

The above solution is obtained through minimizing an algebraic distance which is not physically meaningful. We can refine it through maximum likelihood inference.

We are given N images of the 1D calibration object, and there are three points on the object. Point A is fixed, and points B and C move around A. Assume that the image points are corrupted by independent and identically distributed noise. The maximum likelihood estimate can be obtained by minimizing the following functional:

$$\sum_{i=1}^{N} \left(\|\mathbf{a}_i - \phi(\mathbf{A}, \mathbf{A})\|^2 + \|\mathbf{b}_i - \phi(\mathbf{A}, \mathbf{B}_i)\|^2 + \|\mathbf{c}_i - \phi(\mathbf{A}, \mathbf{C}_i)\|^2 \right) , \qquad (2.46)$$

where $\phi(\mathbf{A}, \mathtt{M})$ $(\mathtt{M} \in \{\mathtt{A}, \mathtt{B}_i, \mathtt{C}_i\})$ is the projection of point $\mathtt{M}$ onto the image, according to equations (2.33) to (2.35). More precisely, $\phi(\mathbf{A}, \mathtt{M}) = \frac{1}{z_M} \mathbf{A} \mathtt{M}$, where z_M is the z-component of $\mathtt{M}$.

The unknowns to be estimated are

- 5 camera intrinsic parameters, α, β, γ, u_0, and v_0, that define matrix $\mathbf{A}$;

- 3 parameters for the coordinates of the fixed point $\mathtt{A}$;

- $2N$ additional parameters to define points $\mathtt{B}_i$ and $\mathtt{C}_i$ at each instant (see below for more details).

Therefore, we have in total $8 + 2N$ unknowns. Regarding the parameterization for $\mathtt{B}$ and $\mathtt{C}$, we use the spherical coordinates ϕ and θ to define the direction of the 1D calibration object, and point $\mathtt{B}$ is then given by

$$\mathtt{B} = \mathtt{A} + L \begin{bmatrix} \sin\theta \cos\phi \\ \sin\theta \sin\phi \\ \cos\theta \end{bmatrix},$$

where L is the known distance between $\mathtt{A}$ and $\mathtt{B}$. In turn, point $\mathtt{C}$ is computed according to (2.32). We therefore need only two additional parameters for each observation.

Minimizing (2.46) is a nonlinear minimization problem, which is solved with the Levenberg-Marquardt algorithm, as implemented in `Minpack` [23]. It requires an initial guess of $\mathbf{A}, \mathtt{A}, \{\mathtt{B}_i, \mathtt{C}_i | i = 1, ..., N\}$, which can be obtained using the technique described in the last subsection.

2.5.6   Estimating the fixed point

In the above discussion, we assumed that the image coordinates, $\mathbf{a}$, of the fixed point $\mathtt{A}$ are known. We now describe how to estimate $\mathbf{a}$ by considering whether the fixed point $\mathtt{A}$ is visible in the image or not.

Invisible fixed point.   The fixed point does not need to be visible in the image, and the camera calibration technique becomes more versatile without the visibility requirement. In that case, we can, for example, hang a string of small balls from the ceiling and calibrate multiple cameras in the room by swinging the string. The fixed point can be estimated by intersecting lines from different images as described below.

Each observation of the 1D object defines an image line. An image line can be represented by a 3D vector $\mathbf{l} = [l_1, l_2, l_3]^T$, defined up to a scale factor such as a point $\mathbf{m} = [u, v]^T$ on the line satisfies $\mathbf{l}^T \widetilde{\mathbf{m}} = 0$. In the sequel, we also use $(\mathbf{n}, q)$ to denote line $\mathbf{l}$, where $\mathbf{n} = [l_1, l_2]^T$ and $q = l_3$. To remove the scale ambiguity, we normalize $\mathbf{l}$ such that $\|\mathbf{l}\| = 1$. Furthermore, each $\mathbf{l}$ is associated with an uncertainty measure represented by a 3×3 covariance matrix $\mathbf{\Lambda}$.

Given N images of the 1D object, we have N lines: $\{(\mathbf{l}_i, \mathbf{\Lambda}_i)|i = 1, \ldots, N\}$. Let the fixed point be $\mathbf{a}$ in the image. Obviously, if there is no noise, we have $\mathbf{l}_i^T \widetilde{\mathbf{a}} = 0$, or $\mathbf{n}_i^T \mathbf{a} + q_i = 0$. Therefore, we can estimate $\mathbf{a}$ by minimizing

$$\mathcal{F} = \sum_{i=1}^{N} w_i \|\mathbf{l}_i^T \widetilde{\mathbf{a}}\|^2 = \sum_{i=1}^{N} w_i \|\mathbf{n}_i^T \mathbf{a} + q_i\|^2 = \sum_{i=1}^{N} w_i (\mathbf{a}^T \mathbf{n}_i \mathbf{n}_i^T \mathbf{a} + 2q_i \mathbf{n}_i^T \mathbf{a} + q_i^2),$$

$$(2.47)$$

where w_i is a weighting factor (see below). By setting the derivative of $\mathcal{F}$ with respect to $\mathbf{a}$ to 0, we obtain the solution, which is given by

$$\mathbf{a} = -\left(\sum_{i=1}^{N} w_i \mathbf{n}_i \mathbf{n}_i^T \right)^{-1} \left(\sum_{i=1}^{N} w_i q_i \mathbf{n}_i \right).$$

The optimal weighting factor w_i in (2.47) is the inverse of the variance of $\mathbf{l}_i^T \widetilde{\mathbf{a}}$, which is $w_i = 1/(\widetilde{\mathbf{a}}^T \mathbf{\Lambda}_i \widetilde{\mathbf{a}})$. Note that the weight w_i involves the unknown $\mathbf{a}$. To overcome this difficulty, we can approximate w_i by $1/\operatorname{trace}(\mathbf{\Lambda}_i)$ for the first iteration and by recomputing w_i with the previously estimated $\mathbf{a}$ in the subsequent iterations. Usually two or three iterations are enough.

Visible fixed point.   Since the fixed point is visible, we have N observations: $\{\mathbf{a}_i|i = 1, \ldots, N\}$. We can therefore estimate $\mathbf{a}$ by minimizing $\sum_{i=1}^{N} \|\mathbf{a} - \mathbf{a}_i\|^2$, assuming that the image points are detected with the same accuracy. The solution is simply $\mathbf{a} = (\sum_{i=1}^{N} \mathbf{a}_i)/N$.

The above estimation does not make use of the fact that the fixed point is also the intersection of the N observed lines of the 1D object. Therefore, a better technique to estimate $\mathbf{a}$ is to minimize the following function:

$$\mathcal{F} = \sum_{i=1}^{N} \left[(\mathbf{a} - \mathbf{a}_i)^T \mathbf{V}_i^{-1} (\mathbf{a} - \mathbf{a}_i) + w_i \|\mathbf{l}_i^T \widetilde{\mathbf{a}}\|^2 \right]$$

$$= \sum_{i=1}^{N} \left[(\mathbf{a} - \mathbf{a}_i)^T \mathbf{V}_i^{-1} (\mathbf{a} - \mathbf{a}_i) + w_i \|\mathbf{n}_i^T \mathbf{a} + q_i\|^2 \right] \qquad (2.48)$$

where $\mathbf{V}_i$ is the covariance matrix of the detected point $\mathbf{a}_i$. The derivative of the above function with respect to $\mathbf{a}$ is given by

$$\frac{\partial \mathcal{F}}{\partial \mathbf{a}} = 2 \sum_{i=1}^{N} \left[\mathbf{V}_i^{-1}(\mathbf{a} - \mathbf{a}_i) + w_i \mathbf{n}_i \mathbf{n}_i^T \mathbf{a} + w_i q_i \mathbf{n}_i \right].$$

Setting it to 0 yields

$$\mathbf{a} = \left(\sum_{i=1}^{N} (\mathbf{V}_i^{-1} + w_i \mathbf{n}_i \mathbf{n}_i^T) \right)^{-1} \left(\sum_{i=1}^{N} (\mathbf{V}_i^{-1} \mathbf{a}_i - w_i q_i \mathbf{n}_i) \right).$$

If more than three points are visible in each image, the known cross-ratio provides an additional constraint in determining the fixed point.

For an accessible description of uncertainty manipulation, refer to [45].

2.5.7   Experimental results

The proposed algorithm has been tested on both computer-simulated data and real data.

Computer Simulations

The simulated camera has the following properties: $\alpha = 1000$, $\beta = 1000$, $\gamma = 0$, $u_0 = 320$, and $v_0 = 240$. The image resolution is 640×480. A stick of 70 cm is simulated with the fixed point A at $[0, 35, 150]^T$. The other endpoint of the stick is B, and C is located at the halfway point between A and B. We have generated 100 random orientations of the stick by sampling θ in $[\pi/6, 5\pi/6]$ and ϕ in $[\pi, 2\pi]$ according to uniform distribution. Points A, B, and C are then projected onto the image.

Gaussian noise with 0 mean and σ standard deviation is added to the projected image points $\mathbf{a}$, $\mathbf{b}$, and $\mathbf{c}$. The estimated camera parameters are compared with the ground truth, and we measure their relative errors with respect to the focal length α. Note that we measure the relative errors in (u_0, v_0) with respect to α, as proposed by Triggs in [32]. He pointed out that the absolute errors in (u_0, v_0) are not geometrically meaningful, while computing the relative error is equivalent to measuring the angle between the true optical axis and the estimated one.

We vary the noise level from 0.1 pixels to 1 pixel. For each noise level, we perform 120 independent trials, and the results shown in Figure 2.8 are the average. Figure 2.8a displays the relative errors of the closed-form solution, and Figure 2.8b displays those of the nonlinear minimization result. Errors

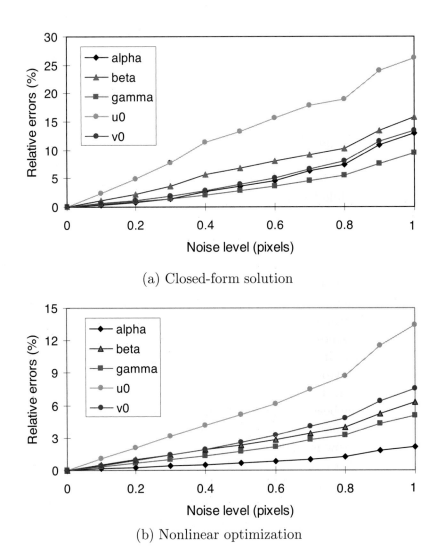

(a) Closed-form solution

(b) Nonlinear optimization

Figure 2.8. Calibration errors with respect to the noise level of the image points.

Table 2.2. Calibration results with real data.

Solution	α	β	γ	u_0	v_0
Closed-form	889.49	818.59	−0.1651 (90.01°)	297.47	234.33
Nonlinear	838.49	799.36	4.1921 (89.72°)	286.74	219.89
Plane-based	828.92	813.33	−0.0903 (90.01°)	305.23	235.17
Relative difference	1.15%	1.69%	0.52% (0.29°)	2.23%	1.84%

increase almost linearly with the noise level. The nonlinear minimization refines the closed-form solution and produces significantly better result (with 50% less errors). At 1 pixel noise level, the errors for the closed-form solution are about 12%, while those for the nonlinear minimization are about 6%.

Real Data

For the experiment with real data, I strung three toy beads together with a stick. The beads are approximately 14 cm apart (i.e., $L = 28$). I then moved the stick around while trying to fix one end with the aid of a book. A video of 150 frames was recorded, and four sample images are shown in Figure 2.9. A bead in the image is modeled as a Gaussian blob in the RGB space, and the centroid of each detected blob is the image point we use for camera calibration. The proposed algorithm is therefore applied to the 150 observations of the beads, and the estimated camera parameters are provided in Table 2.2. The first row is the estimation from the closed-form solution, while the second row is the refined result after nonlinear minimization. For the image skew parameter γ, we also provide the angle between the image axes in parenthesis (it should be very close to 90°).

For comparison, we also used the plane-based calibration technique described in [42] to calibrate the same camera. Five images of a planar pattern were taken, and one of them is shown in Figure 2.10. The calibration result is shown in the third row of Table 2.2. The fourth row displays the relative difference between the plane-based result and the nonlinear solution with respect to the focal length (we use 828.92). As we can observe, the difference is about 2%.

There are several sources contributing to this difference. Besides the image noise and imprecision of the extracted data points, one source is our current rudimentary experimental setup:

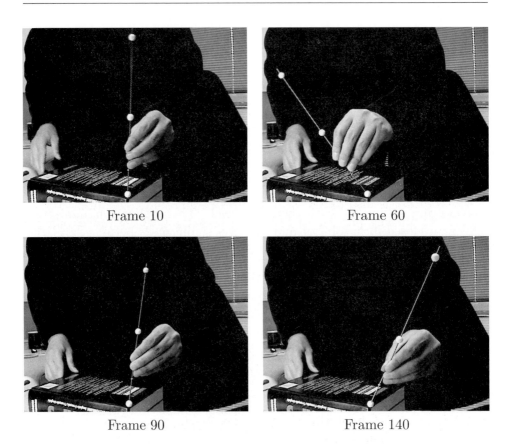

Frame 10   Frame 60

Frame 90   Frame 140

Figure 2.9. Images of a 1D object used for camera calibration.

Figure 2.10. Image of the planar pattern used for camera calibration.

- The supposed-to-be fixed point was not fixed. It slipped around on the surface.

- The positioning of the beads was done with a ruler using eye inspection.

Considering all the factors, the proposed algorithm is very encouraging.

2.6   Self-Calibration

Self-calibration is also called autocalibration. Techniques in this category do not require any particular calibration object. They can be considered 0D approach because only image point correspondences are required. Just by moving a camera in a static scene, the rigidity of the scene provides in general two constraints [22, 21, 20] on the cameras' internal parameters from one camera displacement by using image information alone. Absolute conic, described in Section 2.2.2, is an essential concept in understanding these constraints. Therefore, if images are taken by the same camera with fixed internal parameters, correspondences between three images are sufficient to recover both the internal and external parameters, which allow us to reconstruct 3D structure up to a similarity [20, 17]. Although no calibration objects are necessary, a large number of parameters need to be estimated, resulting in a much harder mathematical problem.

Two recent books [15, 7] provide an excellent recount of those techniques.

2.7   Conclusion

In this chapter, we reviewed several camera calibration techniques. We classified them into four categories, depending whether they use 3D apparatus, 2D objects (planes), 1D objects, or just the surrounding scenes (self-calibration). Recommendations on choosing which technique to use were given in the introduction section.

The techniques described so far are mostly focused on a single-camera calibration. We touched a little bit on stereo calibration in Section 2.4.9. Camera calibration is still an active research area because more and more applications use cameras. In [2], spheres are used to calibrate one or more cameras, which can be considered a 2D approach, since only the surface property is used. In [5], a technique is described to calibrate a camera network consisting of an omni-camera and a number of perspective cameras. In [24], a technique is proposed to calibrate a projector-screen-camera system.

2.8   Appendix: Estimating Homography Between the Model Plane and Its Image

There are many ways to estimate the homography between the model plane and its image. Here, we present a technique based on maximum likelihood criterion. Let M_i and $\mathbf{m}_i$ be the model and image points respectively. Ideally, they should satisfy (2.18). In practice, they do not because of noise in the extracted image points. Let's assume that $\mathbf{m}_i$ is corrupted by Gaussian noise with mean $\mathbf{0}$ and covariance matrix $\Lambda_{\mathbf{m}_i}$. Then, the maximum likelihood estimation of $\mathbf{H}$ is obtained by minimizing the following functional

$$\sum_i (\mathbf{m}_i - \widehat{\mathbf{m}}_i)^T \Lambda_{\mathbf{m}_i}^{-1} (\mathbf{m}_i - \widehat{\mathbf{m}}_i),$$

where $\widehat{\mathbf{m}}_i = \dfrac{1}{\bar{\mathbf{h}}_3^T M_i} \begin{bmatrix} \bar{\mathbf{h}}_1^T M_i \\ \bar{\mathbf{h}}_2^T M_i \end{bmatrix}$, with $\bar{\mathbf{h}}_i$ being the i-th row of $\mathbf{H}$.

In practice, we simply assume $\Lambda_{\mathbf{m}_i} = \sigma^2 \mathbf{I}$ for all i. This is reasonable if points are extracted independently with the same procedure. In this case, the above problem becomes a nonlinear least-squares one: $\min_{\mathbf{H}} \sum_i \|\mathbf{m}_i - \widehat{\mathbf{m}}_i\|^2$. The nonlinear minimization is conducted with the Levenberg-Marquardt algorithm, as implemented in `Minpack` [23]. This requires an initial guess, which can be obtained as follows.

Let $\mathbf{x} = [\bar{\mathbf{h}}_1^T, \bar{\mathbf{h}}_2^T, \bar{\mathbf{h}}_3^T]^T$. Then equation (2.18) can be rewritten as

$$\begin{bmatrix} \widetilde{M}^T & \mathbf{0}^T & -u\widetilde{M}^T \\ \mathbf{0}^T & \widetilde{M}^T & -v\widetilde{M}^T \end{bmatrix} \mathbf{x} = \mathbf{0} .$$

When we are given n points, we have n such equations, which can be written in matrix equation as $\mathbf{L}\mathbf{x} = \mathbf{0}$, where $\mathbf{L}$ is a $2n \times 9$ matrix. As $\mathbf{x}$ is defined up to a scale factor, the solution is well known to be the right singular vector of $\mathbf{L}$ associated with the smallest singular value (or equivalently, the eigenvector of $\mathbf{L}^T\mathbf{L}$ associated with the smallest eigenvalue). In $\mathbf{L}$, some elements are constant 1, some are in pixels, some are in world coordinates, and some are multiplications of both. This makes $\mathbf{L}$ poorly conditioned numerically. Much better results can be obtained by performing a simple data normalization prior to running the above procedure.

Bibliography

[1] Y. I. Abdel-Aziz and H. M. Karara. Direct linear transformation into object space coordinates in close-range photogrammetry. In *Proceedings of the*

Symposium on Close-Range Photogrammetry, University of Illinois at Urbana-Champaign, pages 1–18, January 1971.

[2] M. Agrawal and L. Davis. Camera calibration using spheres: A semi-definite programming approach. In *Proceedings of the 9th International Conference on Computer Vision*, pages 782–789, October 2003. IEEE Computer Society Press.

[3] D. C. Brown. Close-range camera calibration. *Photogrammetric Engineering*, 37(8):855–866, 1971.

[4] B. Caprile and V. Torre. Using Vanishing Points for Camera Calibration. *International Journal of Computer Vision*, 4(2):127–140, March 1990.

[5] X. Chen, J. Yang, and A. Waibel. Calibration of a hybrid camera network. In *Proceedings of the 9th International Conference on Computer Vision*, pages 150–155, October 2003. IEEE Computer Society Press.

[6] W. Faig. Calibration of close-range photogrammetry systems: Mathematical formulation. *Photogrammetric Engineering and Remote Sensing*, 41(12):1479–1486, 1975.

[7] O. Faugeras and Q.-T. Luong. *The Geometry of Multiple Images*. The MIT Press, 2001. With contributions from T. Papadopoulo.

[8] O. Faugeras. *Three-Dimensional Computer Vision: a Geometric Viewpoint*. MIT Press, 1993.

[9] O. Faugeras, T. Luong, and S. Maybank. Camera self-calibration: Theory and experiments. In G. Sandini (Ed.), *Proc. 2nd European Conference on Computer Vision*, volume 588 of *Lecture Notes in Computer Science*, pages 321–334, May 1992. Springer-Verlag.

[10] O. Faugeras and G. Toscani. The calibration problem for stereo. In *Proceedings of the IEEE Conference on Computer Vision and Pattern Recognition*, pages 15–20, June 1986. IEEE.

[11] S. Ganapathy. Decomposition of transformation matrices for robot vision. *Pattern Recognition Letters*, 2:401–412, December 1984.

[12] D. Gennery. Stereo-camera calibration. In *Proceedings of the 10th Image Understanding Workshop*, pages 101–108, 1979.

[13] G. H. Golub and C. F. van Loan. *Matrix Computations*, 3rd ed. Johns Hopkins University Press, Baltimore, Maryland, 1996.

[14] P. Gurdjos, A. Crouzil, and R. Payrissat. Another way of looking at plane-based calibration: The centre circle constraint. In *Proceedings of the 7th European Conference on Computer Vision*, volume IV, pages 252–266, May 2002.

[15] R. Hartley and A. Zisserman. *Multiple View Geometry in Computer Vision*. Cambridge University Press, 2000.

[16] R. Hartley. Self-calibration from multiple views with a rotating camera. In J.-O. Eklundh (Ed.), *Proceedings of the 3rd European Conference on Computer Vision*, volume 800–801 of *Lecture Notes in Computer Science*, pages 471–478, May 1994. Springer-Verlag.

[17] R. Hartley. An algorithm for self calibration from several views. In *Proceedings of the IEEE Conference on Computer Vision and Pattern Recognition*, pages 908–912, June 1994. IEEE.

[18] J. Heikkilä and O. Silvén. A four-step camera calibration procedure with implicit image correction. In *Proceedings of the IEEE Conference on Computer Vision and Pattern Recognition*, pages 1106–1112, June 1997. IEEE Computer Society.

[19] D. Liebowitz and A. Zisserman. Metric rectification for perspective images of planes. In *Proceedings of the IEEE Conference on Computer Vision and Pattern Recognition*, pages 482–488, June 1998. IEEE Computer Society.

[20] Q.-T. Luong and O. D. Faugeras. Self-calibration of a moving camera from point correspondences and fundamental matrices. *International Journal of Computer Vision*, 22(3):261–289, 1997.

[21] Q.-T. Luong. *Matrice Fondamentale et Calibration Visuelle sur l'Environnement-Vers une plus grande autonomie des systèmes robotiques.* PhD thesis, Université de Paris-Sud, Centre d'Orsay, December 1992.

[22] S. J. Maybank and O. D. Faugeras. A theory of self-calibration of a moving camera. *International Journal of Computer Vision*, 8(2):123–152, August 1992.

[23] J. J. More. The Levenberg-Marquardt algorithm, implementation and theory. In G. A. Watson (Ed.), *Numerical Analysis*, Lecture Notes in Mathematics 630. Springer-Verlag, 1977.

[24] T. Okatani and K. Deguchi. Autocalibration of projector-screen-camera system: Theory and algorithm for screen-to-camera homography estimation. In *Proceedings of the 9th International Conference on Computer Vision*, pages 774–781, October 2003. IEEE Computer Society Press.

[25] L. Robert. Camera calibration without feature extraction. *Computer Vision, Graphics, and Image Processing*, 63(2):314–325, March 1995. also available as INRIA Technical Report 2204.

[26] J. G. Semple and G. T. Kneebone. *Algebraic Projective Geometry*. Oxford: Clarendon Press, 1952. Reprinted 1979.

[27] S. W. Shih, Y. P. Hung, and W. S. Lin. Accurate linear technique for camera calibration considering lens distortion by solving an eigenvalue problem. *Optical Engineering*, 32(1):138–149, 1993.

[28] I. Shimizu, Z. Zhang, S. Akamatsu, and K. Deguchi. Head pose determination from one image using a generic model. In *Proceedings of the IEEE Third International Conference on Automatic Face and Gesture Recognition*, pages 100–105, April 1998.

[29] C. C. Slama, editor. *Manual of Photogrammetry*. American Society of Photogrammetry, fourth edition, 1980.

[30] G. Stein. Accurate internal camera calibration using rotation, with analysis of sources of error. In *Proc. Fifth International Conference on Computer Vision*, pages 230–236, June 1995.

[31] P. Sturm and S. Maybank. On plane-based camera calibration: A general algorithm, singularities, applications. In *Proceedings of the IEEE Conference on Computer Vision and Pattern Recognition*, pages 432–437, June 1999. IEEE Computer Society Press.

[32] B. Triggs. Autocalibration from planar scenes. In *Proceedings of the 5th European Conference on Computer Vision*, pages 89–105, June 1998.

[33] R. Y. Tsai. A versatile camera calibration technique for high-accuracy 3D machine vision metrology using off-the-shelf TV cameras and lenses. *IEEE Journal of Robotics and Automation*, 3(4):323–344, August 1987.

[34] T. Ueshiba and F. Tomita. Plane-based calibration algorithm for multi-camera systems via factorization of homography matrices. In *Proceedings of the 9th International Conference on Computer Vision*, pages 966–973, October 2003. IEEE Computer Society Press.

[35] G. Q. Wei and S. D. Ma. A complete two-plane camera calibration method and experimental comparisons. In *Proc. Fourth International Conference on Computer Vision*, pages 439–446, May 1993.

[36] G. Q. Wei and S. D. Ma. Implicit and explicit camera calibration: Theory and experiments. *IEEE Transactions on Pattern Analysis and Machine Intelligence*, 16(5):469–480, 1994.

[37] J. Weng, P. Cohen, and M. Herniou. Camera calibration with distortion models and accuracy evaluation. *IEEE Transactions on Pattern Analysis and Machine Intelligence*, 14(10):965–980, October 1992.

[38] R. Willson. *Modeling and Calibration of Automated Zoom Lenses*. PhD thesis, Department of Electrical and Computer Engineering, Carnegie Mellon University, 1994.

[39] R. Yang and Z. Zhang. Eye gaze correction with stereovision for video teleconferencing. In *Proceedings of the 7th European Conference on Computer Vision*, volume II, pages 479–494, May 2002. Also available as Technical Report MSR-TR-01-119.

[40] R. Yang and Z. Zhang. Model-based head pose tracking with stereovision. In *Proc. Fifth IEEE International Conference on Automatic Face and Gesture Recognition (FG2002)*, pages 255–260, May 2002. Also available as Technical Report MSR-TR-01-102.

[41] Z. Zhang. A flexible new technique for camera calibration. Technical Report MSR-TR-98-71, Microsoft Research, December 1998. Available together with the software at `http://research.microsoft.com/~zhang/Calib/`.

[42] Z. Zhang. A flexible new technique for camera calibration. *IEEE Transactions on Pattern Analysis and Machine Intelligence*, 22(11):1330–1334, 2000.

[43] Z. Zhang. Camera calibration with one-dimensional objects. Technical Report MSR-TR-2001-120, Microsoft Research, December 2001.

[44] Z. Zhang. Camera calibration with one-dimensional objects. In *Proc. European Conference on Computer Vision*, volume IV, pages 161–174, May 2002.

[45] Z. Zhang and O. D. Faugeras. *3D Dynamic Scene Analysis: A Stereo Based Approach*. Springer-Verlag, 1992.

Chapter 3

MULTIPLE VIEW GEOMETRY

Anders Heyden

and Marc Pollefeys

3.1   Introduction

There exist intricate geometric relations between multiple views of a 3D scene. These relations are related to the camera motion and calibration as well as to the scene structure. In this chapter, we introduce these concepts and discuss how they can be applied to recover 3D models from images.

In Section 3.2, a rather thorough description of projective geometry is given. Section 3.3 gives a short introduction to tensor calculus, and Section 3.4 describes in detail the camera model used. In Section 3.5, a modern approach to multiple view geometry is presented and in Section 3.6 simple structure and motion algorithms are presented. In Section 3.7, more advanced algorithms are presented that are suited for automatic processing on real image data. Section 3.8 discusses the possibility of calibrating the camera from images. Section 3.9 describes how the depth can be computed for most image pixels, and Section 3.10 presents how the results of the previous sections can be combined to yield 3D models, render novel views, or combine real and virtual elements in video.

44

3.2 Projective Geometry

Projective geometry is a fundamental tool for dealing with structure from motion problems in computer vision, especially in multiple view geometry. The main reason is that the image formation process can be regarded as a projective transformation from a 3D to a 2D projective space. It is also a fundamental tool for dealing with autocalibration problems and examining critical configurations and critical motion sequences.

This section deals with the fundamentals of projective geometry, including the definitions of projective spaces, homogeneous coordinates, duality, projective transformations, and affine and Euclidean embeddings. For a traditional approach to projective geometry, see [9], and for more modern treatments, see [14, 15, 24].

3.2.1 The central perspective transformation

We start the introduction of projective geometry by looking at a central perspective transformation, which is very natural from an image formation point of view; see Figure 3.1.

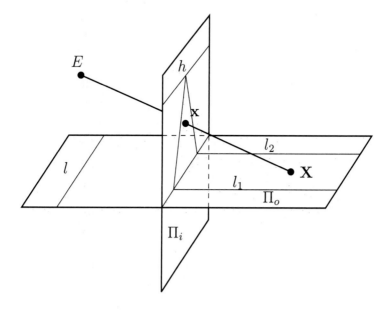

Figure 3.1. A central perspective transformation.

Definition 1. A **central perspective transformation** maps points, $\mathbf{X}$, on the object plane, Π_o, to points on the image plane Π_i, by intersecting the line through E, called the **center**, and $\mathbf{X}$ with Π_i. ∎

We can immediately see the following properties of the planar perspective transformation:

- All points on Π_o map to points on Π_i except for points on l, where l is defined as the intersection of Π_o with the plane incident with E and parallel with Π_i.

- All points on Π_i are images of points on Π_o except for points on h, called the **horizon**, where h is defined as the intersection of Π_i with the plane incident with E and parallel with Π_o.

- Lines in Π_o are mapped to lines in Π_i.

- The images of parallel lines intersect in a point on the horizon, see l_1 and l_2 in Figure 3.1 for an example.

- In the limit where the point E moves infinitely far away, the planar perspective transformation turns into a parallel projection.

Identify the planes Π_o and Π_i with $\mathbb{R}^2$, with a standard cartesian coordinate system $O\mathbf{e}_1\mathbf{e}_2$ in Π_o and Π_i respectively. The central perspective transformation is *nearly* a bijective transformation between Π_o and Π_i, that is, from $\mathbb{R}^2$ to $\mathbb{R}^2$. The problem is the lines $l \in \Pi_o$ and $h \in \Pi_i$. If we remove these lines, we obtain a bijective transformation between $\mathbb{R}^2 \setminus l$ and $\mathbb{R}^2 \setminus h$, but this is not the path that we will follow. Instead, we extend each $\mathbb{R}^2$ with an extra line defined as the images of points on h and points that map to l in the natural way, i.e. maintaining continuity. Thus by adding one artificial line to each plane, it is possible to make the central perspective transformation bijective from ($\mathbb{R}^2$ + an extra line) to ($\mathbb{R}^2$ + an extra line). These extra lines correspond naturally to directions in the other plane, such as the images of the lines l_1 and l_2 intersects in a point on h corresponding to the direction of l_1 and l_2. The intersection point on h can be viewed as the limit of images of a point on l_1 moving infinitely far away. Inspired by this observation, we make the following definition:

Definition 2. Consider the set L of all lines parallel to a given line l in $\mathbb{R}^2$ and assign a point to each such set, $\mathbf{p}_{\text{ideal}}$, called an **ideal point** or **point at infinity**; see Figure 3.2. ∎

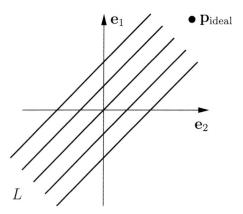

Figure 3.2. The point at infinity corresponding to the set of lines L.

3.2.2 Projective spaces

We are now ready to make a preliminary definition of the 2D projective space, namely, the projective plane.

Definition 3. The **projective plane**, $\mathbb{P}^2$, is defined according to

$$\mathbb{P}^2 = \mathbb{R}^2 \cup \{\text{ideal points}\}.$$

∎

Definition 4. The **ideal line**, l_∞ or **line at infinity** in $\mathbb{P}^2$ is defined according to

$$l_\infty = \{\text{ideal points}\}.$$

∎

The following constructions could easily be made in $\mathbb{P}^2$:

1. Two different points define a line (called the **join** of the points),
2. Two different lines intersect in a point,

with obvious interpretations for ideal points and the ideal line, for example, the line defined by an ordinary point and an ideal point is the line incident with the ordinary point with the direction given by the ideal point. Similarly we define

Definition 5. The **projective line**, $\mathbb{P}^1$, is defined according to

$$\mathbb{P}^1 = \mathbb{R}^1 \cup \{\text{ideal point}\}.$$

∎

Observe that the projective line contains only one ideal point, which could be regarded as the point at infinity.

In order to define 3D projective space, $\mathbb{P}^3$, we start with $\mathbb{R}^3$ and assign an ideal point to each set of parallel lines; that is, to each direction.

Definition 6. The **projective space**, $\mathbb{P}^3$, is defined according to

$$\mathbb{P}^3 = \mathbb{R}^3 \cup \{\text{ideal points}\}.$$

∎

Observe that the ideal points in $\mathbb{P}^3$ constitutes a 2D manifold, which motivates the following definition.

Definition 7. The ideal points in $\mathbb{P}^3$ build up a plane, called the **ideal plane** or **plane at infinity**, also containing ideal lines.   ∎

The plane at infinity contains lines, again called lines at infinity. Every set of parallel planes in $\mathbb{R}^3$ defines an ideal line and all ideal lines build up the ideal plane. A lot of geometrical constructions can be made in $\mathbb{P}^3$. For example,

1. Two different points define a line (called the **join** of the two points).

2. Three different points define a plane (called the **join** of the three points).

3. Two different planes intersect in a line.

4. Three different planes intersect in a point.

3.2.3   Homogeneous coordinates

It is often advantageous to introduce coordinates in the projective spaces, called analytic projective geometry. Introduce a cartesian coordinate system, $Oe_x e_y$ in $\mathbb{R}^2$, and define the line $l : y = 1$; see Figure 3.3. We make the following simple observations:

Lemma 1. *The vectors (p_1, p_2) and (q_1, q_2) determine the same line through the origin iff*

$$(p_1, p_2) = \lambda(q_1, q_2), \quad \lambda \neq 0.$$

Proposition 1. *Every line, $l_p = (p_1, p_2)t$, $t \in \mathbb{R}$, through the origin, except for the x-axis, intersect the line l in one point, $\mathbf{p}$.*

We can now make the following definitions:

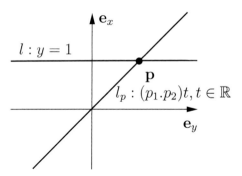

Figure 3.3. Definition of homogeneous coordinates in $\mathbb{P}^1$.

Definition 8. The pairs of numbers (p_1, p_2) and (q_1, q_2) are said to be **equivalent** if

$$(p_1, p_2) = \lambda(q_1, q_2), \quad \lambda \neq 0.$$

We write

$$(p_1, p_2) \sim (q_1, q_2).$$

∎

There is a one-to-one correspondence between lines through the origin and points on the line l if we add an extra point on the line, corresponding to the line $x = 0$; that is, the direction $(1, 0)$. By identifying the line l augmented with this extra point, corresponding to the point at infinity, p_∞, with $\mathbb{P}^1$, we can make the following definitions:

Definition 9. The **1D projective space**, $\mathbb{P}^1$, consists of pairs of numbers (p_1, p_2) (under the equivalence above), where $(p_1, p_2) \neq (0, 0)$. The pair (p_1, p_2) is called **homogeneous coordinates** for the corresponding point in $\mathbb{P}^1$. ∎

Theorem 1. *There is a natural division of $\mathbb{P}^1$ into two disjoint subsets*

$$\mathbb{P}^1 = \{(p_1, 1) \in \mathbb{P}^1\} \cup \{(p_1, 0) \in \mathbb{P}^1\},$$

corresponding to ordinary points and the ideal point.

The introduction of homogeneous coordinates can easily be generalized to $\mathbb{P}^2$ and $\mathbb{P}^3$ using three and four homogeneous coordinates respectively. In the case of $\mathbb{P}^2$ fix a cartesian coordinate system, $Oe_x e_y e_z$ in $\mathbb{R}^3$ and define

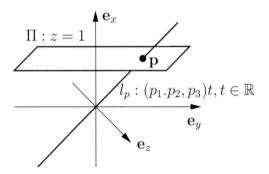

Figure 3.4. Definition of homogeneous coordinates in $\mathbb{P}^2$.

the plane $\Pi : z = 1$; see Figure 3.4. The vectors (p_1, p_2, p_3) and (q_1, q_2, q_3) determines the same line through the origin iff

$$(p_1, p_2, p_3) = \lambda(q_1, q_2, q_3), \quad \lambda \neq 0.$$

Every line through the origin, except for those in the x-y-plane, intersect the plane Π in one point. Again, there is a one-to-one correspondence between lines through the origin and points on the plane Π if we add an extra line corresponding to the lines in the plane $z = 0$, that is, the line at infinity, l_∞, built up by lines of the form $(p_1, p_2, 0)$. We can now identify the plane Π augmented with this extra line, corresponding to the points at infinity, l_∞, with $\mathbb{P}^2$.

Definition 10. The pairs of numbers (p_1, p_2, p_3) and (q_1, q_2, q_3) are said to be equivalent iff

$$(p_1, p_2, p_3) = \lambda(q_1, q_2, q_3), \quad \lambda \neq 0 \quad \text{written} \quad (p_1, p_2, p_3) \sim (q_1, q_2, q_3).$$

                                                                ■

Definition 11. The 2D projective space $\mathbb{P}^2$ consists of all triplets of numbers $(p_1, p_2, p_3) \neq (0, 0, 0)$. The triplet (p_1, p_2, p_3) is called homogeneous coordinates for the corresponding point in $\mathbb{P}^2$.   ■

Theorem 2. *There is a natural division of $\mathbb{P}^2$ into two disjoint subsets*

$$\mathbb{P}^2 = \{(p_1, p_2, 1) \in \mathbb{P}^2\} \cup \{(p_1, p_2, 0) \in \mathbb{P}^2\},$$

corresponding to ordinary points and ideal points (or points at infinity).

The same procedure can be carried out to construct $\mathbb{P}^3$ (and even $\mathbb{P}^n$ for any $n \in \mathbb{N}$), but it is harder to visualize, since we have to start with $\mathbb{R}^4$.

Definition 12. The **3D (nD) projective space** $\mathbb{P}^3$ ($\mathbb{P}^n$) is defined as the set of 1D linear subspaces in a vector space, $\mathbb{V}$ (usually $\mathbb{R}^4$ ($\mathbb{R}^{n+1}$)) of dimension 4 ($n+1$). Points in $\mathbb{P}^3$ ($\mathbb{P}^n$) are represented using **homogeneous coordinates** by vectors $(p_1, p_2, p_3, p_4) \neq (0, 0, 0, 0)$ $((p_1, \ldots, p_{n+1}) \neq (0, \ldots, 0))$, where two vectors represent the same point iff they differ by a global scale factor. There is a natural division of $\mathbb{P}^3$ ($\mathbb{P}^n$) into two disjoint subsets

$$\mathbb{P}^3 = \{(p_1, p_2, p_3, 1) \in \mathbb{P}^3\} \cup \{(p_1, p_2, p_3, 0) \in \mathbb{P}^3\}$$
$$(\mathbb{P}^n = \{(p_1, \ldots, p_n, 1) \in \mathbb{P}^n\} \cup \{(p_1, \ldots, p_n, 0) \in \mathbb{P}^n\},$$

corresponding to ordinary points and ideal points (or points at infinity).   ∎

Finally, geometrical entities are defined similarly in $\mathbb{P}^3$.

3.2.4   Duality

Remember that a line in $\mathbb{P}^2$ is defined by two points $\mathbf{p}_1$ and $\mathbf{p}_2$ according to

$$l = \{\mathbf{x} = (x_1, x_2, x_3) \in \mathbb{P}^2 \mid \mathbf{x} = t_1\mathbf{p}_1 + t_2\mathbf{p}_2, \quad (t_1, t_2) \in \mathbb{R}^2\}.$$

Observe that since (x_1, x_2, x_3) and $\lambda(x_1, x_2, x_3)$ represents the same point in $\mathbb{P}^2$, the parameters (t_1, t_2) and $\lambda(t_1, t_2)$ gives the same point. This gives the equivalent definition:

$$l = \{\mathbf{x} = (x_1, x_2, x_3) \in \mathbb{P}^2 \mid \mathbf{x} = t_1\mathbf{p}_1 + t_2\mathbf{p}_2, \quad (t_1, t_2) \in \mathbb{P}^1\}.$$

By eliminating the parameters t_1 and t_2, we could also write the line in the form

$$l = \{\mathbf{x} = (x_1, x_2, x_3) \in \mathbb{P}^2 \mid n_1x_1 + n_2x_2 + n_3x_3 = 0\}, \qquad (3.1)$$

where the normal vector, $\mathbf{n} = (n_1, n_2, n_3)$, could be calculated as $\mathbf{n} = \mathbf{p}_1 \times \mathbf{p}_2$. Observe that if (x_1, x_2, x_3) fulfills (3.1), then $\lambda(x_1, x_2, x_3)$ also fulfills (3.1), and that if the line, l, is defined by (n_1, n_2, n_3), then the same line is defined by $\lambda(n_1, n_2, n_3)$, which means that $\mathbf{n}$ could be considered as an element in $\mathbb{P}^2$.

The line equation in (3.1) can be interpreted in two different ways; see Figure 3.5:

- Given $\mathbf{n} = (n_1, n_2, n_3)$, the points $\mathbf{x} = (x_1, x_2, x_3)$ that fulfill (3.1) constitute the line defined by $\mathbf{n}$.

   – Given $\mathbf{x} = (x_1, x_2, x_3)$, the lines $\mathbf{n} = (n_1, n_2, n_3)$ that fulfill (3.1) constitute the lines coincident by $\mathbf{x}$.

Definition 13. The set of lines incident with a given point $\mathbf{x} = (x_1, x_2, x_3)$ is called a **pencil** of lines.   ∎

In this way, there is a one-to-one correspondence between points and lines in $\mathbb{P}^2$ given by

$$\mathbf{x} = (a, b, c) \quad \leftrightarrow \quad \mathbf{n} = (a, b, c),$$

as illustrated in Figure 3.5.

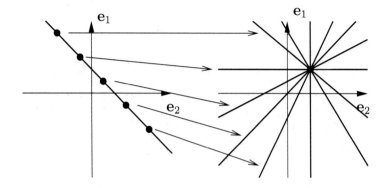

Figure 3.5. Duality of points and lines in $\mathbb{P}^2$.

   Similarly, there exists a duality between points and planes in $\mathbb{P}^3$. A plane π in $\mathbb{P}^3$ consists of the points $\mathbf{x} = (x_1, x_2, x_3, x_4)$ that fulfill the equation

$$\pi = \{\mathbf{x} = (x_1, x_2, x_3, x_4) \in \mathbb{P}^3 \mid n_1 x_1 + n_2 x_2 + n_3 x_3 + n_4 x_4 = 0\}, \quad (3.2)$$

where $\mathbf{n} = (n_1, n_2, n_3, n_4)$ defines the plane. From (3.2) a similar argument leads to a duality between planes and points in $\mathbb{P}^3$. The following theorem is fundamental in projective geometry:

Theorem 3. *Given a statement valid in a projective space, the dual to that statement is also valid, where the dual is obtained by interchanging entities with their duals, intersection with join, and so on.*

For instance, a line in $\mathbb{P}^3$ could be defined as the join of two points. Thus, the dual to a line is the intersection of two planes, which again is a line; that is, the dual to a line in $\mathbb{P}^3$ is a line. We say that lines are **self-dual**. A line in $\mathbb{P}^3$ defined as the join of two points, $\mathbf{p}_1$ and $\mathbf{p}_2$, as in

$$l = \{\mathbf{x} = (x_1, x_2, x_3, x_4) \in \mathbb{P}^3 \mid \mathbf{x} = t_1 \mathbf{p}_1 + t_2 \mathbf{p}_2, \quad (t_1, t_2) \in \mathbb{P}^1\}$$

is said to be given in **parametric form** and (t_1, t_2) can be regarded as homogeneous coordinates on the line. A line in $\mathbb{P}^3$ defined as the intersection of two planes, π and μ, consisting of the common points to the pencil of planes in

$$l : \{s\pi + t\mu \mid (s, t) \in \mathbb{P}^1\}$$

is said to be given in **intersection form**.

Definition 14. A **conic**, c, in $\mathbb{P}^2$ is defined as

$$c = \{\mathbf{x} = (x_1, x_2, x_3) \in \mathbb{P}^2 \mid \mathbf{x}^T C \mathbf{x} = 0\}, \qquad (3.3)$$

where C denotes a 3×3 matrix. If C is nonsingular, the conic is said to be **proper**; otherwise, it is said to be **degenerate**. ∎

The dual to a general curve in $\mathbb{P}^2$ ($\mathbb{P}^3$) is defined as the set of tangent lines (tangent planes) to the curve.

Theorem 4. *The dual, c^*, to a conic $c : \mathbf{x}^T C \mathbf{x}$ is the set of lines*

$$\{\mathbf{l} = (l_1, l_2, l_3) \in \mathbb{P}^2 \mid \mathbf{l}^T C' \mathbf{l} = 0\},$$

where $C' = C^{-1}$.

3.2.5   Projective transformations

The central perspective transformation in Figure 3.1 is an example of a projective transformation. The general definition is as follows:

Definition 15. A **projective transformation** from $\mathbf{p} \in \mathbb{P}^n$ to $\mathbf{p}' \in \mathbb{P}^m$ is defined as a linear transformation in homogeneous coordinates:

$$\mathbf{x}' \sim H\mathbf{x}, \qquad (3.4)$$

where $\mathbf{x}$ and $\mathbf{x}'$ denote homogeneous coordinates for $\mathbf{p}$ and $\mathbf{p}'$ respectively, and H denote a $(m+1) \times (n+1)$ matrix of full rank. ∎

All projective transformations form a group, denoted $\mathcal{G}_P$. For example, a projective transformation from $\mathbf{x} \in \mathbb{P}^2$ to $\mathbf{y} \in \mathbb{P}^2$ is given by

$$\begin{bmatrix} y_1 \\ y_2 \\ y_3 \end{bmatrix} \sim H \begin{bmatrix} x_1 \\ x_2 \\ x_3 \end{bmatrix},$$

where H denote a nonsingular 3×3 matrix. Such a projective transformation from $\mathbb{P}^2$ to $\mathbb{P}^2$ is usually called a **homography**.

It is possible to embed an affine space in the projective space by a simple construction:

Definition 16. The subspaces

$$\mathcal{A}_i^n = \{ (x_1, x_2, \ldots, x_{n+1}) \in \mathbb{P}^n \mid x_i \neq 0 \}$$

of $\mathbb{P}^n$ are called **affine pieces** of $\mathbb{P}^n$. The plane $H_i : x_i = 0$ is called the **plane at infinity**, corresponding to the affine piece $\mathcal{A}_i^n$. Usually, $i = n + 1$ is used and called the **standard affine piece**, and in this case the plane at infinity is denoted H_∞.   ■

We can now identify points in $\mathbb{A}^n$ with points, $\mathbf{x}$, in $\mathcal{A}_i^n \subset \mathbb{P}^n$, by

$$\mathbb{P}^n \ni (x_1, x_2, \ldots, x_n, x_{n+1}) \sim (y_1, y_2, \ldots, y_n, 1) \equiv (y_1, y_2, \ldots, y_n) \in \mathbb{A}^n.$$

There is even an affine structure in this affine piece, given by the following proposition:

Proposition 2. *The subgroup, $\mathcal{H}$, of projective transformations, $\mathcal{G}_P$, that preserves the plane at infinity consists exactly of the projective transformations of the form (3.4), with*

$$H = \begin{bmatrix} A_{n \times n} & b_{n \times 1} \\ 0_{1 \times n} & 1 \end{bmatrix},$$

where the indices denote the sizes of the matrices.

We can now identify the affine transformations in $\mathbb{A}$ with the subgroup $\mathcal{H}$:

$$\mathbb{A} \ni \mathbf{x} \mapsto A\mathbf{x} + b \in \mathbb{A},$$

which gives the affine structure in $\mathcal{A}_i^n \subset \mathbb{P}^n$.

Definition 17. When a plane at infinity has been chosen, two lines are said to be **parallel** if they intersect at a point at the plane at infinity.   ■

We may even go one step further and define a Euclidean structure in $\mathbb{P}^n$.

Definition 18. The (singular, complex) conic, Ω, in $\mathbb{P}^n$ defined by

$$x_1^2 + x_1^2 + \ldots + x_n^2 = 0 \quad \text{and} \quad x_{n+1} = 0$$

is called the **absolute conic**.   ■

Observe that the absolute conic is located at the plane at infinity, it contains only complex points, and it is singular.

Lemma 2. *The dual to the absolute conic, denoted Ω', is given by the set of planes*

$$\Omega' = \{\Pi = (\Pi_1, \Pi_2, \ldots \Pi_{n+1}) \mid \Pi_1^2 + \ldots + \Pi_n^2 = 0.$$

In matrix form, Ω' can be written as $\Pi^T C' \Pi = 0$ with

$$C' = \begin{bmatrix} I_{n \times n} & 0_{n \times 1} \\ 0_{1 \times n} & 0 \end{bmatrix}.$$

Proposition 3. *The subgroup, $\mathcal{K}$, of projective transformations, $\mathcal{G}_P$, that preserves the absolute conic consists exactly of the projective transformations of the form (3.4), with*

$$H = \begin{bmatrix} cR_{n \times n} & t_{n \times 1} \\ 0_{1 \times n} & 1 \end{bmatrix},$$

where $0 \neq c \in \mathbb{R}$ and R denote an orthogonal matrix; that is, $RR^T = R^T R = I$.

Observe that the corresponding transformation in the affine space $\mathbb{A} = \mathcal{A}_n^{n+1}$ can be written as

$$\mathbb{A} \ni \mathbf{x} \mapsto cR\mathbf{x} + t \in \mathbb{A},$$

which is a similarity transformation. This means that we have a Euclidean structure (to be precise, a similarity structure) in $\mathbb{P}^n$ given by the absolute conic.

3.3   Tensor Calculus

Tensor calculus is a natural tool to use when the objects at study are expressed in a specific coordinate system but have physical properties that are independent on the chosen coordinate system. Another advantage is that it gives a simple and compact notation and the rules for tensor algebra make it easy to remember even quite complex formulas. For a more detailed treatment, see [58], and for an engineering approach, see [42].

In this chapter, a simple definition of a tensor and the basic rules for manipulating tensors are given. We start with a straightforward definition:

Definition 19. An **affine tensor** is an object in a linear space, $\mathcal{V}$, that consists of a collection of numbers that are related to a specific choice of coordinate system in $\mathcal{V}$, indexed by one or several indices:

$$A_{j_1, j_2, \cdots, j_m}^{i_1, i_2, \cdots, i_n}.$$

Furthermore, this collection of numbers transforms in a predefined way when a change of coordinate system in $\mathcal{V}$ is made; see Definition 20. The number of indices $(n + m)$ is called the **degree** of the tensor. The indices may take any value from 1 to the dimension of $\mathcal{V}$. The upper indices are called **contravariant** indices and the lower indices are called **covariant** indices. ∎

There are some simple conventions that have to be remembered:

- *The index rule* When an index appears in a formula, the formula is valid for every value of the index: $a_i = 0 \Rightarrow a_1 = 0, a_2 = 0, \ldots$.

- *The summation convention* When an index appears twice in a formula, it is implicitly assumed that a summation takes place over that index: $a_i b^i = \sum_{i=1,\dim \mathcal{V}} a_i b^i$.

- *The compatibility rule* A repeated index must appear once as a sub-index and once as a super-index.

- *The maximum rule* An index cannot be used more than twice in a term.

Definition 20. When the coordinate system in $\mathcal{V}$ is changed from $\mathbf{e}$ to $\widehat{\mathbf{e}}$ and the points with coordinates $\mathbf{x}$ are changed to $\widehat{\mathbf{x}}$, according to

$$\widehat{\mathbf{e}}_j = S^i_j \mathbf{e}_i \quad \Leftrightarrow \quad \mathbf{x}_i = S^i_j \widehat{\mathbf{x}}_j,$$

then the affine tensor components change according to

$$\widehat{\mathbf{u}}_k = S^j_k \mathbf{u}_j \quad \text{and} \quad \mathbf{v}^j = S^j_k \widehat{\mathbf{v}}^k,$$

for lower and upper indices respectively. ∎

From this definition, the terminology for indices can be motivated, since the covariant indices covary with the basis vectors and the contravariant indices contravary with the basis vectors. It turns out that a vector (e.g., the coordinates of a point) is a contravariant tensor of degree one and that a one-form (e.g., the coordinate of a vector defining a line in $\mathbb{R}^2$ or a hyperplane in $\mathbb{R}^n$) is a covariant tensor of degree one.

Definition 21. The second-order tensor

$$\delta_{ij} = \begin{cases} 1 & i = j \\ 0 & i \neq j \end{cases}$$

is called the **Kronecker delta**. When $\dim \mathcal{V} = 3$, the third order tensor

$$\epsilon_{ijk} = \begin{cases} 1 & \text{(i,j,k) an even permutation} \\ -1 & \text{(i,j,k) an odd permutation} \\ 0 & \text{(i,j,k) has a repeated value} \end{cases}$$

is called the **Levi-Cevita epsilon**.

3.4   Modeling Cameras

This chapter deals with the task of building a mathematical model of a camera. We give a mathematical model of the standard pinhole camera and define intrinsic and extrinsic parameters. For more detailed treatment, see [24, 15], and for a different approach see [26].

3.4.1   The pinhole camera

The simplest optical system used for modeling cameras is the **pinhole camera**. The camera is modeled as a box with a small hole in one of the sides and a photographic plate at the opposite side; see Figure 3.6. Introduce a coordinate system as in Figure 3.6. Observe that the origin of the coordinate system is located at the center of projection, the **focal point**, and that the z-axis coincides with the optical axis. The distance from the focal point to the image, f, is called the **focal length**. Similar triangles give

$$\frac{x}{f} = \frac{X}{Z} \quad \text{and} \quad \frac{y}{f} = \frac{Y}{Z}. \tag{3.5}$$

This equation can be written in matrix form, using homogeneous coordinates, as

$$\lambda \begin{bmatrix} x \\ y \\ 1 \end{bmatrix} = \begin{bmatrix} f & 0 & 0 & 0 \\ 0 & f & 0 & 0 \\ 0 & 0 & 1 & 0 \end{bmatrix} \begin{bmatrix} X \\ Y \\ Z \\ 1 \end{bmatrix}, \tag{3.6}$$

where the **depth**, λ, is equal to Z.

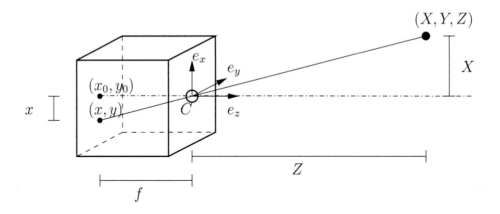

Figure 3.6. The pinhole camera with a coordinate system.

3.4.2 The camera matrix

Introducing the notation

$$K = \begin{bmatrix} f & 0 & 0 \\ 0 & f & 0 \\ 0 & 0 & 1 \end{bmatrix} \quad \mathbf{x} = \begin{bmatrix} x \\ y \\ 1 \end{bmatrix} \quad \mathbf{X} = \begin{bmatrix} X \\ Y \\ Z \\ 1 \end{bmatrix}, \tag{3.7}$$

in (3.6) gives

$$\lambda \mathbf{x} = K[\, I_{3\times 3} \mid \mathbf{0}_{3\times 1}\,]\mathbf{X} = P\mathbf{X}, \tag{3.8}$$

where $P = K\,[\, I_{3\times 3} \mid \mathbf{0}_{3\times 1}\,]$.

Definition 22. A 3×4 matrix P relating extended image coordinates $\mathbf{x} = (x, y, 1)$ to extended object coordinates $\mathbf{X} = (X, Y, Z, 1)$ via the equation

$$\lambda \mathbf{x} = P\mathbf{X}$$

is called a **camera matrix**, and the equation above is called the **camera equation**.   ■

Observe that the focal point is given as the right nullspace to the camera matrix, since $P\mathbf{C} = 0$, where $\mathbf{C}$ denote homogeneous coordinates for the focal point, C.

3.4.3 The intrinsic parameters

In a refined camera model, the matrix K in (3.7) is replaced by

$$K = \begin{bmatrix} \gamma f & sf & x_0 \\ 0 & f & y_0 \\ 0 & 0 & 1 \end{bmatrix}, \tag{3.9}$$

where the parameters have the following interpretations, see Figure 3.7:

- f : **focal length**, also called camera constant
- γ : **aspect ratio**, modeling nonquadratic light-sensitive elements
- s : **skew**, modeling nonrectangular light-sensitive elements
- (x_0, y_0) : **principal point**, orthogonal projection of the focal point onto the image plane; see Figure 3.6.

These parameters are called the **intrinsic parameters**, since they model intrinsic properties of the camera. For most cameras, $s = 0$ and $\gamma \approx 1$ and the principal point is located close to the center of the image.

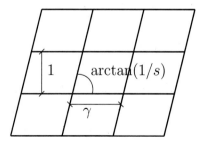

Figure 3.7. The intrinsic parameters.

Definition 23. A camera is said to be **calibrated** if K is known. Otherwise, it is said to be **uncalibrated.**   ■

3.4.4 The extrinsic parameters

It is often advantageous to be able to express object coordinates in a different coordinate system than the camera coordinate system. This is especially the case when the relation between these coordinate systems is not known. For this purpose, it is necessary to model the relation between two different coordinate systems in 3D. The natural way to do this is to model the relation as a Euclidean transformation. Denote the camera coordinate system with e_c and points expressed in this coordinate system with index c–for example, (X_c, Y_c, Z_c)–and similarly denote the object coordinate system with e_o and points expressed in this coordinate system with index o; see Figure 3.8. A

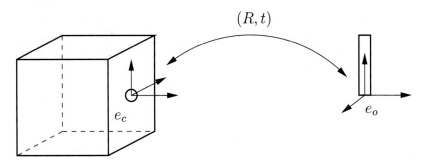

Figure 3.8. Using different coordinate systems for the camera and the object.

Euclidean transformation from the object coordinate system to the camera coordinate system can be written in homogeneous coordinates as

$$\begin{bmatrix} X_c \\ Y_c \\ Z_c \\ 1 \end{bmatrix} = \begin{bmatrix} R^T & 0 \\ 0 & 1 \end{bmatrix} \begin{bmatrix} I & -t \\ 0 & 1 \end{bmatrix} \begin{bmatrix} X_o \\ Y_o \\ Z_o \\ 1 \end{bmatrix} \implies \mathbf{X}_c = \begin{bmatrix} R^T & -R^T t \\ 0 & 1 \end{bmatrix} \mathbf{X}_o, \quad (3.10)$$

where R denotes an orthogonal matrix and t a vector, encoding the rotation and translation in the rigid transformation. Observe that the focal point $(0,0,0)$ in the c-system corresponds to the point t in the o-system. Inserting (3.10) in (3.8), taking into account that $\mathbf{X}$ in (3.8) is the same as $\mathbf{X}_c$ in (3.10), we obtain

$$\lambda \mathbf{x} = K R^T [I \mid -t] \mathbf{X}_o = P\mathbf{X}, \quad (3.11)$$

with $P = K R^T [I \mid -t]$. Usually, it is assumed that object coordinates are expressed in the object coordinate system and the index o in $\mathbf{X}_o$ is omitted. Observe that the focal point, $C_f = t = (t_x, t_y, t_z)$, is given from the right nullspace to P according to

$$P \begin{bmatrix} t_x \\ t_y \\ t_z \\ 1 \end{bmatrix} = K R^T [I \mid -t] \begin{bmatrix} t_x \\ t_y \\ t_z \\ 1 \end{bmatrix} = \mathbf{0}.$$

Given a camera, described by the camera matrix P, this camera could also be described by the camera matrix μP, $0 \neq \mu \in \mathbb{R}$, since these give the same image point for each object point. This means that the camera matrix is only defined up to an unknown scale factor. Moreover, the camera matrix P can be regarded as a projective transformation from $\mathbb{P}^3$ to $\mathbb{P}^2$; see (3.8) and (3.4).

Observe also that replacing t with μt and (X, Y, Z) with $(\mu X, \mu Y, \mu Z)$, $0 \neq \mu \in \mathbb{R}$, gives the same image since

$$KR^T[\, I \mid \, -\mu t\,] \begin{bmatrix} \mu X \\ \mu Y \\ \mu Z \\ 1 \end{bmatrix} = \mu KR^T[\, I \mid \, -t\,] \begin{bmatrix} X \\ Y \\ Z \\ 1 \end{bmatrix}.$$

We refer to this ambiguity as the **scale ambiguity**.

We can now calculate the number of parameters in the camera matrix, P:

- K: 5 parameters (f, γ, s, x_0, y_0)
- R: 3 parameters
- t: 3 parameters

Summing up gives a total of 11 parameters, which is the same as in a general 3×4 matrix defined up to scale. This means that for an uncalibrated camera, the factorization $P = KR^T[\, I \mid \, -t\,]$ has no meaning, and we can instead deal with P as a general 3×4 matrix.

Given a calibrated camera with camera matrix $P = KR^T[\, I \mid \, -t\,]$ and corresponding camera equation

$$\lambda \mathbf{x} = KR^T[\, I \mid \, -t\,]\mathbf{X},$$

it is often advantageous to make a change of coordinates from $\mathbf{x}$ to $\widehat{\mathbf{x}}$ in the image according to $\mathbf{x} = K\widehat{\mathbf{x}}$, which gives

$$\lambda K\widehat{\mathbf{x}} = KR^T[\, I \mid \, -t\,]\mathbf{X} \quad \Rightarrow \quad \lambda \widehat{\mathbf{x}} = R^T[\, I \mid \, -t\,]\mathbf{X} = \widehat{P}\mathbf{X}.$$

Now the camera matrix becomes $\widehat{P} = R^T[\, I \mid \, -t\,]$.

Definition 24. A camera represented with a camera matrix of the form

$$P = R^T[\, I \mid \, -t\,]$$

is called a **normalized camera**. ∎

Note that even when K is only approximatively known, the above normalization can be useful for numerical reasons. This is discussed more in detail in Section 3.7.

3.4.5   Properties of the pinhole camera

We conclude this section with some properties of the pinhole camera.

Proposition 4. *The set of 3D points that project to an image point,* $\mathbf{x}$, *is given by*

$$\mathbf{X} = \mathbf{C} + \mu P^+ \mathbf{x}, \quad 0 \neq \mu \in \mathbb{R},$$

where $\mathbf{C}$ *denotes the focal point in homogeneous coordinates and* P^+ *denotes the pseudo-inverse of* P.

Proposition 5. *The set of 3D-points that projects to a line,* $\mathbf{l}$, *is the points lying on the plane* $\Pi = P^T \mathbf{l}$.

Proof: It is obvious that the set of points lie on the plane defined by the focal point and the line $\mathbf{l}$. A point $\mathbf{x}$ on $\mathbf{l}$ fulfills $\mathbf{x}^T \mathbf{l} = 0$ and a point $\mathbf{X}$ on the plane Π fulfills $\Pi^T \mathbf{X} = 0$. Since $\mathbf{x} \sim P\mathbf{X}$, we have $(P\mathbf{X})^T \mathbf{l} = \mathbf{X}^T P^T \mathbf{l} = 0$ and identification with $\Pi^T \mathbf{X} = 0$ gives $\Pi = P^T \mathbf{l}$.   ■

Lemma 3. *The projection of a quadric,* $\mathbf{X}^T C \mathbf{X} = 0$ *(dually* $\Pi^T C' \Pi = 0$, $C' = C^{-1}$), *is an image conic,* $\mathbf{x}^T c \mathbf{x} = 0$ *(dually* $\mathbf{l}^T c' \mathbf{l} = 0$, $c' = c^{-1}$), *with* $c' = P C' P^T$.

Proof: Use the previous proposition.   ■

Proposition 6. *The image of the absolute conic is given by the conic* $\mathbf{x}^T \omega \mathbf{x} = 0$ *(dually* $\mathbf{l}^T \omega' \mathbf{l} = 0$), *where* $\omega' = KK^T$.

Proof: The result follows from the previous lemma:

$$\omega' \sim P\Omega' P^T \sim KR^T \begin{bmatrix} I & -t \end{bmatrix} \begin{bmatrix} I & \mathbf{0} \\ \mathbf{0} & 0 \end{bmatrix} \begin{bmatrix} I \\ -t^T \end{bmatrix} RK^T = KR^T RK^T = KK^T.$$

     ■

3.5   Multiple View Geometry

Multiple view geometry is the subject in which relations between coordinates of feature points in different views are studied. It is an important tool for understanding the image formation process for several cameras and for designing reconstruction algorithms. For a more detailed treatment, see [27] or [24], and for a different approach, see [26]. For the algebraic properties of multilinear constraints, see [30].

3.5.1   The structure and motion problem

The following problem is central in computer vision:

Problem 1. *Structure and motion. Given a sequence of images with corresponding feature points* $\mathbf{x}_{ij}$*, taken by a perspective camera,*

$$\lambda_{ij}\mathbf{x}_{ij} = P_i\mathbf{X}_j, \quad i = 1,\ldots,m, \quad j = 1,\ldots,n,$$

determine the camera matrices, P_i *(***motion***) and the 3D points,* $\mathbf{X}_j$ *(***structure***), under different assumptions on the intrinsic and/or extrinsic parameters. This is called the* **structure and motion problem***.*

It turns out that there is a fundamental limitation on the solutions to the structure and motion problem when the intrinsic parameters are unknown and possibly varying, namely with an **uncalibrated image sequence**.

Theorem 5. *Given an uncalibrated image sequence with corresponding points, it is only possible to reconstruct the object up to an unknown projective transformation.*

Proof: Assume that $\mathbf{X}_j$ is a reconstruction of n points in m images, with camera matrices P_i according to

$$\mathbf{x}_{ij} \sim P_i\,\mathbf{X}_j, \; i = 1,\ldots m, \; j = 1,\ldots n.$$

Then $H\,\mathbf{X}_j$ is also a reconstruction, with camera matrices $P_i\,H^{-1}$, for every nonsingular 4×4 matrix H, since

$$\mathbf{x}_{ij} \sim P_i\,\mathbf{X}_j \sim P_iH^{-1}H\mathbf{X}_j \sim (P_iH^{-1})\,(H\mathbf{X}_j).$$

The transformation

$$\mathbf{X} \mapsto H\mathbf{X}$$

corresponds to all projective transformations of the object.   ∎

In the same way it can be shown that if the cameras are calibrated, then it is possible to reconstruct the scene up to an unknown similarity transformation.

3.5.2   The two-view case

The epipoles

Consider two images of the same point $\mathbf{X}$ as in Figure 3.9.

Definition 25. The **epipole**, $\mathbf{e}_{i,j}$, is the projection of the focal point of camera i in image j.   ∎

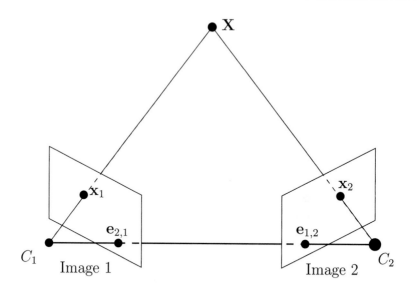

Figure 3.9. Two images of the same point and the epipoles.

Proposition 7. *Let*

$$P_1 = [\, A_1 \mid b_1 \,] \quad and \quad P_2 = [\, A_2 \mid b_2 \,].$$

Then the epipole, $\mathbf{e}_{1,2}$ *is given by*

$$\mathbf{e}_{1,2} = -A_2 A_1^{-1} b_1 + b_2. \tag{3.12}$$

Proof: The focal point of camera 1, C_1, is given by

$$P_1 \begin{bmatrix} C_1 \\ 1 \end{bmatrix} = [\, A_1 \mid b_1 \,] \begin{bmatrix} C_1 \\ 1 \end{bmatrix} = A_1 C_1 + b_1 = 0;$$

that is, $C_1 = -A_1^{-1} b_1$, and then the epipole is obtained from

$$P_2 \begin{bmatrix} C_1 \\ 1 \end{bmatrix} = [\, A_2 \mid b_2 \,] \begin{bmatrix} C_1 \\ 1 \end{bmatrix} = A_2 C_1 + b_2 = -A_2 A_1^{-1} b_1 + b_2.$$

■

It is convenient to use the notation $A_{12} = A_2 A_1^{-1}$. Assume that we have calculated two camera matrices representing the two-view geometry,

$$P_1 = [\, A_1 \mid b_1 \,] \quad and \quad P_2 = [\, A_2 \mid b_2 \,].$$

According to Theorem 5, we can multiply these camera matrices with

$$H = \begin{bmatrix} A_1^{-1} & -A_1^{-1}b_1 \\ 0 & 1 \end{bmatrix}$$

from the right and obtain

$$\bar{P}_1 = P_1 H = [\, I \mid 0 \,] \quad \bar{P}_2 = P_2 H = [\, A_2 A_1^{-1} \mid b_2 - A_2 A_1^{-1} b_1 \,].$$

Thus, we may always assume that the first camera matrix is $[\, I \mid 0 \,]$. Observe that $\bar{P}_2 = [\, A_{12} \mid \mathbf{e} \,]$, where $\mathbf{e}$ denotes the epipole in the second image. Observe also that we may multiply again with

$$\bar{H} = \begin{bmatrix} I & 0 \\ \mathbf{v}^T & 1 \end{bmatrix}$$

without changing $\bar{P}_1$, but

$$\bar{H}\bar{P}_2 = [\, A_{12} + \mathbf{e}\mathbf{v}^T \mid \mathbf{e} \,];$$

that is, the last column of the second camera matrix still represents the epipole.

Definition 26. A pair of camera matrices is said to be in **canonical form** if

$$P_1 = [\, I \mid 0 \,] \quad \text{and} \quad P_2 = [\, A_{12} + \mathbf{e}\mathbf{v}^T \mid \mathbf{e} \,], \tag{3.13}$$

where $\mathbf{v}$ denotes a three-parameter ambiguity.   ∎

The fundamental matrix

The fundamental matrix was originally discovered in the calibrated case in [38] and in the uncalibrated case in [13]. Consider a fixed point, $\mathbf{X}$, in two views:

$$\lambda_1 \mathbf{x}_1 = P_1 \mathbf{X} = [\, A_1 \mid b_1 \,]\mathbf{X}, \quad \lambda_2 \mathbf{x}_2 = P_2 \mathbf{X} = [\, A_2 \mid b_2 \,]\mathbf{X}.$$

Use the first camera equation to solve for X, Y, Z:

$$\lambda_1 \mathbf{x}_1 = P_1 \mathbf{X} = [\, A_1 \mid b_1 \,]\mathbf{X} = A_1 \begin{bmatrix} X \\ Y \\ Z \end{bmatrix} + b_1 \quad \Rightarrow \quad \begin{bmatrix} X \\ Y \\ Z \end{bmatrix} = A_1^{-1}(\lambda_1 \mathbf{x}_1 - b_1)$$

and insert into the second one

$$\lambda_2 \mathbf{x}_2 = A_2 A_1^{-1}(\lambda_1 \mathbf{x}_1 - b_1) + b_2 = \lambda_1 A_{12}\mathbf{x}_1 + (-A_{12}b_1 - b_2).$$

That is, $\mathbf{x}_2$, $A_{12}\mathbf{x}_1$ and $\mathbf{t} = -A_{12}b_1 + b_2 = \mathbf{e}_{1,2}$ are linearly dependent. Observe that $\mathbf{t} = \mathbf{e}_{1,2}$, the epipole in the second image. This condition can be written as $\mathbf{x}_1^T A_{12}^T T_{\mathbf{e}} \mathbf{x}_2 = \mathbf{x}_1^T F \mathbf{x}_2 = 0$, with $F = A_{12}^T T_{\mathbf{e}}$, where $T_{\mathbf{x}}$ denotes the skew-symmetric matrix corresponding to the vector $\mathbf{x}$; that is, $T_{\mathbf{x}}(y) = x \times y$.

Definition 27. The bilinear constraint

$$\mathbf{x}_1^T F \mathbf{x}_2 = 0 \tag{3.14}$$

is called the **epipolar constraint** and

$$F = A_{12}^T T_{\mathbf{e}}$$

is called the **fundamental matrix**.   ■

Theorem 6. *The epipole in the second image is obtained as the right nullspace to the fundamental matrix and the epipole in the left image is obtained as the left nullspace to the fundamental matrix.*

Proof: Follows from $F\mathbf{e} = A_{12}^T T_{\mathbf{e}} \mathbf{e} = A_{12}^T(\mathbf{e} \times \mathbf{e}) = 0$. The statement about the epipole in the left image follows from symmetry.   ■

Corollary 1. *The fundamental matrix is singular:* $\det F = 0$.

Given a point, $\mathbf{x}_1$, in the first image, the coordinates of the corresponding point in the second image fulfill

$$0 = \mathbf{x}_1^T F \mathbf{x}_2 = (\mathbf{x}_1^T F)\mathbf{x}_2 = \mathbf{l}(\mathbf{x}_1)^T \mathbf{x}_2 = 0,$$

where $\mathbf{l}(\mathbf{x}_1)$ denotes the line represented by $\mathbf{x}_1^T F$.

Definition 28. The line $\mathbf{l} = F^T \mathbf{x}_1$ is called the **epipolar line** corresponding to $\mathbf{x}_1$.   ■

The geometrical interpretation of the epipolar line is the geometric construction in Figure 3.10. The points $\mathbf{x}_1$, C_1, and C_2 define a plane, Π, intersecting the second image plane in the line $\mathbf{l}$, containing the corresponding point.

From the previous considerations, we have the following pair:

$$F = A_{12}^T T_{\mathbf{e}} \quad \Leftrightarrow \quad P_1 = [\, I \,|\, 0 \,], \quad P_2 = [\, A_{12} \,|\, \mathbf{e}\,]. \tag{3.15}$$

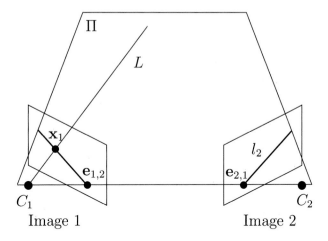

Figure 3.10. The epipolar line.

Observe that

$$F = A_{12}^T T_e = (A_{12} + \mathbf{ev}^T)^T T_{\mathbf{e}}$$

for every vector $\mathbf{v}$, since

$$(A_{12} + \mathbf{ev})^T T_{\mathbf{e}}(\mathbf{x}) = A_{12}^T (\mathbf{e} \times \mathbf{x}) + \mathbf{ve}^T (\mathbf{e} \times \mathbf{x}) = A_{12}^T T_{\mathbf{e}} \mathbf{x},$$

since $\mathbf{e}^T (\mathbf{e} \times \mathbf{x}) = \mathbf{e} \cdot (\mathbf{e} \times \mathbf{x}) = 0$. This ambiguity corresponds to the transformation

$$\bar{H}\bar{P}_2 = [\, A_{12} + \mathbf{ev}^T \mid \mathbf{e} \,].$$

We conclude that there are three free parameters in the choice of the second camera matrix when the first is fixed to $P_1 = [\, I \mid 0 \,]$.

The infinity homography

Consider a plane in the 3D object space, Π, defined by a vector $\mathbf{V}$: $\mathbf{V}^T \mathbf{X} = 0$ and the following construction; see Figure 3.11. Given a point in the first image, construct the intersection with the optical ray and the plane Π and project to the second image. This procedure gives a homography between points in the first and second images that depends on the chosen plane Π.

Proposition 8. *The homography corresponding to the plane* $\Pi : \mathbf{V}^T \mathbf{X} = 0$ *is given by the matrix*

$$H_\Pi = A_{12} - \mathbf{ev}^T,$$

where $\mathbf{e}$ *denotes the epipole and* $\mathbf{V} = [\, \mathbf{v}\, 1\,]^T$.

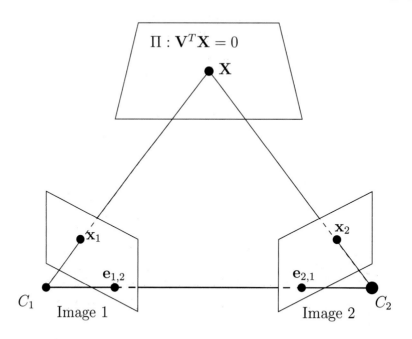

Figure 3.11. The homography corresponding to the plane Π.

Proof: Assume that

$$P_1 = [\, I \mid 0 \,], \quad P_2 = [\, A_{12} \mid \mathbf{e} \,].$$

Write $\mathbf{V} = [\, v_1\, v_2\, v_3\, 1\,]^T = [\,\mathbf{v}\, 1\,]^T$ (assuming $v_4 \neq 0$; that is, the plane is not incident with the origin, or the focal point of the first camera) and $\mathbf{X} = [\, X\, Y\, Z\, W \,]^T = [\,\mathbf{w}\, W\,]^T$, which gives

$$\mathbf{V}^T\mathbf{X} = \mathbf{v}^T\mathbf{w} + W, \tag{3.16}$$

which implies that $\mathbf{v}^T\mathbf{w} = -W$ for points in the plane Π. The first camera equation gives

$$\mathbf{x}_1 \sim [\, I \mid 0\,]\mathbf{X} = \mathbf{w},$$

and using (3.16) gives $\mathbf{v}^T\mathbf{x}_1 = -W$. Finally, the second camera matrix gives

$$\mathbf{x}_2 \sim [\, A_{12} \mid \mathbf{e} \,] \begin{bmatrix} \mathbf{x}_1 \\ -\mathbf{v}^T\mathbf{x}_1 \end{bmatrix} = A_{12}\mathbf{x}_1 - \mathbf{e}\mathbf{v}^T\mathbf{x}_1 = (A_{12} - \mathbf{e}\mathbf{v}^T)\mathbf{x}_1.$$

■

Observe that when $\mathbf{V} = (0, 0, 0, 1)$, that is, $\mathbf{v} = (0, 0, 0)$, the plane Π is the plane at infinity.

Definition 29. The homography

$$H_\infty = H_{\Pi_\infty} = A_{12}$$

is called the **homography corresponding to the plane at infinity** or **infinity homography**. ∎

Note that the epipolar line through the point $\mathbf{x}_2$ in the second image can be written as $\mathbf{x}_2 \times \mathbf{e}$, implying

$$(\mathbf{x}_2 \times \mathbf{e})^T H \mathbf{x}_1 = \mathbf{x}_1^T H^T T_\mathbf{e} \mathbf{x}_2 = 0,$$

that is, the epipolar constraint, and we get

$$F = H^T T_e.$$

Proposition 9. *There is a one-to-one correspondence between planes in 3D, homographies between two views, and factorization of the fundamental matrix as $F = H^T T_e$.*

Finally, we note that the matrix $H_\Pi^T T_e H_\Pi$ is skew symmetric, implying that

$$F H_\Pi + H_\Pi^T F^T = 0. \tag{3.17}$$

3.5.3 Multiview constraints and tensors

Consider one object point, $\mathbf{X}$, and its m images, $\mathbf{x}_i$, according to the camera equations $\lambda_i \mathbf{x}_i = P_i \mathbf{X}$, $i = 1 \ldots m$. These equations can be written as

$$\underbrace{\begin{bmatrix} P_1 & \mathbf{x}_1 & 0 & 0 & \cdots & 0 \\ P_2 & 0 & \mathbf{x}_2 & 0 & \cdots & 0 \\ P_3 & 0 & 0 & \mathbf{x}_3 & \cdots & 0 \\ \vdots & \vdots & \vdots & \vdots & \ddots & \vdots \\ P_m & 0 & 0 & 0 & \cdots & \mathbf{x}_m \end{bmatrix}}_{M} \begin{bmatrix} \mathbf{X} \\ -\lambda_1 \\ -\lambda_2 \\ -\lambda_3 \\ \vdots \\ -\lambda_m \end{bmatrix} = \begin{bmatrix} 0 \\ 0 \\ 0 \\ \vdots \\ 0 \end{bmatrix}. \tag{3.18}$$

We immediately get the following proposition:

Proposition 10. *The matrix, M, in (3.18) is rank deficient; that is,*

$$\operatorname{rank} M < m + 4,$$

*which is referred to as the **rank condition**.*

The rank condition implies that all $(m+4) \times (m+4)$ minors of M are equal to 0. These can be written using Laplace expansions as sums of products of determinants of *four rows* taken from the first four columns of M and of image coordinates. There are three different categories of such minors depending on the number of rows taken from each image, since one row has to be taken from each image, and then the remaining four rows can be distributed freely. The three different types are:

1. Taking the two remaining rows from one camera matrix and the two remaining rows from another camera matrix, gives 2-view constraints.

2. Taking the two remaining rows from one camera matrix, one row from another and one row from a third camera matrix, gives three-view constraints.

3. Taking one row from each of four different camera matrices, gives four-view constraints.

Observe that the minors of M can be factorized as products of the two-, three-, or four-view constraint and image coordinates in the other images. In order to get a notation that connects to the tensor notation, we use (x^1, x^2, x^3) instead of (x, y, z) for homogeneous image coordinates. We also denote row number i of a camera matrix P by P^i.

The monofocal tensor

Before we proceed to the multiview tensors, we make the following observation:

Proposition 11. *The epipole in image 2 from camera 1, $\mathbf{e} = (e^1, e^2, e^3)$ in homogeneous coordinates, can be written as*

$$e^j = \det \begin{bmatrix} P_1^1 \\ P_1^2 \\ P_1^3 \\ P_2^j \end{bmatrix}. \tag{3.19}$$

Proposition 12. *The numbers e^j constitute a first-order contravariant tensor, where the transformations of the tensor components are related to projective transformations of the image coordinates.*

Definition 30. The first-order contravariant tensor, e^j, is called the **monofocal tensor**.

The bifocal tensor

Considering minors obtained by taking three rows from one image, and three rows from another image:

$$\det \begin{bmatrix} P_1 & \mathbf{x}_1 & 0 \\ P_2 & 0 & \mathbf{x}_2 \end{bmatrix} = \det \begin{bmatrix} P_1^1 & x_1^1 & 0 \\ P_1^2 & x_1^2 & 0 \\ P_1^3 & x_1^3 & 0 \\ P_2^1 & 0 & x_2^1 \\ P_2^2 & 0 & x_2^2 \\ P_2^3 & 0 & x_2^3 \end{bmatrix} = 0,$$

which gives a *bilinear constraint*:

$$\sum_{i,j=1}^{3} F_{ij}\mathbf{x}_1^i\mathbf{x}_2^j = F_{ij}\mathbf{x}_1^i\mathbf{x}_2^j = 0, \tag{3.20}$$

where

$$F_{ij} = \sum_{i',i'',j',j''=1}^{3} \epsilon_{ii'i''}\epsilon_{jj'j''} \det \begin{bmatrix} P_1^{i'} \\ P_1^{i''} \\ P_2^{j'} \\ P_2^{j''} \end{bmatrix}.$$

The following proposition follows from (3.20).

Proposition 13. *The numbers F_{ij} constitute a second-order covariant tensor.*

Here the transformations of the tensor components are related to projective transformations of the image coordinates.

Definition 31. The second-order covariant tensor, F_{ij}, is called the **bifocal tensor**, and the bilinear constraint in (3.20) is called the **bifocal constraint**. ∎

Observe that the indices tell us which row to *exclude* from the corresponding camera matrix when forming the determinant. The geometric interpretation of the bifocal constraint is that corresponding view-lines in two images intersect in 3D; see Figure 3.9. The bifocal tensor can also be used to transfer a point to the corresponding epipolar line (see Figure 3.10), according to $l_j^2 = F_{ij}\mathbf{x}_1^i$. This transfer can be extended to a homography between epipolar lines in the first view and epipolar lines in the second view according to

$$l_i^1 = F_{ij}\epsilon_{j''}^{jj'}l_{j'}^2\mathbf{e}^{j''},$$

since $\epsilon_{j''}^{jj'}l_{j'}^{2}\mathbf{e}^{j''}$ gives the cross product between the epipole $\mathbf{e}$ and the line $\mathbf{l}^2$, which gives a point on the epipolar line.

The trifocal tensor

The trifocal tensor was originally discovered in the calibrated case in [60] and in the uncalibrated case in [55]. Considering minors obtained by taking three rows from one image, two rows from another image, and two rows from a third image, for example,

$$\det \begin{bmatrix} P_1^1 & x_1^1 & 0 & 0 \\ P_1^2 & x_1^2 & 0 & 0 \\ P_1^3 & x_1^3 & 0 & 0 \\ P_2^1 & 0 & x_2^1 & 0 \\ P_2^2 & 0 & x_2^2 & 0 \\ P_3^1 & 0 & 0 & x_3^1 \\ P_3^3 & 0 & 0 & x_3^3 \end{bmatrix} = 0,$$

gives a *trilinear constraints*:

$$\sum_{i,j,j',k,k'=1}^{3} T_i^{jk} \mathbf{x}_1^i \epsilon_{jj'j''} \mathbf{x}_2^{j'} \epsilon_{kk'k''} \mathbf{x}_3^{k'} = 0, \tag{3.21}$$

where

$$T_i^{jk} = \sum_{i',i''=1}^{3} \epsilon_{ii'i''} \det \begin{bmatrix} P_1^{i'} \\ P_1^{i''} \\ P_2^{j} \\ P_3^{k} \end{bmatrix}. \tag{3.22}$$

Note that there are in total nine constraints indexed by j'' and k'' in (3.21).

Proposition 14. *The numbers T_i^{jk} constitute a third-order mixed tensor that is covariant in i and contravariant in j and k.*

Definition 32. The third order mixed tensor, T_i^{jk} is called the **trifocal tensor**, and the trilinear constraint in (3.21) is called the **trifocal constraint**.   ∎

Again, the lower index tells us which row to *exclude* from the first camera matrix, and the upper indices tell us which rows to *include* from the second and third camera matrices respectively, and these indices becomes covariant and contravariant respectively. Observe that the order of the images is important, since the first image is treated differently. If the images are permuted, another set of coefficients is obtained. The geometric interpretation of the trifocal constraint is that the view-line in the first image and the planes corresponding to arbitrary lines coincident with the corresponding points in the second and third images (together with the focal points) respectively intersect in 3D; see Figure 3.12. The following theorem is straightforward to prove.

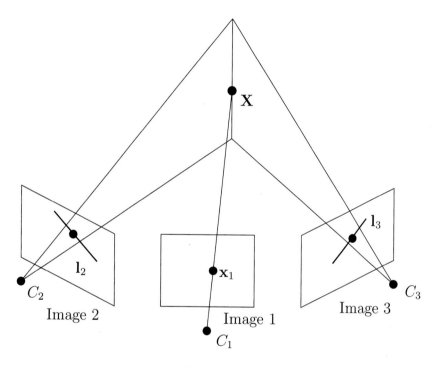

Figure 3.12. Geometrical interpretation of the trifocal constraint.

Theorem 7. *Given three corresponding lines, $\mathbf{l}^1$, $\mathbf{l}^2$, and $\mathbf{l}^3$, in three images, represented by the vectors $(\mathbf{l}_1^1, \mathbf{l}_2^1, \mathbf{l}_3^1)$ and so on, then*

$$\mathbf{l}_k^3 = T_k^{ij} \mathbf{l}_i^1 \mathbf{l}_j^2. \tag{3.23}$$

From this theorem it is possible to transfer the images of a line seen in two images to a third image, called **tensorial transfer**. The geometrical interpretation is that two corresponding lines define two planes in 3D that intersect in a line, that can be projected onto the third image. There are also other transfer equations, such as

$$\mathbf{x}_2^j = T_i^{jk}\mathbf{x}_1^i\mathbf{l}_k^3 \quad \text{and} \quad \mathbf{x}_3^k = T_i^{jk}\mathbf{x}_2^j\mathbf{l}_k^3,$$

with obvious geometrical interpretations.

The quadrifocal tensor

The quadrifocal tensor was independently discovered in several papers, including [64, 26]. Considering minors obtained by taking two rows from each one of four different images gives a *quadrilinear constraint*,

$$\sum_{i,i',j,j',k,k',l,l'=1}^{3} Q^{ijkl}\epsilon_{ii'i''}\mathbf{x}_{i'}^1\epsilon_{jj'j''}\mathbf{x}_{j'}^2\epsilon_{kk'k''}\mathbf{x}_{k'}^3\epsilon_{ll'l''}\mathbf{x}_{l'}^4 = 0, \tag{3.24}$$

where

$$Q^{ijkl} = \det\begin{bmatrix} P_1^i \\ P_2^j \\ P_3^k \\ P_4^l \end{bmatrix}.$$

Note that there are in total 81 constraints indexed by i'', j'', k'', and l'' in (3.24).

Proposition 15. *The numbers Q^{ijkl} constitute a fourth-order contravariant tensor.*

Definition 33. The fourth-order contravariant tensor, Q^{ijkl} is called the **quadrifocal tensor** and the quadrilinear constraint in (3.24), is called the **quadrifocal constraint**. ∎

Note that there are in total 81 constraints indexed by i'', j'', k'', and l''. Again, the upper indices tell us which rows to *include* from each camera matrix respectively. They become contravariant indices. The geometric interpretation of the quadrifocal constraint is that the four planes corresponding to arbitrary lines coincident with the corresponding points in the images intersect in 3D.

3.6 Structure and Motion I

We now study the structure and motion problem in detail. First, we solve the problem when the structure in known, called *resection*, then when the motion is known, called *intersection*. Then we present a linear algorithm to solve for both structure and motion using the multifocal tensors, and finally a factorization algorithm is presented. Again, refer to [24] for a more detailed treatment.

3.6.1   Resection

Problem 2 (Resection). *Assume that the structure is given; that is, the object points, $\mathbf{X}_j$, $j = 1, \ldots n$, are given in some coordinate system. Calculate the camera matrices P_i, $i = 1, \ldots m$ from the images, that is, from $\mathbf{x}_{i,j}$.*

The simplest solution to this problem is the classical DLT algorithm based on the fact that the camera equations

$$\lambda_j \mathbf{x}_j = P\mathbf{X}_j, \quad j = 1 \ldots n$$

are linear in the unknown parameters, λ_j and P.

3.6.2   Intersection

Problem 3 (Intersection). *Assume that the motion is given; that is, the camera matrices, P_i, $i = 1, \ldots m$, are given in some coordinate system. Calculate the structure $\mathbf{X}_j$, $j = 1, \ldots n$ from the images, that is, from $\mathbf{x}_{i,j}$.*

Consider the image of $\mathbf{X}$ in camera 1 and 2

$$\begin{cases} \lambda_1 \mathbf{x}_1 = P_1 \mathbf{X}, \\ \lambda_2 \mathbf{x}_2 = P_2 \mathbf{X}, \end{cases} \tag{3.25}$$

which can be written in matrix form as (compared to (3.18)):

$$\begin{bmatrix} P_1 & \mathbf{x}_1 & \mathbf{0} \\ P_2 & \mathbf{0} & \mathbf{x}_2 \end{bmatrix} \begin{bmatrix} \mathbf{X} \\ -\lambda_1 \\ -\lambda_2 \end{bmatrix} = \mathbf{0}, \tag{3.26}$$

which again is linear in the unknowns, λ_i and $\mathbf{X}$. This linear method can of course be extended to an arbitrary number of images.

3.6.3   Linear estimation of tensors

We are now able to solve the structure and motion problem given in Problem 1. The general scheme is as follows:

1. Estimate the components of a multiview tensor linearly from image correspondences.
2. Extract the camera matrices from the tensor components.
3. Reconstruct the object using intersection, that is, (3.26).

The eight-point algorithm

Each point correspondence gives one linear constraint on the components of the bifocal tensor according to the bifocal constraint:

$$F_{ij}\mathbf{x}_1^i\mathbf{x}_2^j = 0.$$

Each pair of corresponding points gives a linear homogeneous constraint on the nine tensor components F_{ij}. Thus, given at least eight corresponding points, we can solve linearly (e.g., by SVD) for the tensor components. After the bifocal tensor (fundamental matrix) has been calculated, it has to be factorized as $F = A_{12}T_{\mathbf{e}}^T$, which can be done by first solving for $\mathbf{e}$ using $F\mathbf{e} = 0$ (i.e., finding the right nullspace to F) and then for A_{12} by solving a linear system of equations. One solution is

$$A_{12} = \begin{bmatrix} 0 & 0 & 0 \\ F_{13} & F_{23} & F_{33} \\ -F_{12} & -F_{22} & -F_{32} \end{bmatrix},$$

which can be seen from the definition of the tensor components. In the case of noisy data, it might happen that $\det F \neq 0$, and the right nullspace does not exist. One solution is to solve $F\mathbf{e} = 0$ in least-squares sense using SVD. Another possibility is to project F to the closest rank-2 matrix, again using SVD. Then the camera matrices can be calculated from (3.26) and finally using intersection, (3.25), to calculate the structure.

The seven-point algorithm

A similar algorithm can be constructed for the case of corresponding points in three images.

Proposition 16. *The trifocal constraint in (3.21) contains four linearly independent constraints in the tensor components T_i^{jk}.*

Corollary 2. *At least seven corresponding points in three views are needed in order to estimate the 27 homogeneous components of the trifocal tensor.*

The main difference from the eight-point algorithm is that it is not obvious how to extract the camera matrices from the trifocal tensor components. Start with the transfer equation

$$\mathbf{x}_2^j = T_i^{jk}\mathbf{x}_1^i l_k^3,$$

which can be seen as a homography between the first two images by fixing a line in the third image. The homography is the one corresponding to the plane Π defined by the focal point of the third camera and the fixed line in the third camera. Thus, we know from (3.17) that the fundamental matrix between image 1 and image 2 obeys

$$FT^{.J} + (T^{.J})^T F^T = 0,$$

where $T^{.J}$ denotes the matrix obtained by fixing the index J. Since this is a linear constraint on the components of the fundamental matrix, it can easily be extracted from the trifocal tensor. Then the camera matrices P_1 and P_2 could be calculated, and finally, the entries in the third camera matrix P_3 can be recovered linearly from the definition of the tensor components in (3.22); see [27].

An advantage of using three views is that lines could be used to constrain the geometry, using (3.23), giving two linearly independent constraints for each corresponding line.

The six-point algorithm

Again, a similar algorithm can be constructed for the case of corresponding points in four images.

Proposition 17. *The quadrifocal constraint in (3.24) contains 16 linearly independent constraints in the tensor components Q^{ijkl}.*

From this proposition, it seems that five corresponding points would be sufficient to calculate the 81 homogeneous components of the quadrifocal tensor. However, the following proposition says that this is not possible.

Proposition 18. *The quadrifocal constraint in (3.24) for two corresponding points contains 31 linearly independent constraints in the tensor components Q^{ijkl}.*

Corollary 3. *At least six corresponding points in three views are needed in order to estimate the 81 homogeneous components of the quadrifocal tensor.*

Since one independent constraint is lost for each pair of corresponding points in four images, we get $6 \cdot 16 - \binom{6}{2} = 81$ linearly independent constraints.

Again, it is not obvious how to extract the camera matrices from the trifocal tensor components. First, a trifocal tensor has to be extracted, and then a fundamental matrix and the camera matrices must be extracted. It is outside the scope of this work to give the details for this; see [27]. Also, in this case, corresponding lines can be used by looking at transfer equations for the quadrifocal tensor.

3.6.4   Factorization

A disadvantage with using multiview tensors to solve the structure and motion problem is that when many images ($\gg 4$) are available, the information in all images cannot be used with equal weight. An alternative is to use a factorization method, see [59].

Write the camera equations

$$\lambda_{i,j}\mathbf{x}_{i,j} = P_i\mathbf{X}_j, \quad i = 1,\dots,m, \quad j = 1,\dots,n$$

for a fixed image i in matrix form as

$$\mathfrak{X}_i\Lambda_i = P_i\mathbb{X}, \tag{3.27}$$

where

$$\mathfrak{X}_i = \begin{bmatrix} \mathbf{x}_{i,1}^T & \mathbf{x}_{i,2}^T & \dots & \mathbf{x}_{i,n}^T \end{bmatrix}, \quad \mathbb{X} = \begin{bmatrix} \mathbf{X}_1^T & \mathbf{X}_2^T & \dots & \mathbf{X}_n^T \end{bmatrix},$$
$$\Lambda_i = \mathrm{diag}(\lambda_{i,1}, \lambda_{i,2}, \dots, \lambda_{i,n}).$$

The camera matrix equations for all images can now be written as

$$\widehat{\mathfrak{X}} = \mathbb{P}\mathbb{X}, \tag{3.28}$$

where

$$\widehat{\mathfrak{X}} = \begin{bmatrix} \mathfrak{X}_1\Lambda_1 \\ \mathfrak{X}_2\Lambda_2 \\ \vdots \\ \mathfrak{X}_m\Lambda_m \end{bmatrix}, \quad \mathbb{P} = \begin{bmatrix} P_1 \\ P_2 \\ \vdots \\ P_3 \end{bmatrix}.$$

Observe that $\widehat{\mathfrak{X}}$ only contains image measurements apart from the unknown depths.

Proposition 19.

$$\mathrm{rank}\,\widehat{\mathfrak{X}} \leq 4$$

This follows from (3.28), since $\widehat{\mathfrak{X}}$ is a product of a $3m \times 4$ and a $4 \times n$ matrix. Assume that the depths, Λ_i, are known, corresponding to affine cameras, we may use the following simple **factorization** algorithm:

1. Build up the matrix $\widehat{\mathfrak{X}}$ from image measurements.

2. Factorize $\widehat{\mathfrak{X}} = U\Sigma V^T$ using SVD.

3. Extract $\mathbb{P}$ = the first four columns of $U\Sigma$ and $\mathbb{X}$ = the first four rows of V^T.

In the perspective case, this algorithm can be extended to the **iterative factorization** algorithm:

1. Set $\lambda_{i,j} = 1$.

2. Build up the matrix $\widehat{\mathfrak{X}}$ from image measurements and the current estimate of $\lambda_{i,j}$.

3. Factorize $\widehat{\mathfrak{X}} = U\Sigma V^T$ using SVD.

4. Extract $\mathbb{P}$ = the first four columns of $U\Sigma$ and $\mathbb{X}$ = the first four rows of V^T.

5. Use the current estimate of $\mathbb{P}$ and $\mathbb{X}$ to improve the estimate of the depths from the camera equations $\mathfrak{X}_i \Lambda_i = P_i \mathbb{X}$.

6. If the error (reprojection errors or σ_5) is too large, go to step 2.

Definition 34. The fifth singular value, σ_5, in the SVD above is called the **proximity measure** and is a measure of the accuracy of the reconstruction.   ∎

Theorem 8. *The algorithm above minimizes the proximity measure.*

Figure 3.13 shows an example of a reconstruction using the iterative factorization method applied on four images of a toy block scene. Observe that the proximity measure decreases quite fast and the algorithm converges in about 20 steps.

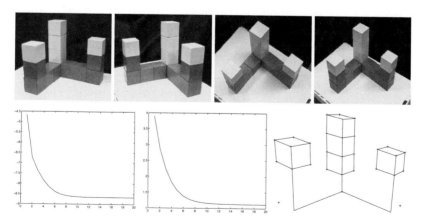

Figure 3.13.   Top: Four images. Bottom: The proximity measure for each iteration, the standard deviation of reprojection errors, and the reconstruction.

3.7   Structure and Motion II

In this section, we discuss the problem of structure and motion recovery from a more practical point of view. We present an approach to automatically build up the projective structure and motion from a sequence of images. The presented approach is sequential, which offers the advantage that corresponding features are not required to be visible in all views.

We assume that for each pair of consecutive views we are given a set of potentially corresponding feature points. The feature points are typically obtained by using a corner detector [19]. If the images are not too widely separated, corresponding features can be identified by comparing the local intensity neighborhoods of the feature points on a pixel-by-pixel basis. In this case, typically only feature points that have similar coordinates are compared. In case of video, it might be more appropriate to use a feature tracker [56] that follows features from frame to frame. For more widely separated views, more complex features would have to be used [39, 53, 69, 40].

3.7.1   Two-view geometry computation

The first step of our sequential structure and motion computation approach consists of computing the geometric relation between two consecutive views. As seen in Section 3.5.2, this consists of recovering the fundamental matrix. In principle, the linear method presented in Section 3.6.3 could be used. In this section, a practical algorithm is presented to compute the fundamental matrix from a set of corresponding points perturbed with noise and containing a significant proportion of outliers.

Linear algorithms

Before presenting the complete robust algorithm, we revisit the linear algorithm. Given a number of corresponding points, (3.14) can be used to compute F. This equation can be rewritten in the following form:

$$\begin{bmatrix} x_1 x_2 & y_1 x_2 & x_2 & x_1 y_2 & y_1 y_2 & y_2 & x_1 & y_1 & 1 \end{bmatrix} \mathbf{f} = 0, \qquad (3.29)$$

with $\mathbf{x}_1 = [x_1\, y_1\, 1]^\top$, $\mathbf{x}_2 = [x_2\, y_2\, 1]^\top$ and $\mathbf{f}$ a vector containing the elements of the fundamental matrix. As discussed before, stacking eight or more of these equations allows for a linear solution. Even for seven corresponding points, the one parameter family of solutions obtained by solving the linear equations can be restricted to one or three solutions by enforcing the cubic rank-2 constraint $\det(F_1 + \lambda F_2) = 0$. Note also that, as pointed out by

Hartley [22], it is important to normalize the image coordinates before solving the linear equations. Otherwise, the columns of (3.29) would differ by several orders of magnitude and the error would be concentrated on the coefficients corresponding to the smaller columns. This normalization can be achieved by transforming the image center to the origin and scaling the images so that the coordinates have a standard deviation of unity.

Nonlinear algorithms

The result of the linear equations can be refined by minimizing the following criterion [70]:

$$ \mathcal{C}(F) = \sum \left(d(\mathbf{x}_2, F\mathbf{x}_1)^2 + d(\mathbf{x}_1, F^\top \mathbf{x}_2)^2 \right), \tag{3.30} $$

with $d(.,.)$ representing the Euclidean distance in the image. This criterion can be minimized through a Levenberg-Marquardt algorithm [50]. An even better approach consists of computing the maximum likelihood estimation (for Gaussian noise) by minimizing the following criterion:

$$ \mathcal{C}(F, \widehat{\mathbf{x}}_1, \widehat{\mathbf{x}}_2) = \sum \left(d(\widehat{\mathbf{x}}_1, \mathbf{x}_1)^2 + d(\widehat{\mathbf{x}}_2, \mathbf{x}_2)^2 \right) \text{ with } \widehat{\mathbf{x}}_2^\top F \widehat{\mathbf{x}}_1 = 0. \tag{3.31} $$

Although in this case the minimization has to be carried out over a much larger set of variables, this can be achieved efficiently by taking advantage of the sparsity of the problem.

Robust algorithm

To compute the fundamental matrix from a set of matches that were automatically obtained from a pair of real images, it is important to explicitly deal with outliers. If the set of matches is contaminated with even a small set of outliers, the result of the above methods can become unusable. This is typical for all types of least-squares approaches (even nonlinear ones). The problem is that the quadratic penalty allows for a single outlier that is very far away from the true solution to completely bias the final result.

An approach that can be used to cope with this problem is the RANSAC algorithm that was proposed by Fischler and Bolles [17]. A minimal subset of the data, in this case seven point correspondences, is randomly selected from the set of potential correspondences, and the solution obtained from it is used to segment the remainder of the dataset in *inliers* and *outliers*. If the initial subset contains no outliers, most of the correct correspondences will support the solution. However, if one or more outliers are contained in the initial subset, it is highly improbable that the computed solution will find a lot of support among the remainder of the potential correspondences, yielding a low "inlier" ratio. This procedure is repeated until a satisfying solution is obtained. This is typically defined as a probability in excess of 95% that a good subsample was selected. The expression for this probability is $\Gamma = 1-(1-\rho^p)^m$ with ρ the fraction of inliers, p the number of features in each sample, seven in this case, and m the number of trials (see Rousseeuw [51]).

Two-view geometry computation

The different algorithms described above can be combined to yield a practical algorithm to compute the two-view geometry from real images:

1. Compute initial set of potential correspondences (and set $\rho_{max} = 0, m = 0$).

2. While $(1 - (1 - \rho_{max}^7)^m) < 95\%$,

    (a) Randomly select a minimal sample (seven pairs of corresponding points),

    (b) Compute the solution(s) for F (yielding one or three solutions),

    (c) Determine percentage of inliers ρ (for all solutions),

    (d) Increment m, update ρ_{max} if $\rho_{max} < \rho$.

3. Refine F based on all inliers.

4. Look for additional matches along epipolar lines.

5. Refine F based on all correct matches (preferably using (3.31)).

3.7.2   Structure and motion recovery

Once the epipolar geometry has been computed between all consecutive views, the next step consists of reconstructing the structure and motion for the whole sequence. Contrary to the factorization approach of Section 3.6.4, here a sequential approach is presented. First, the structure and motion is initialized for two views and then gradually extended toward the whole sequence. Finally, the solution is refined through a global minimization over all the unknown parameters.

Initializing the structure and motion

Initial motion computation   Two images of the sequence are used to determine a reference frame. The world frame is aligned with the first camera.

The second camera is chosen so that the epipolar geometry corresponds to the retrieved fundamental matrix F:

$$P_1 = \begin{bmatrix} I_{3\times3} & | & \mathbf{0}_3 \end{bmatrix}$$
$$P_2 = \begin{bmatrix} T_e F + \mathbf{ev}^\top & | & \sigma\mathbf{e} \end{bmatrix}. \tag{3.32}$$

Eq. (3.32) is not completely determined by the epipolar geometry (i.e., F and $\mathbf{e}$), but has 4 more degrees of freedom (i.e., $\mathbf{v}$ and σ). The vector $\mathbf{v}$ determines the position of the reference plane (i.e., the plane at infinity in an affine or metric frame), and σ determines the global scale of the reconstruction. The location of the reference plane should not make any difference if the algorithm is projectively invariant. To achieve this, it is important to use homogeneous representations for all 3D entities and to only use image measurements for minimizations. The value for the parameter σ has no importance and can be fixed to one.

Initial structure computation   Once two projection matrices have been fully determined, the matches can be reconstructed through triangulation. Due to noise, the lines of sight will not intersect perfectly. In the projective case, the minimizations should be carried out in the images and not in projective 3D space. Therefore, the distance between the reprojected 3D point and the image points should be minimized:

$$d(\mathbf{x}_1, P_1\mathbf{X})^2 + d(\mathbf{x}_2, P_2\mathbf{X})^2. \tag{3.33}$$

It was noted by Hartley and Sturm [23] that the only important choice is to select in which epipolar plane the point is reconstructed. Once this choice is made, it is trivial to select the optimal point from the plane. Since a bundle of epipolar planes has only one parameter, the dimension of the problem is reduced from three to one. Minimizing the following equation is thus equivalent to minimizing equation (3.33).

$$d(\mathbf{x}_1, \mathbf{l}_1(\lambda))^2 + d(\mathbf{x}_2, \mathbf{l}_2(\lambda))^2, \tag{3.34}$$

with $\mathbf{l}_1(\lambda)$ and $\mathbf{l}_2(\lambda)$ the epipolar lines obtained in function of the parameter λ describing the bundle of epipolar planes. It turns out (see [23]) that this equation is a polynomial of degree 6 in λ. The global minimum of equation (3.34) can thus easily be computed directly. In both images, the points on the epipolar line $\mathbf{l}_1(\lambda)$ and $\mathbf{l}_2(\lambda)$ closest to the points $\mathbf{x}_1$ and $\mathbf{x}_2$ respectively are selected. Since these points are in epipolar correspondence, their lines of sight intersect exactly in a 3D point. In the case where (3.31) is minimized to obtain the fundamental matrix F, the procedure described here is unnecessary and the pairs $(\widehat{\mathbf{x}}_1, \widehat{\mathbf{x}}_2)$ can be reconstructed directly.

Updating the structure and motion

The previous section dealt with obtaining an initial reconstruction from two views. This section discusses how to add a view to an existing reconstruction. First, the pose of the camera is determined, then the structure is updated based on the added view, and finally new points are initialized.

Projective pose estimation   For every additional view, the pose toward the preexisting reconstruction is determined. This is illustrated in Figure 3.14. It is assumed that the epipolar geometry has been computed between view $i - 1$ and i. The matches that correspond to already reconstructed points are used to infer correspondences between 2D and 3D. Based on these, the projection matrix P_i is computed using a robust procedure similar to the

Figure 3.14. Image matches $(\mathbf{x}_{i-1}, \mathbf{x}_i)$ are found as described before. Since some image points, $\mathbf{x}_{i-1}$, relate to object points, $\mathbf{X}$, the pose for view i can be computed from the inferred matches $(\mathbf{X}, \mathbf{x}_i)$. A point is accepted as an inlier if a solution for $\widehat{\mathbf{X}}$ exists for which $d(P\widehat{\mathbf{X}}, \mathbf{x}_i) < 1$ for each view k in which $\mathbf{X}$ has been observed.

one laid out for computing the fundamental matrix. In this case, a minimal sample of six matches is needed to compute P_i. A point is considered an inlier if there exists a 3D point that projects sufficiently close to all associated image points. This requires refining the initial solution of $\mathbf{X}$ based on all observations, including the last. Because this is computationally expensive (remember that this has to be done for each generated hypothesis), it is advised to use a modified version of RANSAC that cancels the verification of unpromising hypothesis [5]. Once P_i has been determined, the projection of already reconstructed points can be predicted so that some additional matches can be obtained. This means that the search space is gradually reduced from the full image to the epipolar line to the predicted projection of the point.

This procedure relates only the image to the previous image. In fact, it is implicitly assumed that once a point gets out of sight, it will not come back. Although this is true for many sequences, this assumption does not always hold. Assume that a specific 3D point got out of sight, but that it becomes visible again in the two most recent views. This type of point could be interesting to avoid error accumulation. However, the naive approach would just reinstantiate a new independent 3D point. A possible solution to this problem was proposed in [35].

Refining and extending structure   The structure is refined using an iterated linear reconstruction algorithm on each point. The scale factors can also be eliminated from (3.25) so that homogeneous equations in $\mathbf{X}$ are obtained:

$$
\begin{aligned}
P_3 \mathbf{X} x - P_1 \mathbf{X} &= 0 \\
P_3 \mathbf{X} y - P_2 \mathbf{X} &= 0
\end{aligned}
\tag{3.35}
$$

with P_i the i-th row of P and (x, y) being the image coordinates of the point. An estimate of $\mathbf{X}$ is computed by solving the system of linear equations obtained from all views where a corresponding image point is available. To obtain a better solution, the criterion $\sum d(P\mathbf{X}, \mathbf{x})^2$ should be minimized. This can be approximately obtained by iteratively solving the following weighted linear least-squares problem:

$$
\frac{1}{P_3 \widetilde{\mathbf{X}}}
\begin{bmatrix}
P_3 x - P_1 \\
P_3 y - P_2
\end{bmatrix}
\mathbf{X} = 0,
\tag{3.36}
$$

where $\widetilde{\mathbf{X}}$ is the previous solution for $\mathbf{X}$. This procedure can be repeated a few times. By solving this system of equations through SVD, a normalized homogeneous point is automatically obtained. If a 3D point is not observed, the position is not updated. In this case one can check if the point was seen in a sufficient number of views to be kept in the final reconstruction. This minimum number of views can, for example, be put to three. This avoids having an important number of outliers due to spurious matches.

Of course, in an image sequence some new features will appear in every new image. If point matches are available that were not related to an existing point in the structure, then a new point can be initialized, as described in Section 3.7.2.

Refining structure and motion

Once the structure and motion has been obtained for the whole sequence, it is recommended to refine it through a global minimization step so that a bias toward the initial views is avoided. A maximum likelihood estimation can be obtained through *bundle adjustment* [67, 57]. The goal is to find the parameters of the camera view P_k and the 3D points $\mathbf{X}_i$ for which the sum of squared distances between the observed image points $\mathbf{m}_{ki}$ and the reprojected image points $P_k(\mathbf{X}_i)$ is minimized. It is advised to extend the camera projection model to also take radial distortion into account. For m views and n points, the following criterion should be minimized:

$$\min_{P_k, \mathbf{X}_i} \sum_{k=1}^{m} \sum_{i=1}^{n} d(\mathbf{x}_{ki}, P_k(\mathbf{X}_i))^2. \tag{3.37}$$

If the errors on the localization of image features are independent and satisfy a zero-mean Gaussian distribution, then it can be shown that bundle adjustment corresponds to a maximum likelihood estimator. This minimization problem is huge; for example, for a sequence of 20 views and 100 points/view, a minimization problem in more than 6000 variables has to be solved (most of them related to the structure). A straightforward computation is obviously not feasible. However, the special structure of the problem can be exploited to solve the problem much more efficiently [67, 57]. The key reason for this is that a specific residual is only dependent on one point and one camera, which results in a very sparse structure for the normal equations.

Structure and motion recovery algorithm

To conclude this section, an overview of the structure and motion recovery algorithm is given. The whole procedure consists of the following steps:

    1. Match or track points over the whole image sequence (see Section 3.5.2).

2. Initialize the structure and motion recovery:

   (a) Select two views that are suited for initialization,

   (b) Set up the initial frame,

   (c) Reconstruct the initial structure.

3. For every additional view,

   (a) Infer matches to the existing 3D structure,

   (b) Compute the camera pose using a robust algorithm,

   (c) Refine the existing structure,

   (d) Initialize new structure points.

4. Refine the structure and motion through bundle adjustment.

The results of this algorithm are the camera poses for all the views and the reconstruction of the interest points. For most applications, such as MatchMoving (aligning a virtual camera with the motion of a real camera; see Section 3.10.3), the camera poses are the most useful.

3.8   Autocalibration

As shown by Theorem 5, for a completely uncalibrated image sequence, the reconstruction is only determined up to a projective transformation. While it is true that the full calibration is often not available, some knowledge of the camera intrinsics is usually available. This knowledge can be used to recover the structure and motion up to a similarity transformation. This type of approach is called *autocalibration* or self-calibration in the literature. A first class of algorithms assumes constant, but unknown, intrinsic camera parameters [16, 21, 44, 28, 65]. Another class of algorithms assumes some intrinsic camera parameters to be known, while others can vary [47, 29]. Specific algorithms have also been proposed for restricted camera motion, such as pure rotation [20, 11], or restricted scene structure, such as planar scenes [66].

The absolute conic and its image

The central concept for autocalibration is the absolute conic. As stated in Proposition 3, the absolute conic allows us to identify the similarity structure in a projective space. In other words, if, given a projective reconstruction, we could locate the conic corresponding to the absolute conic in the real world, this would be equivalent to recovering the structure of the observed scene up to a similarity. In this case, a transformation that transforms the absolute conic to its canonical representation in Euclidean space–that is, $\Omega' = diag(1, 1, 1, 0)$–would yield a reconstruction similar to the original (i.e., identical up to orientation, position, and scale).

As was seen in Proposition 6, the image of the absolute conic is directly related to the intrinsic camera parameters, and this independently of the choice of projective basis:

$$P\Omega'P^\top \sim KK^\top. \tag{3.38}$$

Therefore, constraints on the intrinsics can be used to constrain the location of the conic corresponding to the absolute conic. Most autocalibration algorithms are based on (3.38).

Critical motion sequences

Autocalibration is not always guaranteed to yield a unique solution. Depending on the available constraints on the intrinsics and on the camera motion, the remaining ambiguity on the reconstruction might be larger than a similarity. This problem was identified as the problem of *critical motion sequences*. The first complete analysis of the problem for constant intrinsic camera parameters was made by Sturm [61]. Analysis for some other cases can be found in [62, 43, 32]. It was also shown that in some cases, the ambiguity notwithstanding, correct novel views could be generated [45].

Linear autocalibration

In this section, we present a simple linear algorithm for autocalibration of cameras. The approach, published in [49], is related to the initial approach published in [46], but avoids most of the problems due to critical motion sequences by incorporating more a priori knowledge. As input, it requires a projective representation of the camera projection matrices.

As discussed in Section 3.4.3, for most cameras it is reasonable to assume that the pixels are close to square and that the principal point is close to the center of the image. The focal length (measured in pixel units) is typically of the same order of magnitude as the image size. It is therefore a good idea to perform the following normalization:

$$P_N = K_N^{-1} P \text{ with } K_N = \begin{bmatrix} w+h & 0 & \frac{w}{2} \\ & w+h & \frac{h}{2} \\ & & 1 \end{bmatrix}, \qquad (3.39)$$

where w and h are the width and height respectively of the image. After normalization, the focal length should be of the order of unity and the principal point should be close to the origin. The above normalization would scale a focal length of a 60mm lens to 1, and thus focal lengths in the range of 20mm to 180mm would end up in the range $[1/3, 3]$, and the principal point should now be close to the origin. The aspect ratio is typically around 1, and the skew can be assumed 0 for all practical purposes. Making this a priori knowledge more explicit and estimating reasonable standard deviations yields $f \approx rf \approx 1 \pm 3$, $u \approx v \approx 0 \pm 0.1$, $r \approx 1 \pm 0.1$, and $s = 0$, which approximately translates to the following expectations for ω':

$$\omega' \sim K K^\top = \begin{bmatrix} \gamma^2 f^2 + x_0^2 & x_0 y_0 & x_0 \\ x_0 y_0 & f^2 + y_0^2 & y_0 \\ x_0 & y_0 & 1 \end{bmatrix} \approx \begin{bmatrix} 1 \pm 9 & \pm 0.01 & \pm 0.1 \\ \pm 0.01 & 1 \pm 9 & \pm 0.1 \\ \pm 0.1 & \pm 0.1 & 1 \end{bmatrix},$$
$$(3.40)$$

and $\omega'_{22}/\omega'_{11} \approx 1 \pm 0.2$. These constraints can also be used to constrain the left-hand side of (3.38). The uncertainty can be accounted for by weighting the equations, yielding the following set of constraints:

$$\begin{aligned} \frac{\nu}{9} \left(P_1 \Omega' P_1{}^\top - P_3 \Omega' P_3{}^\top \right) &= 0 \\ \frac{\nu}{9} \left(P_2 \Omega' P_2{}^\top - P_3 \Omega' P_3{}^\top \right) &= 0 \\ \frac{\nu}{0.2} \left(P_1 \Omega' P_1{}^\top - P_2 \Omega' P_2{}^\top \right) &= 0 \\ \frac{\nu}{0.1} \left(P_1 \Omega' P_2{}^\top \right) &= 0 \\ \frac{\nu}{0.1} \left(P_1 \Omega' P_3{}^\top \right) &= 0 \\ \frac{\nu}{0.01} \left(P_2 \Omega' P_3{}^\top \right) &= 0 \end{aligned} \qquad (3.41)$$

with P_i the i-th row of P and ν a scale factor that can be set to 1. If for the solution $P_3 \Omega' P_3{}^\top$ varies widely for the different views, we might want to iterate with $\nu = (P_3 \widetilde{\Omega}' P_3{}^\top)^{-1}$, with $\widetilde{\Omega}'$ the result of the previous iteration. Since Ω' is a symmetric 4×4 matrix, it is linearly parametrized by 10 coefficients. An estimate of the dual absolute quadric Ω' can be obtained by solving the above set of equations for all views through homogeneous linear least-squares. The rank-3 constraint can be imposed by forcing the smallest singular value to zero. The upgrading transformation T can be obtained from diag $(1, 1, 1, 0) = T \Omega' T^\top$ by decomposition of Ω'.

Autocalibration refinement

This result can be further refined through bundle adjustment (see Section 3.7.2). In this case, the constraints on the intrinsics should be enforced during the minimization process. Constraints on the intrinsics can be enforced either exactly through parametrization or approximately by adding a residual for the deviation from the expected value in the global minimization process.

3.9   Dense Depth Estimation

With the camera calibration given for all viewpoints of the sequence, we can proceed with methods developed for calibrated structure from motion algorithms. The feature tracking algorithm already delivers a sparse surface model based on distinct feature points. This, however, is not sufficient to reconstruct geometrically correct and visually pleasing surface models. This task is accomplished by a dense disparity matching that estimates correspondences from the gray level images directly by exploiting additional geometrical constraints. The dense surface estimation is done in a number of steps. First, image pairs are rectified to the standard stereo configuration. Then, disparity maps are computed through a stereo matching algorithm. Finally, a multiview approach integrates the results obtained from several view pairs.

3.9.1   Rectification

Since the calibration between successive image pairs was computed, the epipolar constraint that restricts the correspondence search to a 1D search range can be exploited. Image pairs can be warped so that epipolar lines coincide with image scan lines. The correspondence search is then reduced to a matching of the image points along each image scanline. This results in a dramatic increase of the computational efficiency of the algorithms by enabling several optimizations in the computations.

For some motions (i.e., when the epipole is located in the image), standard rectification based on planar homographies [2] is not possible, and a more advanced procedure should be used. The approach proposed in [48] avoids this problem. The method works for all possible camera motions. The key idea is to use polar coordinates with the epipole as origin. Corresponding lines are given through the epipolar geometry. By taking the orientation [36] into account, the matching ambiguity is reduced to half epipolar lines. A minimal image size is achieved by computing the angle between two consec-

utive epipolar lines that correspond to rows in the rectified images to have
the worst case pixel on the line preserve its area.

Some examples   A first example comes from the *castle* sequence. In Figure 3.15, an image pair and the associated rectified image pair are shown.
A second example was filmed with a handheld digital video camera in the
Béguinage in Leuven.   Due to the narrow streets, only forward motion is
feasible.   In this case, the full advantage of the polar rectification scheme
becomes clear, since this sequence could not have been handled through traditional planar rectification. An example of a rectified image pair is given in
Figure 3.16. Note that the whole left part of the rectified images corresponds
to the epipole. On the right side of this figure, a model that was obtained
by combining the results from several image pairs is shown.

Figure 3.15. Original image pair (left) and rectified image pair (right).

3.9.2   Stereo matching

The goal of a dense stereo algorithm is to compute the corresponding pixel for
every pixel of an image pair. After rectification, the correspondence search
is limited to corresponding scanlines. As illustrated in Figure 3.17, finding

Figure 3.16. Rectified image pair (left) and some views of the reconstructed scene (right).

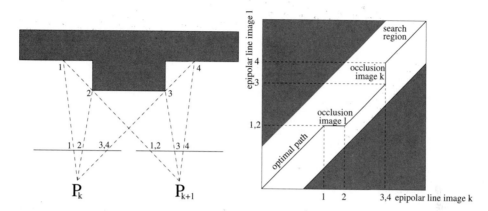

Figure 3.17. The ordering constraint (left) and dense matching as a path search problem (right).

the correspondences for a pair of scanlines can be seen as a path search problem. Besides the epipolar geometry, other constraints, like preserving the order of neighboring pixels, bidirectional uniqueness of the match, and detection of occlusions, can be exploited. In most cases, it is also possible to limit the search to a certain disparity range (an estimate of this range can be obtained from the reconstructed 3D feature points). Besides these constraints, a stereo algorithm should also take into account the similarity between corresponding points and the continuity of the surface. It is possible to compute the optimal path, taking all the constraints into account, using dynamic programming [8, 12, 41]. Other computationally more expensive approaches also take continuity across scanlines into account. Real-time approaches, on the other hand, estimate the best match independently for every pixel. A complete taxonomy of stereo algorithms can be found in [52].

3.9.3   Multiview linking

The pairwise disparity estimation allows us to compute image-to-image correspondences between adjacent rectified image pairs and independent depth estimates for each camera viewpoint. An optimal joint estimate is achieved by fusing all independent estimates into a common 3D model. The fusion can be performed in an economical way through controlled correspondence linking (see Figure 3.18). A point is transferred from one image to the next as follows:

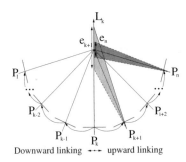

Figure 3.18. Depth fusion and uncertainty reduction from correspondence linking.

$$\mathbf{x}_2 = R'^{-1}(R(\mathbf{x}_1) + D(R(\mathbf{x}_1))), \qquad (3.42)$$

with $R(.)$ and $R'(.)$ functions that map points from the original image into the rectified image and $D(.)$ a function that corresponds to the disparity map. When the depth obtained from the new image point $\mathbf{x}_2$ is outside the confidence interval, the linking is stopped; otherwise, the result is fused with the previous values through a Kalman filter. This approach is discussed in more detail in [34]. This approach combines the advantages of small baseline and wide baseline stereo. The depth resolution is increased through the combination of multiple viewpoints and large global baseline, while the matching is simplified through the small local baselines. It can provide a very dense depth map by avoiding most occlusions. Due to multiple observations of a single surface point, the texture can be enhanced and noise and highlights can be removed.

3.10   Visual Modeling

In the previous sections, we explained how the camera motion and calibration and the depth estimates for (almost) every pixel could be obtained. This yields all the necessary information to build different types of visual models. In this section, several types of models are considered. First, the construction of texture-mapped 3D surface models is discussed. Then, a combined image- and geometry-based approach is presented that can render models ranging from pure plenoptic to view-dependent texture and geometry models. Finally, the possibility of combining real and virtual scenes is treated.

3.10.1   3D surface reconstruction

The 3D surface is approximated by a triangular mesh to reduce geometric complexity and to tailor the model to the requirements of computer graphics visualization systems. A simple approach consists of overlaying a 2D triangular mesh on top of one of the images and then building a corresponding 3D mesh by placing the vertices of the triangles in 3D space according to the values found in the corresponding depth map. To reduce noise, it is recommended to first smooth the depth image (the kernel can be chosen of the same size as the mesh triangles). The image itself can be used as a texture map. While projective texture mapping would normally be required, the small size of the triangles allows us to use standard (affine) texture mapping (the texture coordinates are trivially obtained as the 2D coordinates of the vertices).

It can happen that for some vertices no depth value is available or that the confidence is too low. In these cases, the corresponding triangles are not reconstructed. The same happens when triangles are placed over discontinuities. This is achieved by selecting a maximum angle between the normal of a triangle and the line-of-sight through its center (e.g., 85 degrees). This simple approach works very well on the dense depth maps as obtained through multiview linking. The surface reconstruction approach is illustrated in Figure 3.19.

A further example is shown in Figure 3.20. The video sequence was recorded with a handheld camcorder on an archaeological site in Sagalasso, Turkey (courtesy of Marc Waelkens). It shows a decorative medusa head that was part of a monumental fountain. The video sequence was processed fully automatically by using the algorithms discussed in sections 3.7, 3.8, 3.9, and 3.10. From the bundle adjustment and the multiview linking, the accuracy

Figure 3.19. Surface reconstruction approach (top): Triangular mesh is overlaid on top of the image. The vertices are back-projected in space according to the depth values. From this, a 3D surface model is obtained (bottom).

was estimated to be of $\frac{1}{500}$ (compared to the size of the reconstructed object). This has to be compared with the image resolution of 720×576. Note that the camera was uncalibrated and, besides the unknown focal length and principal point, has significant radial distortion and an aspect ratio different from one (i.e., 1.09), which were all automatically recovered from the video sequence.

To reconstruct more complex shapes, it is necessary to combine results from multiple depth maps. The simplest approach consists of generating separate models independently and then loading them together in the graphics system. Since all depth maps are located in a single coordinate frame, registration is not an issue. Often it is interesting to integrate the different meshes into a single mesh. A possible approach is given in [10].

Figure 3.20. 3D-from-video: one of the video frames (upper-left), recovered structure and motion (upper-right), textured and shaded 3D model (middle), and more views of textured 3D model (bottom).

3.10.2   Image-based rendering

In the previous section, we presented an approach to construct 3D models. If the goal is to generate novel views, other approaches are available. In recent years, a multitude of image-based approaches have been proposed that render images from images without the need for an explicit intermediate 3D model. The best known approaches are lightfield and lumigraph rendering [37, 18] and image warping [4, 54, 1].

Here we briefly introduce an approach to render novel views directly from images recorded with a handheld camera. If available, some depth information can also be used to refine the underlying geometric assumption. A more extensive discussion of this work can be found in [25, 33]. A related approach was presented in [3]. A lightfield is the collection of the lightrays corresponding to all the pixels in all the recorded images. Therefore, rendering from a lightfield consists of looking up the "closest" ray(s) passing through every pixel of the novel view. Determining the closest ray consists of two steps: (1) determining in which views the closest rays are located, and (2) within that view selecting the ray that intersects the implicit geometric assumption in the same point. For example, if the assumption is that the scene is far away, the corresponding implicit geometric assumption might be Π_∞ so that parallel rays would be selected.

In our case, the view selection works as follows. All the camera projection centers are projected in the novel view and Delaunay triangulated. For every pixel within a triangle, the recorded views corresponding to the three vertices are selected as "closest" views. If the implicit geometric assumption is planar, a homography relates the pixels in the novel view with those in a recorded view. Therefore, a complete triangle in the novel view can be efficiently drawn using texture mapping. The contributions of the three cameras can be combined using alpha blending. The geometry can be approximated by one plane for the whole scene, one plane per camera triple, or by several planes for one camera triple. The geometric construction is illustrated in Figure 3.21.

This approach is illustrated in Figure 3.22 with an image sequence of 187 images recorded by waving a camera over a cluttered desk. In the lower part of Figure 3.22, a detail of a view is shown for the different methods. In the case of one global plane (left image), the reconstruction is sharp where the approximating plane intersects the actual scene geometry. The reconstruction is blurred where the scene geometry diverges from this plane. In the case of local planes (middle image), at the corners of the triangles, the reconstruction is almost sharp, because there the scene geometry is considered

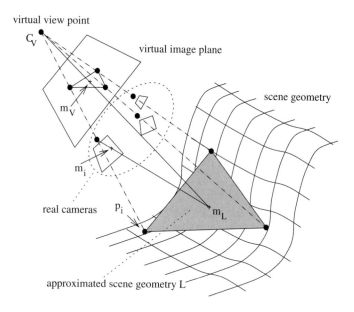

virtual view point

C_v

virtual image plane

scene geometry

m_v

m_i

real cameras

p_i

m_L

approximated scene geometry L

Figure 3.21. Drawing triangles of neighboring projected camera centers and approximated scene geometry.

directly. Within a triangle, ghosting artifacts occur where the scene geometry diverges from the particular local plane. If these triangles are subdivided (right image), these artifacts are reduced further.

3.10.3 Match-moving

Another interesting application of the presented algorithms consists of adding virtual elements to real video. This has important applications in the entertainment industry and several products, such as 2d3's *boujou* and RealViz's *MatchMover*, exist that are based on the techniques described in Section 3.7 and Section 3.8. The key issue consists of registering the motion of the real and the virtual camera. The presented techniques can be used to compute the motion of the camera in the real world. This allows us to restrict the problem of introducing virtual objects in video to determine the desired position, orientation, and scale with respect to the reconstructed camera motion. More details on this approach can be found in [7]. Note that to achieve a seamless integration, other effects such as occlusion and lighting must also be taken care of.

An example is shown in Figure 3.23. The video shows the remains of one of the ancient monumental fountains of Sagalassos. A virtual reconstruction

Figure 3.22. Top: image of the *desk* sequence and sparse structure-and-motion result (left), artificial view rendered using one plane per image triple (right). Details of rendered images showing the differences between the approaches (bottom): one global plane of geometry (left), one local plane for each image triple (middle), and refinement of local planes (right).

of the monument was overlaid on the original frame. The virtual camera was setup to mimic exactly the computed motion and calibration of the original camera.

3.11 Conclusion

In this chapter, the relations between multiple views of a 3D scene were discussed. The discussion started by introducing the basic concepts of projective geometry and tensor calculus. Then, the pinhole camera model, the epipolar geometry, and the multiple view tensors were discussed. Next, approaches

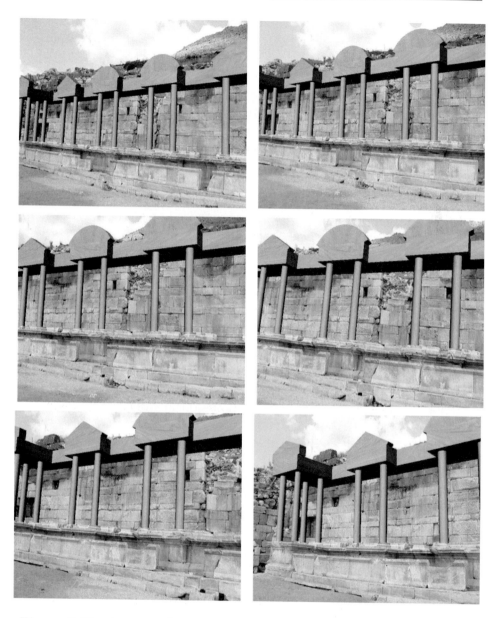

Figure 3.23. Augmented video: six frames (out of 250) from a video where a virtual reconstruction of an ancient monument has been added.

that rely on those concepts to recover both the structure and motion from a sequence of images were presented and illustrated with some real-world examples and applications.

Acknowledgment

The authors wish to acknowledge the financial support of the EU projects InViews, Attest, Vibes, and Murale, as well as the contributions of their (former) colleagues in Leuven and Lund, where most of the presented work was carried out. Luc Van Gool, Maarten Vergauwen, Frank Verbiest, Kurt Cornelis, Jan Tops, Reinhard Koch, and Benno Heigl have contributed to generate the results that illustrate this chapter.

Bibliography

[1] S. Avidan and A. Shashua. Novel view synthesis in tensor space. In *Proc. of IEEE Conference on Computer Vision and Pattern Recognition*, pages 1034–1040, 1997.

[2] N. Ayache. *Artificial Vision for Mobile Robots: Stereo Vision and Multi-sensory Perception*. MIT Press, 1991.

[3] C. Buehler, M. Bosse, L. McMillan, S. Gortler, and M. Cohen. Unstructured lumigraph rendering. In *Proc. SIGGRAPH*, 2001.

[4] S. Chen and L. Williams. View interpolation for image synthesis. *Computer Graphics*, 27(Annual Conference Series):279–288, 1993.

[5] O. Chum and J. Matas. Randomized ransac and t(d,d) test. In *Proc. British Machine Vision Conference*, 2002.

[6] K. Cornelis, M. Pollefeys, M. Vergauwen, and L. Van Gool. Augmented reality from uncalibrated video sequences. In A. Zisserman, M. Pollefeys, L. Van Gool, and A. Fitzgibbon (Eds.), *3D Structure from Images–SMILE 2000*, volume 2018, pages 150–167. Springer-Verlag, 2001.

[7] K. Cornelis, M. Pollefeys, M. Vergauwen, F. Verbiest, and L. Van Gool. Tracking based structure and motion recovery for augmented video productions. In *Proceedings ACM Symposium on Virtual Reality Software and Technology–VRST 2001*, pages 17–24, November 2001.

[8] I. Cox, S. Hingoraini, and S. Rao. A maximum likelihood stereo algorithm. *Computer Vision and Image Understanding*, 63(3):542–567, May 1996.

[9] H. S. M. Coxeter. *Projective Geometry*. Blaisdell Publishing Company, 1964.

[10] B. Curless and M. Levoy. A volumetric method for building complex models from range images. In *Proc. SIGGRAPH '96*, pages 303–312, 1996.

[11] L. de Agapito, R. Hartley, and E. Hayman. Linear selfcalibration of a rotating and zooming camera. In *Proc. IEEE Conf. Comp. Vision Patt. Recog.*, volume I, pages 15–21, June 1999.

[12] L. Falkenhagen. Depth estimation from stereoscopic image pairs assuming piecewise continuous surfaces. In *Proc. of European Workshop on Combined Real and Synthetic Image Processing for Broadcast and Video Production*, November 1994.

[13] O. Faugeras. What can be seen in three dimensions with an uncalibrated stereo rig? In *Proc. European Conf. on Computer Vision*, pages 563–578, 1992.

[14] O. Faugeras. *Three-Dimensional Computer Vision*. MIT Press, 1993.

[15] O. Faugeras and Q.-T. Luong. *The Geometry of Multiple Images*. MIT Press, 2001.

[16] O. Faugeras, Q.-T. Luong, and S. Maybank. Camera self-calibration: Theory and experiments. In *Computer Vision–ECCV'92*, volume 588 of *Lecture Notes in Computer Science*, pages 321–334. Springer-Verlag, 1992.

[17] M. Fischler and R. Bolles. Random sampling consensus: A paradigm for model fitting with application to image analysis and automated cartography. *Commun. Assoc. Comp. Mach.*, 24:381–395, 1981.

[18] S. Gortler, R. Grzeszczuk, R. Szeliski, and M. F. Cohen. The Lumigraph. In *Proc. SIGGRAPH '96*, pages 43–54. ACM Press, New York, 1996.

[19] C. Harris and M. Stephens. A combined corner and edge detector. In *Fourth Alvey Vision Conference*, pages 147–151, 1988.

[20] R. Hartley. An algorithm for self calibration from several views. In *Proc. IEEE Conf. Comp. Vision Patt. Recog.*, pages 908–912, 1994.

[21] R. Hartley. Euclidean reconstruction from uncalibrated views. In J. L. Mundy, A. Zisserman, and D. Forsyth (Eds.), *Applications of Invariance in Computer Vision*, volume 825 of *Lecture Notes in Computer Science*, pages 237–256. Springer-Verlag, 1994.

[22] R. Hartley. In defense of the eight-point algorithm. *IEEE Trans. on Pattern Analysis and Machine Intelligence*, 19(6):580–593, June 1997.

[23] R. Hartley and P. Sturm. Triangulation. *Computer Vision and Image Understanding*, 68(2):146–157, 1997.

[24] R. Hartley and A. Zisserman. *Multiple View Geometry in Computer Vision*. Cambridge University Press, 2000.

[25] B. Heigl, R. Koch, M. Pollefeys, J. Denzler, and L. Van Gool. Plenoptic modeling and rendering from image sequences taken by hand-held camera. In *Proc. DAGM*, pages 94–101, 1999.

[26] A. Heyden. *Geometry and Algebra of Multiple Projective Transformations*. PhD thesis, Lund University, Sweden, 1995.

[27] A. Heyden. Tensorial properties of multilinear constraints. *Mathematical Methods in the Applied Sciences*, 23:169–202, 2000.

[28] A. Heyden and K. Åström. Euclidean reconstruction from constant intrinsic parameters. In *Proc. 13th International Conference on Pattern Recognition*, pages 339–343. IEEE Computer Soc. Press, 1996.

[29] A. Heyden and K. Åström. Euclidean reconstruction from image sequences with varying and unknown focal length and principal point. In *Proc. IEEE Conference on Computer Vision and Pattern Recognition*, pages 438–443. IEEE Computer Soc. Press, 1997.

[30] A. Heyden and K. Åström. Algebraic properties of multilinear constraints. *Mathematical Methods in the Applied Sciences*, 20:1135–1162, 1997.

[31] F. Kahl. Critical motions and ambiguous euclidean reconstructions in autocalibration. In *Proc. International Conference on Computer Vision*, pages 469–475, 1999.

[32] F. Kahl, B. Triggs, and K. Åström. Critical motions for autocalibration when some intrinsic parameters can vary. *Journal of Mathematical Imaging and Vision*, 13(2):131–146, 2000.

[33] R. Koch, B. Heigl, and M. Pollefeys. Image-based rendering from uncalibrated lightfields with scalable geometry. In G. Gimel'farb, R. Klette, and T. Huang (Eds.), *Multi-Image Analysis*, volume 2032 of *Lecture Notes in Computer Science*, pages 51–66. Springer-Verlag, 2001.

[34] R. Koch, M. Pollefeys, and L. Van Gool. Multi viewpoint stereo from uncalibrated video sequences. In *Proc. European Conference on Computer Vision*, pages 55–71, 1998.

[35] R. Koch, M. Pollefeys, B. Heigl, L. Van Gool, and H. Niemann. Calibration of hand-held camera sequences for plenoptic modeling. In *Proc. International Conference on Computer Vision*, pages 585–591, 1999.

[36] S. Laveau and O. Faugeras. Oriented projective geometry for computer vision. In B. Buxton and R. Cipolla (Eds.), *Proc. European Conference on Computer Vision, Lecture Notes in Computer Science*, Vol. 1064, pages 147–156. Springer-Verlag, 1996.

[37] M. Levoy and P. Hanrahan. Lightfield rendering. In *Proc. SIGGRAPH '96*, pages 31–42. ACM Press, 1996.

[38] H. C. Longuet-Higgins. A computer algorithm for reconstructing a scene from two projections. *Nature*, 293:133–135, 1981.

[39] D. Lowe. Object recognition from local scale-invariant features. In *Proc. International Conference on Computer Vision*, pages 1150–1157, 1999.

[40] J. Matas, S. Obdrzalek, and O. Chum. Local affine frames for wide-baseline stereo. In *Proc. 16th International Conference on Pattern Recognition*, volume 4, pages 363–366, 2002.

[41] G. Van Meerbergen, M. Vergauwen, M. Pollefeys, and L. Van Gool. A hierarchical symmetric stereo algorithm using dynamic programming. *International Journal on Computer Vision*, 47(1/2/3):275–285, 2002.

[42] N. Myklestad. *Cartesian Tensors–the Mathematical Language of Engineering*. Van Nostrand, Princeton, 1967.

[43] M. Pollefeys. *Self-calibration and metric 3D reconstruction from uncalibrated image sequences*. PhD thesis, Katholieke Universiteit Leuven, 1999.

[44] M. Pollefeys and L. Van Gool. Stratified self-calibration with the modulus constraint. *IEEE Transactions on Pattern Analysis and Machine Intelligence*, 21(8):707–724, 1999.

[45] M. Pollefeys and L. Van Gool. Do ambiguous reconstructions always give ambiguous images? In *Proc. International Conference on Computer Vision*, pages 187–192, 2001.

[46] M. Pollefeys, R. Koch, and L. Van Gool. Self-calibration and metric reconstruction in spite of varying and unknown internal camera parameters. In *Proc. International Conference on Computer Vision*, pages 90–95. Narosa Publishing House, 1998.

[47] M. Pollefeys, R. Koch, and L. Van Gool. Self-calibration and metric reconstruction in spite of varying and unknown internal camera parameters. *International Journal of Computer Vision*, 32(1):7–25, 1999.

[48] M. Pollefeys, R. Koch, and L. Van Gool. A simple and efficient rectification method for general motion. In *Proc. International Conference on Computer Vision*, pages 496–501, 1999.

[49] M. Pollefeys, F. Verbiest, and L. Van Gool. Surviving dominant planes in uncalibrated structure and motion recovery. In A. Heyden, G. Sparr, M. Nielsen, and P. Johansen (Eds.), *7th European Conference on Computer Vision*, volume 2351 of *Lecture Notes in Computer Science*, pages 837–851, 2002.

[50] W. Press, S. Teukolsky, and W. Vetterling. *Numerical Recipes in C: The Art of Scientific Computing*. Cambridge University Press, 1992.

[51] P. Rousseeuw. *Robust Regression and Outlier Detection*. Wiley, 1987.

[52] D. Scharstein and R. Szeliski. A taxonomy and evaluation of dense two-frame stereo correspondence algorithms. *International Journal of Computer Vision*, 47(1/2/3):7–42, April–June 2002.

[53] C. Schmid and R. Mohr. Local grayvalue invariants for image retrieval. *IEEE PAMI*, 19(5):530–534, 1997.

[54] S. Seitz and C. Dyer. View morphing. *Computer Graphics*, Annual Conference Series, 30:21–30, 1996.

[55] A. Shashua. Trilinearity in visual recognition by alignment. In *Proc. European Conf. on Computer Vision*, 1994.

[56] J. Shi and C. Tomasi. Good features to track. In *Proc. IEEE Conference on Computer Vision and Pattern Recognition*, pages 593–600, 1994.

[57] C. Slama. *Manual of Photogrammetry*. 4th ed., American Society of Photogrammetry, USA, 1980.

[58] B. Spain. *Tensor Calculus*. University Mathematical Texts, Oliver and Boyd, Edingburgh, 1953.

[59] G. Sparr. Simultaneous reconstruction of scene structure and camera locations from uncalibrated image sequences. In *Proc. International Conf. on Pattern Recognition*, 1996.

[60] M. E. Spetsakis and J. Aloimonos. A unified theory of structure from motion. In *Proc. DARPA IU Workshop*, 1990.

[61] P. Sturm. Critical motion sequences for monocular self-calibration and uncalibrated Euclidean reconstruction. In *Proc. Conference on Computer Vision and Pattern Recognition*, pages 1100–1105. IEEE Computer Society Press, 1997.

[62] P. Sturm. Critical motion sequences for the self-calibration of cameras and stereo systems with variable focal length. In *Proc. 10th British Machine Vision Conference*, pages 63–72, 1999.

[63] P. Sturm. Critical motion sequences for the self-calibration of cameras and stereo systems with variable focal length. In *Proc. BMVC*, 1999.

[64] B. Triggs. Matching constraints and the joint image. In *Proc. Int. Conf. on Computer Vision*, 1995.

[65] B. Triggs. The absolute quadric. In *Proc. 1997 Conference on Computer Vision and Pattern Recognition*, pages 609–614. IEEE Computer Society Press, 1997.

[66] B. Triggs. Autocalibration from planar scenes. In *ECCV*, volume I, pages 89–105, June 1998.

[67] B. Triggs, P. McLauchlan, R. Hartley, and A. Fiztgibbon. Bundle adjustment–A modern synthesis. In R. Szeliski, B. Triggs, A. Zisserman (Eds.), *Vision Algorithms: Theory and Practice*, LNCS Vol. 1883, pages 298–372. Springer-Verlag, 2000.

[68] G. Turk and M. Levoy. Zippered polygon meshes from range images. In *Proceedings of SIGGRAPH '94*, pages 311–318, 1994.

[69] T. Tuytelaars and L. Van Gool. Wide baseline stereo based on local, affinely invariant regions. In *British Machine Vision Conference*, pages 412–422, 2000.

[70] Z. Zhang, R. Deriche, O. Faugeras, and Q.-T. Luong. A robust technique for matching two uncalibrated images through the recovery of the unknown epipolar geometry. *Artificial Intelligence Journal*, 78:87–119, October 1995.

Chapter 4

ROBUST TECHNIQUES
FOR COMPUTER VISION

Peter Meer

4.1 Robustness in Visual Tasks

Visual information makes up about 75 percent of all the sensorial information received by a person during a lifetime. This information is processed not only efficiently but also transparently. Our awe of visual perception was perhaps the best captured by the 17th-century British essayist Joseph Addison in an essay on imagination [1].

> Our sight is the most perfect and most delightful of all our senses.
> It fills the mind with the largest variety of ideas, converses with
> its objects at the greatest distance, and continues the longest in
> action without being tired or satiated with its proper enjoyments.

The ultimate goal of computer vision is to mimic human visual perception. Therefore, in the broadest sense, robustness of a computer vision algorithm is judged against the performance of a human observer performing an equivalent task. In this context, robustness is the ability to extract the visual information of relevance for a specific task, even when this information is carried only by a small subset of the data and/or is significantly different from an already stored representation.

To understand why the performance of generic computer vision algorithms is still far away from that of human visual perception, we should consider the hierarchy of computer vision tasks. They can be roughly classified into three large categories:

- *low level*, dealing with extraction from a single image of salient simple features, such as edges, corners, homogeneous regions, curve fragments;

- *intermediate level*, dealing with extraction of semantically relevant characteristics from one or more images, such as grouped features, depth, motion information;

- *high level*, dealing with the interpretation of the extracted information.

A similar hierarchy is difficult to distinguish in human visual perception, which appears as a single integrated unit. In the visual tasks performed by a human observer, an extensive top-down information flow carrying representations derived at higher levels seems to control the processing at lower levels. See [84] for a discussion on the nature of these interactions.

A large amount of psychophysical evidence supports this "closed-loop" model of human visual perception. Preattentive vision phenomena in which salient information pops out from the image (e.g., [55, 109]) or perceptual constancies in which changes in the appearance of a familiar object are attributed to external causes [36, Chap. 9] are only some of the examples. Similar behavior is yet to be achieved in generic computer vision techniques. For example, preattentive vision-type processing seems to imply that a region of interest is delineated *before* extracting its salient features.

To approach the issue of robustness in computer vision, we start by mentioning one of the simplest perceptual constancies, the shape constancy. Consider a door opening in front of an observer. As the door opens, its image changes from a rectangle to a trapezoid, but the observer will report only the movement. That is, *additional information* not available in the input data was also taken into account. We *know* that a door is a rigid structure, and therefore it is very unlikely that its image changed due to a nonrigid transformation. Since the perceptual constancies are based on rules embedded in the visual system, they can be also deceived. A well-known example is the Ames room in which the rules used for perspective foreshortening compensation are violated [36, p. 241].

The previous example does not seem to reveal much. Any computer vision algorithm of rigid motion recovery is based on a similar approach. However, the example emphasizes that the employed rigid motion model is only associated with the data and is not intrinsic to it. We could use a completely different model, say of nonrigid doors, but the result would not be satisfactory. Robustness thus is closely related to the availability of a model adequate for the goal of the task.

In today's computer vision algorithms, the information flow is almost exclusively bottom-up. Feature extraction is followed by grouping into semantical primitives, which in turn is followed by a task-specific interpretation of the ensemble of primitives. The lack of top-down information flow is arguably the main reason computer vision techniques cannot yet autonomously handle visual data under a wide range of operating conditions. This fact is well understood in the vision community, and different approaches were proposed to simulate the top-down information stream.

The increasingly popular Bayesian paradigm is such an attempt. By using a probabilistic representation for the possible outcomes, multiple hypotheses are incorporated into the processing, which in turn guide the information recovery. The dependence of the procedure on the accuracy of the employed representation is relaxed in the semiparametric or nonparametric Bayesian methods, such as particle filtering for motion problems [51]. Incorporating a learning component into computer vision techniques (e.g., [3, 29]) is another somewhat similar approach to using higher level information during the processing.

Comparison with human visual perception is not a practical way to arrive at a definition of robustness for computer vision algorithms. For example, robustness in the context of the human visual system extends to abstract concepts. We can recognize a chair independent of its design, size, or the period in which it was made. However, in a somewhat similar experiment, when an object recognition system was programmed to decide if a simple drawing represents a chair, the results were rather mixed [96].

We will not consider high-level processes when examining the robustness of vision algorithms, nor will we discuss the role of top-down information flow. A computer vision algorithm will be called robust if it can tolerate outliers (i.e., data that does not obey the assumed model). This definition is similar to the one used in statistics for robustness [40, p. 6]

> In a broad informal sense, robust statistics is a body of knowledge, partly formalized into "theories of statistics," relating to deviations from idealized assumptions in statistics.

Robust techniques have been used in computer vision for at least 30 years. In fact, those most popular today are related to old methods proposed to solve specific image understanding or pattern recognition problems. Some of them were rediscovered only in the last few years.

The best known example is the Hough transform, a technique to extract multiple instances of a low-dimensional manifold from a noisy background. The Hough transform is a U.S. patent granted in 1962 [47] for the detection of linear trajectories of subatomic particles in a bubble chamber. In the rare cases when Hough transform is explicitly referenced, this patent is used, though an earlier publication also exists [46]. Similarly, the most popular robust regression methods today in computer vision belong to the family of random sample consensus (RANSAC), proposed in 1980 to solve the perspective n-point problem [25]. The usually employed reference is [26]. An old pattern recognition technique for density gradient estimation proposed in 1975 [32], the mean shift, recently became a widely used method for feature space analysis. See also [31, p. 535].

In theoretical statistics, investigation of robustness started in the early 1960s, and the first robust estimator, the M-estimator, was introduced by Huber in 1964. See [49] for the relevant references. Another popular family of robust estimators, including the least median of squares (LMedS), was introduced by Rousseeuw in 1984 [86]. By the end of 1980s, these robust techniques became known in the computer vision community.

Application of robust methods to vision problems was restricted at the beginning to replacing a nonrobust parameter estimation module with its robust counterpart; see [4, 41, 59, 102]. See also the review paper [74]. While this approach was successful in most of the cases, some failures were reported [77]. Today we know that these failures are due to the inability of most robust estimators to handle data in which more than one structure is present [9, 97], a situation frequently met in computer vision but almost never in statistics. For example, a window operator often covers an image patch which contains two homogeneous regions of almost equal sizes, or there can be several independently moving objects in a visual scene.

A large part of today's robust computer vision toolbox is indigenous. There are good reasons for this. The techniques imported from statistics were designed for data with characteristics significantly different from that of the data in computer vision. If the data does not obey the assumptions implied by the method of analysis, the desired performance may not be achieved. The development of robust techniques in the vision community (such as RANSAC) were motivated by applications. In these techniques, the user has more freedom to adjust the procedure to the specific data than in a similar technique taken from the statistical literature (such as LMedS). Thus, some of the theoretical limitations of a robust method can be alleviated by data-specific tuning, which sometimes resulted in attributing better performance to a technique than is theoretically possible in the general case.

A decade ago, when a vision task was solved with a robust technique, the focus of the research was on the methodology and not on the application. Today the emphasis has changed, and often the employed robust techniques are barely mentioned. It is no longer of interest to have an exhaustive survey of "robust computer vision." For some representative results, see the review paper [98] or the special issue [94].

This chapter focuses on the theoretical foundations of the robust methods in the context of computer vision applications. We provide a unified treatment for most estimation problems and put the emphasis on the underlying concepts, not on the details of implementation of a specific technique. We describe the assumptions embedded in the different classes of robust methods and clarify some misconceptions often arising in the vision literature. Based on this theoretical analysis, new robust methods, better suited for the complexity of computer vision tasks, can be designed.

4.2   Models and Estimation Problems

In this section, we examine the basic concepts involved in parameter estimation. We describe the different components of a model and show how to find the adequate model for a given computer vision problem. Estimation is analyzed as a generic problem, and the differences between nonrobust and robust methods are emphasized. We also discuss the role of the optimization criterion in solving an estimation problem.

4.2.1   Elements of a model

The goal of data analysis is to provide, for data spanning a very high-dimensional space, an equivalent low-dimensional representation. A set of measurements consisting of n data vectors $\mathbf{y}_i \in \mathcal{R}^q$ can be regarded as a point in $\mathcal{R}^{nq}$. If the data can be described by a model with only $p \ll nq$ parameters, we have a much more compact representation. Should new data points become available, their relation to the initial data then can be established using only the model. A model has two main components:

- the constraint equation;

- the measurement equation.

The constraint describes our a priori knowledge about the nature of the process generating the data, while the measurement equation describes the way the data was obtained.

In the general case, a constraint has two levels. The first level is that of the quantities providing the input into the estimation. These *variables* $\{y_1, \ldots, y_q\}$ can be obtained by direct measurement or can be the output of another process. The variables are grouped together in the context of the process to be modeled. Each ensemble of values for the q variables provides a single input data point, a q-dimensional vector $\mathbf{y} \in \mathcal{R}^q$.

At the second level of a constraint, the variables are combined into *carriers*, also called *basis functions*:

$$x_j = \varphi_j(y_1, \ldots, y_q) = \varphi_j(\mathbf{y}) \qquad j = 1, \ldots, m \ . \tag{4.1}$$

A carrier is usually a simple, nonlinear function in a subset of the variables. In computer vision, most carriers are monomials.

The *constraint* is a set of algebraic expressions in the carriers and the parameters $\theta_0, \theta_1, \ldots, \theta_p$:

$$\psi_k(x_1, \ldots, x_m; \theta_0, \theta_1, \ldots, \theta_p) = 0 \qquad k = 1, \ldots, K \ . \tag{4.2}$$

One of the goals of the estimation process is to find the values of these parameters, that is, to mold the constraint to the available measurements.

The constraint captures our a priori knowledge about the physical and/or geometrical relations underlying the process in which the data was generated, Thus, the constraint is valid only for the true (uncorrupted) values of the variables. In general, these values are not available. The estimation process replaces in the constraint the true values of the variables with their *corrected* values, and the true values of the parameters with their estimates. We return to this issue in Section 4.2.2.

The expression of the constraint (4.2) is too general for our discussion, and we use only a scalar (univariate) constraint, $K = 1$, which is *linear in the carriers and the parameters*:

$$\alpha + \mathbf{x}^\top \boldsymbol{\theta} = 0 \qquad \mathbf{x}^\top = [\varphi_1(\mathbf{y}) \quad \cdots \quad \varphi_p(\mathbf{y})], \tag{4.3}$$

where the parameter θ_0 associated with the constant carrier was renamed α, all the other carriers were gathered into the vector $\mathbf{x}$, and the parameters into the vector $\boldsymbol{\theta}$. The linear structure of this model implies that $m = p$. Note that the constraint (4.3) in general is *nonlinear in the variables*.

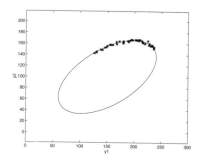

Figure 4.1. A typical nonlinear regression problem. Estimate the parameters of the ellipse from the noisy data points.

The parameters α and $\boldsymbol{\theta}$ are defined in (4.3) only up to a multiplicative constant. This ambiguity can be eliminated in many different ways. We show in Section 4.2.6 that often it is advantageous to impose $\|\boldsymbol{\theta}\| = 1$. Any condition additional to (4.3) is called an *ancillary constraint*.

In some applications, one of the variables has to be singled out. This variable, denoted z, is called the *dependent variable*; all the others are *independent variables*, which enter into the constraint through the carriers. The constraint becomes

$$z = \alpha + \mathbf{x}^\top \boldsymbol{\theta}, \qquad (4.4)$$

and the parameters are no longer ambiguous.

To illustrate the role of the variables and carriers in a constraint, we consider the case of the ellipse (Figure 4.1). The constraint can be written as

$$(\mathbf{y} - \mathbf{y}_c)^\top Q(\mathbf{y} - \mathbf{y}_c) - 1 = 0, \qquad (4.5)$$

where the two variables are the coordinates of a point on the ellipse, $\mathbf{y}^\top = [y_1 \; y_2]$. The constraint has five parameters: the two coordinates of the ellipse center $\mathbf{y}_c$, and the three distinct elements of the 2×2 symmetric, positive definite matrix Q. The constraint is rewritten under the form (4.3) as

$$\alpha + \theta_1 y_1 + \theta_2 y_2 + \theta_3 y_1^2 + \theta_4 y_1 y_2 + \theta_5 y_2^2 = 0, \qquad (4.6)$$

where

$$\alpha = \mathbf{y}_c^\top Q \mathbf{y}_c - 1 \qquad \boldsymbol{\theta}^\top = [-2\mathbf{y}_c^\top Q \quad Q_{11} \quad 2Q_{12} \quad Q_{22}]. \qquad (4.7)$$

Three of the five carriers,

$$\mathbf{x}^{\top} = [y_1 \quad y_2 \quad y_1^2 \quad y_1 y_2 \quad y_2^2], \tag{4.8}$$

are nonlinear functions in the variables.

Ellipse estimation uses the constraint (4.6). This constraint, however, has not five but six parameters, which are again defined only up to a multiplicative constant. Furthermore, the same constraint can also represent two other conics: a parabola or a hyperbola. The ambiguity of the parameters is therefore eliminated by using the ancillary constraint, which enforces that the quadratic expression (4.6) represents an ellipse:

$$4\theta_3\theta_5 - \theta_4^2 = 1 . \tag{4.9}$$

The nonlinearity of the constraint in the variables makes ellipse estimation a difficult problem. See [27, 57, 72, 117] for different approaches and discussions.

For most of the variables, only the noise-corrupted version of their true value is available. Depending on the nature of the data, the noise is due to the measurement errors or to the inherent uncertainty at the output of another estimation process. While for convenience we use the term measurement for any input into an estimation process, the above distinction about the origin of the data should be kept in mind.

The general assumption in computer vision problems is that the noise is additive. Thus, the *measurement equation* is

$$\mathbf{y}_i = \mathbf{y}_{io} + \delta\mathbf{y}_i \qquad \mathbf{y}_i \in \mathcal{R}^q \qquad i = 1, \dots, n, \tag{4.10}$$

where $\mathbf{y}_{io}$ is the true value of $\mathbf{y}_i$, the i-th measurement. The subscript o denotes the true value of a measurement. Since the constraints (4.3) or (4.4) capture our a priori knowledge, they are valid for the true values of the measurements or parameters, and should have been written as

$$\alpha + \mathbf{x}_o^{\top}\boldsymbol{\theta} = 0 \qquad \text{or} \qquad z_o = \alpha + \mathbf{x}_o^{\top}\boldsymbol{\theta}, \tag{4.11}$$

where $\mathbf{x}_o = \mathbf{x}(\mathbf{y}_o)$. In the ellipse example, y_{1o} and y_{2o} should have been used in (4.6).

The noise corrupting the measurements is assumed to be independent and identically distributed (i.i.d.)

$$\delta\mathbf{y}_i \sim GI(\mathbf{0}, \sigma^2 \mathbf{C}_y), \tag{4.12}$$

where $GI(\cdot)$ stands for a general symmetric distribution of independent outcomes. Note that this distribution does not necessarily have to be normal.
A warning is in order, though. By characterizing the noise only with its first
two central moments we implicitly agree to normality, since only the normal
distribution is defined uniquely by these two moments.

The independency assumption usually holds when the input data points
are physical measurements, but may be violated when the data is the output
of another estimation process. It is possible to take into account the correlation between two data points $\mathbf{y}_i$ and $\mathbf{y}_j$ in the estimation (e.g., [75]), but it
is rarely used in computer vision algorithm. Most often this is not a crucial
omission, since the main source of performance degradation is the failure of
the constraint to adequately model the structure of the data.

The covariance of the noise is the product of two components in (4.12).
The *shape* of the noise distribution is determined by the matrix $\mathbf{C}_y$. This
matrix is assumed to be known and can also be singular. Indeed, for those
variables that are available without error, there is no variation along their dimensions in $\mathcal{R}^q$. The shape matrix is normalized to have $\det[\mathbf{C}_y] = 1$, where
in the singular case the determinant is computed as the product of nonzero
eigenvalues (which are also the singular values for a covariance matrix). For
independent variables, the matrix $\mathbf{C}_y$ is diagonal, and if all the independent
variables are corrupted by the same measurement noise, $\mathbf{C}_y = \mathbf{I}_q$. This is often the case when variables of the same nature (e.g., spatial coordinates) are
measured in the physical world. Note that the independency of the n measurements $\mathbf{y}_i$ and the independency of the q variables y_k are not necessarily
related properties.

The second component of the noise covariance is the *scale* σ, which in
general is not known. The main message of this chapter is that

> robustness in computer vision cannot be achieved without having
> access to a reasonably correct value of the scale.

The importance of scale is illustrated through the simple example in
Figure 4.2. All the data points except the one marked with the star belong
to the same (linear) model in Figure 4.2a. The points obeying the model are
called *inliers*, and the point far away is an *outlier*. The shape of the noise
corrupting the inliers is circular symmetric, that is, $\sigma^2\mathbf{C}_y = \sigma^2\mathbf{I}_2$. The data in
Figure 4.2b differs from the data in Figure 4.2a only by the value of the scale
σ. Should the value of σ from the first case be used when analyzing the data
in the second case, many inliers will be discarded with severe consequences
on the performance of the estimation process.

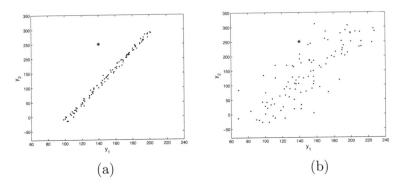

Figure 4.2. The importance of scale. The difference between the data in (a) and (b) is only in the scale of the noise.

The true values of the variables are not available, and instead of $\mathbf{y}_{io}$ at the beginning of the estimation process, the measurement $\mathbf{y}_i$ has to be used to compute the carriers. The first two central moments of the noise associated with a carrier can be approximated by error propagation.

Let $x_{ij} = \varphi_j(\mathbf{y}_i)$ be the j-th element, $j = 1, \ldots, p$, of the carrier vector $\mathbf{x}_i = \mathbf{x}(\mathbf{y}_i) \in \mathcal{R}^p$, computed for the i-th measurement $\mathbf{y}_i \in \mathcal{R}^q$, $i = 1, \ldots, n$. Since the measurement vectors $\mathbf{y}_i$ are assumed to be independent, the carrier vectors $\mathbf{x}_i$ are also independent random variables.

The second-order Taylor expansion of the carrier x_{ij} around the corresponding true value $x_{ijo} = \varphi_j(\mathbf{y}_{io})$ is

$$x_{ij} \approx x_{ijo} + \left[\frac{\partial \varphi_j(\mathbf{y}_{io})}{\partial \mathbf{y}}\right]^\top (\mathbf{y}_i - \mathbf{y}_{io}) + \frac{1}{2}(\mathbf{y}_i - \mathbf{y}_{io})^\top \frac{\partial^2 \varphi_j(\mathbf{y}_{io})}{\partial \mathbf{y} \partial \mathbf{y}^\top}(\mathbf{y}_i - \mathbf{y}_{io}),$$

$$(4.13)$$

where $\frac{\partial \varphi_j(\mathbf{y}_{io})}{\partial \mathbf{y}}$ is the gradient of the carrier with respect to the vector of the variables $\mathbf{y}$, and $H_j(\mathbf{y}_{io}) = \frac{\partial^2 \varphi_j(\mathbf{y}_{io})}{\partial \mathbf{y} \partial \mathbf{y}^\top}$ is its Hessian matrix, both computed in the true value of the variables $\mathbf{y}_{io}$. From the measurement equation (4.10) and (4.13), the second-order approximation for the expected value of the noise corrupting the carrier x_{ij} is

$$\mathrm{E}[x_{ij} - x_{ijo}] = \frac{\sigma^2}{2}\mathrm{trace}[C_y H_j(\mathbf{y}_{io})],$$

$$(4.14)$$

which shows that this noise is not necessarily zero-mean. The first-order approximation of the noise covariance is obtained by straightforward error propagation:

$$\text{cov}[\mathbf{x}_i - \mathbf{x}_{io}] = \sigma^2 C_{x_i} = \sigma^2 J_{x|y}(\mathbf{y}_{io})^\top C_y J_{x|y}(\mathbf{y}_{io}), \qquad (4.15)$$

where $J_{x|y}(\mathbf{y}_{io})$ is the Jacobian of the carrier vector $\mathbf{x}$ with respect to the vector of the variables $\mathbf{y}$, computed in the true values $\mathbf{y}_{io}$. In general, the moments of the noise corrupting the carriers are functions of $\mathbf{y}_{io}$ and thus are point-dependent. A point-dependent noise process is called *heteroscedastic*. Note that the dependence is through the true values of the variables, which in general are not available. In practice, the true values are substituted with the measurements.

To illustrate the heteroscedasticity of the carrier noise, we return to the example of the ellipse. From (4.8) we obtain the Jacobian

$$J_{x|y} = \begin{bmatrix} 1 & 0 & 2y_1 & y_2 & 0 \\ 0 & 1 & 0 & y_1 & 2y_2 \end{bmatrix} \qquad (4.16)$$

and the Hessians

$$H_1 = H_2 = 0 \quad H_3 = \begin{bmatrix} 2 & 0 \\ 0 & 0 \end{bmatrix} \quad H_4 = \begin{bmatrix} 0 & 1 \\ 1 & 0 \end{bmatrix} \quad H_5 = \begin{bmatrix} 0 & 0 \\ 0 & 2 \end{bmatrix}. \qquad (4.17)$$

Assume that the simplest measurement noise distributed $GI(\mathbf{0}, \sigma^2 I_2)$ is corrupting the two spatial coordinates (the variables). The noise corrupting the carriers, however, has nonzero mean and a covariance that is a function of $\mathbf{y}_o$:

$$E[\mathbf{x} - \mathbf{x}_o] = \begin{bmatrix} 0 & 0 & \sigma^2 & 0 & \sigma^2 \end{bmatrix}^\top \qquad \text{cov}[\mathbf{x} - \mathbf{x}_o] = \sigma^2 J_{x|y}(\mathbf{y}_o)^\top J_{x|y}(\mathbf{y}_o) \ . \qquad (4.18)$$

To accurately estimate the parameters of the general model, the heteroscedasticity of the carrier noise has to be taken into account, as discussed in Section 4.2.5.

4.2.2 Estimation of a model

We can proceed now to a formal definition of the estimation process.

Given the model:

    – the noisy measurements $\mathbf{y}_i$, which are the additively corrupted versions of the true values $\mathbf{y}_{io}$

$$\mathbf{y}_i = \mathbf{y}_{io} + \delta\mathbf{y}_i \quad \mathbf{y}_i \in \mathcal{R}^q \quad \delta\mathbf{y}_i \sim GI(\mathbf{0}, \sigma^2 C_y) \qquad i = 1, \ldots, n$$

    – the covariance of the errors $\sigma^2 C_y$, known only up to the scale σ

- the constraint obeyed by the true values of the measurements

$$\alpha + \mathbf{x}_{io}^{\top}\boldsymbol{\theta} = 0 \quad \mathbf{x}_{io} = \mathbf{x}(\mathbf{y}_{io}) \qquad i = 1, \ldots, n$$

and some ancillary constraints.

Find the estimates:

- for the model parameters, $\widehat{\alpha}$ and $\widehat{\boldsymbol{\theta}}$
- for the true values of the measurements, $\widehat{\mathbf{y}}_i$
- such that they satisfy the constraint

$$\widehat{\alpha} + \widehat{\mathbf{x}}_i^{\top}\widehat{\boldsymbol{\theta}} = 0 \quad \widehat{\mathbf{x}}_i = \mathbf{x}(\widehat{\mathbf{y}}_i) \qquad i = 1, \ldots, n$$

and all the ancillary constraints.

The true values of the measurements $\mathbf{y}_{io}$ are called *nuisance parameters*, since they have only a secondary role in the estimation process. We treat the nuisance parameters as unknown constants, in which case we have a *functional* model [33, p. 2]. When the nuisance parameters are assumed to obey a known distribution whose parameters also have to be estimated, we have a *structural* model. For robust estimation the functional models are more adequate, since they require fewer assumptions about the data.

The estimation of a functional model has two distinct parts. First, the parameter estimates are obtained in the main *parameter estimation* procedure, followed by the computation of the nuisance parameter estimates in the *data correction* procedure. The nuisance parameter estimates $\widehat{\mathbf{y}}_i$ are called the *corrected data* points. The data correction procedure is usually not more than the projection of the measurements $\mathbf{y}_i$ on the already estimated constraint surface.

The parameter estimates are most often obtained by seeking the global minima of an *objective function*. The variables of the objective function are the normalized distances between the measurements and their true values. They are defined from the squared Mahalanobis distances

$$d_i^2 = \frac{1}{\sigma^2}(\mathbf{y}_i - \mathbf{y}_{io})^{\top}\mathbf{C}_y^{+}(\mathbf{y}_i - \mathbf{y}_{io}) = \frac{1}{\sigma^2}\delta\mathbf{y}_i^{\top}\mathbf{C}_y^{+}\delta\mathbf{y}_i \qquad i = 1, \ldots, n, \quad (4.19)$$

where $+$ stands for the pseudoinverse operator, since the matrix $\mathbf{C}_y$ can be singular, in which case (4.19) is only a pseudodistance. Note that $d_i \geq 0$. Through the estimation procedure, the $\mathbf{y}_{io}$ are replaced with $\widehat{\mathbf{y}}_i$ and the distance d_i becomes the absolute value of the *normalized residual*.

The objective function $\mathcal{J}(d_1, \ldots, d_n)$ is always a positive semidefinite function taking value zero only when all the distances are zero. We should distinguish between homogeneous and nonhomogeneous objective functions. A homogeneous objective function has the property

$$\mathcal{J}(d_1, \ldots, d_n) = \frac{1}{\sigma} \mathcal{J}(\|\delta \mathbf{y}_1\|_{\mathbf{C}_y}, \ldots, \|\delta \mathbf{y}_n\|_{\mathbf{C}_y}), \tag{4.20}$$

where $\|\delta \mathbf{y}_i\|_{\mathbf{C}_y} = [\delta \mathbf{y}_i^\top \mathbf{C}_y^+ \delta \mathbf{y}_i]^{1/2}$ is the covariance weighted norm of the measurement error. The homogeneity of an objective function is an important property in the estimation. Only for homogeneous objective functions we have

$$[\widehat{\alpha}, \widehat{\boldsymbol{\theta}}] = \operatorname*{argmin}_{\alpha, \boldsymbol{\theta}} \mathcal{J}(d_1, \ldots, d_n) = \operatorname*{argmin}_{\alpha, \boldsymbol{\theta}} \mathcal{J}\left(\|\delta \mathbf{y}_1\|_{\mathbf{C}_y}, \ldots, \|\delta \mathbf{y}_n\|_{\mathbf{C}_y}\right), \tag{4.21}$$

meaning that the scale σ does not play any role in the main estimation process. Since the value of the scale is not known a priori, by removing it an important source for performance deterioration is eliminated. All the following objective functions are homogeneous

$$\mathcal{J}_{LS} = \frac{1}{n} \sum_{i=1}^{n} d_i^2 \qquad \mathcal{J}_{LAD} = \frac{1}{n} \sum_{i=1}^{n} d_i \qquad \mathcal{J}_{LkOS} = d_{k:n}, \tag{4.22}$$

where $\mathcal{J}_{LS}$ yields the family of least squares estimators, $\mathcal{J}_{LAD}$ the least absolute deviations estimator, and $\mathcal{J}_{LkOS}$ the family of least k-th order statistics estimators. In an LkOS estimator, the distances are assumed sorted in ascending order, and the k-th element of the list is minimized. If $k = n/2$, the least median of squares (LMedS) estimator, discussed in detail in Section 4.4.4, is obtained.

The most important example of nonhomogeneous objective functions is that of the M-estimators

$$\mathcal{J}_M = \frac{1}{n} \sum_{i=1}^{n} \rho(d_i), \tag{4.23}$$

where $\rho(u)$ is a nonnegative, even-symmetric *loss function*, nondecreasing with $|u|$. The class of $\mathcal{J}_M$ includes as particular cases $\mathcal{J}_{LS}$ and $\mathcal{J}_{LAD}$, for $\rho(u) = u^2$ and $\rho(u) = |u|$ respectively, but in general this objective function is not homogeneous. The family of M-estimators, discussed in Section 4.4.2, have the loss function

$$\rho(u) = \begin{cases} 1 - (1 - u^2)^d & |u| \leq 1 \\ 1 & |u| > 1 \end{cases}, \qquad (4.24)$$

where $d = 0, 1, 2, 3$. It will be shown later in the chapter that all the robust techniques popular today in computer vision can be described as M-estimators.

The definitions introduced so far implicitly assume that *all* the n data points obey the model; that is, they are inliers. In this case, nonrobust estimation techniques provide a satisfactory result. In the presence of outliers, only $n_1 \leq n$ measurements are inliers and obey (4.3). The number n_1 is not known. The measurement equation (4.10) becomes

$$\begin{aligned} \mathbf{y}_i &= \mathbf{y}_{io} + \delta \mathbf{y}_i & \delta \mathbf{y}_i \sim GI(\mathbf{0}, \sigma^2 \mathbf{C}_y) & \qquad i = 1, \ldots, n_1 & \qquad (4.25) \\ \mathbf{y}_i & & & \qquad i = (n_1 + 1), \ldots, n, \end{aligned}$$

where nothing is assumed known about the $n - n_1$ outliers. Sometimes in robust methods proposed in computer vision, such as [99, 106, 113], the outliers were modeled as obeying a uniform distribution.

A robust method has to determine n_1 simultaneously with the estimation of the inlier model parameters. Since n_1 is unknown, at the beginning of the estimation process, the model is still defined for $i = 1, \ldots, n$. Only through the optimization of an adequate objective function are the data points classified into inliers or outliers. The result of the robust estimation is the *inlier/outlier dichotomy* of the data.

The estimation process maps the input, the set of measurements $\mathbf{y}_i$, $i = 1, \ldots, n$ into the output, the estimates $\widehat{\alpha}$, $\widehat{\boldsymbol{\theta}}$ and $\widehat{\mathbf{y}}_i$. The measurements are noisy and the uncertainty about their true value is mapped into the uncertainty about the true value of the estimates. The computational procedure employed to obtain the estimates is called the *estimator*. To describe the properties of an estimator, the estimates are treated as random variables. The estimate $\widehat{\boldsymbol{\theta}}$ is used generically in the next two sections to discuss these properties.

4.2.3 Robustness of an estimator

Depending on n, the number of available measurements, we should distinguish between small (finite) sample and large (asymptotic) sample properties of an estimator [75, secs. 6, 7]. In the latter case, n becomes large enough that further increase in its value no longer has a significant influence on the estimates. Many of the estimator properties proven in theoretical statistics are asymptotic and are not necessarily valid for small data sets. Rigorous analysis of small sample properties is difficult. See [85] for examples in pattern recognition.

What is a small or a large sample depends on the estimation problem at hand. Whenever the model is not accurate even for a large number of measurements, the estimate remains highly sensitive to the input. This situation is frequently present in computer vision, where only a few tasks would qualify as large-scale behavior of the employed estimator. We will not discuss here asymptotic properties, such as the *consistency*, which describes the relation of the estimate to its true value when the number of data points grows unbounded. Our focus is on the *bias* of an estimator, the property which is also central in establishing whether the estimator is robust or not.

Let $\boldsymbol{\theta}$ be the true value of the estimate $\widehat{\boldsymbol{\theta}}$. The estimator mapping the measurements $\mathbf{y}_i$ into $\widehat{\boldsymbol{\theta}}$ is unbiased if

$$\mathrm{E}[\widehat{\boldsymbol{\theta}}] = \boldsymbol{\theta}, \tag{4.26}$$

where the expectation is taken over all possible sets of measurements of size n (i.e., over the joint distribution of the q variables). Assume now that the input data contains n_1 inliers and $n - n_1$ outliers. In a "thought" experiment, we keep all the inliers fixed and allow the outliers to be placed anywhere in $\mathcal{R}^q$, the space of the measurements $\mathbf{y}_i$. Clearly, some of these arrangements will have a larger effect on $\widehat{\boldsymbol{\theta}}$ than others. We define the *maximum bias* as

$$b_{max}(n_1, n) = \max_{\mathcal{O}} \|\widehat{\boldsymbol{\theta}} - \boldsymbol{\theta}\|, \tag{4.27}$$

where $\mathcal{O}$ stands for the arrangements of the $n - n_1$ outliers. We say that an estimator exhibits a *globally robust* behavior in a given task if and only if

$$\text{for} \quad n_1 < n \qquad b_{max}(n_1, n) < t_b, \tag{4.28}$$

where $t_b \geq 0$ is a threshold depending on the task. That is, the presence of outliers cannot introduce an estimation error beyond the tolerance deemed acceptable for that task. To *qualitatively* assess the robustness of the estimator, we can define

$$\eta(n) = 1 - \min_{n_1} \frac{n_1}{n} \qquad \text{while (4.28) holds}, \tag{4.29}$$

which measures its outlier rejection capability. Note that the definition is based on the worst-case situation, which may not appear in practice.

The robustness of an estimator is assured by the employed objective function. Among the three homogeneous objective functions in (4.22), minimization of two criteria, the least squares $\mathcal{J}_{LS}$ and the least absolute deviations $\mathcal{J}_{LAD}$, does not yield robust estimators. A striking example for the (less known) nonrobustness of the latter is discussed in [89, p. 20]. The LS and LAD estimators are not robust, since their homogeneous objective function (4.20) is also *symmetric*. The value of a symmetric function is invariant under the permutations of its variables, the distances d_i in our case. Thus, in a symmetric function, all the variables have equal importance.

To understand why these two objective functions lead to a nonrobust estimator, consider the data containing a single outlier located far away from all the other points, the inliers (Figure 4.2a). The scale of the inlier noise, σ, has no bearing on the minimization of a homogeneous objective function (4.21). The symmetry of the objective function, on the other hand, implies that during the optimization *all* the data points, including the outlier, are treated in the same way. For a parameter estimate close to the true value, the outlier yields a very large measurement error $\|\delta \mathbf{y}_i\|_{\mathbf{C}_y}$. The optimization procedure therefore tries to compensate for this error and biases the fit toward the outlier. For any threshold t_b on the tolerated estimation errors, the outlier can be placed far enough from the inliers such that (4.28) is not satisfied. This means $\eta(n) = 0$.

In a robust technique, the objective function cannot be both symmetric and homogeneous. For the M-estimators, $\mathcal{J}_M$ (4.23) is only symmetric, while the least k-th order statistics objective function $\mathcal{J}_{LkOS}$ (4.20) is only homogeneous.

Consider $\mathcal{J}_{LkOS}$. When at least k measurements in the data are inliers and the parameter estimate is close to the true value, the k-th error is computed based on an inlier, and it is small. The influence of the outliers is avoided, and if (4.28) is satisfied, for the LkOS estimator, $\eta(n) = (n-k)/n$. As shown in the next section, the condition (4.28) depends on the level of noise corrupting the inliers. When the noise is large, the value of $\eta(n)$ decreases. Therefore, it is important to realize that $\eta(n)$ measures the global robustness of the employed estimator only in the context of the task. However, this is what we really care about in an application!

Several strategies can be adopted to define the value of k. Prior to the estimation process, k can be set to a given percentage of the number of points n. For example, if $k = n/2$, the least median of squares (LMedS) estimator [86] is obtained. Similarly, the value of k can be defined implicitly by setting the level of the allowed measurement noise and maximizing the number of data points within this tolerance. This is the approach used in the random sample consensus (RANSAC) estimator [26], which solves

$$\widehat{\boldsymbol{\theta}} = \arg\max_{\boldsymbol{\theta}}\; d_{k:n} \qquad \text{subject to} \quad \|\delta\mathbf{y}_{k:n}\|_{\mathbf{C}_y} < s(\sigma), \qquad (4.30)$$

where $s(\sigma)$ is a *user-set threshold* related to the scale of the inlier noise. In a third, less generic strategy, an auxiliary optimization process is introduced to determine the best value of k by analyzing a sequence of scale estimates [63, 76].

Besides the global robustness property discussed until now, the *local robustness* of an estimator also has to be considered when evaluating performance. Local robustness is measured through the *gross error sensitivity*, which describes the worst influence a single measurement can have on the value of the estimate [89, p. 191]. Local robustness is a central concept in the theoretical analysis of robust estimators and has a complex relation to global robustness (e.g., [40, 69]). It also has important practical implications.

Large gross error sensitivity (poor local robustness) means that for a critical arrangement of the n data points, a slight change in the value of a measurement $\mathbf{y}_i$ yields an unexpectedly large change in the value of the estimate $\widehat{\boldsymbol{\theta}}$. Such behavior is certainly undesirable. Several robust estimators in computer vision, such as LMedS and RANSAC, have large gross error sensitivity, as shown in Section 4.4.4.

4.2.4   Definition of robustness

We have defined global robustness in a task-specific manner. An estimator is considered robust only when the estimation error is guaranteed to be less than what can be tolerated in the application (4.28). This definition is different from the one used in statistics, where global robustness is closely related to the *breakdown point* of an estimator. The (explosion) breakdown point is the minimum percentage of outliers in the data for which the value of maximum bias *becomes unbounded* [89, p. 117]. Also, the maximum bias is defined in statistics relative to a *typically good* estimate computed with all the points being inliers and not relative to the true value, as in (4.27).

For computer vision problems, the statistical definition of robustness is too narrow. First, a finite maximum bias can still imply unacceptably large estimation errors. Second, in statistics the estimators of models linear in the variables are often required to be affine equivariant; that is, an affine transformation of the input (measurements) should change the output (estimates) by the inverse of that transformation [89, p. 116]. It can be shown that the breakdown point of an affine equivariant estimator cannot exceed 0.5, i.e., the inliers must be the absolute majority in the data [89, p. 253], [69]. According to the definition of robustness in statistics, once the number of outliers exceeds that of inliers, the former can be arranged into a false structure, thus compromising the estimation process.

Our definition of robust behavior is better suited for estimation in computer vision where often the information of interest is carried by less than half of the data points and/or the data may also contain multiple structures. Data with multiple structures is characterized by the presence of several instances of the *same* model, each corresponding in (4.11) to a different set of parameters $\alpha_k, \boldsymbol{\theta}_k, k = 1, \ldots, K$. Independently moving objects in a scene is just one example in which such data can appear. (The case of simultaneous presence of different models is too rare to be considered here.)

The data in Figure 4.3 is a simple example of the multistructured case. Outliers not belonging to any of the model instances can also be present. During the estimation of any of the individual structures, all the other data points act as outliers. Multistructured data is very challenging, and once the measurement noise becomes large (Figure 4.3b), none of the current robust estimators can handle it. Theoretical analysis of robust processing for data containing two structures can be found in [9, 97], and we discuss it in Section 4.4.7.

The definition of robustness employed here, besides being better suited for data in computer vision, also has the advantage of highlighting the complex relation between σ, the scale of the inlier noise, and $\eta(n)$, the amount of outlier tolerance. To avoid misconceptions, we do not recommend the use of the term *breakdown point* in the context of computer vision.

Assume for the moment that the data contains only inliers. Since the input is corrupted by measurement noise, the estimate $\widehat{\boldsymbol{\theta}}$ will differ from the true value $\boldsymbol{\theta}$. The larger the scale of the inlier noise, the higher the probability of a significant deviation between $\boldsymbol{\theta}$ and $\widehat{\boldsymbol{\theta}}$. The inherent uncertainty of an estimate computed from noisy data thus sets a lower bound on t_b (4.28). Several such bounds can be defined, the best known being the Cramer-Rao bound [75, p. 78]. Most bounds are computed under strict assumptions about the distribution of the measurement noise. Given the complexity of the visual data, the significance of a bound in a real application is often questionable. For a discussion of the Cramer-Rao bound in the context of computer vision, see [58, Chap. 14], and for an example, see [95].

Next, assume that the employed robust method can handle the percentage of outliers present in the data. After the outliers are removed, the esti-

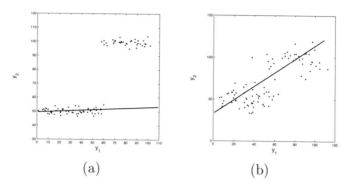

(a)   (b)

Figure 4.3. Multistructured data. The measurement noise is small in (a) and large in (b). The line is the fit obtained with the least median of squares (LMedS) estimator.

mate $\widehat{\theta}$ is computed from fewer data points and therefore it is less reliable (a small sample property). The probability of a larger deviation from the true value increases, which is equivalent to an increase of the lower bound on t_b. Thus, for a given level of the measurement noise (the value of σ), as the employed estimator has to remove more outliers from the data, the chance of larger estimation errors (the lower bound on t_b) also increases. The same effect is obtained when the number of removed outliers is kept the same but the level of the measurement noise increases.

In practice, the tolerance threshold t_b is set by the application to be solved. When the level of the measurement noise corrupting the inliers increases, eventually we are no longer able to keep the estimation errors below t_b. Based on our definition of robustness, the estimator no longer can be considered robust! Note that by defining robustness through the breakdown point, as it is done in statistics, the failure of the estimator would not have been recorded. Our definition of robustness also covers the *numerical robustness* of a nonrobust estimator when all the data obeys the model. In this case, the focus is exclusively on the size of the estimation errors, and the property is related to the *efficiency* of the estimator.

The loss of robustness is best illustrated with multistructured data. For example, the LMedS estimator was designed to reject up to half the points being outliers. When used to robustly fit a line to the data in Figure 4.3a, the LMedS estimator correctly recovers the lower structure, which contains 60 percent of the points. However, when applied to the similar but heavily corrupted data in Figure 4.3b, LMedS completely fails and the obtained fit is not different from that of the nonrobust least squares [9, 77, 97]. As will be shown in Section 4.4.7, the failure of LMedS is part of a more general deficiency of robust estimators.

4.2.5 Taxonomy of estimation problems

The model described at the beginning of Section 4.2.2, the measurement equation

$$\mathbf{y}_i = \mathbf{y}_{io} + \delta\mathbf{y}_i \quad \mathbf{y}_i \in \mathcal{R}^q \quad \delta\mathbf{y}_i \sim GI(\mathbf{0}, \sigma^2 \mathbf{C}_y) \qquad i = 1, \dots, n \quad (4.31)$$

and the constraint

$$\alpha + \mathbf{x}_{io}^\top \boldsymbol{\theta} = 0 \quad \mathbf{x}_{io} = \mathbf{x}(\mathbf{y}_{io}) \qquad i = 1, \dots, n, \qquad (4.32)$$

is general enough to apply to almost all computer vision problems. The constraint is linear in the parameters α and $\boldsymbol{\theta}$, but nonlinear in the variables $\mathbf{y}_i$. A model in which all the variables are measured with errors is called in statistics an *errors-in-variables* (EIV) model [111, 115].

We discussed in Section 4.2.1 the problem of ellipse fitting using such a nonlinear EIV model (Figure 4.1). Nonlinear EIV models also appear in any computer vision problem in which the constraint has to capture an incidence relation in projective geometry. For example, consider the epipolar constraint between the affine coordinates of corresponding points in two images A and B:

$$[y_{B1o} \quad y_{B2o} \quad 1]^\top \mathbf{F} [y_{A1o} \quad y_{A2o} \quad 1] = 0, \qquad (4.33)$$

where F is a rank-2 matrix [43, Chap. 8]. When this bilinear constraint is rewritten as (4.32), four of the eight carriers,

$$\mathbf{x}_o^\top = [y_{A1o} \quad y_{A2o} \quad y_{B1o} \quad y_{B2o} \quad y_{A1o}y_{B1o} \quad y_{A2o}y_{B1o} \quad y_{A2o}y_{B1o} \quad y_{A2o}y_{B2o}], \qquad (4.34)$$

are nonlinear functions in the variables. Several nonlinear EIV models used in recovering 3D structure from uncalibrated image sequences are discussed in [34].

To obtain an unbiased estimate, the parameters of a nonlinear EIV model have to be computed with nonlinear optimization techniques such as the Levenberg-Marquardt method. See [43, Appen. 4] for a discussion. However, the estimation problem can be also approached as a *linear model in the carriers* and taking into account the heteroscedasticity of the noise process associated with the carriers (Section 4.2.1). Several such techniques were

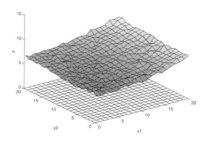

Figure 4.4. A typical traditional regression problem. Estimate the parameters of the surface defined on a sampling grid.

proposed in the computer vision literature: the renormalization method [58], the heteroscedastic errors-in-variables (HEIV) estimator [64, 71, 70], and the fundamental numerical scheme (FNS) [12]. All of them return estimates unbiased in a first-order approximation.

Since the focus of this chapter is on robust estimation, we use only the less general *linear errors-in-variables regression* model. In this case, the carriers are linear expressions in the variables, and the constraint (4.32) becomes

$$\alpha + \mathbf{y}_{io}^{\top}\boldsymbol{\theta} = 0 \qquad i = 1, \ldots, n \ . \tag{4.35}$$

An important case of the general EIV model is obtained by considering the constraint (4.4). This is the *traditional regression* model where only a single variable, denoted z, is measured with error, and therefore the measurement equation becomes

$$
\begin{aligned}
z_i &= z_{io} + \delta z_i && \delta z_i \sim GI(0, \sigma^2) && i = 1, \ldots, n \\
\mathbf{y}_i &= \mathbf{y}_{io} && && i = 1, \ldots, n,
\end{aligned}
\tag{4.36}
$$

while the constraint is expressed as

$$z_{io} = \alpha + \mathbf{x}_{io}^{\top}\boldsymbol{\theta} \qquad \mathbf{x}_{io} = \mathbf{x}(\mathbf{y}_{io}) \qquad i = 1, \ldots, n \ . \tag{4.37}$$

Note that the nonlinearity of the carriers is no longer relevant in the traditional regression model, since now their value is known.

In traditional regression, the covariance matrix of the variable vector

$$\mathbf{v}^{\top} = [z \quad \mathbf{y}] \qquad \sigma^2 \mathbf{C}_v = \sigma^2 \begin{bmatrix} 1 & \mathbf{0}^{\top} \\ \mathbf{0} & \mathrm{O} \end{bmatrix} \tag{4.38}$$

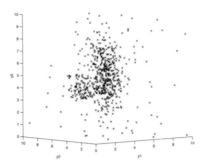

Figure 4.5. A typical location problem. Determine the center of the cluster.

has rank one, and the normalized distances, d_i (4.19), used in the objective functions become

$$d_i^2 = \frac{1}{\sigma^2}(\mathbf{v}_i - \mathbf{v}_{io})^{\top} \mathbf{C}_v^+ (\mathbf{v}_i - \mathbf{v}_{io}) = \frac{(z_i - z_{io})^2}{\sigma^2} = \left(\frac{\delta z_i}{\sigma}\right)^2. \qquad (4.39)$$

The two regression models, the linear EIV (4.35) and the traditional (4.37), must be estimated with different least squares techniques, as shown in Section 4.4.1. Using the method optimal for traditional regression when estimating an EIV regression model yields biased estimates. In computer vision, the traditional regression model appears almost exclusively only when an image defined on the sampling grid is to be processed. In this case, the pixel coordinates are the independent variables and can be considered available uncorrupted (Figure 4.4).

All the models discussed so far are related the class of *regression problems*. A second, equally important class of estimation problems also exist. They are the *location problems* in which the goal is to determine an estimate for the "center" of a set of noisy measurements. The location problems are closely related to clustering in pattern recognition.

In practice, a location problem is of interest only in the context of robust estimation. The measurement equation is

$$\mathbf{y}_i = \mathbf{y}_{io} + \delta\mathbf{y}_i \qquad\qquad i = 1, \ldots, n_1 \qquad\qquad (4.40)$$
$$\mathbf{y}_i \qquad\qquad\qquad\qquad i = (n_1 + 1), \ldots, n$$

with the constraint

$$\mathbf{y}_{io} = \boldsymbol{\theta} \qquad\qquad i = 1, \ldots, n_1 \qquad\qquad (4.41)$$

with n_1, the number of inliers, unknown.

The important difference from the regression case (4.25) is that now we do not assume that the noise corrupting the inliers can be characterized by a single covariance matrix; that is, the cloud of inliers has an elliptical shape. This allows us to handle data such as in Figure 4.5.

The goal of the estimation process in a location problem is twofold.

– Find a robust estimate $\widehat{\boldsymbol{\theta}}$ for the center of the n measurements.

– Select the n_1 data points associated with this center.

The discussion in Section 4.2.4 about the definition of robustness also applies to location estimation.

While handling multistructured data in regression problems is an open research question, clustering multistructured data is the main application of the location estimators. The feature spaces derived from visual data are complex and usually contain several clusters. The goal of feature space analysis is to delineate each significant cluster through a robust location estimation process. We return to location estimation in Section 4.3.

4.2.6   Linear EIV regression model

To focus on the issue of robustness in regression problems, only the simplest linear EIV regression model (4.35) will be used. The measurements are corrupted by i.i.d. noise,

$$\mathbf{y}_i = \mathbf{y}_{io} + \delta\mathbf{y}_i \quad \mathbf{y}_i \in \mathcal{R}^p \quad \delta\mathbf{y}_i \sim GI(\mathbf{0}, \sigma^2 \mathbf{I}_p) \qquad i = 1,\dots,n, \quad (4.42)$$

where the number of variables q was aligned with p, the dimension of the parameter vector $\boldsymbol{\theta}$. The constraint is rewritten under the more convenient form

$$g(\mathbf{y}_{io}) = \mathbf{y}_{io}^\top \boldsymbol{\theta} - \alpha = 0 \qquad i = 1,\dots,n \ . \qquad (4.43)$$

To eliminate the ambiguity up to a constant of the parameters, the following two ancillary constraints are used

$$\|\boldsymbol{\theta}\| = 1 \qquad \alpha \geq 0 \ . \qquad (4.44)$$

The three constraints together define the Hessian normal form of a plane in $\mathcal{R}^p$. Figure 4.6 shows the interpretation of the two parameters. The unit vector $\boldsymbol{\theta}$ is the direction of the normal, while α is the distance from the origin.

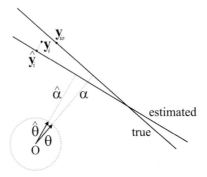

Figure 4.6. The concepts of the linear EIV regression model. The constraint is in the Hessian normal form.

In general, given a surface $f(\mathbf{y}_o) = 0$ in $\mathcal{R}^p$, the first-order approximation of the shortest Euclidean distance from a point $\mathbf{y}$ to the surface is [110, p. 101]

$$\|\mathbf{y} - \widehat{\mathbf{y}}\| \simeq \frac{|f(\mathbf{y})|}{\|\nabla f(\widehat{\mathbf{y}})\|}, \tag{4.45}$$

where $\widehat{\mathbf{y}}$ is the orthogonal projection of the point onto the surface, and $\nabla f(\widehat{\mathbf{y}})$ is the gradient computed in the location of that projection. The quantity $f(\mathbf{y})$ is called the *algebraic distance*, and it can be shown that it is zero only when $\mathbf{y} = \mathbf{y}_o$; that is, the point is on the surface.

Taking into account the linearity of the constraint (4.43) and that $\boldsymbol{\theta}$ has unit norm, (4.45) becomes

$$\|\mathbf{y} - \widehat{\mathbf{y}}\| = |g(\mathbf{y})|. \tag{4.46}$$

That is, the Euclidean distance from a point to a hyperplane written under the Hessian normal form is the absolute value of the algebraic distance.

When all the data points obey the model, the least squares objective function $\mathcal{J}_{LS}$ (4.22) is used to estimate the parameters of the linear EIV regression model. The i.i.d. measurement noise (4.42) simplifies the expression of the distances d_i (4.19) and the minimization problem (4.21) can be written as

$$[\widehat{\alpha}, \widehat{\boldsymbol{\theta}}] = \operatorname*{argmin}_{\alpha, \boldsymbol{\theta}} \frac{1}{n} \sum_{i=1}^{n} \|\mathbf{y}_i - \mathbf{y}_{io}\|^2 . \tag{4.47}$$

Combining (4.46) and (4.47), we obtain

$$[\widehat{\alpha}, \widehat{\boldsymbol{\theta}}] = \operatorname*{argmin}_{\alpha, \boldsymbol{\theta}} \frac{1}{n} \sum_{i=1}^{n} g(\mathbf{y}_i)^2 . \tag{4.48}$$

To solve (4.48), the true values $\mathbf{y}_{io}$ are replaced with the orthogonal projection of the $\mathbf{y}_i$-s onto the hyperplane. The orthogonal projections $\widehat{\mathbf{y}}_i$ associated with the solution $\widehat{\alpha}, \widehat{\boldsymbol{\theta}}$ are the corrected values of the measurements $\mathbf{y}_i$ and satisfy (Figure 4.6)

$$\widehat{g}(\widehat{\mathbf{y}}_i) = \widehat{\mathbf{y}}_i^\top \widehat{\boldsymbol{\theta}} - \widehat{\alpha} = 0 \qquad i = 1, \ldots, n . \qquad (4.49)$$

The estimation process (discussed in Section 4.4.1) returns the parameter estimates, after which the ancillary constraints (4.44) can be imposed. The employed parameterization of the linear model

$$\boldsymbol{\omega}_1 = [\boldsymbol{\theta}^\top \ \alpha]^\top = [\theta_1 \ \theta_2 \ \cdots \ \theta_p \ \alpha]^\top \in \mathcal{R}^{p+1}, \qquad (4.50)$$

however, is redundant. The vector $\boldsymbol{\theta}$, being a unit vector, is restricted to the p-dimensional unit sphere in R^p. This can be taken into account by expressing $\boldsymbol{\theta}$ in polar angles [115] as $\boldsymbol{\theta} = \boldsymbol{\theta}(\boldsymbol{\beta})$

$$\boldsymbol{\beta} = [\beta_1 \ \beta_2 \ \cdots \ \beta_{p-1}]^\top \quad 0 \le \beta_j \le \pi \ \ j = 1, \cdots, p-2 \qquad 0 \le \beta_{p-1} < 2\pi \qquad (4.51)$$

where the mapping is

$$\begin{aligned}
\theta_1(\boldsymbol{\beta}) &= sin\beta_1 \cdots sin\beta_{p-2} \, sin\beta_{p-1} \\
\theta_2(\boldsymbol{\beta}) &= sin\beta_1 \cdots sin\beta_{p-2} \, cos\beta_{p-1} \\
\theta_3(\boldsymbol{\beta}) &= sin\beta_1 \cdots sin\beta_{p-3} \, cos\beta_{p-2} \\
&\vdots \\
\theta_{p-1}(\boldsymbol{\beta}) &= sin\beta_1 \, cos\beta_2 \\
\theta_p(\boldsymbol{\beta}) &= cos\beta_1 .
\end{aligned} \qquad (4.52)$$

The polar angles β_j and α provide the second representation of a hyperplane

$$\boldsymbol{\omega}_2 = [\boldsymbol{\beta}^\top \ \alpha]^\top = [\beta_1 \ \beta_2 \ \cdots \ \beta_{p-1} \ \alpha]^\top \in \mathcal{R}^p . \qquad (4.53)$$

The $\boldsymbol{\omega}_2$ representation, being based in part on the mapping from the unit sphere to $\mathcal{R}^{p-1}$, is inherently discontinuous. See [24, Chap. 5] for a detailed discussion of such representations. The problem is well known in the context of Hough transform, where this parameterization is widely used.

To illustrate the discontinuity of the mapping, consider the representation of a line, $p = 2$. In this case, only a single polar angle β is needed, and the equation of a line in the Hessian normal form is

$$y_1 cos\beta + y_2 sin\beta - \alpha = 0 . \qquad (4.54)$$

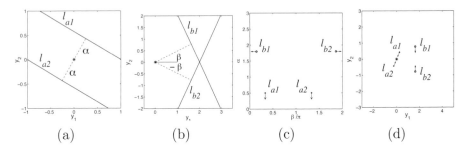

Figure 4.7. Discontinuous mappings due to the polar representation of $\boldsymbol{\theta}$. (a) Two lines with same α and antipodal polar angles β. (b) Two lines with same α and polar angles β differing only in sign. (c) The $\boldsymbol{\omega}_2$ parameter space. (d) The $\boldsymbol{\omega}_3$ parameter space.

In Figures 4.7a and 4.7b, two pairs of lines are shown, each pair having the same α but different polar angles. Take $\beta_1 = \beta$. The lines in Figure 4.7a have the relation $\beta_2 = \beta + \pi$, while those in Figure 4.7b $\beta_2 = 2\pi - \beta$. When represented in the $\boldsymbol{\omega}_2$ parameter space (Figure 4.7c), the four lines are mapped into four points.

Let $\alpha \to 0$ for the first pair, and $\beta \to 0$ for the second pair. In the input space, each pair of lines merges into a single line, but the four points in the $\boldsymbol{\omega}_2$ parameter space remain distinct, as shown by the arrows in Figure 4.7c.

A different parameterization of the hyperplane in $\mathcal{R}^p$ can avoid this problem, though no representation of the Hessian normal form can provide a continuous mapping into a feature space. In the new parameterization, all the hyperplanes not passing through the origin are represented by their point closest to the origin. This point has the coordinates $\alpha\boldsymbol{\theta}$ and is the intersection of the plane with the normal from the origin. The new parameterization is

$$\boldsymbol{\omega}_3 = \alpha\boldsymbol{\theta} = [\alpha\theta_1 \ \alpha\theta_2 \ \cdots \ \alpha\theta_p]^{\top} \in \mathcal{R}^p \ . \tag{4.55}$$

It is important to notice that the space of $\boldsymbol{\omega}_3$ is in fact the space of the input, as can also be seen from Figure 4.6. Thus, when the pairs of lines collapse, so do their representations in the $\boldsymbol{\omega}_3$ space (Figure 4.7d).

Planes that contain the origin have to be treated separately. In practice, this also applies to planes passing near the origin. A plane with small α is translated along the direction of the normal $\boldsymbol{\theta}$ with a known quantity τ. The plane is then represented as $\tau\boldsymbol{\theta}$. When m planes are close to the origin, the direction of translation is $\sum\limits_{i=1}^{m} \boldsymbol{\theta}_i$ and the parameters of each translated plane are adjusted accordingly. After processing in the $\boldsymbol{\omega}_3$ space, it is easy to convert back to the $\boldsymbol{\omega}_1$ representation.

Estimation of the linear EIV regression model parameters by total least squares (Section 4.4.1) uses the $\boldsymbol{\omega}_1$ parameterization. The $\boldsymbol{\omega}_2$ parameterization will be employed in the robust estimation of the model (Section 4.4.5). The parameterization $\boldsymbol{\omega}_3$ is useful when the problem of robust multiple regression is approached as a feature space analysis problem [9].

4.2.7   Objective function optimization

The objective functions used in robust estimation are often nondifferentiable, and analytical optimization methods, like those based on the gradient, cannot be employed. The k-th order statistics, $\mathcal{J}_{LkOS}$ (4.22), is such an objective function. Nondifferentiable objective functions also have many local extrema, and to avoid being trapped in one of these minima, the optimization procedure should be run starting from several initial positions. A numerical technique to implement robust estimators with nondifferentiable objective functions is based on *elemental subsets*.

An elemental subset is the smallest number of data points required to fully instantiate a model. In the linear EIV regression case this means p points in a general position; that is, the points define a basis for a $(p-1)$-dimensional affine subspace in $\mathcal{R}^p$ [89, p. 257]. For example, if $p = 3$, not all three points can lie on a line in 3D.

The p points in an elemental subset thus define a full-rank system of equations from which the model parameters α and $\boldsymbol{\theta}$ can be computed analytically. Note that using p points suffices to solve this homogeneous system. The ancillary constraint $\|\boldsymbol{\theta}\| = 1$ is imposed at the end. The obtained parameter vector $\boldsymbol{\omega}_1 = [\boldsymbol{\theta}^\top \ \alpha]^\top$ will be called, with a slight abuse of notation, a *model candidate*.

The number of possibly distinct elemental subsets in the data $\binom{n}{p}$ can be very large. In practice, an exhaustive search over all the elemental subsets is not feasible, and a *random sampling* of this ensemble has to be used. The sampling drastically reduces the amount of computations at the price of a negligible decrease in the outlier rejection capability of the implemented robust estimator.

Assume that the number of inliers in the data is n_1 and that N elemental subsets, p-tuples, were drawn independently from that data. The probability that *none* of these subsets contains only inliers is (after disregarding the artifacts due to the finite sample size)

$$P_{fail} = \left[1 - \left(\frac{n_1}{n} \right)^p \right]^N . \qquad (4.56)$$

We can choose a small probability P_{error} to bound upward P_{fail}. Then the equation

$$P_{fail} = P_{error} \qquad (4.57)$$

provides the value of N as a function of the percentage of inliers n_1/n, the dimension of the parameter space p, and P_{error}. This probabilistic sampling strategy was applied independently in computer vision for the RANSAC estimator [26] and in statistics for the LMedS estimator [89, p. 198].

Several important observations must be made. The value of N obtained from (4.57) is an absolute lower bound, since it implies that *any* elemental subset which contains only inliers can provide a satisfactory model candidate. However, the model candidates are computed from the smallest possible number of data points and the influence of the noise is the largest possible. Thus, the assumption used to compute N is not guaranteed to be satisfied once the measurement noise becomes significant. In practice, n_1 is not known prior to the estimation, and the value of N has to be chosen large enough to compensate for the inlier noise under a worst-case scenario.

Nevertheless, it is not recommended to increase the size of the subsets. The reason is immediately revealed if we define, in a drawing of subsets of size $q \geq p$, the probability of success as obtaining a subset that contains only inliers

$$P_{success} = \frac{\left(\begin{array}{c} n_1 \\ q \end{array} \right)}{\left(\begin{array}{c} n \\ q \end{array} \right)} = \prod_{k=0}^{q-1} \frac{n_1 - k}{n - k} . \qquad (4.58)$$

This probability is maximized when $q = p$.

Optimization of an objective function using random elemental subsets is *only a computational tool* and has no bearing on the robustness of the corresponding estimator. This fact is not always recognized in the computer vision literature. However, *any* estimator can be implemented using the following numerical optimization procedure.

Objective Function Optimization with Elemental Subsets

- Repeat N times:

  1. choose an elemental subset (p-tuple) by random sampling;

  2. compute the corresponding model candidate;

  3. compute the value of the objective function by assuming the model candidate valid for all the data points.

- The parameter estimate is the model candidate yielding the smallest (largest) objective function value.

This procedure can be applied the same way for the nonrobust least squares objective function $\mathcal{J}_{LS}$ as for the robust least k-th order statistics $\mathcal{J}_{LkOS}$ (4.22). However, while an analytical solution is available for the former (Section 4.4.1), for the latter the above procedure is the only practical way to obtain the estimates.

Performing an exhaustive search over all elemental subsets does not guarantee to find the global extremum of the objective function, since not every location in the parameter space can be visited. Finding the global extremum, however, most often is also not required. When a robust estimator is implemented with the elemental subsets-based search procedure, the goal is only to obtain the inlier/outlier dichotomy, that is, to select the "good" data. The robust estimate corresponding to an elemental subset is then refined by processing the selected inliers with a nonrobust (least squares) estimator. See [87] for an extensive discussion of the related issues from a statistical perspective.

The number of required elemental subsets N can be significantly reduced when information about the reliability of the data points is available. This information can be either provided by the user or derived from the data through an auxiliary estimation process. The elemental subsets are then chosen with a *guided sampling* biased toward the points having a higher probability to be inliers. See [103] and [104] for computer vision examples.

We have emphasized that the random sampling of elemental subsets is not more than a computational procedure. However, guided sampling has a different nature, since it relies on a fuzzy preclassification of the data (derived automatically or supplied by the user). Guided sampling can yield a significant improvement in the performance of the estimator relative to the unguided approach. The better quality of the elemental subsets can be converted either into fewer samples in the numerical optimization (while preserving the outlier rejection capacity $\eta(n)$ of the estimator) or into an increase of $\eta(n)$ (while preserving the same number of elemental subsets N).

We conclude that guided sampling should be regarded as a robust technique, while the random sampling procedure should be not. Their subtle but important difference has to be recognized when designing robust methods for solving complex vision tasks.

In most applications, information reliable enough to guide the sampling is not available. However, the amount of computations still can be reduced by performing in the space of the parameters local searches with optimization techniques that do not rely on derivatives. For example, in [90] line search was proposed to improve the implementation of the LMedS estimator. Let $\omega_1^b = [\theta_b^\top \ \alpha_b]^\top$ be the currently best model candidate, as measured by the value of the objective function. From the next elemental subset, the model candidate $\omega_1 = [\theta^\top \ \alpha]^\top$ is computed. The objective function is then assessed at several locations along the line segment $\omega_1^b - \omega_1$, and if an improvement relative to ω_1^b is obtained, the best model candidate is updated.

In Section 4.4.5, we use a more effective multidimensional unconstrained optimization technique, the *simplex-based direct search*. The simplex search is a heuristic method proposed in 1965 by Nelder and Mead [78]. See also [82, Sec. 10.4]. Simplex search is a heuristic with no theoretical foundations. Recently, direct search methods gained renewed interest, and significant progress was reported in the literature [66, 114], but in our context there is no need to use these computationally more intensive techniques.

To take into account the fact that θ is a unit vector, the simplex search should be performed in the space of the polar angles $\beta \in R^{p-1}$. A simplex in R^{p-1} is the volume delineated by p vertices in a nondegenerate position; the points define an affine basis in R^{p-1}. For example, in R^2 the simplex is a triangle, in R^3 it is a tetrahedron. In our case, the vertices of the simplex are the polar angle vectors $\beta_k \in R^{p-1}$, $k = 1, \ldots, p$, representing p unit vectors $\theta_k \in R^p$. Each vertex is associated with the value of a scalar function $f_k = f(\beta_k)$. For example, $f(u)$ can be the objective function of an estimator. The goal of the search is to find the global (say) maximum of this function.

We can always assume that at the beginning of an iteration the vertices are labeled such that $f_1 \leq f_2 \leq \cdots \leq f_p$. In each iteration, an attempt is made to improve the least favorable value of the function, f_1 in our case, by trying to find a new location β_1' for the vertex β_1 such that $f_1 < f(\beta_1')$.

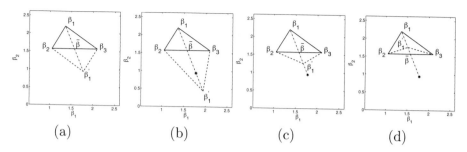

Figure 4.8. Basic operations in simplex-based direct search. (a) Reflection. (b) Expansion. (c) Outside contraction. (d) Inside contraction.

Simplex-Based Direct Search Iteration

First $\bar{\boldsymbol{\beta}}$, the centroid of the nonminimum vertices, $\boldsymbol{\beta}_k$, $k = 2, \ldots, p$, is obtained. The new location is then computed with one of the following operations along the direction $\bar{\boldsymbol{\beta}} - \boldsymbol{\beta}_1$: reflection, expansion, and contraction.

1. The *reflection* of $\boldsymbol{\beta}_1$, denoted $\boldsymbol{\beta}'$ (Figure 4.8a) is defined as

$$\boldsymbol{\beta}' = c_r \boldsymbol{\beta}_1 + (1 - c_r)\bar{\boldsymbol{\beta}}, \tag{4.59}$$

where $c_r < 0$ is the reflection coefficient. If $f_2 < f(\boldsymbol{\beta}') \le f_p$, then $\boldsymbol{\beta}'_1 = \boldsymbol{\beta}'$, and the next iteration is started.

2. If $f(\boldsymbol{\beta}') > f_p$ (i.e., the reflection has produced a new maximum), the simplex is *expanded* by moving $\boldsymbol{\beta}'$ to $\boldsymbol{\beta}^*$ (Figure 4.8b):

$$\boldsymbol{\beta}^* = c_e \boldsymbol{\beta}' + (1 - c_e)\bar{\boldsymbol{\beta}}, \tag{4.60}$$

where the expansion coefficient $c_e > 1$. If $f(\boldsymbol{\beta}^*) > f(\boldsymbol{\beta}')$, the expansion is successful and $\boldsymbol{\beta}'_1 = \boldsymbol{\beta}^*$. Else, $\boldsymbol{\beta}'_1 = \boldsymbol{\beta}'$. The next iteration is started.

3. If $f(\boldsymbol{\beta}') \le f_2$, the vector $\boldsymbol{\beta}_{1n}$ is defined as either $\boldsymbol{\beta}_1$ or $\boldsymbol{\beta}'$, whichever has the larger associated function value, and a *contraction* is performed:

$$\boldsymbol{\beta}^* = c_c \boldsymbol{\beta}_{1n} + (1 - c_c)\bar{\boldsymbol{\beta}} \,. \tag{4.61}$$

First, a contraction coefficient $0 < c_c < 1$ is chosen for outside contraction (Figure 4.8c). If $f(\boldsymbol{\beta}^*) > f(\boldsymbol{\beta}_{1n})$, then $\boldsymbol{\beta}'_1 = \boldsymbol{\beta}^*$, and the next iteration is started. Otherwise, an inside contraction is performed (Figure 4.8d) in which c_c is replaced with $-c_c$, and the condition $f(\boldsymbol{\beta}^*) > f(\boldsymbol{\beta}_{1n})$ is again verified.

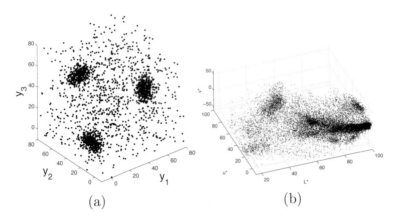

Figure 4.9. Multistructured data in the location estimation problem. (a) The "traditional" case. (b) A typical computer vision example.

4. Should both contractions fail, all the vertices are updated:

$$\boldsymbol{\beta}_k \longleftarrow \frac{1}{2}\left(\boldsymbol{\beta}_k + \boldsymbol{\beta}_p\right) \qquad\qquad k = 1,\ldots,(p-1), \qquad (4.62)$$

and the next iteration is started.

Recommended values for the coefficients are $c_r = -1$, $c_e = 1.5$, $c_c = 0.5$.

To assess the convergence of the search, several stopping criteria can be employed. For example, the variance of the p function values f_k should fall below a threshold which is exponentially decreasing with the dimension of the space, or the ratio of the smallest and largest function values f_1/f_p should be close to one. Similarly, the volume of the simplex should shrink below a dimension-dependent threshold. In practice, the most effective stopping criteria are application-specific, incorporating additional information that was not used during the optimization.

In the previous sections, we analyzed the problem of robust estimation from a generic point of view. We can now examine the two classes of estimation problems: location and regression. In each case, we introduce a new, robust technique whose improved behavior was achieved by systematically exploiting the principles discussed so far.

4.3   Location Estimation

In this section, we show that in the context of computer vision tasks, often only nonparametric approaches can provide a robust solution for the location estimation problem. We employ a class of nonparametric techniques in which the data points are regarded as samples from an unknown probability density. The location estimates are then defined as the modes of this density. Explicit computation of the density is avoided by using the mean shift procedure.

4.3.1   Why nonparametric methods?

The most general model of the location problem is that of multiple structures,

$$k = 1, \ldots, K \qquad\qquad m_1 = 1 \quad \cdots \quad m_{K+1} = n_1 + 1$$

$$
\begin{aligned}
\mathbf{y}_i^{(k)} &= \mathbf{y}_{io}^{(k)} + \delta\mathbf{y}_i \qquad \mathbf{y}_{io}^{(k)} = \boldsymbol{\theta}^{(k)} \qquad\qquad i = m_k, \ldots, (m_{k+1} - 1) \\
\mathbf{y}_i & \qquad\qquad\qquad\qquad\qquad\qquad\qquad\quad i = (n_1 + 1), \ldots, n, \qquad (4.63)
\end{aligned}
$$

with no information being available about the nature of the inlier noise $\delta\mathbf{y}_i$, the $n - n_1$ outliers, or the number of structures present in the data K. The model (4.63) is also used in *cluster analysis*, the equivalent pattern recognition problem. Clustering under its most general form is an unsupervised learning method of unknown categories from incomplete prior information [52, p. 242]. The books [52], [21, Chap. 10], [44, Sec. 14.3] provide a complete coverage of the related pattern recognition literature.

   Many of the pattern recognition methods are not adequate for data analysis in computer vision. To illustrate their limitations, we compare the two data sets shown in Figure 4.9. The data in Figure 4.9a obeys what is assumed in traditional clustering methods when the proximity to a cluster center is measured as a function of Euclidean or Mahalanobis distances. In this case, the shape of the clusters is restricted to elliptical, and the inliers are assumed to be normally distributed around the true cluster centers. A different metric will impose a different shape on the clusters. The number of the structures (clusters) K is a parameter to be supplied by the user and has a large influence on the quality of the results. While the value of K can be also derived from the data by optimizing a cluster validity index, this approach is not robust, since it is based on (possibly erroneous) data partitions.

   Expectation maximization (EM) is a frequently used technique today in computer vision to model the data. See [44, Sec. 8.5.2] for a short description. The EM algorithm also relies on strong prior assumptions. A likelihood function, defined from a mixture of predefined (most often normal) probability densities, is maximized. The obtained partition of the data thus employs "tiles" of given shape. The number of required mixture components is often difficult to determine, and the association of these components with true cluster centers may not be obvious.

Examine now the data in Figure 4.9b in which the pixels of a color image were mapped into the 3D L*u*v* color space. The significant clusters correspond to similarly colored pixels in the image. The clusters have a large variety of shapes and their number is not obvious. Any technique that imposes a preset shape on the clusters will have difficulties accurately separating K significant structures from the background clutter while simultaneously also having to determine the value of K.

Following our goal-oriented approach toward robustness (Section 4.2.3), a location estimator should be declared robust only if it returns a satisfactory result. From the above discussion, we can conclude that robustly solving location problems in computer vision often requires techniques that use the least possible amount of prior assumptions about the data. Such techniques belong to the family of *nonparametric* methods.

In nonparametric methods, the n data points are regarded as outcomes from an (unknown) probability distribution $f(\mathbf{y})$. Each data point is assumed to have an equal probability

$$\mathrm{Prob}[\mathbf{y} = \mathbf{y}_i] = \frac{1}{n} \qquad i = 1, \ldots, n \ . \qquad (4.64)$$

When several points have the same value, the probability is n^{-1} times the multiplicity. The ensemble of points defines the *empirical distribution* $f(\mathbf{y}|\mathbf{y}_1 \cdots \mathbf{y}_n)$ of the data. The empirical distribution is the nonparametric maximum likelihood estimate of the distribution from which the data was drawn [22, p. 310]. It is also the "least committed" description of the data.

Every clustering technique exploits the fact that the clusters are the denser regions in the space. However, this observation can be pushed further in the class of nonparametric methods considered here, for which a region of higher density implies more probable outcomes of the random variable $\mathbf{y}$. Therefore, in each dense region, the location estimate (cluster center) should be associated with the most probable value of $\mathbf{y}$, that is, with the *local mode* of the empirical distribution

$$\widehat{\boldsymbol{\theta}}^{(k)} = \underset{\mathbf{y}}{\mathrm{argmax}}_k \, f(\mathbf{y}|\mathbf{y}_1 \cdots \mathbf{y}_n) \qquad k = 1, \ldots, K \ . \qquad (4.65)$$

Note that by detecting all the significant modes of the empirical distribution, the number of clusters K is automatically determined. The mode-based clustering techniques make extensive use of density estimation during data analysis.

4.3.2   Kernel density estimation

The modes of a random variable $\mathbf{y}$ are the local maxima of its probability density function $f(\mathbf{y})$. However, only the empirical distribution, the data points $\mathbf{y}_i$, $i = 1, \ldots, n$, are available. To accurately determine the locations of the modes, first a *continuous* estimate of the underlying density $\widehat{f}(\mathbf{y})$ has to be defined. Later we will see that this step can be eliminated by directly estimating the gradient of the density (Section 4.3.3).

To estimate the probability density in $\mathbf{y}$, a small neighborhood is defined around $\mathbf{y}$. The neighborhood usually has a simple shape: cube, sphere, or ellipsoid. Let its volume be V_y, and m_y be the number of data points inside. Then the density estimate is [21, Sec. 4.2]

$$\widehat{f}(\mathbf{y}) = \frac{m_y}{nV_y}, \tag{4.66}$$

which can be employed in two different ways.

- In the *nearest neighbors* approach, the neigborhoods (the volumes V_y) are scaled to keep the number of points m_y constant. A mode corresponds to a location in which the neighborhood has the smallest volume.
- In the *kernel density* approach, the neigborhoods have the same volume V_y and the number of points m_y inside are counted. A mode corresponds to a location in which the neighborhood contains the largest number of points.

The *minimum volume ellipsoid* (MVE) robust location estimator proposed in statistics [89, p. 258] is a technique related to the nearest neighbors approach. The ellipsoids are defined by elemental subsets obtained through random sampling, and the numerical optimization procedure discussed in Section 4.2.7 is employed. The location estimate is the center of the smallest ellipsoid, which contains a given percentage of the data points. In a robust clustering method proposed in computer vision, the MVE estimator was used to sequentially remove the clusters from the data, starting from the largest [53]. However, by imposing an elliptical shape for the clusters, severe artifacts were introduced and the method was never successful in real vision applications.

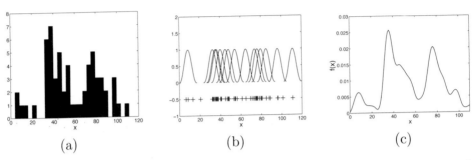

(a)   (b)   (c)

Figure 4.10. Kernel density estimation. (a) Histogram of the data. (b) Some of the employed kernels. (c) The estimated density.

For our goal of finding the local maxima of $\widehat{f}(\mathbf{y})$, the kernel density methods are more suitable. Kernel density estimation is a widely used technique in statistics and pattern recognition, where it is also called the Parzen window method. See [92, 112] for a description in statistics, and [21, Sec. 4.3] [44, Sec. 6.6] for a description in pattern recognition.

We start with the simplest case of 1D data. Let y_i, $i = 1, \ldots, n$, be scalar measurements drawn from an arbitrary probability distribution $f(y)$. The kernel density estimate $\widehat{f}(y)$ of this distribution is obtained based on a *kernel function* $K(u)$ and a *bandwidth h* as the average

$$\widehat{f}(y) = \frac{1}{nh} \sum_{i=1}^{n} K\left(\frac{y - y_i}{h}\right) . \tag{4.67}$$

Only the class of symmetric kernel functions with bounded support will be considered. They satisfy the following properties:

$$K(u) = 0 \quad \text{for } |u| > 1 \qquad \int_{-1}^{1} K(u) = 1 \tag{4.68}$$

$$K(u) = K(-u) \geq 0 \qquad K(u_1) \geq K(u_2) \quad \text{for } |u_1| \leq |u_2| .$$

Other conditions on the kernel function or on the density to be estimated [112, p. 18], are of less significance in practice. The even symmetry of the kernel function allows us to define its *profile* $k(u)$:

$$K(u) = c_k k(u^2) \qquad k(u) \geq 0 \quad \text{for } 0 \leq u \leq 1, \tag{4.69}$$

where c_k is a normalization constant determined by (4.68). The shape of the kernel implies that the profile is a monotonically decreasing function.

The kernel density estimate is a *continuous* function derived from the discrete data, the empirical distribution. An example is shown in Figure 4.10. When instead of the histogram of the n points (Figure 4.10a) the data is represented as an ordered list (Figure 4.10b, bottom), we are in fact using the empirical distribution. By placing a kernel in each point (Figure 4.10b), the data is convolved with the symmetric kernel function. The density estimate in a given location is the average of the contributions from each kernel (Figure 4.10c). Since the employed kernel has a finite support, not all the points contribute to a density estimate. The bandwidth h scales the size of the kernels, i.e., the number of points whose contribution is averaged when computing the estimate. The bandwidth thus controls the amount of smoothing present in $\widehat{f}(y)$.

For multivariate measurements $\mathbf{y}_i \in R^p$, in the most general case, the bandwidth h is replaced by a symmetric, positive definite bandwidth matrix H. The estimate of the probability distribution at location $\mathbf{y}$ is still computed as the average

$$\widehat{f}(\mathbf{y}) = \frac{1}{n} \sum_{i=1}^{n} K_{\mathrm{H}} \left(\mathbf{y} - \mathbf{y}_i \right), \tag{4.70}$$

where the bandwidth matrix H scales the kernel support to be radial symmetric, that is, to have the desired elliptical shape and size

$$K_{\mathrm{H}}(\mathbf{u}) = [\det[\mathrm{H}]]^{-1/2} K(\mathrm{H}^{-1/2}\mathbf{u}) . \tag{4.71}$$

Since only circular symmetric prototype kernels $K(\mathbf{u})$ are considered, we have, using the profile $k(u)$,

$$K(\mathbf{u}) = c_{k,p} k(\mathbf{u}^\top \mathbf{u}) . \tag{4.72}$$

From (4.70), taking into account (4.72) and (4.71) results in

$$\begin{aligned}
\widehat{f}(\mathbf{y}) &= \frac{c_{k,p}}{n[\det[\mathrm{H}]]^{1/2}} \sum_{i=1}^{n} k\left((\mathbf{y} - \mathbf{y}_i)^\top \mathrm{H}^{-1}(\mathbf{y} - \mathbf{y}_i) \right) \\
&= \frac{c_{k,p}}{n[\det[\mathrm{H}]]^{1/2}} \sum_{i=1}^{n} k\left(\mathrm{d}[\mathbf{y}, \mathbf{y}_i, \mathrm{H}]^2 \right),
\end{aligned} \tag{4.73}$$

where the expression $\mathrm{d}[\mathbf{y}, \mathbf{y}_i, \mathrm{H}]^2$ denotes the squared Mahalanobis distance from $\mathbf{y}$ to $\mathbf{y}_i$. The case $\mathrm{H} = h^2 \mathrm{I}_p$ is the most often used. The kernels then have a circular support whose radius is controlled by the bandwidth h, and (4.70) becomes

$$\widehat{f}_K(\mathbf{y}) = \frac{1}{nh^p} \sum_{i=1}^{n} K\left(\frac{\mathbf{y} - \mathbf{y}_i}{h}\right) = \frac{c_{k,p}}{nh^p} \sum_{i=1}^{n} k\left(\left\|\frac{\mathbf{y} - \mathbf{y}_i}{h}\right\|^2\right), \qquad (4.74)$$

where the dependence of the density estimate on the kernel was made explicit.

The quality of a density estimate $\widehat{f}(\mathbf{y})$ is assessed in statistics using the *asymptotic mean integrated error* (AMISE), that is, the integrated mean square error

$$\text{MISE}(\mathbf{y}) = \int \text{E}\left[f(\mathbf{y}) - \widehat{f}(\mathbf{y})\right]^2 d\mathbf{y} \qquad (4.75)$$

between the true density and its estimate for $n \to \infty$, while $h \to 0$ at a slower rate. The expectation is taken over all data sets of size n. Since the bandwidth h of a circular symmetric kernel has a strong influence on the quality of $\widehat{f}(\mathbf{y})$, the bandwidth minimizing an approximation of the AMISE error is of interest. Unfortunately, this bandwidth depends on $f(\mathbf{y})$, the unknown density [112, Sec. 4.3].

For the univariate case, several practical rules are available [112, Sec. 3.2]. For example, the information about $f(y)$ is substituted with $\widehat{\sigma}$, a robust scale estimate derived from the data

$$\widehat{h} = \left[\frac{243R(K)}{35\mu_2(K)^2 n}\right]^{1/5} \widehat{\sigma}, \qquad (4.76)$$

where

$$\mu_2(K) = \int_{-1}^{1} u^2 K(u) du \qquad R(K) = \int_{-1}^{1} K(u)^2 du. \qquad (4.77)$$

The scale estimate $\widehat{\sigma}$ is discussed in Section 4.4.3.

For a given bandwidth, the AMISE measure is minimized by the *Epanechnikov* kernel [112, p. 104] having the profile

$$k_E(u) = \begin{cases} 1 - u & 0 \le u \le 1 \\ 0 & u > 1 \end{cases}, \qquad (4.78)$$

which yields the kernel

$$K_E(\mathbf{y}) = \begin{cases} \frac{1}{2} c_p^{-1}(p+2)(1 - \|\mathbf{y}\|^2) & \|\mathbf{y}\| \le 1 \\ 0 & \text{otherwise} \end{cases}, \qquad (4.79)$$

where c_p is the volume of the p-dimensional unit sphere. Other kernels can also be defined. The *truncated normal* has the profile

$$k_N(u) = \begin{cases} e^{-au} & 0 \le u \le 1 \\ 0 & u > 1 \end{cases}, \qquad (4.80)$$

where a is chosen such that e^{-a} is already negligibly small. Neither of the two profiles defined above has continuous derivatives at the boundary $u = 1$. This condition is satisfied (for the first two derivatives) by the *biweight* kernel having the profile

$$k_B(u) = \begin{cases} (1 - u)^3 & 0 \le u \le 1 \\ 0 & u > 1 \end{cases}. \qquad (4.81)$$

Its name here is taken from robust statistics; in the kernel density estimation literature, it is called the triweight kernel [112, p. 31].

The bandwidth matrix H is the critical parameter of a kernel density estimator. For example, if the region of summation (bandwidth) is too large, significant features of the distribution, like multimodality, can be missed by oversmoothing. Furthermore, locally the data can have very different densities, and using a single bandwidth matrix often is not enough to obtain a satisfactory estimate.

There are two ways to adapt the bandwidth to the local structure. In each case, the adaptive behavior is achieved by first performing a pilot density estimation. The bandwidth matrix can be associated with the location $\mathbf{y}$ in which the distribution is to be estimated, or else each measurement $\mathbf{y}_i$ can be taken into account in (4.70) with its own bandwidth matrix:

$$\widehat{f}_K(\mathbf{y}) = \frac{1}{n} \sum_{i=1}^n K_{\mathbf{H}_i}(\mathbf{y} - \mathbf{y}_i) . \qquad (4.82)$$

It can be shown that (4.82), called the *sample point* density estimator, has superior statistical properties [39].

The local maxima of the density $f(\mathbf{y})$ are by definition the roots of the equation

$$\nabla f(\mathbf{y}) = 0, \qquad (4.83)$$

that is, the zeros of the density gradient. Note that the converse is not true, since any *stationary point* of $f(\mathbf{y})$ satisfies (4.83). The true density, however, is not available, and in practice, the estimate of the gradient $\widehat{\nabla} f(\mathbf{y})$ has to be used.

In the next section, we describe the *mean shift* technique, which avoids explicit computation of the density estimate when solving (4.83). The mean shift procedure also associates each data point to the nearest density maximum and thus performs a nonparametric clustering in which the shape of the clusters is not set a priori.

4.3.3 Adaptive mean shift

The mean shift method was described in several publications [16, 18, 17]. Here we consider its most general form in which each measurement $\mathbf{y}_i$ is associated with a known bandwidth matrix H_i, $i = 1,\dots,n$. Taking the gradient of the sample point density estimator (4.82), we obtain, after recalling (4.73) and exploiting the linearity of the expression,

$$\widehat{\nabla} f_K(\mathbf{y}) \equiv \nabla \widehat{f}_K(\mathbf{y}) \tag{4.84}$$

$$= \frac{2c_{k,p}}{n} \sum_{i=1}^{n} [\det[\mathrm{H}_i]]^{-1/2} \, \mathrm{H}_i^{-1}(\mathbf{y} - \mathbf{y}_i) \, k'\left(d[\mathbf{y},\mathbf{y}_i,\mathrm{H}_i]^2\right) \ .$$

The function $g(x) = -k'(x)$ satisfies the properties of a profile, and thus we can define the kernel $G(\mathbf{u}) = c_{g,p} g(\mathbf{u}^\top \mathbf{u})$. For example, for the Epanechnikov kernel, the corresponding new profile is

$$g_E(u) = \begin{cases} 1 & 0 \le u \le 1 \\ 0 & u > 1 \end{cases}, \tag{4.85}$$

and thus $G_E(\mathbf{u})$ is the uniform kernel. For convenience, we introduce the notation

$$Q_i(\mathbf{y}) = \det[\mathrm{H}_i]^{-1/2} \, \mathrm{H}_i^{-1} g\left(d[\mathbf{y},\mathbf{y}_i,\mathrm{H}_i]^2\right) \ . \tag{4.86}$$

From the definition of $g(u)$ and (4.69),

$$Q_i(\mathbf{y}) = O \qquad\qquad d[\mathbf{y},\mathbf{y}_i,\mathrm{H}_i] > 1 \ . \tag{4.87}$$

Then (4.84) can be written as

$$\widehat{\nabla} f_K(\mathbf{y}) = \frac{2c_{k,p}}{n} \left(\sum_{i=1}^{n} Q_i(\mathbf{y})\right) \left[\left(\sum_{i=1}^{n} Q_i(\mathbf{y})\right)^{-1} \sum_{i=1}^{n} Q_i(\mathbf{y})\mathbf{y}_i - \mathbf{y}\right], \tag{4.88}$$

and the roots of the equation (4.83) are the solutions of

$$\mathbf{y} = \left(\sum_{i=1}^{n} Q_i(\mathbf{y})\right)^{-1} \sum_{i=1}^{n} Q_i(\mathbf{y})\mathbf{y}_i, \tag{4.89}$$

which can be solved only iteratively

$$\mathbf{y}^{[l+1]} = \left(\sum_{i=1}^{n} Q_i\left(\mathbf{y}^{[l]}\right)\right)^{-1} \sum_{i=1}^{n} Q_i\left(\mathbf{y}^{[l]}\right)\mathbf{y}_i \qquad l = 0, 1, \dots \qquad (4.90)$$

The meaning of an iteration becomes apparent if we consider the particular case $\mathrm{H}_i = h_i^2 \mathrm{I}$ yielding

$$\mathbf{y} = \frac{\sum_{i=1}^{n}\mathbf{y}_i\, h_i^{-(p+2)}\, g\left(\left\|\frac{\mathbf{y}-\mathbf{y}_i}{h_i}\right\|^2\right)}{\sum_{i=1}^{n} h_i^{-(p+2)}\, g\left(\left\|\frac{\mathbf{y}-\mathbf{y}_i}{h_i}\right\|^2\right)}, \qquad (4.91)$$

which becomes, when all $h_i = h$,

$$\mathbf{y} = \frac{\sum_{i=1}^{n}\mathbf{y}_i\, g\left(\left\|\frac{\mathbf{y}-\mathbf{y}_i}{h}\right\|^2\right)}{\sum_{i=1}^{n} g\left(\left\|\frac{\mathbf{y}-\mathbf{y}_i}{h}\right\|^2\right)}. \qquad (4.92)$$

From (4.87) we see that at every step only a *local* weighted mean is computed. The robustness of the mode detection method is the direct consequence of this property. In the next iteration, the computation is repeated centered on the previously computed mean. The difference between the current and previous locations, the vector

$$\mathbf{m}_G^{[l+1]} = \mathbf{y}^{[l+1]} - \mathbf{y}^{[l]} \qquad l = 0, 1, \dots, \qquad (4.93)$$

is called the *mean shift* vector, where the fact that the weighted averages are computed with the kernel G was made explicit. Adapting (4.88) to the two particular cases above, it can be shown that

$$\mathbf{m}_G^{[l+1]} = c\frac{\widehat{\nabla} f_K(\mathbf{y}^{[l]})}{\widehat{f}_G(\mathbf{y}^{[l]})}, \qquad (4.94)$$

where c is a positive constant. Thus, the mean shift vector is aligned with the gradient estimate of the density, and the window of computations is always moved toward regions of higher density. See [17] for the details. A relation similar to (4.94) still holds in the general case, but now the mean shift and gradient vectors are connected by a linear transformation.

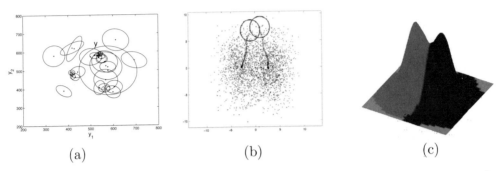

$$(a) \qquad\qquad (b) \qquad\qquad (c)$$

Figure 4.11. The main steps in mean shift-based clustering. (a) Computation of the weighted mean in the general case. (b) Mean shift trajectories of two points in a bimodal data. (c) Basins of attraction.

In the mean shift procedure, the user controls the resolution of the data analysis by providing the bandwidth information. Since most often circular symmetric kernels are used, only the bandwidth parameters h_i are needed.

Mean Shift Procedure

1. Choose a data point $\mathbf{y}_i$ as the initial $\mathbf{y}^{[0]}$.

2. Compute $\mathbf{y}^{[l+1]}$, $\;l = 0, 1, \ldots$ the weighted mean of the points at less than unit Mahalanobis distance from $\mathbf{y}^{[l]}$. Each point is considered with its own metric.

3. Determine if $\left\| \mathbf{m}_G^{[l+1]} \right\|$ is less than the tolerance. If yes, stop.

4. Replace $\mathbf{y}^{[l]}$ with $\mathbf{y}^{[l+1]}$; that is, move the processing toward a region with higher point density. Return to Step 2.

The most important properties of the mean shift procedure are illustrated graphically in Figure 4.11. In Figure 4.11a, the setup of the weighted mean computation in the general case is shown. The kernel associated with a data point is nonzero only within the elliptical region centered on that point. Thus, only those points contribute to the weighted mean in $\mathbf{y}$ whose kernel support contains $\mathbf{y}$.

The evolution of the iterative procedure is shown in Figure 4.11b for the simplest case of identical circular kernels (4.92). When the locations of the points in a window are averaged, the result is biased toward the region of

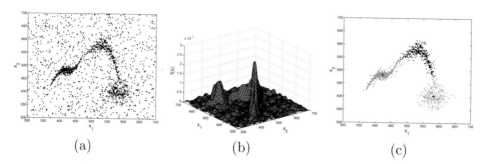

Figure 4.12. An example of clustering using the mean shift procedure. (a) The 2D input. (b) Kernel density estimate of the underlying distribution. (c) The basins of attraction of the three significant modes (marked +).

higher point density in that window. By moving the window into the new position, we move uphill on the density surface. The mean shift procedure is a gradient ascent technique. The processing climbs toward the highest point on the side of the density surface on which the initial position $\mathbf{y}^{[0]}$ was placed. At convergence (which can be proven), the local maximum of the density, the sought mode, is detected.

The two initializations in Figure 4.11b are on different components of this mixture of two Gaussians. Therefore, while the two mean shift procedures start from nearby locations, they converge to different modes, both of which are accurate location estimates.

A nonparametric classification of the data into clusters can be obtained by starting a mean shift procedure from every data point. A set of points converging to nearby locations defines the *basin of attraction* of a mode. Since the points are processed independently, the shape of the basin of attraction is not restricted in any way. The basins of attraction of the two modes of a Gaussian mixture (Figure 4.11c) were obtained without using the nature of the distributions.

The 2D data in Figure 4.12a illustrates the power of the mean shift-based clustering. The three clusters have arbitrary shapes, and the background is heavily cluttered with outliers. Traditional clustering methods would have difficulty yielding satisfactory results. The three significant modes in the data are clearly revealed in a kernel density estimate (Figure 4.12b). The mean shift procedure detects all three modes, and the associated basins of attraction provide a good delineation of the individual clusters (Figure 4.12c). In practice, using only a subset of the data points suffices for an accurate delineation. See [16] for details of the mean shift-based clustering.

The original mean shift procedure was proposed in 1975 by Fukunaga and Hostetler [32]. See also [31, p. 535]. It came again into attention with the paper [10]. In spite of its excellent qualities, mean shift is less known in the statistical literature. The book [92, Sec. 6.2.2] discusses [32], and a similar technique is proposed in [11] for bias reduction in density estimation.

The simplest, *fixed-bandwidth* mean shift procedure in which all $\mathbf{H}_i = h^2 \mathbf{I}_p$ is the one most frequently used in computer vision applications. The adaptive mean shift procedure discussed in this section, however, is not difficult to implement with circular symmetric kernels: $\mathbf{H}_i = h_i^2 \mathbf{I}_p$. The bandwidth value h_i associated with the data point $\mathbf{y}_i$ can be defined as the distance to the k-th neighbor; that is, for the pilot density estimation, the nearest neighbors approach is used. An implementation for high dimensional spaces is described in [35]. Other, more sophisticated methods for local bandwidth selection are described in [15, 18]. Given the complexity of the visual data, such methods, which are based on assumptions about the local structure, may not provide any significant gain in performance.

4.3.4 Applications

We now sketch two applications of the fixed-bandwidth mean shift procedure, that is, circular kernels with $\mathbf{H}_i = h^2 \mathbf{I}_p$.

- discontinuity preserving filtering and segmentation of color images;

- tracking of nonrigid objects in a color image sequence.

These applications are the subject of [17] and [19] respectively, which should be consulted for details.

An image can be regarded as a vector field defined on the 2D lattice. The dimension of the field is one in the gray-level case and three for color images. The image coordinates belong to the *spatial* domain, while the gray level or color information is in the *range* domain. To be able to use in the mean shift procedure circular symmetric kernels, the validity of an Euclidean metric must be verified for both domains. This is most often true in the spatial domain and for gray-level images in the range domain. For color images, mapping the RGB input into the L*u*v* (or L*a*b*) color space provides the closest possible Euclidean approximation for the perception of color differences by human observers.

The goal in image filtering and segmentation is to generate an accurate, piecewise-constant representation of the input. The constant parts should correspond in the input image to contiguous regions with similarly colored pixels, while the discontinuities should correspond to significant changes in color. This is achieved by considering the spatial and range domains jointly. In the joint domain, the basin of attraction of a mode corresponds to a contiguous, homogeneous region in the input image, and the valley between two modes most often represents a significant color discontinuity in the input. The joint mean shift procedure uses a product kernel

$$K(\mathbf{y}) = \frac{c}{h_s^2 h_r^q} k\left(\left\|\frac{\mathbf{y}^s}{h_s}\right\|^2\right) k\left(\left\|\frac{\mathbf{y}^r}{h_r}\right\|^2\right), \tag{4.95}$$

where $\mathbf{y}^s$ and $\mathbf{y}^r$ are the spatial and the range parts of the feature vector, $k(u)$ is the profile of the kernel used in both domains (though they can also differ), h_s and h_r are the employed bandwidths parameters, and c is the normalization constant. The dimension of the range domain q is one for the gray level and three for the color images. The user sets the value of the two bandwidth parameters according to the desired resolution of the image analysis.

In *discontinuity preserving filtering*, every pixel is allocated to the nearest mode in the joint domain. All the pixels in the basin of attraction of the mode get the range value of that mode. From the spatial arrangement of the basins of attraction, the region adjacency graph (RAG) of the input image is then derived. A transitive closure algorithm is performed on the RAG, and the basins of attraction of adjacent modes with similar range values are fused. The result is the *segmented* image.

The gray-level image example in Figure 4.13 illustrates the role of the mean shift procedure. The small region of interest (ROI) in Figure 4.13a is shown in a wireframe representation in Figure 4.13b. The 3D G kernel used in the mean shift procedure (4.92) is in the top left corner. The kernel is the product of two uniform kernels: a circular, symmetric 2D kernel in the spatial domain and a 1D kernel for the gray values.

At every step of the mean shift procedure, the average of the 3D data points is computed and the kernel is moved to the next location. When the kernel is defined at a pixel on the high plateau on the right in Figure 4.13b, adjacent pixels (neighbors in the spatial domain) have very different gray-level values and will *not* contribute to the average. This is how the mean shift procedure achieves the discontinuity preserving filtering. Note that the probability density function whose local mode is sought cannot be visualized, since it would require a 4D space, the fourth dimension being that of the density.

The result of the segmentation for the ROI is shown in Figure 4.13c and for the entire image in Figure 4.13d. A more accurate segmentation is

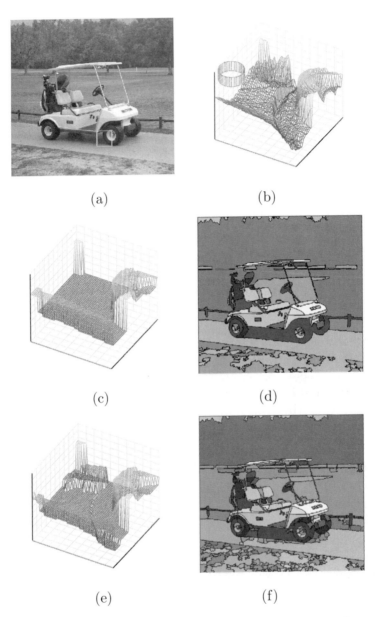

(a)   (b)

(c)   (d)

(e)   (f)

Figure 4.13. The image segmentation algorithm. (a) The gray-level input image with a region of interest (ROI) marked. (b) The wireframe representation of the ROI and the 3D window used in the mean shift procedure. (c) The segmented ROI. (d) The segmented image. (e) The segmented ROI when local discontinuity information is integrated into the mean shift procedure. (f) The segmented image.

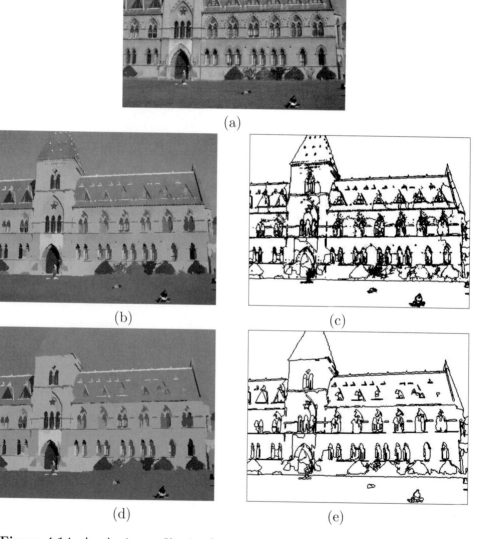

(a)

(b)   (c)

(d)   (e)

Figure 4.14. A color image filtering/segmentation example. (a) The input image. (b) The filtered image. (c) The boundaries of the delineated regions. (d) The segmented image. (e) The boundaries of the delineated regions.

obtained if edge information is incorporated into the mean shift procedure (Figures 4.13e and 4.13f). The technique is described in [13].

A color image example is shown in Figure 4.14. The input has large homogeneous regions, and after filtering (Figures 4.14b and 4.14c), many of the delineated regions already correspond to semantically meaningful parts of the image. However, this is more the exception than the rule in filtering. A more realistic filtering process can be observed around the windows, where many small regions (basins of attraction containing only a few pixels) are present. These regions are either fused or attached to a larger neighbor during the transitive closure process on the RAG, and the segmented image (Figures 4.14d and 4.14e) is less cluttered. The quality of any segmentation, however, can be assessed only through the performance of subsequent processing modules for which it serves as input.

The discontinuity preserving filtering and the image segmentation algorithm were integrated with a novel edge detection technique [73] in the *Edge Detection and Image SegmentatiON* (EDISON) system [13]. The C++ source code of EDISON is available on the web at `www.caip.rutgers.edu/riul/`.

The second application of the mean shift procedure is *tracking* of a dynamically changing neighborhood in a sequence of color images. This is a critical module in many object recognition and surveillance tasks. The problem is solved by analyzing the image sequence as pairs of two consecutive frames. See [19] for a complete discussion.

The neighborhood to be tracked (i.e., the *target model* in the first image) contains n_a pixels. We are interested only in the amount of relative translation of the target between the two frames. Therefore, without loss of generality, the target model can be considered centered on $\mathbf{y}_a = \mathbf{0}$. In the next frame, the *target candidate* is centered on $\mathbf{y}_b$ and contains n_b pixels.

In both color images, kernel density estimates are computed in the joint 5D domain. In the spatial domain, the estimates are defined in the center of the neighborhoods, while in the color domain, the density is sampled at m locations $\mathbf{c}$. Let $c = 1, \ldots, m$ be a scalar hashing index of these 3D sample points. A kernel with profile $k(u)$ and bandwidths h_a and h_b is used in the spatial domain. The sampling in the color domain is performed with the Kronecker delta function $\delta(u)$ as kernel.

The result of the two kernel density estimations are the two discrete color densities associated with the target in the two images. For $c = 1, \ldots, m$

$$\text{model:} \qquad \widehat{f_a}(c) = A \sum_{i=1}^{n_a} k\left(\left\| \frac{\mathbf{y}_{a,i}}{h_a} \right\|^2 \right) \delta\left[\mathbf{c}(\mathbf{y}_{a,i}) - \mathbf{c} \right] \qquad (4.96)$$

$$\text{candidate:} \qquad \widehat{f_b}(c, \mathbf{y}_b) = B \sum_{i=1}^{n_b} k\left(\left\| \frac{\mathbf{y}_b - \mathbf{y}_{b,i}}{h_b} \right\|^2 \right) \delta\left[\mathbf{c}(\mathbf{y}_{b,i}) - \mathbf{c} \right] \quad (4.97)$$

where $\mathbf{c}(\mathbf{y})$ is the color vector of the pixel at $\mathbf{y}$. The normalization constants A, B are determined such that

$$\sum_{c=1}^{m} \widehat{f_a}(c) = 1 \qquad\qquad \sum_{c=1}^{m} \widehat{f_b}(c, \mathbf{y}_b) = 1 \ . \qquad\qquad (4.98)$$

The normalization assures that the template matching score between these two discrete signals is

$$\rho(\mathbf{y}_b) = \sum_{c=1}^{m} \sqrt{\widehat{f_a}(c) \widehat{f_b}(c, \mathbf{y}_b)}, \qquad\qquad (4.99)$$

and it can be shown that

$$d(\mathbf{y}_b) = \sqrt{1 - \rho(\mathbf{y}_b)} \qquad\qquad (4.100)$$

is a metric distance between $\widehat{f_a}(c)$ and $\widehat{f_b}(c, \mathbf{y}_b)$

To find the location of the target in the second image, the distance (4.100) has to be minimized over $\mathbf{y}_b$, or equivalently (4.99) has to be maximized. That is, the local maximum of $\rho(\mathbf{y}_b)$ has to be found by performing a search in the second image. This search is implemented using the mean shift procedure.

The local maximum is a root of the template matching score gradient

$$\nabla\rho(\mathbf{y}_b) = \frac{1}{2} \sum_{c=1}^{m} \nabla\widehat{f_b}(c, \mathbf{y}_b) \sqrt{\frac{\widehat{f_a}(c)}{\widehat{f_b}(c, \mathbf{y}_b)}} = 0 \ . \qquad\qquad (4.101)$$

Taking into account (4.97) yields

$$\sum_{c=1}^{m} \sum_{i=1}^{n_b} (\mathbf{y}_b - \mathbf{y}_{b,i}) \, k'\left(\left\| \frac{\mathbf{y}_b - \mathbf{y}_{b,i}}{h} \right\|^2 \right) \delta\left[\mathbf{c}(\mathbf{y}_{b,i}) - \mathbf{c} \right] \sqrt{\frac{\widehat{f_a}(c)}{\widehat{f_b}(c, \mathbf{y}_b)}} = 0 \ .$$

$$(4.102)$$

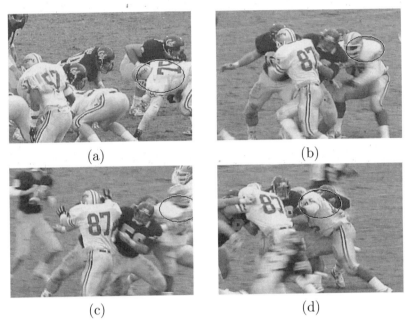

Figure 4.15. An example of the tracking algorithm. (a) The first frame of a color image sequence with the target model manually defined as the marked elliptical region. (b) to (d) Localization of the target in different frames.

As in Section 4.3.3, we can introduce the profile $g(u) = -k'(u)$ and define the weights

$$q_i(\mathbf{y}_b) = \sum_{c=1}^{m} \sqrt{\frac{\widehat{f}_a(c)}{\widehat{f}_b(c, \mathbf{y}_b)}} \, \delta\left[\mathbf{c}(\mathbf{y}_{b,i}) - \mathbf{c}\right] \qquad (4.103)$$

and obtain the iterative solution of (4.101) from

$$\mathbf{y}_b^{[l+1]} = \frac{\sum_{i=1}^{n_b} \mathbf{y}_{b,i}^{[l]} \, q_i\left(\mathbf{y}_b^{[l]}\right) \, g\left(\left\|\frac{\mathbf{y}_b^{[l]} - \mathbf{y}_{b,i}}{h_b}\right\|^2\right)}{\sum_{i=1}^{n_b} q_i\left(\mathbf{y}_b^{[l]}\right) \, g\left(\left\|\frac{\mathbf{y}_b^{[l]} - \mathbf{y}_{b,i}}{h_b}\right\|^2\right)}, \qquad (4.104)$$

which is a mean shift procedure, the only difference being that at each step the weights (4.103) are also computed.

In Figure 4.15, four frames of an image sequence are shown. The target model, defined in the first frame (Figure 4.15a), is successfully tracked throughout the sequence. As can be seen, the localization is satisfactory in spite of the target candidates' color distribution being significantly different from that of the model. While the model can be updated as we move along the sequence, the main reason for the good performance is the small amount of translation of the target region between two consecutive frames. The search in the second image always starts from the location of the target model center in the first image. The mean shift procedure then finds the *nearest* mode of the template matching score, and with high probability this is the target candidate location we are looking for. See [19] for more examples and extensions of the tracking algorithm, and [14] for a version with automatic bandwidth selection.

The robust solution of the location estimation problem presented in this section put the emphasis on employing the least possible amount of a priori assumptions about the data and belongs to the class of nonparametric techniques. Nonparametric techniques require a larger number of data points supporting the estimation process than their parametric counterparts. In parametric methods, the data is more constrained, and as long as the model is obeyed, the parametric methods are better in extrapolating over regions where data is not available. However, if the model is not correct, a parametric method will still impose it at the price of severe estimation errors. This important tradeoff must be kept in mind when feature space analysis is used in a complex computer vision task.

4.4   Robust Regression

The linear EIV regression model (Section 4.2.6) is employed for the discussion of the different regression techniques. In this model, the inliers are measured as

$$\mathbf{y}_i = \mathbf{y}_{io} + \delta\mathbf{y}_i \quad \mathbf{y}_i \in \mathcal{R}^p \quad \delta\mathbf{y}_i \sim GI(\mathbf{0}, \sigma^2 \mathbf{I}_p) \qquad i = 1, \dots, n_1, \quad (4.105)$$

and their true values obey the constraints

$$g(\mathbf{y}_{io}) = \mathbf{y}_{io}^\top \boldsymbol{\theta} - \alpha = 0 \quad i = 1, \dots, n_1 \qquad \|\boldsymbol{\theta}\| = 1 \qquad \alpha \geq 0 . \quad (4.106)$$

The number of inliers must be much larger than the number of free parameters of the model, $n_1 \gg p$. Nothing is assumed about the $n - n_1$ outliers.

After a robust method selects the inliers, they are often postprocessed with a nonrobust technique from the least squares (LS) family to obtain the final parameter. Therefore, we start by discussing the LS estimators. Next, the family of M-estimators is introduced, and the importance of the scale parameter related to the noise of the inliers is emphasized.

All the robust regression methods popular today in computer vision can be described within the framework of M-estimation, and thus their performance also depends on the accuracy of the scale parameter. To avoid this deficiency, we approach M-estimation in a different way and introduce the pbM-estimator, which does not require the user to provide the value of the scale.

In Section 4.2.5, it was shown that when a nonlinear EIV regression model is processed as a linear model in the carriers, the associated noise is heteroscedastic. Since the robust methods discussed in this section assume the model (4.105) and (4.106), they return biased estimates if employed for solving nonlinear EIV regression problems. However, this does not mean they should not be used! The role of any robust estimator is only to establish a satisfactory inlier/outlier dichotomy. As long as most of the inliers were recovered from the data, postprocessing with the proper nonlinear (and nonrobust) method will provide the correct estimates.

Regression in the presence of multiple structures in the data are not considered beyond the particular case of two structures in the context of structured outliers. We show why all the robust regression methods fail to handle such data once the measurement noise becomes large.

Each of the regression techniques in this section is related to one of the objective functions described in Section 4.2.2. Using the same objective function, location models can also be estimated, but we do not discuss these location estimators. For example, many of the traditional clustering methods belong to the least squares family [52, Sec. 3.3.2], or there is a close connection between the mean shift procedure and M-estimators of location [17].

4.4.1   Least squares family

We saw in Section 4.2.3 that the least squares family of estimators is not robust, since its objective function $\mathcal{J}_{LS}$ (4.22) is a symmetric function in *all* the measurements. Therefore, in this section, we assume that the data contains only inliers; that is, $n = n_1$.

The parameter estimates of the linear EIV regression model are obtained by solving the minimization

$$[\widehat{\alpha}, \widehat{\boldsymbol{\theta}}] = \underset{\alpha, \boldsymbol{\theta}}{\operatorname{argmin}} \frac{1}{n} \sum_{i=1}^{n} \|\mathbf{y}_i - \mathbf{y}_{io}\|^2 = \underset{\alpha, \boldsymbol{\theta}}{\operatorname{argmin}} \frac{1}{n} \sum_{i=1}^{n} g(\mathbf{y}_i)^2 \qquad (4.107)$$

subject to (4.106). The minimization yields the *total least squares* (TLS) estimator. For an in-depth analysis of the TLS estimation, see [111]. Related problems were discussed in the 19th century [33, p. 30], though the method most frequently used today, based on the singular value decomposition (SVD) was proposed only in 1970 by Golub and Reinsch [37]. See [38] for the linear algebra background.

To solve the minimization problem (4.107), we define the $n \times p$ matrices of the measurements Y and of the true values Y_o:

$$ Y = [\mathbf{y}_1 \ \mathbf{y}_2 \ \cdots \ \mathbf{y}_n]^\top \qquad Y_o = [\mathbf{y}_{1o} \ \mathbf{y}_{2o} \ \cdots \ \mathbf{y}_{no}]^\top . \qquad (4.108) $$

Then (4.107) can be rewritten as

$$ [\widehat{\alpha}, \ \widehat{\boldsymbol{\theta}}] = \underset{\alpha, \boldsymbol{\theta}}{\mathrm{argmin}} \, \| Y - Y_o \|_F^2 \qquad (4.109) $$

subject to

$$ Y_o \boldsymbol{\theta} - \alpha \mathbf{1}_n = \mathbf{0}_n, \qquad (4.110) $$

where $\mathbf{1}_n (\mathbf{0}_n)$ is a vector in $\mathcal{R}^n$ of all ones (zeros), and $\|A\|_F$ is the Frobenius norm of the matrix A.

The parameter α is eliminated next. The data is centered by using the orthogonal projector matrix $G = I_n - \frac{1}{n} \mathbf{1}_n \mathbf{1}_n^\top$, which has the property $G \mathbf{1}_n = \mathbf{0}_n$. It is easy to verify that

$$ \widetilde{Y} = GY = [\widetilde{\mathbf{y}}_1 \ \widetilde{\mathbf{y}}_2 \ \cdots \ \widetilde{\mathbf{y}}_n]^\top \qquad \widetilde{\mathbf{y}}_i = \mathbf{y}_i - \frac{1}{n} \sum_{i=1}^n \mathbf{y}_i = \mathbf{y}_i - \bar{\mathbf{y}} . \quad (4.111) $$

The matrix $\widetilde{Y}_o = GY_o$ is similarly defined. The parameter estimate $\widehat{\boldsymbol{\theta}}$ is then obtained from the minimization

$$ \widehat{\boldsymbol{\theta}} = \underset{\boldsymbol{\theta}}{\mathrm{argmin}} \, \| \widetilde{Y} - \widetilde{Y}_o \|_F^2 \qquad (4.112) $$

subject to

$$ \widetilde{Y}_o \boldsymbol{\theta} = \mathbf{0}_n . \qquad (4.113) $$

The constraint (4.113) implies that the rank of the true data matrix $\widetilde{Y}_o$ is only $p - 1$ and that the true $\boldsymbol{\theta}$ spans its nullspace. Indeed, our linear model requires that the true data points belong to a hyperplane in $\mathcal{R}^p$, which is a $(p-1)$-dimensional affine subspace. The vector $\boldsymbol{\theta}$ is the unit normal to this plane.

The available measurements, however, are located nearby the hyperplane and thus the measurement matrix $\widetilde{Y}$ has full-rank p. The solution of the TLS thus is the rank $p-1$ approximation of $\widetilde{Y}$. This approximation is obtained from the SVD of $\widetilde{Y}$ written as a dyadic sum:

$$\widetilde{Y} = \sum_{k=1}^{p} \widetilde{\sigma}_k \widetilde{\mathbf{u}}_k \widetilde{\mathbf{v}}_k^{\top}, \tag{4.114}$$

where the singular vectors $\widetilde{\mathbf{u}}_i$, $i = 1, \ldots, n$, and $\widetilde{\mathbf{v}}_j$, $j = 1, \ldots, p$ provide orthonormal bases for the four linear subspaces associated with the matrix $\widetilde{Y}$ [38, Sec. 2.6.2], and $\widetilde{\sigma}_1 \geq \widetilde{\sigma}_2 \geq \cdots \geq \widetilde{\sigma}_p > 0$ are the singular values of this full-rank matrix.

The optimum approximation yielding the minimum Frobenius norm for the error is the truncation of the dyadic sum (4.114) at $p-1$ terms [111, p. 31]:

$$\widehat{\widetilde{Y}} = \sum_{k=1}^{p-1} \widetilde{\sigma}_k \widetilde{\mathbf{u}}_k \widetilde{\mathbf{v}}_k^{\top}, \tag{4.115}$$

where the matrix $\widehat{\widetilde{Y}}$ contains the centered *corrected measurements* $\widehat{\widetilde{y}}$. These corrected measurements are the orthogonal projections of the available $\widetilde{y}_i$ on the hyperplane characterized by the parameter estimates (Figure 4.6). The TLS estimator is also known as *orthogonal least squares*.

The rank-one nullspace of $\widehat{\widetilde{Y}}$ is spanned by $\widetilde{\mathbf{v}}_p$, the right singular vector associated with the smallest singular value $\widetilde{\sigma}_p$ of $\widetilde{Y}$ [38, p.72]. Since $\widetilde{\mathbf{v}}_p$ is a unit vector,

$$\widehat{\boldsymbol{\theta}} = \widetilde{\mathbf{v}}_p . \tag{4.116}$$

The estimate of α is obtained by reversing the centering operation:

$$\widehat{\alpha} = \bar{\mathbf{y}}^{\top} \widehat{\boldsymbol{\theta}} . \tag{4.117}$$

The parameter estimates of the linear EIV model can also be obtained in a different, though completely equivalent, way. We define the carrier vector $\mathbf{x}$ by augmenting the variables with a constant

$$\mathbf{x} = [\mathbf{y}^{\top} \quad -1]^{\top} \qquad \sigma^2 C = \sigma^2 \begin{bmatrix} \mathbf{I}_p & \mathbf{0} \\ \mathbf{0}^{\top} & 0 \end{bmatrix}, \tag{4.118}$$

which implies that the covariance matrix of the carriers is singular. Using the $n \times (p+1)$ matrices

$$X = [\mathbf{x}_1 \ \mathbf{x}_2 \ \cdots \ \mathbf{x}_n]^\top \qquad X_o = [\mathbf{x}_{1o} \ \mathbf{x}_{2o} \ \cdots \ \mathbf{x}_{no}]^\top, \qquad (4.119)$$

the constraint (4.110) can be written as

$$X_o \boldsymbol{\omega} = \mathbf{0}_n \qquad \boldsymbol{\omega} = [\boldsymbol{\theta}^\top \ \alpha]^\top \qquad \|\boldsymbol{\theta}\| = 1 \quad \alpha \geq 0, \qquad (4.120)$$

where the subscript 1 of this parameterization in Section 4.2.6 was dropped.

Using Lagrangian multipliers, it can be shown that the parameter estimate $\widehat{\boldsymbol{\omega}}$ is the eigenvector of the *generalized* eigenproblem

$$X^\top X \boldsymbol{\omega} = \lambda C \boldsymbol{\omega} \qquad (4.121)$$

corresponding to the smallest eigenvalue λ_{min}. This eigenproblem is equivalent to the definition of the right singular values of the matrix $\widetilde{Y}$ [38, Sec. 8.3]. The condition $\|\widehat{\boldsymbol{\theta}}\| = 1$ is then imposed on the vector $\widehat{\boldsymbol{\omega}}$.

The first-order approximation for the covariance of the parameter estimate is [70, Sec.5.2.2]

$$C_{\widehat{\boldsymbol{\omega}}} = \widehat{\sigma}^2 (X^\top X - \lambda_{min} C)^+, \qquad (4.122)$$

where the pseudoinverse has to be used, since the matrix has rank p following (4.121). The estimate of the noise standard deviation is

$$\widehat{\sigma}^2 = \frac{\sum_{i=1}^n \widehat{g}(\mathbf{y}_i)^2}{n - p + 1} = \frac{\lambda_{min}}{n - p + 1} = \frac{\widetilde{\sigma}_p^2}{n - p + 1}, \qquad (4.123)$$

where $\widehat{g}(\mathbf{y}_i) = \mathbf{y}_i^\top \widehat{\boldsymbol{\theta}} - \widehat{\alpha}$ are the residuals. The covariances for the other parameterizations of the linear EIV model, $\boldsymbol{\omega}_2$ (4.53) and $\boldsymbol{\omega}_3$ (4.55), can be obtained through error propagation.

Note that when computing the TLS estimate with either of the two methods, special care has to be taken to execute *all* the required processing steps. The first approach starts with the data being centered, while in the second approach a generalized eigenproblem has to be solved. These steps are sometimes neglected in computer vision algorithms.

In the traditional linear regression model, only the variable z is corrupted by noise (4.36), and the constraint is

$$z_{io} = \alpha + \mathbf{x}_{io}^\top \boldsymbol{\theta} \qquad \mathbf{x}_{io} = \mathbf{x}(\mathbf{y}_{io}) \qquad i = 1, \ldots, n \ . \qquad (4.124)$$

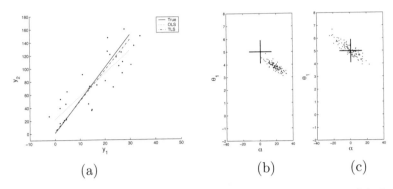

(a)   (b)   (c)

Figure 4.16. OLS versus TLS estimation of a linear EIV model. (a) A typical trial. (b) The scatterplot of the OLS estimates. A significant bias is present. (c) The scatterplot of the TLS estimates. The true parameter values correspond to the location marked +.

This model is actually valid for fewer computer vision problems (Figure 4.4) than is suggested in the literature. The corresponding estimator is the well-known (ordinary) *least squares* (OLS)

$$\widehat{\boldsymbol{\omega}} = (X_o^\top X_o)^{-1} X_o^\top \mathbf{z} \qquad C_{\widehat{\boldsymbol{\omega}}} = \widehat{\sigma}^2 (X_o^\top X_o)^{-1} \qquad \widehat{\sigma}^2 = \frac{\sum_{i=1}^n z_i^2}{n - p},$$

$$(4.125)$$

where

$$X_o = \begin{bmatrix} \mathbf{x}_{1o} & \mathbf{x}_{2o} & \cdots & \mathbf{x}_{no} \\ 1 & 1 & \cdots & 1 \end{bmatrix}^\top \qquad \mathbf{z} = [z_1 \; z_2 \; \cdots \; z_n]^\top . \qquad (4.126)$$

If the matrix X_o is poorly conditioned, the pseudoinverse should be used instead of the full inverse.

In the presence of significant measurement noise, using the OLS estimator when the data obeys the full EIV model (4.105) results in biased estimates [111, p. 232]. This is illustrated in Figure 4.16. The $n = 30$ data points are generated from the model

$$5y_{1o} - y_{2o} + 1 = 0 \qquad \delta \mathbf{y} \sim NI(\mathbf{0}, 5^2 I_2), \qquad (4.127)$$

where $NI(\cdot)$ stands for independent, normally distributed noise. Note that the constraint is not in the Hessian normal form but

$$\alpha + \theta_1 y_{1o} - y_{2o} = 0 \qquad \theta_1 = 5 \quad \alpha = 1, \qquad (4.128)$$

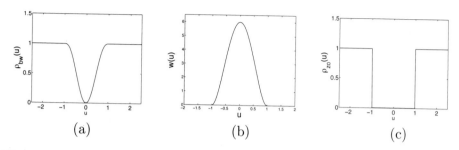

Figure 4.17. Redescending M-estimators. (a) Biweight loss function. (b) The weight function for biweight. (c) Zero-one loss function.

where, in order to compare the performance of the OLS and TLS estimators, the parameter θ_2 was set to -1. When the traditional regression model is associated with this data, it is assumed that

$$y_{2o} \equiv z_o = \theta_1 y_{1o} + \alpha \qquad\qquad \delta z \sim NI(0, 5^2), \qquad\qquad (4.129)$$

and the OLS estimator (4.125) is used to find $\widehat{\theta}_1$ and $\widehat{\alpha}$. The scatterplot of the result of 100 trials is shown in Figure 4.16b, and the estimates are far away from the true values.

Either TLS estimation method discussed above can be employed to find the TLS estimate. However, to eliminate the multiplicative ambiguity of the parameters, the ancillary constraint $\widehat{\theta}_2 = -1$ has to be used. See [111, Sec. 2.3.2]. The TLS estimates are unbiased, and the scatterplot is centered on the true values (Figure 4.16c).

Throughout this section, we have tacitly assumed that the data is not degenerate, that is, the measurement matrix Y has full-rank p. Both the TLS and OLS estimators can be adapted for the rank-deficient case, though then the parameter estimates are no longer unique. Techniques similar to the ones described in this section yield minimum norm solutions. See [111, Chap. 3] for the case of the TLS estimator.

4.4.2   M-estimators

The robust equivalent of the least squares family are the M-estimators, first proposed in 1964 by Huber as a generalization of the maximum likelihood technique in which contaminations in the data distribution are tolerated. See [67] for an introduction to M-estimators and [49] for a more in-depth discussion. We focus only on the class of M-estimators most recommended for computer vision applications.

The robust formulation of (4.48) is

$$[\widehat{a}, \widehat{\boldsymbol{\theta}}] = \underset{\alpha,\boldsymbol{\theta}}{\operatorname{argmin}} \frac{1}{n} \sum_{i=1}^{n} \rho\left(\frac{1}{s}g(\mathbf{y}_i)\right) = \underset{\alpha,\boldsymbol{\theta}}{\operatorname{argmin}} \mathcal{J}_M, \tag{4.130}$$

where s is a parameter that depends on σ, the (unknown) scale of the inlier noise (4.105). With a slight abuse of notation, s will also be called scale. The loss function $\rho(u)$ satisfies the following properties: nonnegative with $\rho(0) = 0$, even symmetric $\rho(u) = \rho(-u)$, and nondecreasing with $|u|$. For $\rho(u) = u^2$, we obtain the LS objective function (4.107).

The different M-estimators introduced in the statistical literature differ through the distribution assumed for the data. See [5] for a discussion in the context of computer vision. However, none of these distributions will provide an accurate model in a real application. Thus, the distinctive theoretical properties of different M-estimators are less relevant in practice.

The *redescending* M-estimators are characterized by bounded loss functions

$$0 \leq \rho(u) \leq 1 \quad |u| \leq 1 \qquad \rho(u) = 1 \quad |u| > 1 \ . \tag{4.131}$$

As will be shown below, in a redescending M-estimator, only those data points that are at distance less than s from the current fit are taken into account. This yields better outlier rejection properties than that of the M-estimators with nonredescending loss functions [69, 115].

The following class of redescending loss functions covers several important M-estimators

$$\rho(u) = \begin{cases} 1 - (1 - u^2)^d & |u| \leq 1 \\ 1 & |u| > 1 \end{cases}, \tag{4.132}$$

where $d = 1, 2, 3$. The loss functions have continuous derivatives up to the $(d-1)$-th order and a unique minimum in $\rho(0) = 0$.

Tukey's *biweight* function $\rho_{bw}(u)$ (Figure 4.17a) is obtained for $d = 3$ [67, p. 295]. This loss function is widely used in the statistical literature and was known at least a century before robust estimation [40, p. 151]. See also [42, vol.I, p. 323]. The loss function obtained for $d = 2$ is denoted $\rho_e(u)$. The case $d = 1$ yields the *skipped mean* loss function, a name borrowed from robust location estimators [89, p. 181],

$$\rho_{sm}(u) = \begin{cases} u^2 & |u| \leq 1 \\ 1 & |u| > 1 \end{cases}, \tag{4.133}$$

which has discontinuous first derivative. It is often used in vision applications, such as [108].

In the objective function of any M-estimator, the geometric distances (4.46) are normalized by the scale s. Since $\rho(u)$ is an even function, we do not need to use absolute values in (4.130). In redescending M-estimators, the scale acts as a hard rejection threshold, and thus its value is of paramount importance. For the moment, we assume that a satisfactory value is already available for s, but we return to this topic in Section 4.4.3.

The M-estimator equivalent to the total least squares is obtained following either TLS method discussed in Section 4.4.1. For example, it can be shown that instead of (4.121), the M-estimate of $\boldsymbol{\omega}$ (4.120) is the eigenvector corresponding to the the smallest eigenvalue of the generalized eigenproblem,

$$X^\top W X \boldsymbol{\omega} = \lambda C \boldsymbol{\omega}, \tag{4.134}$$

where $W \in \mathcal{R}^{n \times n}$ is the diagonal matrix of the nonnegative weights

$$w_i = w(u_i) = \frac{1}{u_i} \frac{d\rho(u_i)}{du} \geq 0 \qquad u_i = \frac{\widehat{g}(\mathbf{y}_i)}{s} \quad i = 1, \ldots, n . \tag{4.135}$$

Thus, in redescending M-estimators, $w(u) = 0$ for $|u| > 1$; that is, the data points whose residual $\widehat{g}(\mathbf{y}_i) = \mathbf{y}_i^\top \widehat{\boldsymbol{\theta}} - \widehat{\alpha}$ relative to the current fit is larger than the scale threshold s are discarded from the computations. The weights $w_{bw}(u) = 6(1 - u^2)^2$ derived from the biweight loss function are shown in Figure 4.17b. The weights derived from the $\rho_e(u)$ loss function are proportional to the Epanechnikov kernel (4.79). For traditional regression, instead of (4.125), the M-estimate is

$$\widehat{\boldsymbol{\omega}} = (X_o^\top W X_o)^{-1} X_o^\top W \mathbf{z} . \tag{4.136}$$

The residuals $\widehat{g}(\mathbf{y}_i)$ in the weights w_i require values for the parameter estimates. Therefore, the M-estimates can be found only by an iterative procedure.

M-estimation with Iterative Weighted Least Squares

Given the scale s,

1. Obtain the initial parameter estimate $\widehat{\boldsymbol{\omega}}^{[0]}$ with total least squares.

2. Compute the weights $w_i^{[l+1]}, l = 0, 1, \ldots$.

3. Obtain the updated parameter estimates, $\widehat{\boldsymbol{\omega}}^{[l+1]}$.

4. Determine if $\|\widehat{\omega}^{[l+1]} - \widehat{\omega}^{[l]}\|$ is less than the tolerance. If yes, stop.

5. Replace $\widehat{\omega}^{[l]}$ with $\widehat{\omega}^{[l+1]}$. Return to Step 2.

For the traditional regression, the procedure is identical. See [67, p. 306]. A different way of computing linear EIV regression M-estimates is described in [115].

The objective function minimized for redescending M-estimators is not convex, and therefore the convergence to a global minimum is not guaranteed. Nevertheless, in practice, convergence is always achieved [67, p. 307], and if the initial fit and the chosen scale value are adequate, the obtained solution is satisfactory. These two conditions are much more influential than the precise nature of the employed loss function. Note that at every iteration, all the data points regarded as inliers are processed, and thus there is no need for postprocessing, as is the case with the elemental subsets-based numerical optimization technique discussed in Section 4.2.7.

In the statistical literature, often the scale threshold s is defined as the product between $\widehat{\sigma}$, the robust estimate for the standard deviation of the inlier noise (4.105), and a tuning constant. The tuning constant is derived from the asymptotic properties of the simplest location estimator, the mean [67, p. 296]. Therefore, its value is rarely meaningful in real applications. Our definition of redescending M-estimators avoids the problem of tuning by using the inlier/outlier classification threshold as the scale parameter s.

The case $d = 0$ in (4.132) yields the *zero-one* loss function

$$\rho_{zo}(u) = \begin{cases} 0 & |u| \leq 1 \\ 1 & |u| > 1 \end{cases}, \tag{4.137}$$

shown in Figure 4.17c. The zero-one loss function is also a redescending M-estimator; however, it is no longer continuous and does not have a unique minimum in $u = 0$. It is only mentioned because, in Section 4.4.4, it is used to link the M-estimators to other robust regression techniques such as LMedS and RANSAC. The zero-one M-estimator is not recommended in applications. The weight function (4.135) is nonzero only at the boundary and the corresponding M-estimator has poor local robustness properties. That is, in a critical data configuration, a single data point can have a very large influence on the parameter estimates.

4.4.3 Median absolute deviation scale estimate

Access to a reliable scale parameter s is a necessary condition for the minimization procedure (4.130) to succeed. The scale s is a strictly monotonically

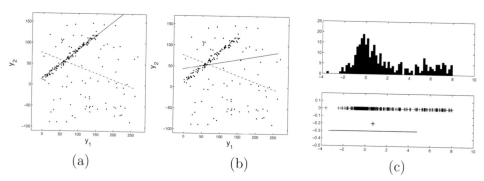

Figure 4.18. Sensitivity of the M-estimation to the $\widehat{s}_{mad}$ scale value. Dashed line: initial TLS fit. Solid line: biweight M-estimate. (a) $c = 1.5$. (b) $c = 3.5$. (c) Overestimation of the scale in the presence of skewness. The median of the residuals is marked $+$ under the sorted data points. The bar below corresponds to $\pm\widehat{s}_{mad}$ computed with $c = 3$.

increasing function of σ, the standard deviation of the inlier noise. Since σ is a nuisance parameter of the model, it can be estimated together with α and $\boldsymbol{\theta}$ at every iteration of the M-estimation process [67, p. 307]. An example of a vision application employing this approach is [7]. However, the strategy is less robust than providing the main estimation process with a fixed-scale value [68]. In the latter case, we talk about an *M-estimator with auxiliary scale* [69, 100].

Two different approaches can be used to obtain the scale prior to the parameter estimation. It can be either *arbitrarily* set by the user, or it can be derived from the data in a pilot estimation procedure. The first approach is widely used in the robust regression techniques developed within the vision community, such as RANSAC or Hough transform. The reason is that it allows an easy way to tune a method to the available data. The second approach is often adopted in the statistical literature for M-estimators and is implicitly employed in the LMedS estimator.

The most frequently used off-line scale estimator is the *median absolute deviation* (MAD), which is based on the residuals $\widehat{g}(\mathbf{y}_i)$ relative to an initial (nonrobust TLS) fit:

$$\widehat{s}_{mad} = c \operatorname*{med}_i |\widehat{g}(\mathbf{y}_i) - \operatorname*{med}_j \widehat{g}(\mathbf{y}_j)|, \qquad (4.138)$$

where c is a constant to be set by the user. The MAD scale estimate measures the spread of the residuals around their median.

In the statistical literature, the constant in (4.138) is often taken as $c = 1.4826$. However, this value is used to obtain a consistent estimate for σ when *all the residuals* obey a normal distribution [67, p. 302]. In computer vision applications where the percentage of outliers is often high, the condition is strongly violated. In the redescending M-estimators, the role of the scale parameter s is to define the inlier/outlier classification threshold. The order of magnitude of the scale can be established by computing the MAD expression, and the rejection threshold is then set as a multiple of this value. There is no need for assumptions about the residual distribution. In [105], the standard deviation of the inlier noise $\hat{\sigma}$ was computed as 1.4826 times a robust scale estimate similar to MAD, the minimum of the LMedS optimization criterion (4.140). The rejection threshold was set at $1.96\hat{\sigma}$ by assuming normally distributed residuals. The result is actually three times the computed MAD value and could be obtained by setting $c = 3$ without any assumption about the distribution of the residuals.

The example in Figures 4.18a and 4.18b illustrates not only the importance of the scale value for M-estimation but also the danger of being locked into the nature of the residuals. The data contains 100 inliers and 75 outliers, and as expected, the initial TLS fit is completely wrong. When the scale parameter is set small by choosing for $\hat{s}_{mad}$ the constant $c = 1.5$, at convergence the final M-estimate is satisfactory (Figure 4.18a). When the scale $\hat{s}_{mad}$ is larger, $c = 3.5$, the optimization process converges to a local minimum of the objective function. This minimum does not correspond to a robust fit (Figure 4.18b). Note that $c = 3.5$ is about the value of the constant that would have been used under the assumption of normally distributed inlier noise.

The location estimator employed for centering the residuals in (4.138) is the median, while the MAD estimate is computed with the second, outer median. However, the median is a reliable estimator only when the distribution underlying the data is unimodal and symmetric [49, p. 29]. It is easy to see that for a heavily skewed distribution (i.e., with a long tail on one side), the median will be biased toward the tail. For such distributions, the MAD estimator severely overestimates the scale, since the 50th percentile of the centered residuals is now shifted toward the boundary of the inlier distribution. The tail is most often due to outliers, and the amount of overestimation increases with both the decrease of the inlier/outlier ratio and the lengthening of the tail. In the example in Figure 4.18c, the inliers (at the left) were obtained from a standard normal distribution. The median is 0.73 instead of zero. The scale computed with $c = 3$ is $\hat{s}_{mad} = 4.08$, which is much larger than 2.5, a reasonable value for the spread of the inliers. Again, c should be chosen smaller.

Scale estimators that avoid centering the data were proposed in the statistical literature [88], but they are computationally intensive and their advantage for vision applications is not immediate. We must conclude that the MAD scale estimate has to be used with care in robust algorithms dealing with real data. Whenever available, independent information provided by the problem at hand should be exploited to validate the obtained scale. The influence of the scale parameter s on the performance of M-estimators can be entirely avoided by a different approach toward this family of robust estimators. This is discussed in Section 4.4.5.

4.4.4   LMedS, RANSAC, and Hough transform

The origin of these three robust techniques was described in Section 4.1. Now we show that they all can be expressed as M-estimators with auxiliary scale.

The LMedS estimator is a least k-th order statistics estimator (4.22) for $k = n/2$ and is the main topic of the book [89]. The LMedS estimates are obtained from

$$[\widehat{\alpha}, \widehat{\boldsymbol{\theta}}] = \operatorname*{argmin}_{\alpha, \boldsymbol{\theta}} \operatorname*{med}_{i} g(\mathbf{y}_i)^2, \qquad (4.139)$$

but in practice we can use

$$[\widehat{\alpha}, \widehat{\boldsymbol{\theta}}] = \operatorname*{argmin}_{\alpha, \boldsymbol{\theta}} \operatorname*{med}_{i} |g(\mathbf{y}_i)| . \qquad (4.140)$$

The difference between the two definitions is largely theoretical and becomes relevant only when the number of data points n is small and even, while the median is computed as the average of the two central values [89, p. 126]. Once the median is defined as the $[n/2]$-th order statistics, the two definitions always yield the same solution. By minimizing the median of the residuals, the LMedS estimator finds in the space of the data the narrowest cylinder containing at least half the points (Figure 4.19a). The minimization is performed with the elemental subsets-based search technique discussed in Section 4.2.7.

The scale parameter s does not appear explicitly in the above definition of the LMedS estimator. Instead of setting an upper bound on the value of the scale—the inlier/outlier threshold of the redescending M-estimator—in the LMedS, a lower bound on the percentage of inliers (50%) is imposed. This eliminates the need for the user to guess the amount of measurement noise, and as long as the inliers are in absolute majority, a somewhat better robust

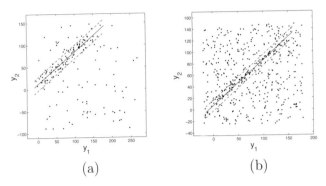

Figure 4.19. The difference between LMedS and RANSAC. (a) LMedS: finds the location of the narrowest band containing half the data. (b) RANSAC: finds the location of the densest band of width specified by the user.

behavior is obtained. For example, the LMedS estimator will successfully process the data in Figure 4.18a.

The relation between the scale parameter and the bound on the percentage of inliers is revealed if the equivalent condition of half the data points being outside of the cylinder is written as

$$\frac{1}{n} \sum_{i=1}^{n} \rho_{zo} \left(\frac{1}{s} g(\mathbf{y}_i) \right) = \frac{1}{2}, \qquad (4.141)$$

where ρ_{zo} is the zero-one loss function (4.137), and the scale parameter is now regarded as a function of the residuals $s\,[g(\mathbf{y}_1), \dots, g(\mathbf{y}_n)]$. By defining $s = \underset{i}{\mathrm{med}}\,|g(\mathbf{y}_i)|$, the LMedS estimator becomes

$$[\widehat{\alpha},\, \widehat{\boldsymbol{\theta}}] = \underset{\alpha,\boldsymbol{\theta}}{\mathrm{argmin}}\, s\,[g(\mathbf{y}_1), \dots, g(\mathbf{y}_n)] \qquad \text{subject to (4.141).} \qquad (4.142)$$

The new definition of LMedS is a particular case of the *S-estimators*, which, while popular in statistics, are not widely know in the vision community. For an introduction to S-estimators, see [89, pp. 135–143], and for a more detailed treatment in the context of EIV models, see [115]. Let $\widehat{s}$ be the minimum of s in (4.142). Then, it can be shown that

$$[\widehat{\alpha},\, \widehat{\boldsymbol{\theta}}] = \underset{\alpha,\boldsymbol{\theta}}{\mathrm{argmin}}\, \frac{1}{n} \sum_{i=1}^{n} \rho_{zo} \left(\frac{1}{\widehat{s}} g(\mathbf{y}_i) \right), \qquad (4.143)$$

and thus the S-estimators are in fact M-estimators with auxiliary scale.

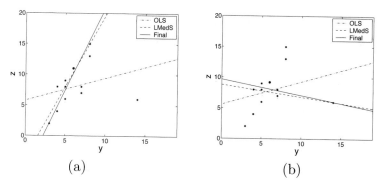

Figure 4.20. The poor local robustness of the LMedS estimator. The difference between the data sets in (a) and (b) is that the point (6, 9.3) was moved to (6, 9.2).

The value of $\widehat{s}$ can also be used as a scale estimator for the noise corrupting the inliers. All the observations made in Section 4.4.3 remain valid. For example, when the inliers are no longer the absolute majority in the data, the LMedS fit is incorrect, and the residuals used to compute $\widehat{s}$ are not reliable.

The RANSAC estimator predates the LMedS [26]. Since the same elemental subsets-based procedure is used to optimize their objective function, sometimes the two techniques were mistakenly considered to be very similar (e.g., [74]). However, their similarity should be judged examining the objective functions and not the way the optimization is implemented. In LMedS, the scale is computed from a condition set on the percentage of inliers (4.142). In RANSAC, the following minimization problem is solved

$$[\widehat{\alpha}, \widehat{\boldsymbol{\theta}}] = \operatorname*{argmin}_{\alpha, \boldsymbol{\theta}} \frac{1}{n} \sum_{i=1}^{n} \rho_{zo}\left(\frac{1}{\widehat{s}} g(\mathbf{y}_i)\right) \qquad \text{given } \widehat{s}. \qquad (4.144)$$

That is, the scale is *provided* by the user. This is a critical difference. Note that (4.144) is the same as (4.30). Since it is relatively easy to tune RANSAC to the data, it can also handle situations in which LMedS would already fail due to the large percentage of outliers (Figure 4.19b). Today RANSAC replaced LMedS in most vision applications, [65, 83, 107], for example.

The use of the zero-one loss function in both LMedS and RANSAC yields very poor local robustness properties, as illustrated in Figure 4.20, an example inspired by [2]. The $n = 12$ data points appear to be a simple case of robust linear regression for which the traditional regression model (4.37) was used. The single outlier on the right corrupts the least squares (OLS) estimator. The LMedS estimator, however, succeeds in recovering the correct fit (Figure 4.20a), and the OLS postprocessing of the points declared inliers (Final) does not yield any further change. The data in Figure 4.20b seems to be the same, but now the LMedS, and therefore the postprocessing, completely failed. Actually, the difference between the two data sets is that the point (6, 9.3) was moved to (6, 9.2).

The configuration of this data, however, is a critical one. The six points in the center can be grouped either with the points that appear to be also inliers (Figure 4.20a) or with the single outlier on the right (Figure 4.20b). In either case, the grouping yields an absolute majority of points, which is preferred by LMedS. There is a hidden bimodality in the data, and as a consequence, a delicate equilibrium exists between the correct and the incorrect fit.

In this example, the LMedS minimization (4.140) seeks the narrowest band containing at least six data points. The width of the band is measured along the z axis, and its boundary is always defined by two of the data points [89, p. 126]. This is equivalent to using the zero-one loss function in the optimization criterion (4.143). A small shift of one of the points thus can change to which fit the value of the minimum in (4.140) corresponds to. The instability of the LMedS is discussed in a practical setting in [45], while more theoretical issues are addressed in [23]. A similar behavior is also present in RANSAC due to (4.144).

For both LMedS and RANSAC, several variants were introduced in which the zero-one loss function is replaced by a smooth function. Since more points then have nonzero weights in the optimization, the local robustness properties of the estimators improve. The *least trimmed squares* (LTS) estimator [89, p. 132]

$$[\widehat{\alpha}, \widehat{\boldsymbol{\theta}}] = \underset{\alpha, \boldsymbol{\theta}}{\operatorname{argmin}} \sum_{i=1}^{k} g(\mathbf{y})_{i:n}^2 \qquad (4.145)$$

minimizes the sum of squares of the k smallest residuals, where k has to be provided by the user. Similar to LMedS, the absolute values of the residuals can also be used.

In the first smooth variant of RANSAC, the zero-one loss function was replaced with the skipped mean (4.133) and was called MSAC [108]. Recently, the same loss function was used in a maximum a posteriori formulation of RANSAC, the MAPSAC estimator [104]. A maximum likelihood motivated variant, the MLESAC [106], uses a Gaussian kernel for the inliers. Guided sampling is incorporated into the IMPSAC version of RANSAC [104]. In every variant of RANSAC, the user has to provide a reasonably accurate scale value for a satisfactory performance.

The use of zero-one loss function is not the only (or main) cause of the failure of LMedS (or RANSAC). In Section 4.4.7, we show that there is a more general problem in applying robust regression methods to multistructured data.

The only robust method designed to handle multistructured data is the *Hough transform*. The idea of Hough transform is to replace the regression problems in the input domain with location problems in the space of the parameters. Then, each significant mode in the parameter space corresponds to an instance of the model in the input space. There is a huge literature dedicated to every conceivable aspect of this technique. The survey papers [50, 62, 81] contain hundreds of references.

Since we are focusing here on the connection between the redescending M-estimators and the Hough transform, only the *randomized* Hough transform (RHT) will be considered [56]. Their equivalence is the most straightforward, but the same equivalence also exists for all the other variants of the Hough transform. The feature space in RHT is built with elemental subsets, and thus we have a mapping from p data points to a point in the parameter space.

Traditionally, the parameter space is quantized into bins; that is, it is an accumulator. The bins containing the largest number of votes yield the parameters of the significant structures in the input domain. This can be described formally as

$$[\widehat{\alpha}, \widehat{\boldsymbol{\beta}}]_k = \operatorname*{argmax}_{[\alpha, \boldsymbol{\beta}]} {}_k \frac{1}{n} \sum_{i=1}^{n} \kappa_{zo} \left(s_\alpha, s_{\beta_1}, \dots, s_{\beta_{p-1}}; g(\mathbf{y}_i) \right), \qquad (4.146)$$

where $\kappa_{zo}(u) = 1 - \rho_{zo}(u)$ and $s_\alpha, s_{\beta_1}, \dots, s_{\beta_{p-1}}$ define the size (scale) of a bin along each parameter coordinate. The index k stands for the different local maxima. Note that the parameterization uses the polar angles, as discussed in Section 4.2.6.

The definition (4.146) is that of a redescending M-estimator with auxiliary scale, where the criterion is a maximization instead of a minimization. The accuracy of the scale parameters is a necessary condition for a satisfactory performance, an issue widely discussed in the Hough transform literature. The advantage of distributing the votes around adjacent bins was recognized early [101]. Later the equivalence with M-estimators was also identified, and the zero-one loss function is often replaced with a continuous function [61, 60, 79].

In this section, we showed that all the robust techniques popular in computer vision can be reformulated as M-estimators. In Section 4.4.3, we emphasized that the scale has a crucial influence on the performance of M-estimators. In the next section, we remove this dependence by approaching the M-estimators in a different way.

4.4.5   The pbM-estimator

The minimization criterion (4.130) of the M-estimators is rewritten as

$$[\widehat{\alpha}, \widehat{\boldsymbol{\theta}}] = \underset{\alpha, \boldsymbol{\theta}}{\text{argmax}} \frac{1}{n} \sum_{i=1}^{n} \kappa \left(\frac{\mathbf{y}_i^\top \boldsymbol{\theta} - \alpha}{s} \right) \qquad \kappa(u) = c_\rho [1 - \rho(u)], \qquad (4.147)$$

where $\kappa(u)$ is called the *M-kernel function*. Note that for a redescending M-estimator, $\kappa(u) = 0$ for $|u| > 1$ (4.131). The positive normalization constant c_ρ assures that $\kappa(u)$ is a proper kernel (4.68).

Consider the unit vector $\boldsymbol{\theta}$ defining a line through the origin in $\mathcal{R}^p$. The projections of the n data points $\mathbf{y}_i$ on this line have the 1D (intrinsic) coordinates $x_i = \mathbf{y}_i^\top \boldsymbol{\theta}$. Following (4.67), the density of the set of points x_i, $i = 1, \ldots, n$, estimated with the kernel $K(u)$ and the bandwidth $\widehat{h}_{\boldsymbol{\theta}}$, is

$$\widehat{f}_{\boldsymbol{\theta}}(x) = \frac{1}{n\widehat{h}_{\boldsymbol{\theta}}} \sum_{i=1}^{n} K \left(\frac{\mathbf{y}_i^\top \boldsymbol{\theta} - x}{\widehat{h}_{\boldsymbol{\theta}}} \right) . \qquad (4.148)$$

Comparing (4.147) and (4.148), we can observe that if $\kappa(u)$ is taken as the kernel function, and $\widehat{h}_{\boldsymbol{\theta}}$ is substituted for the scale s, the M-estimation criterion becomes

$$\widehat{\boldsymbol{\theta}} = \underset{\boldsymbol{\theta}}{\text{argmax}} \left[\widehat{h}_{\boldsymbol{\theta}} \max_x \widehat{f}_{\boldsymbol{\theta}}(x) \right] \qquad (4.149)$$

$$\widehat{\alpha} = \underset{x}{\text{argmax}} \widehat{f}_{\widehat{\boldsymbol{\theta}}}(x) . \qquad (4.150)$$

Given the M-kernel $\kappa(u)$, the bandwidth parameter $\widehat{h}_{\boldsymbol{\theta}}$ can be estimated from the data according to (4.76). Since, as will be shown below, the value of the bandwidth has a weak influence on the the result of the M-estimation, for the entire family of redescending loss functions (4.132), we can use

$$\widehat{h}_{\boldsymbol{\theta}} = n^{-1/5} \, \underset{i}{\text{med}} \, | \mathbf{y}_i^\top \boldsymbol{\theta} - \underset{j}{\text{med}} \, \mathbf{y}_j^\top \boldsymbol{\theta} | . \qquad (4.151)$$

The MAD estimator is employed in (4.151), but its limitations (Section 4.4.3) are of less concern in this context. Also, it is easy to recognize when the

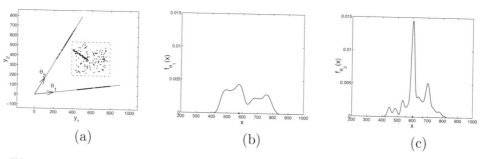

Figure 4.21. M-estimation through projection pursuit. When the data in the rectangle is projected orthogonally on different directions (a), the mode of the estimated density is smaller for an arbitrary direction (b) than for the direction of the normal to the linear structure (c).

data is not corrupted, since the MAD expression becomes too small. In this case, instead of the density estimation, most often a simple search over the projected points suffices.

The geometric interpretation of the new definition of M-estimators is similar to that of the LMedS and RANSAC techniques shown in Figure 4.19. The closer the projection direction is to the normal of the linear structure, the tighter the projected inliers are grouped which increases the mode of the estimated density (Figure 4.21). Again, a cylinder having the highest density in the data has to be located. The new approach is called *projection-based* M-estimator, or pbM-estimator.

The relations (4.149) and (4.150) are the projection pursuit definition of an M-estimator. Projection pursuit was proposed by Friedman and Tukey in 1974 [30] to solve data analysis problems by seeking "interesting" low-dimensional projections of the multidimensional data. The informative value of a projection is measured with a *projection index*, such as the quantity inside the brackets in (4.149). The papers [48, 54] survey all the related topics. It should be emphasized that in the projection pursuit literature, the name projection pursuit regression refers to a technique different from ours. There, a nonlinear additive model is estimated by adding a new term to the model after each iteration; see, for example, [44, Sec. 11.2].

When in the statistical literature a linear regression problem is solved through projection pursuit, either nonrobustly [20] or robustly [89, p. 143], the projection index is a scale estimate. Similar to the S-estimators, the solution is obtained by minimizing the scale, now over the projection directions. The robust scale estimates, like the MAD (4.138) or the median of the absolute value of the residuals (4.142), however, have severe deficiencies for skewed distributions, as discussed in Section 4.4.3. Thus, their use as a projection index will not guarantee a better performance than that of the original implementation of the regression technique.

Projections were employed before in computer vision. In [80], a highly accurate implementation of the Hough transform was achieved by using local projections of the pixels onto a set of directions. Straight edges in the image were then found by detecting the maxima in the numerically differentiated projections. The L_2E estimator, proposed recently in the statistical literature [91], solves a minimization problem similar to the kernel density estimate formulation of M-estimators; however, the focus is on the parametric model of the inlier residual distribution.

The critical parameter of the redescending M-estimators is the scale s, the inlier/outlier selection threshold. The novelty of the pbM-estimator is the way the scale parameter is manipulated. The pbM-estimator avoids the need of M-estimators for an accurate scale prior to estimation by using the bandwidth $\widehat{h}_{\boldsymbol{\theta}}$ as scale during the search for the optimal projection direction. The bandwidth being an approximation of the AMISE optimal solution (4.75) tries to preserve the sensitivity of the density estimation process as the number of data points n becomes large. This is the reason for the $n^{-1/5}$ factor in (4.151). Since $\widehat{h}_{\boldsymbol{\theta}}$ is the outlier rejection threshold at this stage, a too small value increases the probability of incorrectly assigning the optimal projection direction to a local alignment of points. Thus, it is recommended that once n becomes large, say $n > 10^3$, the computed bandwidth value is slightly increased by a factor that is monotonic in n.

After the optimal projection direction $\widehat{\boldsymbol{\theta}}$ is found, the actual inlier/outlier dichotomy of the data is defined by analyzing the shape of the density around the mode. The nearest local minima on the left and on the right correspond in $\mathcal{R}^p$, the space of the data, to the transition between the inliers belonging to the sought structure (which has a higher density) and the background clutter of the outliers (which has a lower density). The locations of the minima define the values $\alpha_1 < \alpha_2$. Together with $\widehat{\boldsymbol{\theta}}$ they yield the two hyperplanes in $\mathcal{R}^p$ separating the inliers from the outliers. Note that the equivalent scale of the M-estimator is $s = \alpha_2 - \alpha_1$ and that the minima may not be symmetrically located relative to the mode.

The 2D data in the example in Figure 4.22a contains 100 inliers and 500 outliers. The density of the points projected on the direction of the true normal (Figure 4.22b) has a sharp mode. Since the pbM-estimator deals only with 1D densities, there is no need to use the mean shift procedure

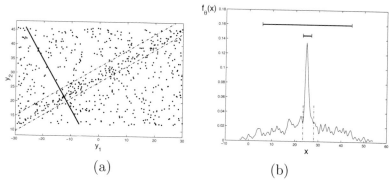

Figure 4.22. Determining the inlier/outlier dichotomy through the density of the projected data. (a) 2D data. Solid line: optimal projection direction. Dashed lines: boundaries of the detected inlier region. (b) The kernel density estimate of the projected points. Vertical dashed lines: the left and right local minima. The bar at the top is the scale $\pm\widehat{s}_{mad}$ computed with $c = 3$. The bar below is $\pm\widehat{h}_{\widehat{\theta}}$, the size of the kernel support. Both are centered on the mode.

(Section 4.3.3) to find the modes, and a simple heuristic suffices to define the local minima if they are not obvious.

The advantage of the pbM-estimator arises from using a more adequate scale in the optimization. In our example, the $\widehat{s}_{mad}$ scale estimate based on the TLS initial fit (to the whole data) and computed with $c = 3$ is about 10 times larger than $\widehat{h}_{\widehat{\theta}}$, the bandwidth computed for the optimal projection direction (Figure 4.22b). When a redescending M-estimator uses $\widehat{s}_{mad}$, the optimization of the objective function is based on a too large band, which almost certainly leads to a nonrobust behavior.

Sometimes the detection of the minima can be fragile. See the right minimum in Figure 4.22b. A slight change in the projected location of a few data points could have changed this boundary to the next, much more significant local minimum. However, this sensitivity is tolerated by the pbM-estimator. First, by the nature of the projection pursuit, many different projections are investigated, and thus it is probable that at least one satisfactory band is found. Second, from any reasonable inlier/outlier dichotomy of the data, postprocessing of the points declared inliers (the region bounded by the two hyperplanes in $\mathcal{R}^p$) can recover the correct estimates. Since the *true* inliers are, with high probability, the absolute majority among the points *declared* inliers, the robust LTS estimator (4.145) can now be used.

The significant improvement in outlier tolerance of the pbM-estimator was obtained at the price of replacing the iterative weighted least squares algorithm of the traditional M-estimation with a search in $\mathcal{R}^p$ for the optimal projection direction $\hat{\boldsymbol{\theta}}$. This search can be efficiently implemented using the simplex-based technique discussed in Section 4.2.7.

A randomly selected p-tuple of points (an elemental subset) defines the projection direction $\boldsymbol{\theta}$, from which the corresponding polar angles $\boldsymbol{\beta}$ are computed (4.52). The vector $\boldsymbol{\beta}$ is the first vertex of the initial simplex in $\mathcal{R}^{p-1}$. The remaining $p-1$ vertices are then defined as

$$\boldsymbol{\beta}_k = \boldsymbol{\beta} + \mathbf{e}_k * \boldsymbol{\gamma} \qquad k = 1, \ldots, (p-1), \tag{4.152}$$

where $\mathbf{e}_k \in R^{p-1}$ is a vector of 0-s except a 1 in the k-th element, and $\boldsymbol{\gamma}$ is a small angle. While the value of $\boldsymbol{\gamma}$ can depend on the dimension of the space, using a constant value such as $\boldsymbol{\gamma} = \pi/12$ seems to suffice in practice. Because $\boldsymbol{\theta}$ is only a projection direction, during the search, the polar angles are allowed to wander outside the limits, assuring a unique mapping in (4.52). The simplex-based maximization of the projection index (4.149) does not have to be extremely accurate, and the number of iterations in the search should be relatively small.

The projection-based implementation of the M-estimators is summarized below.

The pbM-estimator

- Repeat N times:
    1. Choose an elemental subset (p-tuple) by random sampling.

    2. Compute the TLS estimate of $\boldsymbol{\theta}$.

    3. Build the initial simplex in the space of polar angles $\boldsymbol{\beta}$.

    4. Perform a simplex-based direct search to find the local maximum of the projection index.
- Find the left and right local minima around the mode of the density corresponding to the largest projection index.
- Define the inlier/outlier dichotomy of the data. Postprocess the inliers to find the final estimates of α and $\boldsymbol{\theta}$.

4.4.6 Applications

The superior outlier tolerance of the pbM-estimator relative to other robust techniques is illustrated with two experiments. The percentage of inliers in the data is assumed unknown and can be significantly less than that of the outliers. Therefore, the LMedS estimator cannot be applied. It is shown in [106] that MLESAC and MSAC have very similar performance and are superior to RANSAC. We have compared RANSAC and MSAC with the pbM-estimator.

In both experiments, ground truth was available, and the *true* standard deviation of the inliers σ_t could be computed. The output of any robust regression is the inlier/outlier dichotomy of the data. Let the standard deviation of the points *declared* inliers measured relative to the *true fit* be $\widehat{\sigma}_{in}$. The performance of the different estimators was compared through the ratio $\widehat{\sigma}_{in}/\sigma_t$. For a satisfactory result, this ratio should be very close to one.

The same number of computational units is used for all the techniques. A computational unit is either processing of one elemental subset (RANSAC) or one iteration in the simplex-based direct search (pbM). The number of iterations in a search was restricted to 25, but often it ended earlier. Thus, the amount of computation attributed to the pbM-estimator is an upper bound.

In the *first experiment*, the synthetic data contained 100 inlier points obeying an 8D linear EIV regression model (4.106). The measurement noise was normally distributed with covariance matrix $5^2 I_8$. A variable percentage of outliers was uniformly distributed within the bounding box of the region occupied in R^8 by the inliers. The number of computational units was 5000; that is, RANSAC used 5000 elemental subsets while the pbM-estimator initiated 200 local searches. For each experimental condition, 100 trials were run. The *true sample* standard deviation of the inliers σ_t was computed in each trial.

The scale provided to RANSAC was the $\widehat{s}_{mad}$, based on the TLS fit to the data and computed with $c = 3$. The same scale was used for MSAC. However, in an optimal setting, MSAC was also run with the scale $s_{opt} = 1.96\sigma_t$. Note that this information is not available in practice! The graphs in Figure 4.23 show that for any percentage of outliers, the pbM-estimator performs at least as well as MSAC tuned to the optimal scale. This superior performance is obtained in a completely unsupervised fashion. The only parameters used by the pbM-estimator are the generic normalized amplitude values needed for the definition of the local minima. They do not depend on the data or on the application.

In the *second experiment,* two far-apart frames from the *corridor* sequence (Figures 4.24a and 4.24b) were used to estimate the epipolar geometry from point correspondences. As was shown in Section 4.2.5, this is a nonlinear estimation problem, and therefore the role of a robust regression estimator based on the linear EIV model is restricted to selecting the correct matches.

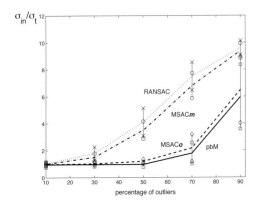

Figure 4.23. RANSAC versus pbM-estimator. The relative standard deviation of the residuals function of the percentage of outliers. Eight-dimensional synthetic data. The employed scale threshold: RANSAC – $\widehat{s}_{mad}$; MSACm – $\widehat{s}_{mad}$; MSACo – s_{opt}. The pbM-estimator has no tuning parameter. The vertical bars mark one standard deviation from the mean.

Subsequent use of a nonlinear (and nonrobust) method can recover the unbiased estimates. Several such methods are discussed in [116].

The Harris corner detector [110, Sec. 4.3] was used to establish the correspondences, from which 265 point pairs were retained. The histogram of the residuals computed as orthogonal distances from the ground-truth plane in 8D is shown in Figure 4.24c. The 105 points in the central peak of the histogram were considered the inliers (Figure 4.24d). Their standard deviation was $\sigma_t = 0.88$.

The number of computational units was 15000; that is, the pbM-estimator used 600 searches. Again, MSAC was tuned to either the optimal scale s_{opt} or to the scale derived from the MAD estimate, $\widehat{s}_{mad}$. The number of true inliers among the points selected by an estimator and the ratio between the standard deviation of the selected points and that of the true inlier noise are shown in the table below.

	selected points/true inliers	$\widehat{\sigma}_{in}/\sigma_t$
MSAC (s_{mad})	219/105	42.32
MSAC (s_{opt})	98/87	1.69
pbM	95/88	1.36

The pbM-estimator successfully recovers the data of interest and behaves like an optimally tuned technique from the RANSAC family. However, in practice, the tuning information is not available.

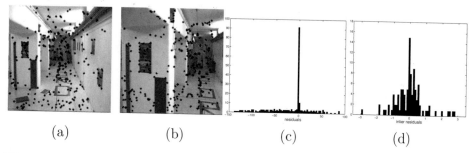

Figure 4.24. Estimating the epipolar geometry for two frames of the *corridor* sequence. (a) and (b) The input images with the points used for correspondences marked. (c) Histogram of the residuals from the ground truth. (d) Histogram of the inliers.

4.4.7 Structured outliers

The problem of multistructured data is not considered in this chapter, but a discussion of robust regression cannot be complete without mentioning the issue of structured outliers. This is a particular case of multistructured data, when only two structures are present and the example shown in Figure 4.3 is a typical case. For such data, once the measurement noise becomes significant, all the robust techniques—M-estimators (including the pbM-estimator), LMedS and RANSAC—behave similarly to the nonrobust least squares estimator. This was first observed for the LMedS [77] and was extensively analyzed in [97]. Here we describe a more recent approach [9].

The true structures in Figure 4.3b are horizontal lines. The lower one contains 60 points and the upper one 40 points. Thus, a robust regression method should return the lower structure as inliers. The measurement noise was normally distributed with covariance $\sigma^2 I_2$, $\sigma = 10$. In Section 4.4.4, it was shown that all robust techniques can be regarded as M-estimators. Therefore, we consider the expression

$$\epsilon(s; \alpha, \beta) = \frac{1}{n} \sum_{i=1}^{n} \rho \left(\frac{y_1 cos\beta + y_2 sin\beta - \alpha}{s} \right), \qquad (4.153)$$

which defines a family of curves parameterized in α and β of the line model (4.54) and in the scale s. The envelope of this family,

$$\epsilon_{min}(s) = \min_{\alpha, \beta} \epsilon(s; \alpha, \beta), \qquad (4.154)$$

represents the value of the M-estimation minimization criterion (4.130) as a function of scale.

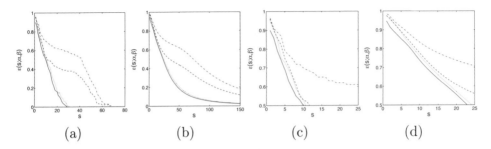

(a)   (b)   (c)   (d)

Figure 4.25. Dependence of $\epsilon(s; \alpha, \beta)$ on the scale s for the data in Figure 4.3b. (a) Zero-one loss function. (b) Biweight loss function. (c) The top left region of (a). (d) The top left region of (b). Solid line: envelope $\epsilon_{min}(s)$. Dashed line: true parameters of the lower structure. Dotdashed line: true parameters of the upper structure. Dotted line: least squares fit parameters.

By definition, for a given value of s, the curve $\epsilon(s; \alpha, \beta)$ can be only above (or touching) the envelope. The comparison of the envelope with a curve $\epsilon(s; \alpha, \beta)$ describes the relation between the employed α, β and the parameter values minimizing (4.153). Three sets of line parameters were investigated using the zero-one (Figure 4.25a) and the biweight (Figure 4.25b) loss functions: the true parameters of the two structures ($\alpha = 50, 100; \beta = \pi/2$) and the least squares parameter estimates (α_{LS}, β_{LS}). The LS parameters yield a line similar to the one in Figure 4.3b, a nonrobust result.

Consider the case of zero-one loss function and the parameters of the lower structure (dashed line in Figure 4.25a). For this loss function, $\epsilon(s; 50, \pi/2)$ is the percentage of data points outside the horizontal band centered on $y_2 = 50$ and with half-width s. As expected, the curve has a plateau around $\epsilon = 0.4$ corresponding to the band having one of its boundaries in the transition region between the two structures. Once the band extends into the second structure, $\epsilon(s; 50, \pi/2)$ further decreases. The curve, however, is not only always above the envelope, but most often also above the curve $\epsilon(s; \alpha_{LS}, \beta_{LS})$. See the magnified area of small scales in Figure 4.25c.

For a given value of the scale (as in RANSAC), a fit similar to least squares will be preferred, since it yields a smaller value for (4.153). The measurement noise being large, a band containing half the data (as in LMedS) corresponds to a scale $s > 12$, the value around which the least squares fit begins to dominate the optimization (Figure 4.25a). As a result, the LMedS will always fail (Figure 4.3b). Note also the very narrow range of scale values (around $s = 10$) for which $\epsilon(s; 50, \pi/2)$ is below $\epsilon(s; \alpha_{LS}, \beta_{LS})$. It shows how accurately the user must tune an estimator in the RANSAC family for a satisfactory performance.

The behavior for the biweight loss function is identical, only the curves are smoother due to the weighted averages (Figures 4.25b and 4.25d). When the noise corrupting the structures is small—in Figure 4.3a it is $\sigma = 2$—the envelope and the curve $\epsilon(s; 50, \pi/2)$ overlap for $s < 8$, which suffices for the LMedS criterion. See [9] for details.

We can conclude that multistructured data has to be processed first by breaking it into parts in which one structure dominates. The technique in [8] combines several of the procedures discussed in this chapter. The sampling was guided by local data density; it was assumed that the structures and the background can be roughly separated by a global threshold on nearest neighbor distances. The pbM-estimator was employed as the estimation module, and the final parameters were obtained by applying adaptive mean shift to a feature space. The technique had a Hough transform flavor, though no scale parameters were required. The density assumption, however, may fail when the structures are defined by linearizing a nonlinear problem, as is often the case in 3D vision. Handling such multistructured data embedded in a significant background clutter remains an open question.

4.5   Conclusion

Our goal in this chapter was to approach robust estimation from the point of view of a practitioner. We used a common statistical framework with solid theoretical foundations to discuss the different types and classes of robust estimators. Therefore, we did not dwell on techniques that have an excellent robust behavior but are of a somewhat ad hoc nature. These techniques, such as tensor voting (see Chapter 5), can provide valuable tools for solving difficult computer vision problems.

Another disregarded topic was the issue of diagnosis. Should an algorithm be able to determine its own failure, one can already talk about robust behavior. When in the late 1980s robust methods became popular in the vision community, the paper [28] was often considered the first robust work in the vision literature. The special issue [93] and the book [6] contain representative collections of papers for the state of the art today.

We emphasized the importance of embedding into the employed model the least possible amount of assumptions necessary for the task at hand. In this way, the developed algorithms are more suitable for vision applications, where the data is often more complex than in the statistical literature.

However, there is a tradeoff to satisfy. As the model becomes less committed (more nonparametric), its power to extrapolate from the available data also decreases. How much is modeled rigorously and how much is purely data driven is an important decision for the designer of an algorithm. The material presented in this chapter is intended to help in making this decision.

Acknowledgments

I must thank several of my current and former graduate students whose work is directly or indirectly present on every page: Haifeng Chen, Dorin Comaniciu, Bogdan Georgescu, Yoram Leedan, and Bogdan Matei. Long discussions with Dave Tyler from the Statistics Department, Rutgers University, helped to crystallize many of the ideas described in this paper. Should they be mistaken, the blame is entirely mine. Preparation of the material was supported by the National Science Foundation under the grant IRI 99-87695.

Bibliography

[1] J. Addison. Pleasures of imagination. *Spectator*, 6(411), June 21, 1712.

[2] G. Antille and H. El May. The use of slices in the LMS and the method of density slices: Foundation and comparison. In Y. Dodge and J. Whittaker (Eds.), *Proc. 10th Symp. Computat. Statist., Neuchatel*, volume I, pages 441–445. Physica-Verlag, 1992.

[3] T. Arbel and F. P. Ferrie. On sequential accumulation of evidence. *Int'l J. of Computer Vision*, 43:205–230, 2001.

[4] P. J. Besl, J. B. Birch, and L. T. Watson. Robust window operators. In *Proc. Int'l Conf. on Computer Vision*, pages 591–600, December 1988.

[5] M. J. Black and A. Rangarajan. On the unification of line processes, outlier rejection, and robust statistics with applications in early vision. *Int'l J. of Computer Vision*, 19:57–91, 1996.

[6] K. J. Bowyer and P. J. Phillips (Eds.). *Empirical evaluation techniques in computer vision.* IEEE Computer Society, 1998.

[7] K. L. Boyer, M. J. Mirza, and G. Ganguly. The robust sequential estimator: A general approach and its application to surface organization in range data. *IEEE Trans. Pattern Anal. Machine Intell.*, 16:987–1001, 1994.

[8] H. Chen and P. Meer. Robust computer vision through kernel density estimation. In *Proc. European Conf. on Computer Vision*, volume I, pages 236–250, May 2002.

[9] H. Chen, P. Meer, and D. E. Tyler. Robust regression for data with multiple structures. In *Proc. IEEE Conf. on Computer Vision and Pattern Recognition*, volume I, pages 1069–1075, December 2001.

[10] Y. Cheng. Mean shift, mode seeking, and clustering. *IEEE Trans. Pattern Anal. Machine Intell.*, 17:790–799, 1995.

[11] E. Choi and P. Hall. Data sharpening as a prelude to density estimation. *Biometrika*, 86:941–947, 1999.

[12] W. Chojnacki, M. J. Brooks, A. van den Hengel, and D. Gawley. On the fitting of surfaces to data with covariances. *IEEE Trans. Pattern Anal. Machine Intell.*, 22:1294–1303, 2000.

[13] C. M. Christoudias, B. Georgescu, and P. Meer. Synergism in low-level vision. In *Proc. 16th International Conference on Pattern Recognition*, volume IV, pages 150–155, August 2002.

[14] R. T. Collins. Mean-shift blob tracking through scale space. In *Proc. IEEE Conf. on Computer Vision and Pattern Recognition*, volume II, pages 234–240, 2003.

[15] D. Comaniciu. An algorithm for data-driven bandwidth selection. *IEEE Trans. Pattern Anal. Machine Intell.*, 25:281–288, 2003.

[16] D. Comaniciu and P. Meer. Distribution free decomposition of multivariate data. *Pattern Analysis and Applications*, 2:22–30, 1999.

[17] D. Comaniciu and P. Meer. Mean shift: A robust approach toward feature space analysis. *IEEE Trans. Pattern Anal. Machine Intell.*, 24:603–619, 2002.

[18] D. Comaniciu, V. Ramesh, and P. Meer. The variable bandwidth mean shift and data-driven scale selection. In *Proc. Int'l Conf. on Computer Vision*, volume I, pages 438–445, July 2001.

[19] D. Comaniciu, V. Ramesh, and P. Meer. Kernel-based object tracking. *IEEE Trans. Pattern Anal. Machine Intell.*, 25:564–577, 2003.

[20] D. Donoho, I. Johnstone, P. Rousseeuw, and W. Stahel. Discussion: Projection pursuit. *Annals of Statistics*, 13:496–500, 1985.

[21] R. O. Duda, P. E. Hart, and D. G. Stork. *Pattern Classification*, 2nd ed. Wiley, 2001.

[22] B. Efron and R. Tibshirani. *An Introduction to the Bootstrap*. Chapman & Hall, 1993.

[23] S. P. Ellis. Instability of least squares, least absolute deviation and least median of squares linear regression. *Statistical Science*, 13:337–350, 1998.

[24] O. Faugeras. *Three-Dimensional Computer Vision*. MIT Press, 1993.

[25] M. A. Fischler and R. C. Bolles. Random sample consensus: A paradigm for model fitting with applications to image analysis and automated cartography. In *DARPA Image Understanding Workshop*, pages 71–88, University of Maryland, College Park, April 1980.

[26] M. A. Fischler and R. C. Bolles. Random sample consensus: A paradigm for model fitting with applications to image analysis and automated cartography. *Comm. Assoc. Comp. Mach*, 24(6):381–395, 1981.

[27] A. W. Fitzgibbon, M. Pilu, and R. B. Fisher. Direct least square fitting of ellipses. *IEEE Trans. Pattern Anal. Machine Intell.*, 21:476–480, 1999.

[28] W. Förstner. Reliability analysis of parameter estimation in linear models with applications to mensuration problems in computer vision. *Computer Vision, Graphics, and Image Processing*, 40:273–310, 1987.

[29] W. T. Freeman, E. G. Pasztor, and O. W. Carmichael. Learning in low-level vision. *Int'l J. of Computer Vision*, 40:25–47, 2000.

[30] J. H. Friedman and J. W. Tukey. A projection pursuit algorithm for exploratory data analysis. *IEEE Trans. Comput.*, 23:881–889, 1974.

[31] K. Fukunaga. *Introduction to Statistical Pattern Recognition*, 2nd ed. Academic Press, 1990.

[32] K. Fukunaga and L. D. Hostetler. The estimation of the gradient of a density function, with applications in pattern recognition. *IEEE Trans. Information Theory*, 21:32–40, 1975.

[33] W. Fuller. *Measurement Error Models*. Wiley, 1987.

[34] B. Georgescu and P. Meer. Balanced recovery of 3D structure and camera motion from uncalibrated image sequences. In *Proc. European Conf. on Computer Vision*, volume II, pages 294–308, 2002.

[35] B. Georgescu, I. Shimshoni, and P. Meer. Mean shift based clustering in high dimensions: A texture classification example. In *Proc. 9th Int'l Conf. on Computer Vision*, pages 456–463, October 2003.

[36] E. B. Goldstein. *Sensation and Perception*, 2nd ed. Wadsworth, 1987.

[37] G. H. Golub and C. Reinsch. Singular value decomposition and least squares solutions. *Number. Math.*, 14:403–420, 1970.

[38] G. H. Golub and C. F. Van Loan. *Matrix Computations*, 2nd ed. Johns Hopkins U. Press, 1989.

[39] P. Hall, T. C. Hui, and J. S. Marron. Improved variable window kernel estimates of probability densities. *Annals of Statistics*, 23:1–10, 1995.

[40] R. Hampel, E. M. Ronchetti, P. J. Rousseeuw, and W. A. Stahel. *Robust Statistics. The Approach Based on Influence Function*. Wiley, 1986.

[41] R. M. Haralick and H. Joo. 2D-3D pose estimation. In *Proceedings of the 9th International Conference on Pattern Recognition*, pages 385–391, November 1988.

[42] R. M. Haralick and L. G. Shapiro. *Computer and Robot Vision*. Addison-Wesley, 1992.

[43] R. Hartley and A. Zisserman. *Multiple View Geometry in Computer Vision*. Cambridge University Press, 2000.

[44] T. Hastie, R. Tibshirani, and J. Friedman. *The Elements of Statistical Learning*. Springer, 2001.

[45] T. P. Hettmansperger and S. J. Sheather. A cautionary note on the method of least median of squares. *The American Statistician*, 46:79–83, 1992.

[46] P. V. C. Hough. Machine analysis of bubble chamber pictures. In *International Conference on High Energy Accelerators and Instrumentation*, Centre Européenne pour la Recherch Nucléaire (CERN), 1959.

[47] P. V. C. Hough. Method and means for recognizing complex patterns. US Patent 3,069,654, December 18, 1962.

[48] P. J. Huber. Projection pursuit (with discussion). *Annals of Statistics*, 13:435–525, 1985.

[49] P. J. Huber. *Robust Statistical Procedures*, 2nd ed. SIAM, 1996.

[50] J. Illingworth and J. V. Kittler. A survey of the Hough transform. *Computer Vision, Graphics, and Image Processing*, 44:87–116, 1988.

[51] M. Isard and A. Blake. Condensation—Conditional density propagation for visual tracking. *Int'l J. of Computer Vision*, 29:5–28, 1998.

[52] A. K. Jain and R. C. Dubes. *Algorithms for Clustering Data.* Prentice Hall, 1988.

[53] J. M. Jolion, P. Meer, and S. Bataouche. Robust clustering with applications in computer vision. *IEEE Trans. Pattern Anal. Machine Intell.*, 13:791–802, 1991.

[54] M. C. Jones and R. Sibson. What is projection pursuit? (with discussion). *J. Royal Stat. Soc. A*, 150:1–37, 1987.

[55] B. Julesz. Early vision and focal attention. *Rev. of Modern Physics*, 63:735–772, 1991.

[56] H. Kälviäinen, P. Hirvonen, L. Xu, and E. Oja. Probabilistic and nonprobabilistic Hough transforms: Overview and comparisons. *Image and Vision Computing*, 13:239–252, 1995.

[57] K. Kanatani. Statistical bias of conic fitting and renormalization. *IEEE Trans. Pattern Anal. Machine Intell.*, 16:320–326, 1994.

[58] K. Kanatani. *Statistical Optimization for Geometric Computation: Theory and Practice.* Elsevier, 1996.

[59] D. Y. Kim, J. J. Kim, P. Meer, D. Mintz, and A. Rosenfeld. Robust computer vision: The least median of squares approach. In *Proceedings 1989 DARPA Image Understanding Workshop*, pages 1117–1134, May 1989.

[60] N. Kiryati and A. M. Bruckstein. What's in a set of points? *IEEE Trans. Pattern Anal. Machine Intell.*, 14:496–500, 1992.

[61] N. Kiryati and A. M. Bruckstein. Heteroscedastic Hough transform (HtHT): An efficient method for robust line fitting in the "errors in the variables" problem. *Computer Vision and Image Understanding*, 78:69–83, 2000.

[62] V. F. Leavers. Survey: Which Hough transform? *Computer Vision, Graphics, and Image Processing*, 58:250–264, 1993.

[63] K. M. Lee, P. Meer, and R. H. Park. Robust adaptive segmentation of range images. *IEEE Trans. Pattern Anal. Machine Intell.*, 20:200–205, 1998.

[64] Y. Leedan and P. Meer. Heteroscedastic regression in computer vision: Problems with bilinear constraint. *Int'l J. of Computer Vision*, 37:127–150, 2000.

[65] A. Leonardis and H. Bischof. Robust recognition using eigenimages. *Computer Vision and Image Understanding*, 78:99–118, 2000.

[66] R. M. Lewis, V. Torczon, and M. W. Trosset. Direct search methods: Then and now. *J. Computational and Applied Math.*, 124:191–207, 2000.

[67] G. Li. Robust regression. In D. C. Hoaglin, F. Mosteller, and J. W. Tukey (Eds.), *Exploring Data Tables, Trends, and Shapes*, pages 281–343. Wiley, 1985.

[68] R. A. Maronna and V. J. Yohai. The breakdown point of simultaneous general M estimates of regression and scale. *J. of Amer. Stat. Assoc.*, 86:699–703, 1991.

[69] R. D. Martin, V. J. Yohai, and R. H. Zamar. Min-max bias robust regression. *Annals of Statistics*, 17:1608–1630, 1989.

[70] B. Matei. *Heteroscedastic Errors-in-Variables Models in Computer Vision.* PhD thesis, Department of Electrical and Computer Engineering, Rutgers University, 2001. Available at `http://www.caip.rutgers.edu/riul/research/theses.html`.

[71] B. Matei and P. Meer. Bootstrapping errors-in-variables models. In B. Triggs, A. Zisserman, and R. Szeliski (Eds.), *Vision Algorithms: Theory and Practice*, pages 236–252. Springer, 2000.

[72] B. Matei and P. Meer. Reduction of bias in maximum likelihood ellipse fitting. In *Proc. Int'l Conf. on Pattern Recognition*, volume III, pages 802–806, September 2000.

[73] P. Meer and B. Georgescu. Edge detection with embedded confidence. *IEEE Trans. Pattern Anal. Machine Intell.*, 23:1351–1365, 2001.

[74] P. Meer, D. Mintz, D. Y. Kim, and A. Rosenfeld. Robust regression methods in computer vision: A review. *Int'l J. of Computer Vision*, 6:59–70, 1991.

[75] J. M. Mendel. *Lessons in Estimation Theory for Signal Processing, Communications, and Control*. Prentice Hall, 1995.

[76] J. V. Miller and C. V. Stewart. MUSE: Robust surface fitting using unbiased scale estimates. In *Proc. IEEE Conf. on Computer Vision and Pattern Recognition*, pages 300–306, June 1996.

[77] D. Mintz, P. Meer, and A. Rosenfeld. Consensus by decomposition: A paradigm for fast high breakdown point robust estimation. In *Proceedings 1991 DARPA Image Understanding Workshop*, pages 345–362, January 1992.

[78] J. A. Nelder and R. Mead. A simplex method for function minimization. *Computer Journal*, 7:308–313, 1965.

[79] P. L. Palmer, J. Kittler, and M. Petrou. An optimizing line finder using a Hough transform algorithm. *Computer Vision and Image Understanding*, 67:1–23, 1997.

[80] D. Petkovic, W. Niblack, and M. Flickner. Projection-based high accuracy measurement of straight line edges. *Machine Vision and Appl.*, 1:183–199, 1988.

[81] P. D. Picton. Hough transform references. *Internat. J. of Patt. Rec and Artific. Intell.*, 1:413–425, 1987.

[82] W. H. Press, S. A. Teukolsky, W. T. Vetterling, and B. P. Flannery. *Numerical Recipes in C*, 2nd ed. Cambridge University Press, 1992.

[83] P. Pritchett and A. Zisserman. Wide baseline stereo matching. In *Proc. Int'l Conf. on Computer Vision*, pages 754–760, January 1998.

[84] Z. Pylyshyn. Is vision continuous with cognition? The case for cognitive impenetrability of visual perception. *Behavioral and Brain Sciences*, 22:341–423, 1999. (with comments).

[85] S. J. Raudys and A. K. Jain. Small sample size effects in statistical pattern recognition: Recommendations for practitioners. *IEEE Trans. Pattern Anal. Machine Intell.*, 13:252–264, 1991.

[86] P. J. Rousseeuw. Least median of squares regression. *J. of Amer. Stat. Assoc.*, 79:871–880, 1984.

[87] P. J. Rousseeuw. Unconventional features of positive-breakdown estimators. *Statistics & Prob. Letters*, 19:417–431, 1994.

[88] P. J. Rousseeuw and C. Croux. Alternatives to the median absolute deviation. *J. of Amer. Stat. Assoc.*, 88:1273–1283, 1993.

[89] P. J. Rousseeuw and A. M. Leroy. *Robust Regression and Outlier Detection.* Wiley, 1987.

[90] D. Ruppert and D. G. Simpson. Comment on "Unmasking Multivariate Outliers and Leverage Points," by P. J. Rousseeuw and B. C. van Zomeren. *J. of Amer. Stat. Assoc.*, 85:644–646, 1990.

[91] D. W. Scott. Parametric statistical modeling by minimum integrated square error. *Technometrics*, 43:247–285, 2001.

[92] B. W. Silverman. *Density Estimation for Statistics and Data Analysis.* Chapman & Hall, 1986.

[93] Special Issue. Performance evaluation. *Machine Vision and Appl.*, 9(5/6), 1997.

[94] Special Issue. Robust statistical techniques in image understanding. *Computer Vision and Image Understanding*, 78, April 2000.

[95] G. Speyer and M. Werman. Parameter estimates for a pencil of lines: Bounds and estimators. In *Proc. European Conf. on Computer Vision*, volume I, pages 432–446, 2002.

[96] L. Stark and K. W. Bowyer. Achieving generalized object recognition through reasoning about association of function to structure. *IEEE Trans. Pattern Anal. Machine Intell.*, 13:1097–1104, 1991.

[97] C. V. Stewart. Bias in robust estimation caused by discontinuities and multiple structures. *IEEE Trans. Pattern Anal. Machine Intell.*, 19:818–833, 1997.

[98] C. V. Stewart. Robust parameter estimation in computer vision. *SIAM Reviews*, 41:513–537, 1999.

[99] C. V. Stewart. Minpran: A new robust estimator for computer vision. *IEEE Trans. Pattern Anal. Machine Intell.*, 17:925–938, 1995.

[100] K. S. Tatsuoka and D. E. Tyler. On the uniqueness of S and constrained M-functionals under non-elliptical distributions. *Annals of Statistics*, 28:1219–1243, 2000.

[101] P. R. Thrift and S. M. Dunn. Approximating point-set images by line segments using a variation of the Hough transform. *Computer Vision, Graphics, and Image Processing*, 21:383–394, 1983.

[102] A. Tirumalai and B. G. Schunk. Robust surface approximation using least median of squares. Technical Report CSE-TR-13-89, Artificial Intelligence Laboratory, 1988. University of Michigan, Ann Arbor.

[103] B. Tordoff and D. W. Murray. Guided sampling and consensus for motion estimation. In *Proc. European Conf. on Computer Vision*, volume I, pages 82–96, May 2002.

[104] P. H. S. Torr and C. Davidson. IMPSAC: Synthesis of importance sampling and random sample consensus. *IEEE Trans. Pattern Anal. Machine Intell.*, 25:354–364, 2003.

[105] P. H. S. Torr and D. W. Murray. The development and comparison of robust methods for estimating the fundamental matrix. *Int'l J. of Computer Vision*, 24:271–300, 1997.

[106] P. H. S. Torr and A. Zisserman. MLESAC: A new robust estimator with application to estimating image geometry. *Computer Vision and Image Understanding*, 78:138–156, 2000.

[107] P. H. S. Torr, A. Zisserman, and S. J. Maybank. Robust detection of degenerate configurations while estimating the fundamental matrix. *Computer Vision and Image Understanding*, 71:312–333, 1998.

[108] P. H. S. Torr and A. Zisserman. Robust computation and parametrization of multiple view relations. In *Proc. Int'l Conf. on Computer Vision*, pages 727–732, January 1998.

[109] A. Treisman. Features and objects in visual processing. *Scientific American*, 254:114–125, 1986.

[110] E. Trucco and A. Verri. *Introductory Techniques for 3-D Computer Vision*. Prentice Hall, 1998.

[111] S. Van Huffel and J. Vandewalle. *The Total Least Squares Problem. Computational Aspects and Analysis*. Society for Industrial and Applied Mathematics, 1991.

[112] M. P. Wand and M. C. Jones. *Kernel Smoothing*. Chapman & Hall, 1995.

[113] I. Weiss. Line fitting in a noisy image. *IEEE Trans. Pattern Anal. Machine Intell.*, 11:325–329, 1989.

[114] M. H. Wright. Direct search methods: Once scorned, now respectable. In D. H. Griffiths and G. A. Watson (Eds.), *Numerical Analysis 1995*, pages 191–208. Addison-Wesley Longman, 1996.

[115] R. H. Zamar. Robust estimation in the errors-in-variables model. *Biometrika*, 76:149–160, 1989.

[116] Z. Zhang. Determining the epipolar geometry and its uncertainty: A review. *Int'l J. of Computer Vision*, 27:161–195, 1998.

[117] Z. Zhang. Parameter-estimation techniques: A tutorial with application to conic fitting. *Image and Vision Computing*, 15:59–76, 1997.

Chapter 5

THE TENSOR VOTING FRAMEWORK

Gérard Medioni

and Philippos Mordohai

5.1 Introduction

The design and implementation of a complete artificial vision system is a daunting challenge. The computer vision community has made significant progress in many areas, but the ultimate goal is still far off. A key component of a general computer vision system is a computational framework that can address a wide range of problems in a unified way. We have developed such a framework over the past several years [40], which is the basis of the augmented framework presented in this chapter. It is based on a data representation formalism that uses *tensors* and an information propagation mechanism termed *tensor voting*.

Most computer vision problems are inverse problems, because the imaging process maps 3D "world" features onto 2D arrays of pixels. The loss of information due to the reduction in dimensionality and the nonlinearity of the imaging process, combined with unavoidable sources of noise, such as limited sensor resolution, quantization of continuous information into pixels, and photometric and projective distortion, guarantee that computer vision algorithms operate on data sets that include numerous outliers and are severely corrupted by noise. Many scene configurations could have generated a given image (see Figure 5.1). An automated computer vision system must impose constraints or prior knowledge about the appearance of familiar objects to select the most likely scene interpretation. In this chapter, we present a methodology for the perceptual organization of *tokens*. The tokens represent the position of elements such as points and curvels, and can also convey other information, such as curve or surface orientation. The organization is achieved by enforcing constraints, as suggested by Gestalt psychology, within the proposed computational framework. All processing is strictly bottom-up without any feedback loops or top-down stages.

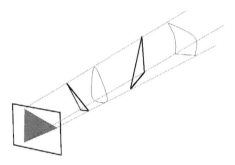

Figure 5.1. An infinite set of scene configurations can produce the same image. Most people, however, would perceive a flat triangle in the above example.

5.1.1 Motivation

The motivation for designing and implementing the *tensor voting framework* came from the observation that even though many computer vision problems are similar, different algorithms were applied to each of them. In most cases, we start from one or more images and attempt to infer descriptions of the depicted scenes in terms of appearance, shape, motion, and so on. The type of primitives, the constraints, and the desired solution may differ; therefore, a general framework must be flexible enough to adjust to different settings and incorporate different constraints. The primitives used in computer vision problems span a wide range of features, including differential image intensity properties, such as edges and corners; elementary structures, such as elementary curves and surface patches; motion vectors; reflectance properties; and texture descriptors. The most generally applicable constraint is the "matter is cohesive" constraint proposed by Marr [39], which states that objects in real scenes are continuous almost everywhere. Other powerful constraints exist, but are usually applicable to specific problems where certain assumptions hold. The desired type of scene descriptions is a large issue, some aspects of which are addressed in the next subsection.

Despite the apparent differences among computer vision problems, the majority of them can be posed as *perceptual organization* problems. In the early 1900s, the Gestalt psychologists [70] formulated a set of principles which guide perceptual organization in the human visual system. These principles include similarity, proximity, good continuation, simplicity, closure, colinearity, and co-curvilinearity, which cooperate, or sometimes compete, to produce salient structures formed by the observed *tokens* (Figure 5.2). We claim that computer vision problems can be addressed within a Gestalt framework, where the primitives are grouped according to the above criteria to give rise to salient structures. For instance, descriptions in terms of shape can be generated by grouping tokens according to similarity, proximity, and good continuation. In other applications, salient groups might appear due to similarity in other properties, such as motion, texture, or surface normal.

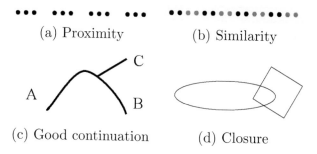

(a) Proximity   (b) Similarity

(c) Good continuation   (d) Closure

Figure 5.2. Simple examples of the Gestalt principles. In (a) the dots are grouped in four groups according to similarity. In (b) the darker dots are grouped in pairs, as are the lighter ones. In (c) the most likely grouping is A to B, and not A to C, due to the smooth continuation of curve tangent from A to B. In (d), the factor of closure generates the perception of an ellipse and a diamond.

The unifying theme in computer vision problems, from a perceptual organization point of view, is the search for salient structures that arise due to nonaccidental alignment of the input tokens and therefore must bear semantic importance. The term *perceptual saliency* indicates the quality of features to be important, stand out conspicuously, be prominent, and attract our attention. In the remainder of this chapter, we demonstrate how a number of computer vision problems can be formulated as perceptual organization problems and how the tensor voting framework provides the machinery to address these problems in a unified way with minor problem-specific adaptations.

5.1.2   Desirable descriptions

An alternative approach to ours, and one that has been widely used, is the formulation of computer vision problems within an optimization framework. Objective functions can be set up according to the requirements and constraints of the problem at hand by imposing penalties to primitives that deviate from the desired models. Local penalties are aggregated into an energy or cost function, which is then optimized using an appropriate method. Due to the inverse nature of computer vision, optimization methods usually need simplifying assumptions to reduce the complexity of the objective functions, as well as careful initialization to avoid convergence at local optima.

The largest difference, however, between optimization methods and the one proposed here is that they arrive at *global* solutions, while we claim that *local* descriptions are more appropriate. Under a global, model-based approach, we cannot discriminate between local model misfits and noise. Furthermore, global descriptions are restricted to features that can be expressed in a parametric form. A local description is more general in the sense that it can encode any smooth feature, with a finite number of discontinuities, as a set of tokens. A hierarchy of more abstract descriptions can be derived from the low-level, local one if that is desired, while the opposite is not always feasible.

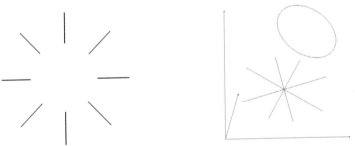

(a) Illusory contour formed by endpoints.   (b) Layered interpretation

Figure 5.3. Illusory contour and its interpretation in terms of layers.

A second requirement on the descriptions we wish to infer is that they should be in terms of layers. Psychological evidence suggests that human perception also supports a layered description of the world. For instance, the illusory contour that clearly appears in Figure 5.3 can be explained by a scene interpretation that consists of a circle on top of a set of lines. Even though we do not aim to emulate the human visual system, this evidence suggests that a layered representation is beneficial for a general computer vision system.

Finally, we choose an object-based representation over a viewpoint-based one. With this choice, we avoid introducing unnecessary viewpoint-dependent elements in intermediate stages of the computation. For instance, if uniqueness along the lines of sight is a reasonable or desired assumption for a given problem, it can be enforced at later stages, thus avoiding the enforcement of constraints too early, when they might be detrimental to the solution.

5.1.3 Our approach

In this section, we briefly review the tensor voting framework, which was originally presented in [40], including extensions that have been developed in the past few years. The two main aspects of the framework are the representation by *tensors* and the information propagation mechanism by *tensor voting*. Its purpose is to serve as a computational mechanism for perceptual grouping of oriented and unoriented tokens generated based on image or other primitives. It has mainly been applied to midlevel vision problems, but it is suitable for any problem, of any dimensionality, that can be formulated as a perceptual organization problem. The novelty of our approach is that there is no objective function that is explicitly defined and optimized according to global criteria. Instead, tensor voting is performed locally, and the saliency of perceptual structures is estimated as a function of the support tokens receive from their neighbors. Tokens with compatible orientations that can form salient structures reinforce each other. The support of a token for its neighbors is expressed by *votes* that are cast according to the Gestalt principles of proximity, colinearity and co-curvilinearity.

Data representation   The representation of a token originally consisted of a symmetric second-order tensor that encodes *perceptual saliency*. The tensor essentially indicates the saliency of each type of perceptual structure the token belongs to and its preferred normal or tangent orientations. Tensors were first used as a signal processing tool for computer vision applications by Granlund and Knutsson [12] and Westin [71]. Our use of tensors differs in that our representation is not signal based, but rather symbolic, where hypotheses for the presence of a perceptual structure at a given location are represented as tokens with associated second-order tensors that encode the most likely type of structure and its preferred tangent or normal orientations. The power of this representation lies in that all types of saliency are encoded by the same tensor.

The second-order symmetric tensors fulfill the requirements set in the previous paragraphs, since they are well suited as descriptors of local structure. As opposed to scalar saliency values, the representation by tensors is richer in information, since it also encodes orientation hypotheses at each location, thus allowing the application of Gestalt principles such as good continuation, colinearity, and co-curvilinearity. Even though this is also feasible with vectors, as in the work of Guy and Medioni [15, 16], the tensors can simultaneously encode all potential types of structures, such as surfaces, curves and regions, as well as the uncertainty of the information.

A representation scheme sufficient for our purposes must be able to encode both smooth perceptual structures as well as discontinuities, which come in two types: orientation and structural discontinuities. The former occur at locations where a perceptual structure is continuous, but its orientation is not, or where multiple salient structures, such as curves or surfaces, meet. Curve orientation discontinuities occur at locations where multiple curve segments intersect, while surface orientation discontinuities occur where multiple surface patches intersect. In other words, whereas there is only one orientation associated with a location within a smooth curve segment or a surface patch or a region boundary, there are multiple orientations associated with locations where a discontinuity occurs. Hence, the desirable data representation is the one that can encode more than one orientation at a given location. It turns out that a second-order symmetric tensor possesses precisely this property, as shown in Sections 5.3 and 5.4.

The second type of discontinuities are structural discontinuities. These occur at locations such as the endpoints B and C of Figure 5.8(a). They are first-order discontinuities, since the perceptual structure is no longer continuous there. Nonaccidental terminations of perceptual structures carry significant weight and should be explicitly detected and represented. The second-order symmetric tensors fail to describe structural discontinuities because they cannot capture first-order properties, and the second-order properties of the structure remain invariant at its boundaries. In order to address this shortcoming of the framework, as it was published in [40], vectors (first-order tensors) were introduced [69]. More specifically, besides the second-order tensor, each token is associated with a *polarity vector* that encodes the likelihood of the token being a termination of a perceptual structure. Polarity vectors are sensitive to first-order properties such as the distribution of neighboring tokens around a given token. Structure terminations can be detected based on their essential property to have all their neighbors, at least locally, on the same side of a half-space.

Tensor Voting   The way to encode primitives in the representation scheme is demonstrated in the following sections. The starting point in our attempt to infer salient structures and scene descriptions is a set of unoriented or oriented tokens. Unoriented tokens express the hypothesis that a perceptual structure of a yet unknown type exists at the token's location. Oriented tokens can be elementary curves (*curvels*), elementary surfaces (*surfels*), and so on. They express the hypothesis that a perceptual structure with the given orientation goes through the location of the token. The question is, how should these hypotheses be combined in order to derive the saliency of each token?

We propose to do this by *tensor voting*, a method of information propagation where tokens convey their orientation preferences to their neighbors in the form of *votes*. First- and second-order votes, which are respectively vectors and second-order tensors, are cast from token to token. The second-order vote is a second-order tensor that has the orientation in terms of normals and tangents the receiver would have if the voter and receiver were part of the same smooth perceptual structure. The first-order vote is a vector that points toward the voter, and thus the interior of the structure, if the receiver were indeed a boundary.

Each vote is an estimate of orientation or termination of a perceptual structure consisting of just two tokens: the voter and the receiver. In essence, simple, smooth, perceptual structures are fitted between the two locations to generate the orientation estimates at the receiver. According to the Gestalt principle of proximity, the strength of the votes attenuates with distance, making the influence from distant tokens, and possible interference from unrelated ones, smaller. The strength of the vote also decreases with increased curvature of the hypothesized structure, making straight continuations preferable to curved ones following the principles of smooth continuation and simplicity.

A large number of first- and second-order votes are accumulated at each location. By analyzing the consistency of the orientation estimations and the amount of support a token receives, we can determine the type of structure that exists at the location and its saliency. The aggregation of support via tensor voting is a generalization to the Hough transform [23] that was first proposed by Guy and Medioni [15, 16] using a vector-based scheme.

5.1.4   Chapter overview

This chapter is organized as follows:

  - Section 5.2 is a review of related work in perceptual organization.

  - Section 5.3 presents the tensor voting framework in 2D.

  - Section 5.4 shows how it can be generalized to 3D.

  - Section 5.5 presents the ND version of the framework and a computational complexity analysis.

  - Section 5.6 addresses how computer vision problems can be posed within the framework and presents our results on stereo and motion analysis.

  - Section 5.7 concludes the chapter after pointing out some issues that still remain open.

5.2   Related Work

Perceptual organization has been an active research area. Important issues include noise robustness, initialization requirements, handling of discontinuities, flexibility in the types that can be represented, and computational complexity. This section reviews related work, which can be classified in the following categories:

  - regularization

  - relaxation labeling

  - computational geometry

  - robust methods

  - level set methods

  - symbolic methods

  - clustering

  - methods based on local interactions

  - methods inspired by psychophysiology and neuroscience

Regularization   Due to the projective nature of imaging, a single image can correspond to different scene configurations. This ambiguity in image formation makes the inverse problem, the inference of structures from images, ill-posed. To address ill-posed problems, constraints have to be imposed on the solution space. Within the regularization theory, this is achieved by selecting an appropriate objective function and optimizing it according to the constraints. Poggio, Torre, and Koch [51] present the application of regularization theory to computer vision problems. Terzopoulos [65] and Robert and Deriche [52] address the issue of preserving discontinuities while enforcing global smoothness in a regularization framework. A Bayesian formulation of the problem based on minimum description length is proposed by Leclerc [32]. Variational techniques are used by Horn and Schunk [22] for the estimation of optical flow, and by Morel and Solimini [43] for image segmentation. In both cases, the goal is to infer functions that optimize the selected criteria while preserving discontinuities.

Relaxation labeling   A different approach to vision problems is relaxation labeling. The problems are cast as the assignment of labels to the elements of the scene from a set of possible labels. Haralick and Shapiro define the consistent labeling problem in [17] and [18]. Labels that violate consistency according to predefined criteria are iteratively removed from the tokens until convergence. Faugeras and Berthod [8] describe a gradient optimization approach to relaxation labeling. A global criterion is defined that combines the concepts of ambiguity and consistency of the labeling process. Geman and Geman discuss how stochastic relaxation can be applied to the task of image restoration in [11]. MAP estimates are obtained by a Gibbs sampler and simulated annealing. Hummel and Zucker [24] develop an underlying theory for the continuous relaxation process. One result is the definition of an explicit function to maximize and guide the relaxation process, leading to a new relaxation operator. The second result is that finding a consistent labeling is equivalent to solving a variational inequality. This work was continued by Parent and Zucker [50] for the inference of trace points in 2D and by Sander and Zucker [54] in 3D.

Computational geometry   Techniques for inferring surfaces from 3D point clouds have been reported in the computer graphics literature. Boissonnat [2] proposes a technique based on computational geometry for object representation by triangulating point clouds in 3D. Hoppe et al. [21] infer surfaces from unorganized point clouds as the zero levels of a signed distance function from the unknown surface. The strength of their approach is that the surface model, topology, and boundaries need not be known a priori. Edelsbrunner and Mücke [7] introduce the *3D alpha shapes* that are based on a 3D Delaunay triangulation of the data. Szeliski, Tonnesen, and Terzopoulos [61] describe a method for modeling surfaces of arbitrary, or changing, topology using a set of oriented dynamic particles that interact according to distance, coplanarity, conormality and cocircularity. Computational geometry methods are limited by their sensitivity to even a very small number of outliers and their computational complexity.

Robust methods   In the presence of noise, robust techniques inspired by RANSAC (random sample consensus) [9] can be applied. Small random samples are selected from the noisy data and are used to derive model hypotheses, which are tested using the remainder of the data set. Hypotheses that are consistent with a large number of the data points are considered valid. Variants of RANSAC include RESC [74] and MIR [31], which are mainly used for segmentation of surfaces from noisy 3D point clouds. The extracted surfaces are limited to planar or quadratic, except for the approach in [33], which can extract high-order polynomial surfaces. Chen et al. [4] introduce robust statistical methods to computer vision. They show how robust M-estimators with auxiliary scale are more general than the other robust estimators previously used in the field. They also point out the difficulties caused by multiple structures in the data and propose a way to detect them. In all cases, an a priori parametric representation of the unknown structure is necessary, thus limiting the applicability of these methods.

Level set methods   The antipode of the explicit representation of surfaces by a set of points is the implicit representation in terms of a function. In [58], Sethian proposes a *level set* approach under which surfaces can be inferred as the zero-level isosurface of a multivariate implicit function. The technique allows for topological changes; thus it can reconstruct surfaces of any genus as well as nonmanifolds. Zhao, Osher, and Fedkiw [20] and Osher and Fedkiw [49] propose efficient ways of handling implicit surfaces as level sets of a function. A combination of points and elementary surfaces and curves can be provided as input to their technique, which can handle local changes locally as well as global deformations and topological changes. All the implicit surface-based approaches are iterative and require careful selection of the implicit function and initialization. The surface in explicit form, as a set of polygons, can be extracted by a technique such as the classic marching cubes algorithm [37]. The simultaneous representation of surfaces, curves, and junctions is impossible.

Symbolic methods   Following the paradigm set by Marr [39], many researchers developed methods for hierarchical grouping of symbolic data. Lowe [38] developed a system for 3D object recognition based on perceptual organization of image edgels. Groupings are selected among the numerous possibilities according to the Gestalt principles, viewpoint invariance, and low likelihood of being accidental formations. Mohan and Nevatia [41] and Dolan and Riseman [6] also propose perceptual organization approaches based on the Gestalt principles. Both are symbolic and operate in a hierarchical bottom-up fashion starting from edgels and increasing the level of abstraction at each iteration. The latter approach aims at inferring curvilinear structures, while the former aims at segmentation and extraction of 3D scene descriptions from collations of features that have high likelihood of being projections of scene objects. Along the same lines is Jacobs' [25] technique for inferring salient convex groups among clutter, since they most likely correspond to world objects. The criteria to determine the nonaccidentalness of the potential structures are convexity, proximity, and contrast of the edgels.

Clustering   A significant current trend in perceptual organization is clustering [26]. Data are represented as nodes of a graph, and the edges between them encode the likelihood that two nodes belong in the same partition of the graph. Clustering is achieved by cutting some of these edges in a way that optimizes global criteria. A landmark approach in the field was the introduction of *normalized cuts* by Shi and Malik [60]. They aim at maximizing the degree of dissimilarity between the partitions normalized by essentially the size of each partition in order to remove the bias for small clusters. Boykov, Veksler, and Zabih [3] use graph-cut based algorithms to approximately optimize energy functions whose explicit optimization is NP-hard. They demonstrate the validity of their approach on a number of computer vision problems. Stochastic clustering algorithms are developed by Cho and Meer [5] and Gdalyahu, Weinshall, and Werman [10]. A consensus of various clusterings of the data is used as a basis of the solution. Finally, Robles-Kelly and Hancock [53] present a perceptual grouping algorithm based on graph cuts and an iterative expectation maximization scheme, which improves the quality of results at the expense of increased computational complexity.

Methods based on local interactions   We now turn our attention to perceptual organization techniques that are based on local interaction between primitives. Shashua and Ullman [59] first addressed the issue of structural saliency and how prominent curves are formed from tokens that are not salient in isolation. They define a locally connected network that assigns a saliency value

to every image location according to the length and smoothness of curvature
of curves going through that location. In [50], Parent and Zucker infer trace
points and their curvature based on spatial integration of local information.
An important aspect of this method is its robustness to noise. This work
was extended to surface inference in three dimensions by Sander and Zucker
[54]. Sarkar and Boyer [55] employ a voting scheme to detect a hierarchy of
tokens. Voting in parameter space has to be performed separately for each
type of feature, thus making the computational complexity prohibitive for
generalization to 3D. The inability of previous techniques to simultaneously
handle surfaces, curves, and junctions was addressed in the precursor of our
research in [15, 16]. A unified framework where all types of perceptual structures can be represented is proposed along with a preliminary version of the
voting scheme presented here. The major advantages of the work of Guy
and Medioni were noise robustness and computational efficiency, since it is
not iterative. How this methodology evolved is presented in the remaining
sections of this chapter.

Methods inspired by psychophysiology and neuroscience   Finally, there is an
important class of perceptual organization methods that are inspired by human perception and research in psychophysiology and neuroscience. Grossberg, Mingolla, and Todorovic [13, 14] developed the *boundary contour system* and the *feature contour system* that can group fragmented and even
illusory edges to form closed boundaries and regions by feature cooperation
in a neural network. Heitger and von der Heydt [19], in a classic paper on
neural contour processing, claim that elementary curves are grouped into
contours via convolution with a set of orientation selective kernels whose
responses decay with distance and difference in orientation. Williams and
Jacobs [72] introduce the *stochastic completion fields* for contour grouping.
Their theory is probabilistic and models the contour from a source to a sink
as the motion of a particle performing a random walk. Particles decay after every step, thus minimizing the likelihood of completions that are not
supported by the data or between distant points. Li [36] presents a contour
integration model based on excitatory and inhibitory cells and a top-down
feedback loop. What is more relevant to our research, that focuses on the
preattentive bottom-up process of perceptual grouping, is that connection
strength decreases with distance and that zero or low curvature alternatives
are preferred to high curvature ones. The model for contour extraction of
Yen and Finkel [73] is based on psychophysical and physiological evidence
that has many similarities to ours. It employs a voting mechanism where
votes, whose strength decays as a Gaussian function of distance, are cast

along the tangent of the osculating circle. An excellent review of perceptual grouping techniques based on cooperation and inhibition fields can be found in [44]. Even though we do not attempt to present a biologically plausible system, the similarities between our framework and the ones presented in this paragraph are nevertheless encouraging.

Relations with our approach   Some important aspects of our approach in the context of the work presented in this section are discussed here. In case of dense, noise-free, uniformly distributed data, we can match the performance of surface extraction methods such as [2, 7, 66, 58, 37]. Furthermore, our results degrade much more gracefully in the presence of noise (see, for example, [15] and [40]). The input can be oriented tokens, unoriented tokens, or a combination of both, while many of the techniques mentioned above require oriented inputs to proceed. In addition, we can extract open and closed forms of all types of structures simultaneously. Our model-free approach allows us to handle arbitrary perceptual structures that adhere just to the "matter is cohesive" principle [39] and not predefined models that restrict the admissible solutions. Model-based approaches cannot easily distinguish between model misfit and noise. Our voting function has many similarities with other voting-based methods, such as the decay with distance and curvature [19, 73, 36] and the use of constant curvature paths [50, 56, 55, 73] that result in an eight-shaped voting field (in 2D) [13, 14, 19, 73, 36]. The major difference is that in our case the votes cast are tensors and not scalars; therefore, they can express much richer information.

5.3   Tensor Voting in 2D

This section introduces the tensor voting framework in 2D. It begins by describing the original second-order representation and voting of [40]. Since the strictly second-order framework cannot handle structure discontinuities, first-order information has been added to the framework. Now tokens are encoded with second-order tensors and polarity vectors. These tokens propagate their information to their neighbors by casting first- and second-order votes. We demonstrate how all types of votes are derived from the same voting function and how voting fields are generated. Votes are accumulated at every token to infer the type of perceptual structure or structures going through it along with their orientation. Voting from token to token is called *sparse voting*. Then dense voting, during which votes are accumulated at all grid positions, can be performed to infer the saliency everywhere. This process allows the continuation of structures and the bridging of gaps. In

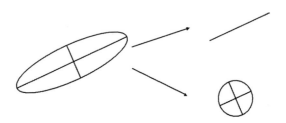

Figure 5.4. Decomposition of a 2D second-order, symmetric tensor into its *stick* and *ball* components.

Table 5.1. Encoding oriented and unoriented 2D inputs as 2D, second-order, symmetric tensors

Input	Second-order tensor	Eigenvalues	Quadratic form
n ⟍ (oriented)	⟍	$\lambda_1 = 1, \lambda_2 = 0$	$\begin{bmatrix} n_1^2 & n_1 n_2 \\ n_1 n_2 & n_2^2 \end{bmatrix}$
● (unoriented)	◯	$\lambda_1 = \lambda_2 = 1$	$\begin{bmatrix} 1 & 0 \\ 0 & 1 \end{bmatrix}$

2D, perceptual structures can be curves, regions, and junctions, as well as their terminations, which are the endpoints of curves and the boundaries of regions. Finally, we present results on synthetic but challenging examples, including illusory contours.

5.3.1   Second-order representation and voting in 2D

The second-order representation is in the form of a second-order symmetric nonnegative definite tensor, which essentially indicates the saliency of each type of perceptual structure (curve, junction, or region in 2D) the token may belong to and its preferred normal and tangent orientations. Tokens cast second-order votes to their neighbors according to the tensors they are associated with.

Second-order representation A 2D, symmetric, nonnegative definite, second-order tensor can be viewed as a 2×2 matrix, or equivalently an ellipse. Intuitively, its shape indicates the type of structure represented and its size the saliency of this information. The tensor can be decomposed as in the following equation:

$$T = \lambda_1 \widehat{e}_1 \widehat{e}_1^T + \lambda_2 \widehat{e}_2 \widehat{e}_2^T = (\lambda_1 - \lambda_2)\widehat{e}_1 \widehat{e}_1^T + \lambda_2 (\widehat{e}_1 \widehat{e}_1^T + \widehat{e}_2 \widehat{e}_2^T), \qquad (5.1)$$

where λ_i are the eigenvalues in decreasing order and $\widehat{e}_i$ are the corresponding eigenvectors (see also Figure 5.4). Note that the eigenvalues are nonnegative, since the tensor is nonnegative definite. The first term in (5.1) corresponds to a degenerate, elongated ellipsoid, termed hereafter the *stick tensor*, that indicates an elementary curve token with $\widehat{e}_1$ as its curve normal. The second term corresponds to a circular disk, termed the *ball tensor*, that corresponds to a perceptual structure that has no preference of orientation or to a location where multiple orientations coexist. The size of the tensor indicates the certainty of the information represented by it. For instance, the size of the stick component $(\lambda_1 - \lambda_2)$ indicates curve saliency.

Based on the above, an elementary curve with normal $\vec{n}$ is represented by a stick tensor parallel to the normal, while an unoriented token is represented by a ball tensor. Note that curves are represented by their normals and not their tangents for reasons that become apparent in higher dimensions. See Table 5.1 for how oriented and unoriented inputs are encoded and the equivalent ellipsoids and quadratic forms.

Second-order voting Now that the inputs, oriented or unoriented, have been encoded with tensors, we examine how the information they contain is propagated to their neighbors. The question we want to answer is: assuming that a token at O with normal $\vec{n}$ and a token at P belong to the same smooth perceptual structure, what information should the token at O cast at P? We first answer the question for the case of a voter with a pure *stick tensor* and show how all other cases can be derived from it. We claim that in the absence of other information, the arc of the *osculating circle* (the circle that shares the same normal as a curve at the given point) at O that goes through P is the most likely smooth path, since it maintains constant curvature. The center of the circle is denoted by C in Figure 5.5. In case of straight continuation from O to P, the osculating circle degenerates to a straight line. Similar use of primitive circular arcs can also be found in [50, 56, 55, 73].

As shown in Figure 5.5, the second-order vote is also a stick tensor and has a normal lying along the radius of the osculating circle at P. What remains to be defined is the magnitude of the vote. According to the Gestalt

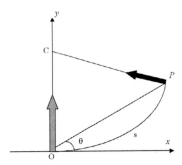

Figure 5.5. Second-order vote cast by a stick tensor located at the origin.

principles, it should be a function of proximity and smooth continuation. The influence from a token to another should attenuate with distance, to minimize interference from unrelated tokens, and with curvature, to favor straight continuation over curved alternatives when both exist. Moreover, no votes are cast if the receiver is at an angle larger than 45° with respect to the tangent of the osculating circle at the voter. Similar restrictions on the fields appear also in [19, 73, 36]. The *saliency decay function* we have selected has the following form:

$$DF(s, \kappa, \sigma) = e^{-(\frac{s^2 + c\kappa^2}{\sigma^2})}, \tag{5.2}$$

where s is the arc length OP, κ is the curvature, c controls the degree of decay with curvature, and σ is the *scale of voting*, which determines the effective neighborhood size. The parameter c is a function of the scale and is optimized to make the extension of two orthogonal line segments to form a right angle equally likely to the completion of the contour with a rounded corner [15]. Its value is given by $c = \frac{-16log(0.1) \times (\sigma - 1)}{\pi^2}$. Scale essentially controls the range within which tokens can influence other tokens. The saliency decay function has infinite extend, but for practical reasons it is cropped at a distance where the votes cast become insignificant. For instance, the field can be defined up to the extend that vote strength becomes less than 1% of the voter's saliency. The scale of voting can also be viewed as a measure of smoothness. A large scale favors long-range interactions and enforces a higher degree of smoothness, aiding noise removal at the same time, while a small scale makes the voting process more local, thus preserving details. Note that σ is the only free parameter in the system.

The 2D, second-order stick voting field for a unit stick voter located at the origin and aligned with the z-axis can be defined as follows as a function of the distance l between the voter and receiver and the angle θ, which is the angle between the tangent of the osculating circle at the voter and the line going through the voter and receiver (see Figure 5.5).

$$\mathbf{S}_{SO}(d, \theta) = DF(s, \kappa, \sigma) \begin{bmatrix} -sin(2\theta) \\ cos(2\theta) \end{bmatrix} [-sin(2\theta) \quad cos(2\theta)]$$

$$s = \frac{\theta l}{sin(\theta)}, \quad \kappa = \frac{2sin(\theta)}{l} \quad (5.3)$$

The votes are also stick tensors. For stick tensors of arbitrary size, the magnitude of the vote is given by (5.2) multiplied by the the size of the stick $\lambda_1 - \lambda_2$. The ball voting field is formally defined in Section 5.3.3. It can be derived from the second-order stick voting field It is used for voting tensors that have nonzero ball components and contain a tensor at every position that expresses the orientation preference and saliency of the receiver of a vote cast by a ball tensor. For arbitrary ball tensors, the magnitude of the votes has to be multiplied by λ_2.

The voting process is identical whether the receiver contains a token or not, but we use the term *sparse voting* to describe a pass of voting where votes are cast to locations that contain tokens only, and the term *dense voting* for a pass of voting from the tokens to all locations within the neighborhood regardless of the presence of tokens. Receivers accumulate the votes cast to them by tensor addition.

Sensitivity to Scale   The scale of the voting field is the single critical parameter of the framework. Nevertheless, the sensitivity to it is low. Similar results should be expected for similar values of scale, and small changes in the output are associated with small changes of scale. We begin analyzing the effects of scale with simple synthetic data for which ground truth can be easily computed. The first example is a set of unoriented tokens evenly spaced on the periphery of a circle of radius 100 (see Figure 5.6a). The tokens are encoded as ball tensors, and tensor voting is performed to infer the most types of structures they form, which are always detected as curves, and their preferred orientations. The theoretic values for the tangent vector at each location can be calculated as $[-sin(\theta) \quad cos(\theta)]^T$.

Table 5.2 reports the angular error in degrees between the ground-truth tangent orientation at each of the 72 locations and the orientation estimated by tensor voting. (Note that the values on the table are for σ^2.) The second example in Figure 5.6b is a set of unoriented tokens lying at equal distances on a square of side 200. After voting, the junctions are detected due to their

(a) 72 unoriented tokens on a circle   (b) 76 unoriented tokens on a square

Figure 5.6. Inputs for quantitative estimation of orientation errors as scale varies.

high ball saliency and the angular error between the ground truth, and the estimated tangent at each curve inlier is reported in Table 5.2. The extreme scale values, especially the larger ones, used here are beyond the range that is typically used in the other experiments presented in this chapter. A σ^2 value of 50 corresponds to a voting neighborhood of 16, while a σ^2 of 5000 corresponds to a neighborhood of 152. An observation from these simple experiments is that, as scale increases, each token receives more votes, leading to a better approximation of the circle by the set of points, while too much influence from the other sides of the square rounds its corners, causing an increase in the reported errors.

A different simple experiment demonstrating the stability of the results with respect to large scale variations is shown in Figure 5.7. The input consists of a set of unoriented tokens from a hand-drawn curve. Approximately half the points of the curve have been randomly removed from the data set, and the remaining points where spaced apart by 3 units of distance. Random outliers whose coordinates follow a uniform distribution have been added to the data. The 166 inliers and the 300 outliers are encoded as ball tensors, and tensor voting is performed with a wide range of scales. The detected inliers at a few scales are shown in Figure 5.7b-d. At the smallest scales, isolated inliers do not receive enough support to appear as salient, but as scale increases, the output becomes stable. In all cases, outliers are successfully removed. Table 5.3 shows the curve saliency of an inlier A and that of an outlier B as the scale varies. (Note that the values on the table again are for σ^2.) Regardless of the scale, the saliency of A is always significantly larger than that of B, making the selection of a threshold that separates inliers from outliers trivial.

Table 5.2. Errors in curve orientation estimation as functions of scale

σ^2	Angular error for circle (degrees)	Angular error for square (degrees)
50	1.01453	1.11601e-007
100	1.14193	0.138981
200	1.11666	0.381272
300	1.04043	0.548581
400	0.974826	0.646754
500	0.915529	0.722238
750	0.813692	0.8893
1000	0.742419	1.0408
2000	0.611834	1.75827
3000	0.550823	2.3231
4000	0.510098	2.7244
5000	0.480286	2.98635

Table 5.3. Errors in curve orientation estimation as functions of scale

σ^2	Saliency of inlier A	Saliency of outlier B
50	0.140	0
100	0.578	0
200	1.533	0
300	3.369	0
400	4.973	0.031
500	6.590	0.061
750	10.610	0.163
1000	14.580	0.297
2000	30.017	1.118
3000	45.499	2.228
4000	60.973	3.569
5000	76.365	5.100

5.3.2   First-order representation and voting in 2D

In this section, the need for integrating first-order information into the framework is justified. The first-order information is conveyed by the polarity

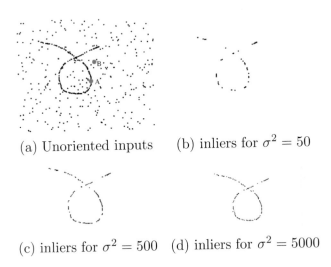

(a) Unoriented inputs   (b) inliers for $\sigma^2 = 50$

(c) inliers for $\sigma^2 = 500$ (d) inliers for $\sigma^2 = 5000$

Figure 5.7. Noisy input for inlier/outlier saliency comparison.

vector that encodes the likelihood of the token being on the boundary of a perceptual structure. Such boundaries in 2D are the endpoints of curves and the end-curves of regions. The direction of the polarity vector indicates the direction of the inliers of the perceptual structure whose potential boundary is the token under consideration. The generation of first-order votes is shown based on the generation of second-order votes.

First-order representation   To illustrate the significance of adding the polarity vectors to the original framework as presented in [40], consider the contour depicted in Figure 5.8a, keeping in mind that we represent curve elements by their normals. The contour consists of a number of points, such as D, that can be considered "smooth inliers," since they can be inferred from their neighbors under the constraint of good continuation. Consider point A, which is an orientation discontinuity where two smooth segments of the contour intersect. A can be represented as having both curve normals simultaneously. This is a second-order discontinuity that is described in the strictly second-order framework by the presence of a salient *ball component* in the tensor structure. The ball component of the second-order tensor captures the uncertainty of orientation, with its size indicating the likelihood that the location is a junction. On the other hand, the endpoints B and C of the contour are smooth continuations of their neighbors in the sense that their normals do not differ significantly from those of the adjacent points.

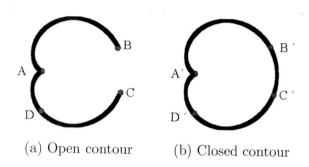

(a) Open contour   (b) Closed contour

Figure 5.8. Curves with orientation discontinuities at A and A' and endpoints B and C. The latter cannot be represented with second-order tensors.

The same is true for their tensors as well. Therefore, they cannot be discriminated from points B' and C' in Figure 5.8b. The property that makes these pairs of points very different is that B and C are terminations of the curve in Figure 5.8a, while their counterparts are not.

The second-order representation is adequate to describe locations where the orientation of the curve varies smoothly and locations where the orientation of the curve is discontinuous, which are second-order discontinuities. The termination of a perceptual structure is a first-order property and is handled by the first-order augmentation to the representation. Polarity vectors are vectors that are directed toward the direction where the majority of the neighbors of a perceptual structure are, and whose magnitude encodes the saliency of the token as a potential termination of the structure. The magnitude of these vectors, termed *polarity* hereafter, is locally maximum for tokens whose neighbors lie on one side of the half-space defined by the polarity vector. In the following sections, the way first-order votes are cast and collected to infer polarity and structure terminations is shown.

First-order voting   Now, we turn our attention to the generation of first-order votes. Note that the first-order votes are cast according to the second-order information of the voter. As shown in Figure 5.9, the first-order vote cast by a unitary stick tensor at the origin is *tangent* to the osculating circle, the smoothest path between the voter and receiver. Its magnitude, since nothing suggests otherwise, is equal to that of the second-order vote according to

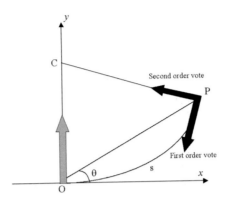

Figure 5.9. Second- and first-order votes cast by a stick tensor located at the origin

(5.2). The first-order voting field for a unit stick voter aligned with the z-axis is

$$\mathbf{S}_{FO}(d, \theta) = DF(s, \kappa, \sigma) \begin{bmatrix} -cos(2\theta) \\ -sin(2\theta) \end{bmatrix}. \qquad (5.4)$$

Note that tokens are cast first- and second-order votes based on their *second-order information only*. This occurs because polarity vectors have to be initialized to zero, since no assumption about structure terminations is available. Therefore, first-order votes are computed based on the second-order representation, which can be initialized (in the form of ball tensors) even with no information other than the presence of a token.

A simple illustration of how first- and second-order votes can be combined to infer curves and their endpoints in 2D appears in Figure 5.10. The input consists of a set of colinear, unoriented tokens, which are encoded as ball tensors. The tokens cast votes to their neighbors and collect the votes cast to them. The accumulated saliency can be seen in Figure 5.10b, where the dashed lines mark the limits of the input. The interior points of the curve receive more support and are more salient than those close to the endpoints. Since second-order voting is an excitatory process, locations beyond the endpoints are also compatible with the inferred line and receive consistent votes from the input tokens. Detection of the endpoints based on the second-order votes is virtually impossible, since there is no systematic way of selecting a threshold that guarantees that the curve will not extend beyond the last input tokens. The strength of the second-order aspect of the framework lies in the ability to extrapolate and fill in gaps caused by missing data. The

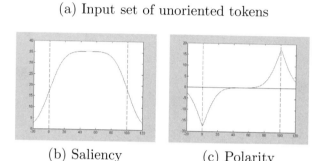

(a) Input set of unoriented tokens

(b) Saliency   (c) Polarity

Figure 5.10. Accumulated saliency and polarity for a simple input.

accumulated polarity can be seen in Figure 5.10c. The endpoints appear clearly as maxima of polarity. The combination of saliency and polarity allows us to infer the curve and terminate it at the correct points. See Section 5.3.4 for a complete analysis of structure inference in 2D.

5.3.3 Voting fields

In this section, we show how votes from ball tensors can be derived with the same saliency decay function and how voting fields can be computed to reduce the computational cost of calculating each vote according to (5.2). Finally, we show how the votes cast by an arbitrary tensor can be computed given the voting fields.

The second-order stick voting field contains at every position a tensor that is the vote cast there by a unitary stick tensor located at the origin and aligned with the y-axis. The shape of the field in 2D can be seen in the upper part of Figure 5.11a. Depicted at every position is the eigenvector corresponding to the largest eigenvalue of the second order tensor contained there. Its size is proportional to the magnitude of the vote. To compute a vote cast by an arbitrary stick tensor, we need to align the field with the orientation of the voter and multiply the saliency of the vote that coincides with the receiver by the saliency of the arbitrary stick tensor, as in Figure 5.11b.

The *ball voting field* can be seen in the lower part of Figure 5.11a. The ball tensor has no preference of orientation, but still can cast meaningful information to other locations. The presence of two proximate unoriented

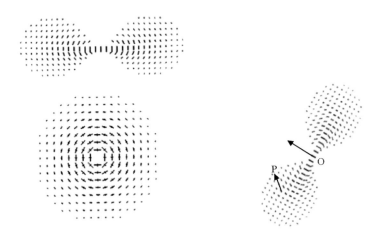

(a) The 2D stick and ball fields   (b) Stick vote cast from O to P

Figure 5.11. Voting fields in 2D and alignment of the stick field with the data for vote generation.

tokens, the voter and the receiver, indicates a potential curve going through the two tokens. The ball voting fields allow us to infer preferred orientations from unoriented tokens, thus minimizing initialization requirements.

The derivation of the ball voting field $\mathbf{B}(P)$ from the stick voting field can be visualized as follows: the vote at P from a unitary ball tensor at the origin O is the integration of the votes of stick tensors that span the space of all possible orientations. In 2D, this is equivalent to a rotating stick tensor that spans the unit circle at O. The 2D ball field can be derived from the stick field $\mathbf{S}(P)$, according to

$$\mathbf{B}(P) = \int_0^{2\pi} R_\theta^{-1} \mathbf{S}(R_\theta P) R_\theta^{-T} d\theta, \qquad (5.5)$$

where R_θ is the rotation matrix to align $\mathbf{S}$ with $\widehat{e}_1$, the eigenvector corresponding to the maximum eigenvalue (the stick component), of the rotating tensor at P.

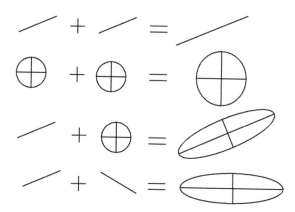

Figure 5.12. Tensor addition in 2D. Purely stick tensors result only from the addition of parallel stick tensors and purely ball tensors from the addition of ball tensors or orthogonal stick tensors.

In practice, the integration is approximated by tensor addition (see also Figure 5.12):

$$T = \sum_{i=0}^{\pi} \sum_{j=0}^{\pi} \vec{v}_i \vec{v}_i^T, \tag{5.6}$$

where V is the accumulated vote and $\vec{v}_i j$ are the stick votes, in vector form (which is equivalent, since a stick tensor has only one nonzero eigenvalue and can be expressed as the outer product of its only significant eigenvector), from O to P cast by the stick tensors $i \in [0, 2\pi)$ that span the unit circle. Normalization has to be performed in order to make the energy emitted by a unitary ball equal to that of a unitary stick. As a result of the integration, the second-order ball field does not contain purely stick or purely ball tensors, but arbitrary second-order symmetric tensors. The field is radially symmetric, as expected, since the voter has no preferred orientation.

The *2D first-order stick voting field* is a vector field, which at every position holds a vector that is equal in magnitude to the stick vote that exists in the same position in the fundamental second-order stick voting field, but is tangent to the smooth path between the voter and receiver instead of normal to it. The *first-order ball voting field* can be derived as in (5.5) by substituting the first-order stick field for the second-order stick field and performing vector instead of tensor addition when integrating the votes at each location.

All voting fields defined here, both first- and second-order, as well as all other fields in higher dimensions, are functions of the position of the receiver relative to the voter, the tensor at the voter, and a single parameter: the scale of the saliency decay function. After these fields have been precomputed at the desired resolution, computing the votes cast by any second-order tensor is reduced to a few look-up operations and linear interpolation. Voting takes place in a finite neighborhood within which the magnitude of the votes cast remains significant. For example, we can find the maximum distance s_{max} from the voter at which the vote cast will have 1% of the voter's saliency, as follows:

$$e^{-(\frac{s_{max}^2}{\sigma^2})} = 0.01. \tag{5.7}$$

The size of this neighborhood is obviously a function of the scale σ. As described in Section 5.3.1, any tensor can be decomposed into the basis components (stick and ball in 2D) according to its eigensystem. Then, the corresponding fields can be aligned with each component. Votes are retrieved by simple look-up operations, and their magnitude is multiplied by the corresponding saliency. The votes cast by the stick component are multiplied by $\lambda_1 - \lambda_2$, and those of the ball component by λ_2.

5.3.4 Vote analysis

Votes are cast from token to token and accumulated by tensor addition in the case of the second-order votes, which are in general arbitrary second-order tensors, and by vector addition in the case of the first-order votes, which are vectors.

Analysis of second-order votes   Analysis of the second-order votes can be performed once the eigensystem of the accumulated second-order 2×2 tensor has been computed. Then the tensor can be decomposed into the stick and ball components:

$$T = (\lambda_1 - \lambda_2)\widehat{e}_1\widehat{e}_1^T + \lambda_2(\widehat{e}_1\widehat{e}_1^T + \widehat{e}_2\widehat{e}_2^T), \tag{5.8}$$

where $\widehat{e}_1\widehat{e}_1^T$ is a *stick tensor*, and $\widehat{e}_1\widehat{e}_1^T + \widehat{e}_2\widehat{e}_2^T$ is a *ball tensor*. The following cases have to be considered. If $\lambda_1 \gg \lambda_2$, this indicates certainty of one normal orientation; therefore, the token most likely belongs on a curve that has the estimated normal at that location. If $\lambda_1 \approx \lambda_2$, the dominant component is the ball and there is no preference of orientation, either because all orientations are equally likely or because multiple orientations coexist at

Figure 5.13. Salient ball tensors at regions and junctions.

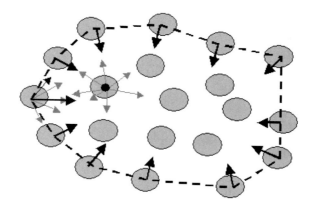

Figure 5.14. First-order votes received at region inliers and boundaries.

the location. This indicates either a token that belongs to a region, which is surrounded by neighbors from the same region from all directions, or a junction where two or more curves intersect and multiple curve orientations are present simultaneously (see Figure 5.13). Junctions can be discriminated from region tokens, since their saliency is a distinct, local maximum of λ_2, whereas the saliency of region inliers is more evenly distributed. An outlier receives only inconsistent votes, so both eigenvalues are small.

Analysis of first-order votes   Vote collection for the first-order case is performed by vector addition. The accumulated result is a vector whose direction points to a weighted center of mass from which votes are cast, and whose magnitude encodes polarity. Since the first-order votes are also weighted by the saliency of the voters and attenuate with distance and curvature, their vector sum points to the direction from which the most salient contributions were received. A relatively low polarity indicates a token that is in the interior of a curve or region, therefore surrounded by neighbors whose votes cancel out each other. On the other hand, a high polarity indicates a token that is on or close to a boundary, thus receiving votes from only one side with respect to the boundary, at least locally. The correct boundaries can be extracted as maxima of polarity along the direction of the polarity vector. See the region of Figure 5.14, where the votes received at interior points cancel each other out, while the votes at boundary points reinforce each other resulting in a large polarity vector there. Table 5.4 illustrates how tokens can be characterized using the collected first- and second-order information.

Table 5.4. Summary of first- and second-order tensor structure for each feature type in 2D

2D feature	Saliency	Second-order tensor orientation	Polarity	Polarity vector
curve interior	high $\lambda_1 - \lambda_2$	normal: $\widehat{e}_1$	low	-
curve endpoint	high $\lambda_1 - \lambda_2$	normal: $\widehat{e}_1$	high	parallel to $\widehat{e}_2$
region interior	high λ_2	-	low	-
region boundary	high λ_2	-	high	normal to boundary
junction	locally max λ_2	-	low	-
outlier	low	-	low	-

Dense structure extraction   In order to extract dense structures based on the sparse set of tokens, an additional step is performed. Votes are collected at *every* location to determine where salient structures exist. The process of collecting votes at all locations regardless of whether they contain an input token or not is called *dense voting*. The tensors at every location are decomposed, and *saliency maps* are built. In 2D, there are two types of saliency that need to be considered: stick and ball saliency. Figure 5.15 shows an example where curve and junction saliency maps are built from a sparse set of oriented tokens.

Curve extraction is performed as a marching process, during which curves are grown starting from seeds. The most salient unprocessed curve token is selected as the seed, and the curve is grown following the estimated normal. The next point is added to the curve with subpixel accuracy as the zero-crossing of the first derivative of saliency along the curve's tangent. Even though saliency values are available at discrete grid positions, the locations

(a) Sparse input tokens   (b) Curve saliency map   (c)Junction saliency map

Figure 5.15. Saliency maps in 2D (darker regions have higher saliency).

of the zero-crossings of the first derivative of saliency can be estimated with subpixel accuracy by interpolation. In practice, to reduce the computational cost of casting unnecessary votes, votes are collected only at grid points toward which the curve can grow, as indicated by the normal of the last point added. As the next point is added to the curve, following the path of maximum saliency, new grid points where saliency has to be computed are determined. Endpoints are critical, since they indicate where the curve should stop.

Regions can be extracted via their boundaries. As described previously, region boundaries can be detected based on their high region saliency and high polarity. Then, they are used as inputs to the curve extraction process described in the previous paragraph. Junctions are extracted as local maxima of the junction saliency map. This is the simplest case because junctions are isolated in space. Examples of the structure extraction processes described here are presented in the following subsection.

5.3.5   Results in 2D

In this section, we present examples of 2D perceptual grouping using the tensor voting framework. Figure 5.16a shows a data set that contains a number of fragmented sinusoidal curves represented by unoriented points contaminated by a large number of outliers. Figure 5.16b shows the output after tensor voting. Curve inliers are colored gray, endpoints black, while junctions appear as gray squares. All the noise has been removed, and the curve segments have been correctly detected and their endpoints and junctions labeled.

The example in Figure 5.17 illustrates the detection of regions and curves. The input consists of a cardioid, represented by a set of unoriented points, and two quadratic curves, also represented as sets of unoriented points, con-

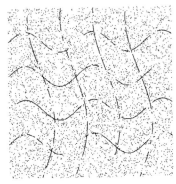
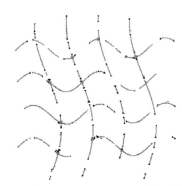

(a) Noisy unoriented data set   (b) Extracted curves, endpoints, and junctions

Figure 5.16. Curve, endpoint, and junction extraction from a noisy data set with sinusoidal curves.

taminated by random uniformly distributed noise (Figure 5.17a). Figures 5.17b-d show the detected region inliers, region boundaries, and curve inliers respectively. Note that some points, where the curves and the cardioid overlap, have been detected as both curve and region inliers.

5.3.6   Illusory contours

An interesting phenomenon is the grouping of endpoints and junctions to form illusory contours. Kanizsa [29] demonstrated the universal and unambiguous perception of illusory contours caused by the alignment of endpoints, distant line segments, junctions, and corners; [19, 13, 14] developed computational models for the inference of illusory contours and point out the differences between them and regular contours. Illusory contours are always convex, and when they are produced by endpoints, their formation occurs in the orthogonal direction of the contours whose endpoints give rise to the illusory ones. A perception of an illusory contour is strong in the lower part of Figure 5.18a, while no contour is formed in the upper part.

The capability to detect endpoints can be put to use for the inference of illusory contours. Figure 5.19a shows a set of line segments whose extensions converge to the same point and whose endpoints form an illusory circle. The segments consist of unoriented points, which are encoded as ball tensors, and propagate first- and second-order votes to their neighbors. The

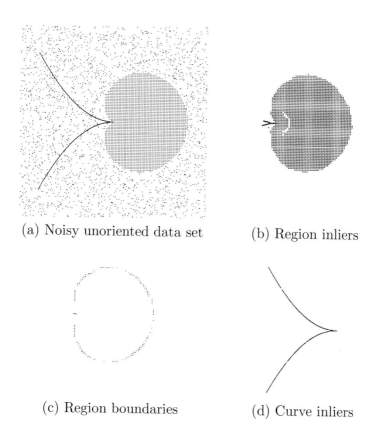

(a) Noisy unoriented data set   (b) Region inliers

(c) Region boundaries   (d) Curve inliers

Figure 5.17. Region and curve extraction from noisy data.

endpoints can be detected based on their high polarity values. They can be seen in Figure 5.19b along with the polarity vectors. The direction of the polarity vectors determines the side of the half-plane where the endpoint should cast second-order votes to ensure that the inferred illusory contours are convex. These votes are cast using a special field, shown in Figure 5.18b. This field is single-sided and is orthogonal to the regular field. Alternatively, the endpoint voting field can be viewed as the positive side of the fundamental stick field rotated by 90 degrees. A dense contour, a circle, is inferred after these votes have been collected and analyzed, as in the previous subsection (Figure 5.19c). Note that first-order fields can be defined in a similar way. An aspect of illusory contour inference that requires further investigation is a mechanism for deciding whether to employ regular voting

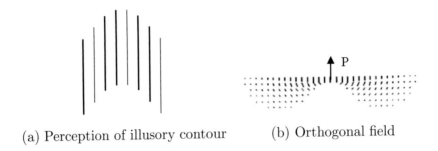

(a) Perception of illusory contour   (b) Orthogonal field

Figure 5.18. Illusory contours are perceived when they are convex (a). The single-sided field used for illusory contour inference. P denotes the polarity vector of the voting endpoint.

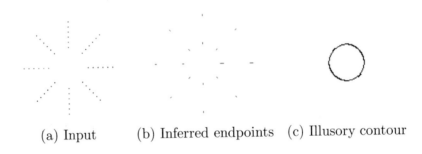

(a) Input   (b) Inferred endpoints   (c) Illusory contour

Figure 5.19. Illusory contour inference from endpoints of linear segments

fields to connect fragmented curves or to employ the orthogonal fields to infer illusory contours.

5.4   Tensor Voting in 3D

We proceed to the generalization of the framework in 3D. The remainder of this section shows that no significant modifications need to be made, apart from taking into account that more types of perceptual structure exist in 3D than in 2D. In fact, the 2D framework is a subset of the 3D framework, which in turn is a subset of the general ND framework. The second-order tensors and the polarity vectors become 3D, while the voting fields are derived from the same *fundamental 2D second-order stick voting field*.

In 3D, the types of perceptual structure that have to be represented are regions (which are now volumes), surfaces, curves, and junctions. Their

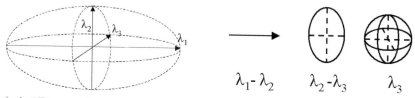

(a) A 3D generic tensor (λ_i are its   (b) Decomposition into the *stick,*
eigenvalues in descending order)   *plate,* and *ball* components

Figure 5.20. A second-order generic tensor and its decomposition in 3D.

terminations are the bounding surfaces of volumes, the bounding curves of
surfaces, and the endpoints of curves. The inputs can be either unoriented
or oriented, in which case there are two types: elementary surfaces (*surfels*)
or elementary curves (*curvels*).

5.4.1   Representation in 3D

The representation of a token consists of a symmetric, second-order tensor
that encodes *saliency* and a vector that encodes *polarity.*

Second-order representation   A 3D, second-order, symmetric, nonnegative
definite tensor is equivalent to a 3×3 matrix and a 3D ellipsoid. The
eigenvectors of the tensor are the axes of the ellipsoid. It can be decomposed
as in the following equation:

$$
T = \lambda_1 \widehat{e}_1 \widehat{e}_1^T + \lambda_2 \widehat{e}_2 \widehat{e}_2^T + \lambda_3 \widehat{e}_3 \widehat{e}_3^T =
$$
$$
= (\lambda_1 - \lambda_2)\widehat{e}_1 \widehat{e}_1^T + (\lambda_2 - \lambda_3)(\widehat{e}_1 \widehat{e}_1^T + \widehat{e}_2 \widehat{e}_2^T) + \lambda_3(\widehat{e}_1 \widehat{e}_1^T + \widehat{e}_2 \widehat{e}_2^T + \widehat{e}_3 \widehat{e}_3^T),
$$
$$(5.9)$$

where λ_i are the eigenvalues in decreasing order and $\widehat{e}_i$ are the corresponding eigenvectors (see also Figure 5.20). Note that the eigenvalues are nonnegative, since the tensor is nonnegative definite, and the eigenvectors are
orthogonal. The first term in (5.9) corresponds to a degenerate, elongated
ellipsoid, the *3D stick tensor*, that indicates an elementary surface token
with $\widehat{e}_1$ as its surface normal. The second term corresponds to a degenerate disk-shaped ellipsoid, termed hereafter the *plate tensor*, that indicates a
curve or a surface intersection with $\widehat{e}_3$ as its tangent, or, equivalently, with
$\widehat{e}_1$ and $\widehat{e}_2$ spanning the plane normal to the curve. Finally, the third term
corresponds to a sphere, the *3D ball tensor*, that corresponds to a structure

that has no preference of orientation. Table 5.5 shows how oriented and unoriented inputs are encoded and the equivalent ellipsoids and quadratic forms.

Table 5.5. Encoding oriented and unoriented 2D inputs as 2D second-order symmetric tensors

Input	Tensor	Eigenvalues	Quadratic form
surfel		$\lambda_1 = 1,$ $\lambda_2 = \lambda_3 = 0$	$\begin{bmatrix} n_1^2 & n_1n_2 & n_1n_3 \\ n_1n_2 & n_2^2 & n_2n_3 \\ n_1n_3 & n_2n_3 & n_3^2 \end{bmatrix}$
curvel		$\lambda_1 = \lambda_2 = 1,$ $\lambda_3 = 0$	$\mathbf{P}$ (see below)
unoriented		$\lambda_1 = \lambda_2 = \lambda_3 = 1$	$\begin{bmatrix} 1 & 0 & 0 \\ 0 & 1 & 0 \\ 0 & 0 & 1 \end{bmatrix}$

$$\mathbf{P} = \begin{bmatrix} n_{11}^2 + n_{21}^2 & n_{11}n_{12} + n_{21}n_{22} & n_{11}n_{13} + n_{21}n_{23} \\ n_{11}n_{12} + n_{21}n_{22} & n_{12}^2 + n_{22}^2 & n_{12}n_{13} + n_{22}n_{23} \\ n_{11}n_{13} + n_{21}n_{23} & n_{12}n_{13} + n_{22}n_{23} & n_{13}^2 + n_{23}^2 \end{bmatrix}$$

The representation using normals instead of tangents can be justified more easily in 3D, where surfaces are arguably the most frequent type of structure. In 2D, normal or tangent representations are equivalent. A surface patch in 3D is represented by a stick tensor parallel to the patch's normal. A curve, which can also be viewed as a surface intersection, is represented by a plate tensor that is normal to the curve. All orientations orthogonal to the curve belong in the 2D subspace defined by the plate tensor. Any two of these orientations that are orthogonal to each other can be used to initialize the plate tensor (see also Table 5.5). Adopting this representation allows a structure with $N - 1$ degrees of freedom in ND (a curve in 2D, a surface in 3D) to be represented by a single vector, while a tangent representation would require the definition of $N-1$ vectors that form a basis for an $(N-1)$-D subspace. Assuming that this is the most frequent structure in the ND space, our choice of representation makes the *stick voting field*, which corresponds to the elementary $(N-1)$-D variety, the basis from which all other voting fields are derived.

First-order representation   The first-order part of the representation is, again, a polarity vector whose size indicates the likelihood of the token being a boundary of a perceptual structure and which points toward the half-space where the majority of the token's neighbors are.

5.4.2   Voting in 3D

Identically to the 2D case, voting begins with a set of oriented and unoriented tokens. Again, both second- and first-order votes are cast according to the second-order tensor of the voter.

Second-order voting   In this section, we begin by showing how a voter with a purely stick tensor generates and casts votes, and then we derive the voting fields for the plate and ball cases. We chose to maintain voting as a function of only the position of the receiver relative to the voter and of the voter's preference of orientation. Therefore, we again address the problem of finding the smoothest path between the voter and receiver by fitting arcs of the osculating circle, as described in Section 5.3.1.

Note that the voter, the receiver, and the stick tensor at the voter define a plane. The voting procedure is restricted in this plane, thus making it identical to the 2D case. The second-order vote, which is the surface normal at the receiver under the assumption that the voter and receiver belong to the same smooth surface, is also a purely stick tensor in the plane (see also Figure 5.5). The magnitude of the vote is defined by the same *saliency decay function*, duplicated here for completeness.

$$DF(s, \kappa, \sigma) = e^{-(\frac{s^2 + c\kappa^2}{\sigma^2})}, \tag{5.10}$$

where s is the arc length OP, κ is the curvature, c is a constant that controls the decay with high curvature, and σ is the scale of voting. Sparse, token-to-token voting is performed to estimate the preferred orientation of tokens, followed by dense voting during which saliency is computed at every grid position.

First-order voting   Following the above line of analysis, the first-order vote is also on the plane defined by the voter, the receiver, and the stick tensor.

As shown in Figure 5.9, the first-order vote cast by a unitary stick tensor at the origin is *tangent* to the osculating circle. Its magnitude is equal to that of the second-order vote.

Voting Fields Voting by any 3D tensor takes place by decomposing the tensor into its three components: the stick, the plate, and the ball. Second-order voting fields are used to compute the second-order votes and first-order voting fields for the first-order ones. Votes are retrieved from the appropriate voting field by look-up operations and are multiplied by the saliency of each component. Stick votes are weighted by $\lambda_1 - \lambda_2$, plate votes by $\lambda_2 - \lambda_3$, and ball votes by λ_3.

The 3D first- and second-order fields can be derived from the corresponding 2D fields by rotation about the voting stick, which is their axis of symmetry. The visualization of the 2D second order stick field in Figure 5.11a is also a cut of the 3D field that contains the stick tensor at the origin.

To show the derivation of the ball voting fields $\mathbf{B}_i(P)$ from the stick voting fields, we can visualize the vote at P from a unitary ball tensor at the origin O as the integration of the votes of stick tensors that span the space of all possible orientations. In 2D, this is equivalent to a rotating stick tensor that spans the unit circle at O, while in 3D the stick tensor spans the unit sphere. The 3D ball field can be derived from the stick field $\mathbf{S}(P)$, as follows. The first-order ball field is derived from the first-order stick field, and the second-order ball field from the second-order stick field.

$$\mathbf{B_i}(P) = \int_0^{2\pi} \int_0^{2\pi} R_{\theta\phi\psi}^{-1} \mathbf{S_i}(R_{\theta\phi\psi}P) R_{\theta\phi\psi}^{-T} d\phi d\psi |_{\theta=0} \qquad (5.11)$$

$$i = 1, 2, \text{ the order of the field,}$$

where $R_{\theta\phi\psi}$ is the rotation matrix to align $\mathbf{S}$ with $\hat{e}_1$, the eigenvector corresponding to the maximum eigenvalue (the stick component), of the rotating tensor at P, and θ, ϕ, ψ are rotation angles about the x-, y-, and z- axes respectively.

In the second-order case, the integration is approximated by tensor addition, $T = \sum \vec{v}_i \vec{v}_i^T$, while in the first-order case by vector addition, $V = \sum \vec{v}_i$. Note that normalization has to be performed in order to make the energy emitted by a unitary ball equal to that of a unitary stick. Both fields are radially symmetric, as expected, since the voter has no preferred orientation.

To complete the description of the voting fields for the 3D case, we need to describe the *plate voting fields* $\mathbf{P}_i(P)$. Since the plate tensor encodes uncertainty of orientation around one axis, it can be derived by integrating the votes of a rotating stick tensor that spans the unit circle: in other words, the plate tensor. The formal derivation is analogous to that of the ball voting fields and can be written as follows:

$$\mathbf{P_i}(P) = \int_0^{2\pi} R_{\theta\phi\psi}\mathbf{S_i}(R_{\theta\phi\psi}^{-1}P)R_{\theta\phi\psi}^T d\psi|_{\theta=\phi=0} \qquad (5.12)$$

$$i = 1, 2, \text{ the order of the field,}$$

where θ, ϕ, ψ, and $R_{\theta\phi\psi}$ have the same meaning as in the previous equation.

Table 5.6. Summary of first- and second-order tensor structure for each feature type in 3D

3D feature	Saliency	Second-order tensor orientation	Polarity	Polarity vector
surface interior	high $\lambda_1 - \lambda_2$	normal: $\widehat{e}_1$	low	-
surface end-curve	high $\lambda_1 - \lambda_2$	normal: $\widehat{e}_1$	high	orthogonal to $\widehat{e}_1$ and end-curve
curve interior	high $\lambda_2 - \lambda_3$	tangent: $\widehat{e}_3$	low	-
curve endpoint	high $\lambda_2 - \lambda_3$	tangent: $\widehat{e}_3$	high	parallel to $\widehat{e}_3$
region interior	high λ_3	-	low	-
region boundary	high λ_3	-	high	normal to bounding surface
junction	locally max λ_3	-	low	-
outlier	low	-	low	-

5.4.3 Vote analysis

After voting from token to token has been completed, we can determine which tokens belong to perceptual structures, as well as the type and preferred orientation of these structures. The eigensystem of each tensor is computed, and the tensor is decomposed according to (5.9). Tokens can be

classified according to the accumulated first- and second-order information
according to Table 5.6.

Dense structure extraction   Now that the most likely type of feature at each
token has been estimated, we want to compute the dense structures (curves,
surfaces, and volumes in 3D) that can be inferred from the tokens. Continuous structures are again extracted through the computation of *saliency
maps*. Three saliency maps need to be built: one that contains $\lambda_1 - \lambda_2$ at
every location, one for $\lambda_2 - \lambda_3$, and one for λ_3. This can be achieved by
casting votes to *all* locations, whether they contain a token or not (dense
voting).

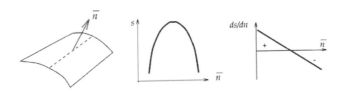

Figure 5.21. Dense surface extraction in 3D. (a) Elementary surface patch with
normal $\vec{n}$. (b) 3D surface saliency along normal direction. (c) First derivative of
surface saliency along normal direction.

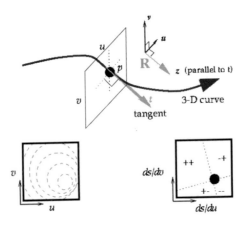

Figure 5.22. Dense curve extraction in 3D. (a) 3D curve with tangent $\vec{t}$ and
the normal plane. (b) Curve saliency isocontours on the normal plane. (c) First
derivatives of surface saliency on the normal plane.

Surfaces and curves are extracted using a modified marching algorithm [62] (based on the marching cubes algorithm [37]). They correspond to zero crossings of the first derivative of the corresponding saliency. Junctions are isolated and can be detected as distinct local maxima of λ_3. Volume inliers are characterized by high λ_3 values. Their bounding surfaces also have high λ_3 values and high polarity, with polarity vectors being normal to the surface and pointing toward the interior of the region. Volumes are extracted as regions of high λ_3, terminated at their boundaries, which have high λ_3 and high polarity. Junctions are isolated, distinct maxima of λ_3.

In order to reduce the computational cost, the accumulation of votes at locations with no prior information and structure extraction are integrated and performed as a marching process. Beginning from seeds, locations with highest saliency, we perform a dense vote only toward the directions dictated by the orientation of the features. Surfaces are extracted with subvoxel accuracy, as the zero-crossings of the first derivative of surface saliency (Figure 5.21). As in the 2D case, discrete saliency values are computed at grid points, and the continuous function is approximated via interpolation. Locations with high *surface* saliency are selected as seeds for surface extraction, while locations with high *curve* saliency are selected as seeds for curve extraction. The marching direction in the former case is perpendicular to the surface normal, while in the latter case, the marching direction is along the curve's tangent (Figure 5.22).

5.4.4   Results in 3D

In this section, we present results on synthetic 3D data sets corrupted by random noise. The first example is on a data set that contains a surface in the from of a spiral inside a cloud of noisy points (Figure 5.23). Both the surface and the noise are encoded as unoriented points. The spiral consists of 19,000 points, while the outliers are 30,000. Figure 5.23b shows the surface boundaries detected after voting. As the spiral is tightened to resemble a cylinder (Figure 5.23c), the surfaces merge and the inferred boundaries are those of a cylinder (Figure 5.23d).

The example in Figure 5.24 illustrates the detection of regions and curves. The input consists of a "peanut" and a plane, encoded as unoriented points contaminated by random uniformly distributed noise (Figure 5.24a). The "peanut" is empty inside, except for the presence of noise, which has an equal probability of being anywhere in space. Figure 5.24b shows the detected surface inliers after tokens with low saliency have been removed. Figure

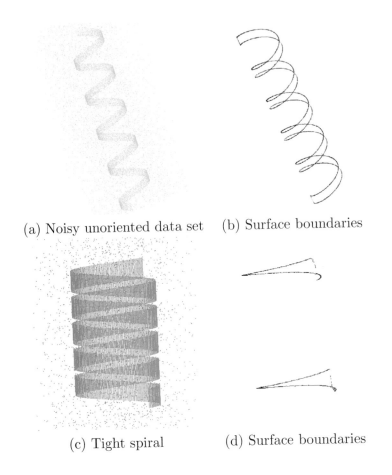

(a) Noisy unoriented data set   (b) Surface boundaries

(c) Tight spiral   (d) Surface boundaries

Figure 5.23. Surface boundary detection from noisy data.

5.24c shows the curve inliers, that is the tokens that lie at the intersection of the two surfaces. Finally, Figure 5.24d shows the extracted dense surfaces.

Figure 5.25a shows two solid generalized cylinders with different parabolic sweep functions. The cylinders are generated by a uniform random distribution of unoriented points in their interior, while the noise is also uniformly distributed but with a lower density. After sparse voting, volume inliers are detected due to their high ball saliency and low polarity, while region boundaries are detected due to their high ball saliency and polarity. The polarity vectors are normal to the bounding surface of the cylinders. Using the detected boundaries as inputs, we perform a dense vote and extract the bounding surface in continuous form in Figure 5.25b.

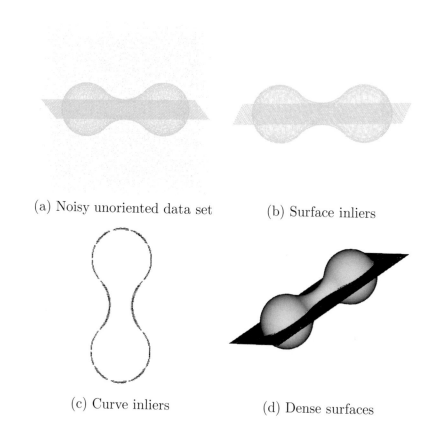

(a) Noisy unoriented data set   (b) Surface inliers

(c) Curve inliers   (d) Dense surfaces

Figure 5.24. Surface and surface intersection extraction from noisy data.

5.5   Tensor Voting in ND

The framework can easily be generalized to higher dimensions to tackle perceptual organization problems in ND spaces. At first glance, it might not be apparent how computer vision problems can be formulated in domains with more than three dimensions, but some computer vision problems are very naturally represented in such domains. For instance, in Section 5.6, we show how optical flow can be formulated as the organization of tokens in smooth layers in a 4D space, where the four dimensions are the image coordinates and the horizontal and vertical velocity.

Both the representation and voting can be generalized to N dimensions along the lines of the previous section, where the generalization from 2D to 3D was shown. The symmetric second order tensor in ND is equivalent to

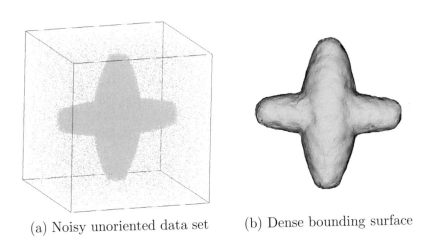

(a) Noisy unoriented data set   (b) Dense bounding surface

Figure 5.25. Region inlier and boundary detection in 3D.

an $N \times N$ matrix and an ND hyperellipsoid. Each axis of the hyperellipsoid can be viewed as a vector that is locally normal to the underlying structure. An elementary hyperplane is represented by a stick tensor, which has only one nonzero eigenvalue and, therefore, one normal. An elementary curve has one tangent and $N - 1$ normals and is represented by a second-order tensor with $N - 1$ nonzero equal eigenvalues, whose corresponding eigenvectors are the normals of the elementary curve. The polarity vector is an ND vector, and its interpretation is the same as in the previous sections. Its magnitude is proportional to the token's likelihood for being a boundary of a perceptual structure, it is locally orthogonal to the boundary and is directed towards the interior of the structure.

As mentioned in Section 5.4.2, first- and second-order voting by a purely stick voter takes place on a plane defined by the voter, the receiver, and the stick tensor of the voter, regardless of the dimensionality of the space. Any cut of a second-order stick voting field in any dimension that contains the the origin and the stick is identical to the *fundamental second-order 2D stick voting field*. The same holds for the first-order stick field as well. The ball voting fields are radially symmetric, since the voter has no preference of orientation. What changes is the number of degrees of freedom of the elementary stick voter at the origin that spans the unit sphere to simulate the effect of the ball voter. The total number of first- and second-order voting fields is $2N$.

Analysis of the results of voting can be performed under the principles summarized in Tables 5.4 and 5.6. An aspect of the framework that is harder to generalize is the extraction of dense structures in ND. The exponential increase in the number of vertices per hypercube requires a large number of votes to be cast and collected, and generates a prohibitively large number of possible configurations for an ND variant of the marching cubes algorithm.

Caution must be applied to certain issues in ND, which are not as prominent in two or three dimensions. Since voting is a function of distance and curvature, these measures have to be meaningful. The space has to be Euclidean, or at least the axes have to be scaled in a way that distances are meaningful, thus making the effect of more relevant neighbors larger than the effect of irrelevant ones. The data are stored in an approximate nearest neighbor (ANN) k-d tree [1], and thus the computational complexity of searching for a token's neighbors is linear with the dimensionality of the space. Finally, depending on the number of input tokens and the dimensionality of the space, the precomputation of the voting fields may become impractical. If the data is relatively sparse in a high-dimensional space, it is more efficient to compute the required votes when they are cast, instead of generating complete voting fields and using them as look-up tables. For sparse data in high dimensions, repetition of votes is less probable, and many of the precomputed votes may never be used.

5.5.1   Computational complexity

In this section, the computational complexity of tensor voting is analyzed. We show that it is a function of the number of tokens and, in the case of dense structure extraction, the size of the extracted structures.

The initialization step consists of sorting the tokens according to their spacial coordinates and storing them in a data structure from which they can be retrieved efficiently. We use the ANN k-d tree as the data structure that holds the tokens. Its construction is of $O(dn\,logn)$ complexity, where d is the dimension of the space and n the number of tokens. This data structure meets the requirements of tensor voting operations, since it is optimal when the locations of the data remain constant and multiple queries that seek the data points within a certain distance from a given location are generated. During voting, the tokens that are within the neighborhood of the voter, as defined by the scale (see equation 5.7 in Section 5.3.4), can be retrieved with $O(logn)$ operations.

The number of votes cast by a voter is not fixed and depends on the distribution of data in space and the scale of voting. For most practical purposes, however, we can assume that each voter casts votes on a fraction of the data set and that votes being cast to all tokens is the undesirable consequence of an extremely large scale. On average, the number of votes cast by each voter can be approximated as the product of data density times the size of the voting field. For each voter and receiver pair, the N components of the voter cast first- and second-order votes, which are computed by look-up operations from the voting fields and linear interpolation. The complexity of voting, after the receivers have been localized, is linear in the number of tokens and the dimension of the space.

The complexity of dense structure extraction, which is implemented in 2D and 3D only, is a function of grid resolution, since it defines the number of curvels or surfels extracted. The surface or curve saliency of a fixed number of grid points has to be computed for each surfel or curvel. For instance, surface saliency values at the eight vertices of a grid cube have to be computed for a surfel to be extracted at subvoxel accuracy inside the cube. The search for tokens within the voting neighborhood of each vertex from which votes are collected is $O(m)$, where m is the number of vertices. The marching cubes algorithm [37] is also linear, since it essentially consists of look-up operations. The number of voxels that are visited, but where no surface is extracted, is typically small (zero for a closed surface). Therefore, the complexity of dense surface extraction in 3D is $O(mlogn)$, where n is the number of tokens and m the number of surfels. The same holds for curve extraction.

5.6   Application to Computer Vision Problems

Thus far, we have demonstrated the capability of the tensor voting framework to group oriented and unoriented tokens into perceptual structures, as well as to remove noise. In practice, however, we rarely have input data in the form we used for the examples of sections 5.3.5 and 5.4.4, except maybe for range or medical data. In this section, the focus is on vision problems, where the input is images, while the desired output varies from application to application, but is some form of scene interpretation in terms of objects and layers.

The missing piece that bridges the gap from the images to the type of input required by the framework is a set of application-specific processes that can generate the tokens from the images. These relatively simple tools enable us to address a variety of computer vision problems within the tensor voting framework, alleviating the need to design software packages dedicated to specific problems. The premise is that solutions to these problems must comprise coherent structures that are salient due to the nonaccidental alignment of simpler tokens. The fact that these solutions might come in different forms, such as image segments, 3D curves, or motion layers, does not pose an additional difficulty to the framework, since they are all characterized by nonaccidentalness and good continuation. As mentioned in Section 5.1.1, problem-specific constraints can easily be incorporated as modules into the framework. For instance, uniqueness along the lines of sight can be enforced in stereo after token saliencies have been computed via tensor voting.

The remainder of this section demonstrates how stereo and motion segmentation can be treated in a very similar way within the framework, with the only difference being the dimensionality of the space. Stereo processing is performed in 3D, either in world coordinates if calibration information is available or in disparity space. Motion segmentation is done in a 4D space with the axes being the two image axes, x and y, and two velocity axes, v_x and v_y, since the epipolar constraint does not apply and motion can occur in any direction.

Four modules have been added to the framework tailored to the problems of stereo and motion segmentation: the initial matching module that generates tokens based on potential pixel correspondences; the enforcement of the uniqueness constraint, since the desired outputs are usually in the form of a disparity or velocity estimate for every pixel; the "discrete densification" module that generates disparity or velocity estimates at pixels for which no salient token exists; and the integration of monocular cues for discontinuity localization.

5.6.1   Initial matching

The goal of the initial matching module is the generation of tokens from potential pixel correspondences. Each correspondence between two pixels in two images gives rise to a token in the 3D (x, y, d) space in the case of stereo, where d is disparity, or in the 4D (x, y, v_x, v_y) space in the case of visual motion, where v_x and v_y are the velocities along the x- and y-axes. The correctness of the potential matches is assessed after tensor voting, when saliency information is available. Since smooth scene surfaces generate smooth layers in 3D or 4D, the decisions on the correctness of potential matches can be made based on their saliency.

The problem of matching is fundamental for computer vision areas dealing with more than one image and has been addressed in a variety of ways. It has been shown that no single technique can overcome all difficulties, which are mainly caused by occlusion and lack of texture. For instance, small matching windows are necessary to estimate the correct disparity or velocity close to region boundaries, but they are very noisy in textureless regions. The opposite is true for large windows, and there is no golden mean solution to the problem of window size selection. To overcome this, the initial matching is performed by normalized cross-correlation using square windows with multiple sizes. Since the objective is dense disparity and velocity map estimation, we adopt an area-based method and apply the windows to all pixels. Peaks in the cross-correlation function of each window size are retained as potential correspondences. The emphasis is toward detecting as many correct correspondences as possible, at the expense of a possibly large number of wrong ones, which can be rejected after tensor voting.

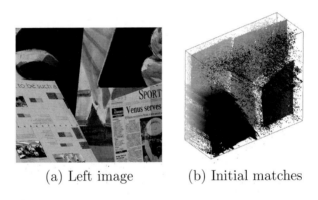

(a) Left image   (b) Initial matches

Figure 5.26. One of the two input frames and the initial matches for the Venus data set. The view of the initial matches is rotated in the 3D (x, y, d) space.

The candidate matches are encoded in the appropriate space as ball tensors, having no bias for a specific orientation, or as stick tensors, biased toward locally fronto-parallel surfaces. Since the correct orientation can be estimated with unoriented inputs, and moreover, a totally wrong initialization with stick tensors would have adverse effects on the solution, we initialize the tokens with ball tensors. Even though the matching score can be used as the initial saliency of the tensor, we prefer to ignore it and initialize all tensors with unit saliency. This allows the seamless combination of multiple matchers, including noncorrelation-based ones. The output of the initial matching module is a cloud of points in which the inliers form salient surfaces. Figure 5.26 shows one of the input images of the stereo pair and the initial matches for the "Venus" data set from [57]).

5.6.2   Uniqueness

Another constraint that is specific to the problems of stereo and motion is that of uniqueness. The required output is usually a disparity or velocity estimate for every pixel in the reference image. Therefore, after sparse voting, the most salient token for each pixel position is retained and the other candidates are rejected. In the case of stereo, we are interested in surface saliency, since even thin objects can be viewed as thin surface patches that reflect light to the cameras. Therefore, the criterion for determining the correct match for a pixel is surface saliency $\lambda_1 - \lambda_2$. In the case of visual motion, smooth objects in the scene that undergo smooth motions appear as 2D "surfaces" in the 4D space. For each (x, y) pair, we are interested in determining a unique velocity $[v_x \ v_y]$, or, equivalently, we are looking for a structure that has two tangents and two normals in the 4D space. The appropriate saliency is $\lambda_2 - \lambda_3$, which encodes the saliency of a 2D manifold in 4D space. Tokens with very low saliency are also rejected, even if they satisfy uniqueness, since the lack of support from the neighborhood indicates that they are probably wrong matches.

5.6.3   Discrete densification

Due to failures in the initial matching stage to generate the correct candidate for every pixel, there are pixels without disparity estimates after the uniqueness enforcement module (Figure 5.27a). In 3D, we could proceed with a dense vote and extract dense surfaces as in Section 5.4.3. Alternatively, since the objective is a disparity or velocity estimate for each pixel, we can limit the processing time by filling in the missing estimates only, based on their neighbors. For lack of a better term, we call this stage *discrete densification*. With a reasonable amount of missing data, this module has a small fraction of the computational complexity of the continuous alternative and is feasible in 4D or higher dimensions.

The first step is the determination of the range of possible disparities or velocities for a given pixel. This can be accomplished by examining the estimates from the previous stage in a neighborhood around the pixel. The size of this neighborhood is not critical and can be taken equal to the size of the voting neighborhood. Once the range of the neighbors is found, it is extended at both ends to allow the missing pixel to have disparity or velocity somewhat smaller or larger than all its neighbors. New candidates are generated for the pixel with disparities or velocities that span the extended range just estimated. Votes are collected at these candidates from their neighbors as before, but the new candidates do not cast any votes. Figure 5.27b shows the dense disparity map computed from the incomplete map of Figure 5.27a.

(a) Sparse disparity map   (b) Dense disparity map

Figure 5.27. Disparity map for Venus after outlier rejection and uniqueness enforcement (missing estimates colored white); and the disparity map after discrete densification.

The incomplete map of Figure 5.27a shows that most holes appear where the image of Figure 5.26a is textureless, and thus the initial matching is ambiguous. The discrete densification stage completes missing data by enforcing good continuation and smoothly extending the inferred structures. Smooth continuation is more likely when there are no cues, such as image intensity edges, suggesting otherwise.

5.6.4 Discontinuity localization

The last stage of processing attempts to correct the errors caused by occlusion. The input is a set disparity or motion layers produced by grouping the tokens of the previous stage. The boundaries of these layers, however, are not correctly localized. The errors are due to occlusion and the shortcomings of the initial matching stage. Occluded pixels that are visible in one image only very often produce high matching scores at the disparity or velocity of the occluding layer. For example, point A, which is occluded in the left image of Figure 5.28, appears as a better match for B' in the right image than does B, which is the correct match. Also, the difference in the level of texture in two adjacent regions, in terms of contrast and frequency, can bias the initial matching toward the more textured region.

Since it is hard to detect the problematic regions during initial matching, we propose a postprocessing step to correct such errors, which was first published in [48] for the case of motion. Since binocular cues are unreliable next

(a) Left image   (b) Right image

Figure 5.28. Problems in matching caused by occlusion. A appears a better match for B' than B, which is the true corresponding pixel.

to layer boundaries due to occlusion, we have to resort to monocular information. A reasonable assumption is that the intensity distributions of two surfaces that are distant in the scene, but appear adjacent from a particular viewpoint, should be quite different. Therefore, the presence of a discontinuity in the scene should also be accompanied by an edge in the image. Instead of detecting edges in the entire image and having to discriminate between true discontinuities and texture edges, we propose to focus on "uncertainty zones" around the boundaries inferred at the previous stage. The width of the uncertainty zones depends on the half-size of the matching window, since this is the maximum surface overextension as well as the width of the occluded region, which can be estimated from the difference in disparity or velocity at the boundary. In practice, we mark as uncertain the entire occluded area extended by half the size of the maximum matching window on either side.

Once the uncertainty zones have been labeled as such, we look for edges in them rather than in the entire image. This is done by tensor voting in 2D, since some of the edges we are looking for may be fragmented or consist of aligned edgels of low contrast. The initialization is performed by running a simple gradient operator and initializing tokens at each pixel position as stick tensors whose orientation is given by the gradient and whose saliency is given by

$$Sal(x,y) = G(x,y) \times e^{-\frac{(x-x_o)^2}{\sigma_e^2}}, \tag{5.13}$$

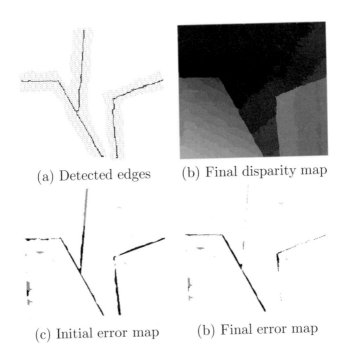

(a) Detected edges   (b) Final disparity map

(c) Initial error map   (b) Final error map

Figure 5.29. Discontinuity localization for the initial disparity map for Venus. Uncertainty zones are marked in gray and the final edges in black. The error maps before and after the correction are in the bottom row. Gray denotes disparity errors between one-half and one level and black errors larger than one disparity level.

where $G(x, y)$ is the magnitude of the gradient at (x, y), x_o is the position of the original layer boundary, and σ_e is a function of the width of the uncertainty zone. The exponential term serves as a prior that biases the extracted edges toward the original layer boundaries and, more importantly, toward the direction of the original boundaries. That is, a vertical boundary is more salient in a region where an almost vertical boundary was originally detected. This makes the edge detection scheme more robust to texture edges of different orientations. Voting is performed in 2D with these tokens as inputs, and edges are extracted starting from seeds, tokens with maximum curve saliency, and grown. Voting enables the completion of fragmented edges while the interference by spurious strong responses of the gradient is not catastrophic. Results for the example of Figure 5.28 can be seen in Figure 5.29.

5.6.5   Stereo

We pose the problem of stereo as the extraction of salient surfaces in 3D disparity space. After the initial matches are generated, as described in Section 5.6.1, the correct ones, which correspond to the actual disparities of scene points, form coherent surfaces in disparity space, which can be inferred by tensor voting. The initial candidate matches are encoded as balls, and tensor voting is performed. Then, the uniqueness constraint is enforced and tokens with low saliency are rejected as outliers. The advantage of the

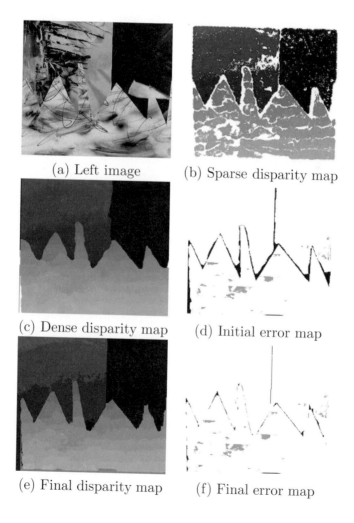

(a) Left image   (b) Sparse disparity map

(c) Dense disparity map   (d) Initial error map

(e) Final disparity map   (f) Final error map

Figure 5.30. Results for the "sawtooth" data set of [57].

tensor voting-based approach is that interaction among tokens occurs in 3D instead of the 2D image space or the 1D epipolar line space, thus minimizing the interference between points that appear adjacent in the image but are projections of distant points in the scene. The output at this stage is a sparse disparity map, which for the Venus example can be seen in Figure 5.26c. Preliminary results on binocular stereo, before the development of many of the modules presented here, have been published in [34, 35].

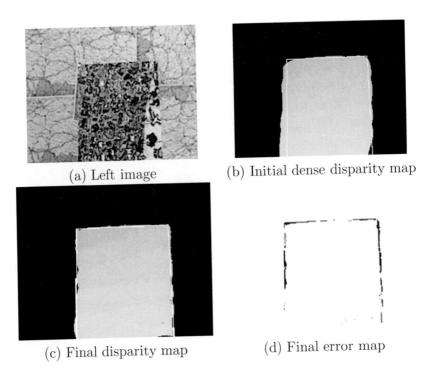

(a) Left image   (b) Initial dense disparity map

(c) Final disparity map   (d) Final error map

Figure 5.31. Results for the "map" data set of [57], where the ground truth depth discontinuities have been superimposed on the disparity maps.

Discrete densification is performed to fill in the holes of the sparse depth map. The results for the "sawtooth" data set of [57] can be seen in Figure 5.30c. Errors mostly exist near depth discontinuities (Figure 5.30d, where errors greater than one disparity level are marked in black and errors between one-half and one disparity level are marked in gray), and they are corrected according to Section 5.6.4. The final disparity and error maps appear in Figure 5.30e and f. Other results can be seen in Figures 5.31 and 5.32. Note that in the last example, the surfaces are not planar and the images are

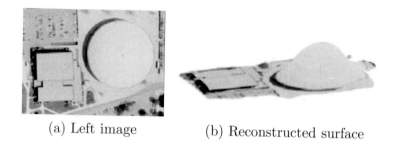

(a) Left image   (b) Reconstructed surface

Figure 5.32. Results for the "arena" data set where the surfaces have been reconstructed in disparity space and texture-mapped.

rather textureless. Despite these difficulties, we can infer the correct surfaces according to Section 5.4.3, which are shown texture-mapped in disparity space in Figure 5.32c-d.

5.6.6   Multiple view stereo

We have also applied the stereo algorithm to the problem of multiple view stereo. The novelty of our approach is the processing of all views simultaneously instead of processing them pairwise and merging the results. The algorithm and results can be found in [42]. The additional difficulty stemming from the increase in the number of views is a considerable increase in the number of tokens that need to be processed. Methods that operate on $2\frac{1}{2}D$ maps are not applicable, since uniqueness with respect to a grid does not hold in this case. This poses no additional difficulties for our method, since our representation is object-centered and not view-centered.

Candidate matches are generated from image pairs, as in the binocular case, and are reconstructed in 3D space using the calibration information that needs to be provided. If two or more candidate matches fall in the same voxel, their saliencies are added, thus making tokens confirmed by more views more salient. The number of tokens is a little below 2 million for the example in Figure 5.33. Processing of large data sets is feasible due to the local nature of the voting process. Tensor voting is performed on the entire set of candidate matches from all images, and then uniqueness is enforced with respect to rays emanating from pixels in the images. Four of the 36 input frames of the lighthouse sequence, a view of the candidate matches, and the inferred surface inliers and boundaries can be seen in Figure 5.33.

(a) Four of the input images   (b) Initial matches

(c) Surface inliers   (d) Surface boundaries

Figure 5.33. Results for the "lighthouse" sequence that consists of 36 images captured using a turntable.

5.6.7 Visual motion from motion cues

In this section, we address the problem of perceptual organization using motion cues only, which was initially published in [46]. The input is a pair of images, such as the ones in Figure 5.34a, where monocular cues provide no information, but when viewed one after another, human observers perceive groupings based on motion. In the case of Figure 5.34a the perceived groups are a translating disk and a static background. As discussed in the introduction of Section 5.6, the appropriate space in the case of visual motion is 4D, with the axes being the two image axes and the two velocity axes v_x and v_y. Processing in this space alleviates problems associated with pixels that are adjacent in the image but belong in different motion layers. For instance, pixels A and B in Figure 5.34b that are close in image space and have similar velocities are also neighbors in the 4D space. On the other hand, pixels A and C that are neighbors in the image but have different velocities are distant in 4D space, and pixel D, which has been assigned an erroneous velocity, appears isolated.

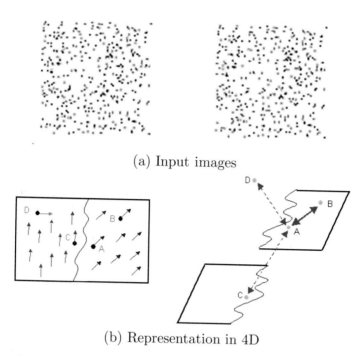

(a) Input images

(b) Representation in 4D

Figure 5.34. Random dot images generated by a translating disk over a static background and advantages of processing in 4D.

Candidate matches are generated by assuming every dot in the first image corresponds to every other dot within a velocity range in the other image. A 3D view (where the v_y component has been dropped) of the point cloud generated by this process for the images of Figure 5.34a can be seen in Figure 5.35a where v_y is not displayed. The correct matches form salient surfaces surrounded by a large number of outliers. The surfaces appear flat in this case because the motion is a pure translation with constant velocity. These matches are encoded as 4D ball tensors and cast votes to their neighbors. Since the correct matches form larger and denser layers, they receive more support from their neighbors and emerge as the most salient velocity candidates for their respective pixels. The relevant saliency is 2D surface saliency in 4D, which is encoded by $\lambda_2 - \lambda_3$ (see also Section 5.6.2). This solves the problem of matching for the dots of the input images. The estimated sparse velocity field can be seen in Figure 5.35b.

People, however, perceive not only the dots as moving but also their neighboring background pixels, the white pixels in the examples shown here. This phenomenon is called motion capture and is addressed at the discrete

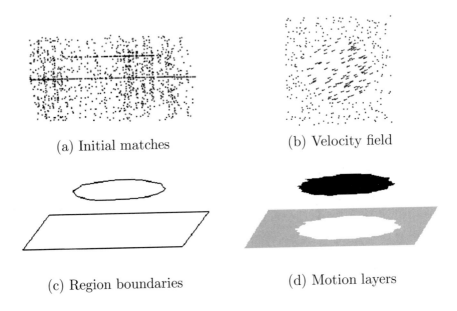

(a) Initial matches   (b) Velocity field

(c) Region boundaries   (d) Motion layers

Figure 5.35. Results on the translating disk over a static background example. The 3D views were generated by dropping v_y.

densification stage, where velocity candidates are generated for all pixels without velocity estimates based on the velocities of their neighbors. Votes are collected at these candidates, and the most salient are retained. Finally, the tokens that have been produced by both stages are grouped into smooth motion layers in 4D space. Regions and boundaries can be seen in Figure 5.35c and d. An example with nonrigid motion can be seen in Figure 5.36, where a disk is undergoing an expanding motion.

5.6.8   Visual motion on real images

After demonstrating the validity of our approach on synthetic data, we applied it to real images [45, 47, 48]. Candidate matches are generated by multiple cross-correlation windows applied to all pixels, as in the case of stereo, but since the epipolar constraint does not hold, the search for matches is done in 2D neighborhoods in the other image. The tokens are initialized as 4D ball tensors, and tensor voting is performed to compute the saliency of the tokens. The token with the largest surface saliency is selected as the correct match for each pixel after outliers with low saliency are removed from the data set. Results on the Yosemite sequence, which is a typical bench-

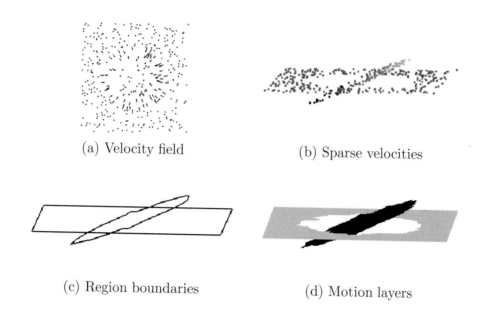

(a) Velocity field

(b) Sparse velocities

(c) Region boundaries

(d) Motion layers

Figure 5.36. Results on an expanding disk. Again, the 3D views were generated by dropping v_y.

mark example for motion estimation for which ground truth exists, can be seen in Figure 5.37. The average angular error is 3.74 degrees with standard deviation 4.3 degrees with velocities computed for 100% of the image. Somewhat smaller errors have been achieved by methods that compute velocities for less than 10% of the pixels [47].

Since the input is real images, the problems described in Section 5.6.4 decrease the quality of the results, as can be seen in Figure 5.38c-d. Discontinuity detection as in Section 5.6.4 and [48] is performed to correct the velocity maps produced after discrete densification. Results can be seen in Figures 5.39 and 5.40.

5.7 Conclusion and Future Work

We have presented the current state of the tensor voting framework, which is a product of a number of years of research performed mostly at the University of Southern California. In this section, we present the contributions of the framework to perceptual organization and computer vision problems, as well as the axes of our ongoing and future research.

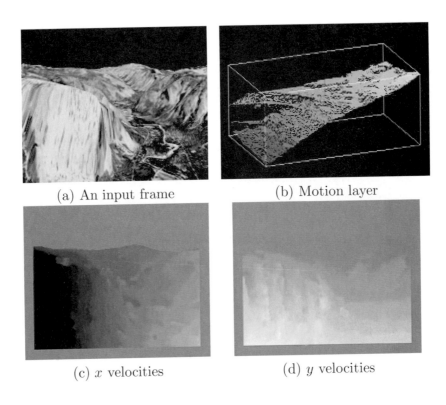

(a) An input frame   (b) Motion layer

(c) x velocities   (d) y velocities

Figure 5.37. Results on the Yosemite sequence.

The tensor voting framework provides a general methodology that can be applied to a large range of problems as long as they can be posed as the inference of salient structures in a metric space of any dimension. The benefits from our representation and voting schemes are that no models need to be known a priori, nor do the data have to fit a parametric model. In addition, all types of perceptual structures can be represented and inferred at the same time. Processing can begin with unoriented inputs, is noniterative, and there is only one free parameter, the scale of the voting field. Tensor voting facilitates the propagation of information locally and enforces smoothness while explicitly detecting and preserving discontinuities with very little initialization requirements. The local nature of the operations makes the framework efficient and applicable to very large data sets. Robustness to noise is an important asset of the framework due to the large amounts of noise that are inevitable in computer vision problems.

(a) Input image   (b) Initial matches

(c) Initial motion layers   (d) Initial boundaries

(e) Corrected motion layers   (f) Corrected boundaries

Figure 5.38. Results on the "candybox" example. The initial matches are shown rotated in (x, y, v_x) space. The second row shows results before discontinuity localization, while the third row is after correction.

Results, besides the organization of generic tokens, have been shown in real computer vision problems such as stereo and motion analysis. Performance equivalent or superior to state-of-the-art algorithms has been achieved

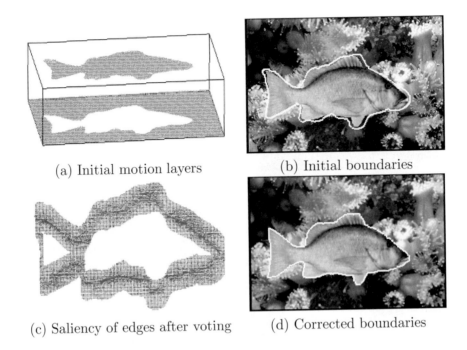

(a) Initial motion layers   (b) Initial boundaries

(c) Saliency of edges after voting   (d) Corrected boundaries

Figure 5.39. Results on the synthetic "fish" example. The initial layers are shown rotated in (x, y, v_x) space. The uncertainty zones and curve saliency after tensor voting in 2D are shown in (c), and the corrected boundaries in (d).

without the use of algorithms that are specific to each problem, but rather with general, simple, and reusable modules. Results in other computer vision problems using tensor voting have also been published in [63, 30, 64, 69, 68, 67, 27, 28]. A major axis of our future work is the generation of tokens from images in a more direct and natural way. This will expand the range of problems we currently address without having to develop new algorithms for each problem.

Arguably the largest remaining issue is that of automatic scale selection or of a multiscale implementation of the tensor voting framework. It has been addressed in [69] under circumstances where additional assumptions, such as the nonexistence of junctions, could be made. Scale affects the level of details that are captured, the degree of smoothness of the output, the completion of missing information, and the robustness to noise. Since these factors are often conflicting, the use of a single scale over a complicated data set is a compromise that fails to produce the best results everywhere. One

(a) Input image   (b) Initial matches

(c) Initial boundaries   (d) Corrected boundaries

Figure 5.40. Results on the "barrier" example. The initial matches are shown rotated in (x, y, v_x) space. Discontinuity detection significantly improves the quality of the layer boundaries.

way to address this is to adapt the scale locally according to some criterion such as data density. Alternatively, a number of scales can be used and the problem can be transformed to one of finding the way to integrate the results over the different scales. A fine-to-coarse scheme was adopted in [69], which is consistent with the majority of the literature, but one could envision coarse-to-fine schemes as well. In any case, the issue remains open and is further complicated by that fact that even human perception is a function of scale.

5.8   Acknowledgments

The authors would like to thank all those who contributed to the research in tensor voting at the University of Southern California: from the original vector voting scheme of Gideon Guy, to the first implementation of the framework as presented in [40] by Chi-Keung Tang and Mi Suen Lee, to the other contributors: Pierre Kornprobst, Amin Massad, Lee Gaucher, Mircea Nicolescu, Eun-Young Kang, Jinman Kang, René Dencker Eriksen, and Wai-Shun Tong.

Bibliography

[1] S. Arya, D. M. Mount, N. S. Netanyahu, R. Silverman, and A. Y. Wu. An optimal algorithm for approximate nearest neighbor searching fixed dimensions. *Journal of the ACM*, 45(6):891–923, 1998.

[2] J. D. Boissonnat. Representing 2D and 3D shapes with the Delaunay triangulation. In *ICPR84*, pages 745–748, 1984.

[3] Y. Boykov, O. Veksler, and R. Zabih. Fast approximate energy minimization via graph cuts. *PAMI*, 23(11):1222–1239, November 2001.

[4] H. Chen, P. Meer, and D. E. Tyler. Robust regression for data with multiple structures. In *CVPR01*, volume I, pages 1069–1075, 2001.

[5] K. J. Cho and P. Meer. Image segmentation from consensus information. *CVIU*, 68(1):72–89, October 1997.

[6] J. Dolan and E. M. Riseman. Computing curvilinear structure by token-based grouping. In *CVPR92*, pages 264–270, 1992.

[7] H. Edelsbrunner and E. P. Mücke. Three-dimensional alpha shapes. *ACM Transactions on Graphics*, 13(1):43–72, 1994.

[8] O. D. Faugeras and M. Berthod. Improving consistency and reducing ambiguity in stochastic labeling: An optimization approach. *PAMI*, 3(4):412–424, July 1981.

[9] M. A. Fischler and R. C. Bolles. Random sample consensus: A paradigm for model fitting with applications to image analysis and automated cartography. *Comm. of the ACM*, 24(6):381–395, 1981.

[10] Y. Gdalyahu, D. Weinshall, and M. Werman. Self-organization in vision: Stochastic clustering for image segmentation, perceptual grouping, and image database organization. *PAMI*, 23(10):1053–1074, October 2001.

[11] S. Geman and D. Geman. Stochastic relaxation, Gibbs distributions, and the Bayesian restoration of images. *PAMI*, 6(6):721–741, November 1984.

[12] G. H. Granlund and H. Knutsson. *Signal Processing for Computer Vision*. Kluwer, December 1995.

[13] S. Grossberg and E. Mingolla. Neural dynamics of form perception: Boundary completion. *Psychological Review*, pages 173–211, 1985.

[14] S. Grossberg and D. Todorovic. Neural dynamics of 1D and 2D brightness perception: A unified model of classical and recent phenomena. *Perception and Psychophysics*, 43:723–742, 1988.

[15] G. Guy and G. Medioni. Inferring global perceptual contours from local features. *IJCV*, 20(1/2):113–133, 1996.

[16] G. Guy and G. Medioni. Inference of surfaces, 3D curves, and junctions from sparse, noisy, 3D data. *PAMI*, 19(11):1265–1277, November 1997.

[17] R. M. Haralick and L. G. Shapiro. The consistent labeling problem: Part i. *PAMI*, 1(2):173–184, April 1979.

[18] R. M. Haralick and L. G. Shapiro. The consistent labeling problem: Part ii. *PAMI*, 2(3):193–203, May 1980.

[19] F. Heitger and R. von der Heydt. A computational model of neural contour processing: Figure-ground segregation and illusory contours. In *ICCV93*, pages 32–40, 1993.

[20] S. Osher, H. K. Zhao, and R. Fedkiw. Fast surface reconstruction using the level set method. In *UCLA Computational and Applied Mathematics Reports*, pages 32–40, 2001.

[21] H. Hoppe, T. DeRose, T. Duchamp, J. McDonald, and W. Stuetzle. Surface reconstruction from unorganized points. *Computer Graphics*, 26(2):71–78, 1992.

[22] B. K. P. Horn and B. G. Schunck. Determining optical flow. *AI*, 17(1-3): 185–203, August 1981.

[23] P. V. C. Hough. Method and means for recognizing complex patterns. In *US Patent*, 1962.

[24] R. A. Hummel and S. W. Zucker. On the foundations of relaxation labeling processes. *PAMI*, 5(3):267–287, May 1983.

[25] D. W. Jacobs. Robust and efficient detection of salient convex groups. *PAMI*, 18(1):23–37, January 1996.

[26] A. K. Jain and R. C. Dubes. *Algorithms for clustering data*. Prentice-Hall, 1988.

[27] J. Jia and C. K. Tang. Image repairing: Robust image synthesis by adaptive ND tensor voting. In *CVPR03*, volume I, pages 643–650, 2003.

[28] J. Kang, I. Cohen, and G. Medioni. Continuous tracking within and across camera streams. In *CVPR03*, volume I, pages 267–272, 2003.

[29] G. K. Kanizsa. Subjective contours. *Scientific American*, 234:48–54, 1976.

[30] P. Kornprobst and G. Medioni. Tracking segmented objects using tensor voting. In *CVPR00*, volume II, pages 118–125, 2000.

[31] K. Koster and M. Spann. Mir: An approach to robust clustering-application to range image segmentation. *PAMI*, 22(5):430–444, May 2000.

[32] Y. G. Leclerc. Constructing simple stable descriptions for image partitioning. *IJCV*, 3(1):73–102, May 1989.

[33] K. M. Lee, P. Meer, and R. H. Park. Robust adaptive segmentation of range images. *PAMI*, 20(2):200–205, February 1998.

[34] M. S. Lee and G. Medioni. Inferring segmented surface description from stereo data. In *CVPR98*, pages 346–352, 1998.

[35] M. S. Lee, G. Medioni, and P. Mordohai. Inference of segmented overlapping surfaces from binocular stereo. *PAMI*, 24(6):824–837, June 2002.

[36] Z. Li. A neural model of contour integration in the primary visual cortex. *Neural Computation*, 10:903–940, 1998.

[37] W. E. Lorensen and H. E. Cline. Marching cubes: A high resolution 3D surface reconstruction algorithm. *Computer Graphics*, 21(4):163–169, 1987.

[38] D. G. Lowe. *Perceptual Organization and Visual Recognition*. Kluwer, June 1985.

[39] D. Marr. *Vision*. Freeman Press, 1982.

[40] G. Medioni, M. S. Lee, and C. K. Tang. *A Computational Framework for Segmentation and Grouping*. Elsevier, 2000.

[41] R. Mohan and R. Nevatia. Perceptual organization for scene segmentation and description. *PAMI*, 14(6):616–635, June 1992.

[42] P. Mordohai and G. Medioni. Perceptual grouping for multiple view stereo using tensor voting. In *ICPR02*, volume III, pages 639–644, 2002.

[43] J. M. Morel and S. Solimini. *Variational Methods in Image Segmentation*. Birkhauser, 1995.

[44] H. Neummann and E. Mingolla. Computational neural models of spatial integration in perceptual grouping. *From Fragments to Objects: Grouping and Segmentation in Vision*, T. F. Shipley and P. J. Kellman (Eds.), pages 353–400, 2001.

[45] M. Nicolescu and G. Medioni. 4D voting for matching, densification and segmentation into motion layers. In *ICPR02*, volume III, pages 303–308, 2002.

[46] M. Nicolescu and G. Medioni. Perceptual grouping from motion cues using tensor voting in 4D. In *ECCV02*, volume III, pages 423–428, 2002.

[47] M. Nicolescu and G. Medioni. Layered 4D representation and voting for grouping from motion. *PAMI*, 25(4):492–501, April 2003.

[48] M. Nicolescu and G. Medioni. Motion segmentation with accurate boundaries – a tensor voting approach. In *CVPR03*, volume I, pages 382–389, 2003.

[49] S. Osher and R. P. Fedkiw. *The Level Set Method and Dynamic Implicit Surfaces*. Springer Verlag, 2002.

[50] P. Parent and S. W. Zucker. Trace inference, curvature consistency, and curve detection. *PAMI*, 11(8):823–839, August 1989.

[51] T. A. Poggio, V. Torre, and C. Koch. Computational vision and regularization theory. *Nature*, 317:314–319, 1985.

[52] L. Robert and R. Deriche. Dense depth map reconstruction: A minimization and regularization approach which preserves discontinuities. In *ECCV96*, volume I, pages 439–451, 1996.

[53] A. Robles-Kelly and E. R. Hancock. An expectation-maximisation framework for perceptual grouping. In *IWVF4, LNCS 2059*, pages 594–605. Springer Verlag, 2001.

[54] P. T. Sander and S. W. Zucker. Inferring surface trace and differential structure from 3D images. *PAMI*, 12(9):833–854, September 1990.

[55] S. Sarkar and K. L. Boyer. A computational structure for preattentive perceptual organization: Graphical enumeration and voting methods. *SMC*, 24: 246–267, 1994.

[56] E. Saund. Labeling of curvilinear structure across scales by token grouping. In *CVPR92*, pages 257–263, 1992.

[57] D. Scharstein and R. Szeliski. A taxonomy and evaluation of dense two-frame stereo correspondence algorithms. *IJCV*, 47(1-3):7–42, April 2002.

[58] J. A. Sethian. *Level Set Methods: Evolving Interfaces in Geometry, Fluid Mechanics, Computer Vision and Materials Science.* Cambridge University Press, 1996.

[59] A. Shashua and S. Ullman. Structural saliency: The detection of globally salient structures using a locally connected network. In *ICCV88*, pages 321–327, 1988.

[60] J. Shi and J. Malik. Normalized cuts and image segmentation. *PAMI*, 22(8): 888–905, August 2000.

[61] R. Szeliski, D. Tonnesen, and D. Terzopoulos. Modeling surfaces of arbitrary topology with dynamic particles. In *CVPR93*, pages 82–87, 1993.

[62] C. K. Tang and G. Medioni. Inference of integrated surface, curve, and junction descriptions from sparse 3D data. *PAMI*, 20(11):1206–1223, November 1998.

[63] C. K. Tang, G. Medioni, and M. S. Lee. Epipolar geometry estimation by tensor voting in 8D. In *ICCV99*, pages 502–509, 1999.

[64] C. K. Tang, G. Medioni, and M. S. Lee. N-dimensional tensor voting and application to epipolar geometry estimation. *PAMI*, 23(8):829–844, August 2001.

[65] D. Terzopoulos. Regularization of inverse visual problems involving discontinuities. *PAMI*, 8(4):413–424, July 1986.

[66] D. Terzopoulos and D. Metaxas. Dynamic 3D models with local and global deformations: Deformable superquadrics. *PAMI*, 13(7):703–714, July 1991.

[67] W. S. Tong and C. K. Tang. Rod-tv: Reconstruction on demand by tensor voting. In *CVPR03*, volume II, pages 391–398, 2003.

[68] W. S. Tong, C. K. Tang, and G. Medioni. Epipolar geometry estimation for non-static scenes by 4D tensor voting. In *CVPR01*, volume I, pages 926–933, 2001.

[69] W. S. Tong, C. K. Tang, and G. Medioni. First-order tensor voting and application to 3D scale analysis. In *CVPR01*, volume I, pages 175–182, 2001.

[70] M. Wertheimer. Laws of organization in perceptual forms. *Psycologische Forschung*, Translation by W. Ellis, *A source book of Gestalt psychology (1938)*, 4:301–350, 1923.

[71] C. F. Westin. *A Tensor Framework for Multidimensional Signal Processing.* Ph.D. thesis, Linkoeping University, Sweden, 1994.

[72] L. R. Williams and D. W. Jacobs. Stochastic completion fields: A neural model of illusory contour shape and salience. *Neural Computation*, 9(4):837–858, 1997.

[73] S. C. Yen and L. H. Finkel. Extraction of perceptually salient contours by striate cortical networks. *Vision Research*, 38(5):719–741, 1998.

[74] X. M. Yu, T. D. Bui, and A. Krzyzak. Robust estimation for range image segmentation and reconstruction. *PAMI*, 16(5):530–538, May 1994.

PART II
APPLICATIONS IN
COMPUTER VISION

The chapters in this section describe a variety of interesting applications in computer vision, ranging from the more traditional (content-based image retrieval, face detection, human tracking) to the more graphics-oriented (image-based lighting and visual effects).

In Chapter 6, Debevec describes how scenes and objects can be illuminated using images of light from the real world. While this operation, also known as image-based lighting, has its roots in computer graphics, it requires computer vision techniques to extract high dynamic range images and resample the captured light.

Many of the special effects seen in movies rely on computer vision techniques to facilitate their production. In Chapter 7, Roble describes some vision techniques that have been used successfully in the movie industry.

A natural extension to current text-based search engines would be image retrieval. In Chapter 8, Gevers and Smeulders present a survey on the theory and techniques for content-based image retrieval. The issues covered include interactive query formulation, image feature extraction, representation and indexing, search techniques, and learning based on feedback.

Li and Lu describe techniques for face detection, alignment, and recognition in Chapter 9. They show how the difficult problems of changing head pose and different illumination can be handled.

In Chapter 10, Turk and Kölsch describe the area of perceptual interfaces, which involves the use of multiple perceptual modalities (e.g., vision, speech, haptic) to enable human-machine interaction. The authors motivate the need for such a study and discuss issues related to vision-based interfaces.

Chapter 6

IMAGE-BASED LIGHTING

Paul E. Debevec

Image-based lighting (IBL) is the process of illuminating scenes and objects—be they real or synthetic—with images of light from the real world. IBL evolved from the techniques known as environment mapping and reflection mapping [3, 20], in which panoramic images are used as texture maps on computer generated (CG) models to show shiny objects reflecting real and synthetic environments. When used effectively, image-based lighting can produce highly realistic renderings of people, objects, and environments, and it can be used to convincingly integrate computer generated objects into real-world scenes.

The basic steps in image-based lighting are

1. Capturing real-world illumination as an omnidirectional, high dynamic range image.

2. Mapping the illumination on to a representation of the environment.

3. Placing the 3D object inside the environment.

4. Simulating the light from the environment illuminating the CG object.

Figure 6.1 shows an object illuminated entirely using IBL. The light was captured in a kitchen and includes light from the fixture on the ceiling, the blue sky from the windows, and the indirect light from the walls, ceiling, and cabinets in the room. The light from this room was mapped onto a large sphere, and the model of the microscope on the table was placed in the middle of the sphere. Then, a global illumination renderer was used to

Figure 6.1. A microscope, modeled by Gary Butcher in 3D Studio Max, rendered using Marcos Fajardo's "Arnold" rendering system with image-based lighting captured within a kitchen.

simulate the object's appearance as illuminated by the light coming from the sphere of incident illumination. The model was created by Gary Butcher in 3D Studio Max, and the renderer used was the "Arnold" global illumination system written by Marcos Fajardo.

Because it is illuminated with captured lighting, the image of the computer-generated microscope should look much like a real microscope would appear in the kitchen environment. IBL simulates not just the direct illumination from the ceiling light and windows, but also the indirect illumination from the rest of the surfaces in the room. The reflections in the smooth curved bottles reveal the appearance of the kitchen, and the shadows on the table reveal the colors and spread of the area light sources. The objects also reflect each other, owing to the ray-tracing component of the global illumination techniques being employed.

The first part of this chapter describes the theory and practice of image-based lighting and provides concrete examples of the technique in action using Greg Ward's RADIANCE lighting simulation system. The second part of the chapter presents advanced image-based lighting techniques, such as having the objects cast shadows into the scene and capturing lighting that varies within space. The final part of this chapter discusses image-based *re*lighting, which is the process of synthetically illuminating real objects and people with light captured from the real world.

6.1   Basic Image-Based Lighting

6.1.1   Capturing light

The first step of image-based lighting is to obtain a measurement of real-world illumination, called a *light probe image* in [7]. Several such images are available in the Light Probe Image Gallery at http://www.debevec.org/Probes/. The web site includes the kitchen environment used to render the microscope as well as lighting captured in various other interior and outdoor environments. A few of these environments are shown in Figure 6.2.

Light probe images are photographically acquired images of the real world with two important properties: first, they are omnidirectional—for every direction in the world, there is a pixel in the image that corresponds to that direction. Second, their pixel values are linearly proportional to the amount of light in the real world. The rest of this section describes techniques for taking images that satisfy both of these properties.

Taking omnidirectional images can be done in a number of ways. The simplest way is to use a regular camera to take a photograph of a mirrored ball placed in a scene. A mirrored ball actually reflects the entire world around it, not just the hemisphere toward the camera: light rays reflecting off the outer circumference of the ball glance toward the camera from the back half of the environment. Another method to obtain omnidirectional images using is to shoot a mosaic of many pictures looking in different directions [4, 26] and combine them using an image stitching program such as RealViz Stitcher. A good way to cover a particularly large area in each shot is to use a fisheye lens [14], allowing the full field to be covered in as few as two

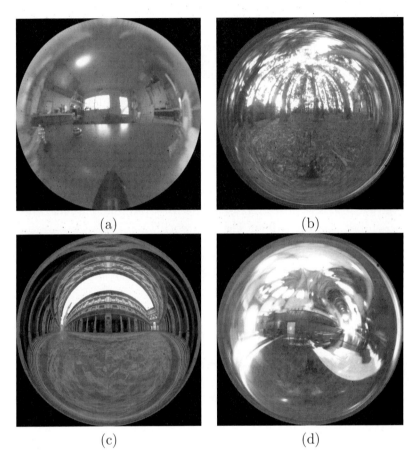

(a)   (b)

(c)   (d)

Figure 6.2. Several light probe images from the Light Probe Image Gallery at `http://www.debevec.org/Probes`. The light is from (a) a residential kitchen, (b) the eucalyptus grove at UC Berkeley, (c) the Uffizi gallery in Florence, Italy, and (d) Grace Cathedral in San Francisco.

images. A final technique is to use a special scanning panoramic camera, such as ones made by Panoscan or Spheron, which uses a vertical row of CCD elements on a rotating camera head to scan across a 360-degree field of view.

Most digital images do not have the property that pixel values are proportional to light levels. Usually, light levels are encoded nonlinearly in order to appear correctly or more pleasingly on nonlinear display devices such as computer monitors. Furthermore, standard digital images typically repre-

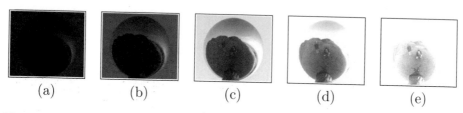

Figure 6.3. A series of differently exposed images of a mirrored ball photographed at Funston Beach near San Francisco. The exposures, ranging in shutter speed from 1/1000 second in (a) to 1/4 second in (e), are merged into a high dynamic range image to be used as an image-based lighting environment.

sent only a small fraction of the dynamic range[1] present in most real-world lighting environments. When a part of the scene is too bright, the pixels saturate to their maximum value (which is 255 for 8-bit images) no matter how bright they really are.

Ensuring that the pixel values in the omnidirectional images are truly proportional to quantities of light can be accomplished using high dynamic range (HDR) photography techniques, as described in [11, 21]. The process typically involves taking a series of pictures of the scene with varying exposure levels, and then using these images both to solve for the response curve of the imaging system and to form a linear-response composite image covering the entire range of illumination values present in the scene. Software for assembling images in this way includes the command-line `mkhdr` program at `http://www.debevec.org/Research/HDR/` and the Windows-based HDR Shop program at `http://www.debevec.org/HDRShop/`.

HDR images can be represented well using a single-precision floating-point number for each of red, green, and blue, allowing the full range of light from thousandths to millions of $W/(srm^2)$ to be represented. HDR image data can be stored in a variety of file formats, including the floating-point version of the TIFF file format and the Portable Floatmap [9] variant of the PBM Plus Portable Pixmap format. Several other representations that use less storage are available, including Greg Ward's RGBE format [29] that uses one byte for each of red, green, blue, and a common 8-bit exponent, and Ward's 24-bit and 32-bit LogLuv formats recently included in the TIFF standard [18]. The light probe images in the Light Probe Image Gallery are available in the RGBE format, allowing them to be easily used in Ward's

---

[1]The ratio between the dimmest and brightest accurately represented regions of an image.

RADIANCE global illumination renderer. We'll see how to do this in the next section.

Figure 6.3 shows a series of images used in creating a light probe image. To acquire these images, a 3-inch polished ball bearing was placed on top of a tripod at Funston Beach near San Francisco, and a digital camera with a telephoto zoom lens was used to take a series of exposures of the ball. Being careful not to disturb the camera, pictures were taken at shutter speeds ranging from 1/4 second to 1/10000 second, allowing the camera to properly image everything from the dark cliffs to the bright sky and the setting sun. The images were assembled using code similar to that now found in `mkhdr` and HDR Shop, yielding a high dynamic range, linear-response image.

6.1.2   Illuminating synthetic objects with real light

Image-based lighting is the process of using light probe images as sources of illumination for CG scenes and objects. IBL is now supported by several commercial renderers, including LightWave 3D, Entropy, and Final Render. For this tutorial, we use the freely downloadable RADIANCE lighting simulation package written by Greg Ward at Lawrence Berkeley Laboratories. RADIANCE is a UNIX package, which means that to use it you must use a computer such as a PC running Linux or a Silicon Graphics or Sun workstation. Here are the steps to using image-based lighting to illuminate synthetic objects in RADIANCE:

1. Download and install RADIANCE.

   First, test to see if you already have RADIANCE installed by typing `which rpict` at a UNIX command prompt. If the shell returns "Command not found," you must install RADIANCE. To do this, visit the RADIANCE web site at `http://radsite.lbl.gov/radiance` and click on the download option. As of this writing, the current version is 3.1.8, and it is precompiled for SGI and Sun workstations. For other operating systems, such as Linux, you can download the source files and then compile the executable programs using the `makeall` script. Once installed, make sure that the RADIANCE binary directory is in your `$PATH` and that your `$RAYPATH` environment variable includes the RADIANCE library directory. Your system administrator should be able to help you if you are not familiar with installing software packages on UNIX.

Table 6.1. The material specifiers in `scene.rad`

```
# Materials
void plastic red_plastic 0 0 5 .7 .1 .1 .06 .1
void metal steel 0 0 5 0.6 0.62 0.68 1 0
void metal gold 0 0 5 0.75 0.55 0.25 0.85 0.2
void plastic white_matte 0 0 5 .8 .8 .8 0 0
void dielectric crystal 0 0 5 .5 .5 .5 1.5 0
void plastic black_matte 0 0 5 .02 .02 .02 .00 .00
void plastic gray_plastic 0 0 5 0.25 0.25 0.25 0.06
0.0
```

2. Create the scene.

The first thing we do is create a RADIANCE scene file. RADIANCE scene files contain the specifications for the geometry, reflectance properties, and lighting in your scene. We create a very simple scene with a few spheres sitting on a platform. First, let us specify the material properties for the spheres. Create a new directory and then call up your favorite text editor to type in the material specifications in Table 6.1 to the file `scene.rad`.

These lines specify the diffuse and specular characteristics of the materials we use in our scene, including crystal, steel, and red plastic. In the case of the red plastic, the diffuse RGB color is (.7, .1, .1), the proportion of light reflected specularly is .06, and the specular roughness is .1. The 0 0 5 just after the names tell RADIANCE how many alphanumeric, integer, and floating-point parameters to expect in each specifier.

Now let us add some objects with these material properties to our scene. The objects we choose are some spheres sitting on a pedestal. Add the lines in Table 6.2 to the end of `scene.rad`.

The lines in Table 6.2 specify five spheres made from the various materials sitting in an arrangement on the pedestal. The first sphere, `ball0`, is made of the `red_plastic` material and located in the scene at (2,0.5,2) with a radius of 0.5. The pedestal itself is composed of two beveled boxes made with the RADIANCE `genbox` generator program.

Table 6.2. The geometric shapes in `scene.rad`

```
# Objects
red_plastic sphere ball0 0 0 4 2 0.5 2 0.5
steel sphere ball1 0 0 4 2 0.5 -2 0.5
gold sphere ball2 0 0 4 -2 0.5 -2 0.5
white_matte sphere ball3 0 0 4 -2 0.5 2 0.5
crystal sphere ball4 0 0 4 0 1 0 1
!genworm black_matte twist "cos(2*PI*t)*(1+0.1*cos(30*PI*t))"
"0.06+0.1+0.1*sin(30*PI*t)" "sin(2*PI*t)*(1+0.1*cos(30*PI*t))"
"0.06" 200 | xform -s 1.1 -t 2 0 2 -a 4 -ry 90 -i 1
!genbox gray_plastic pedestal_top 8 0.5 8 -r 0.08 | xform -t
-4 -0.5 -4
!genbox gray_plastic pedestal_shaft 6 16 6 | xform -t -3
-16.5 -3
```

Table 6.3. A traditional light source for `scene.rad`

```
# Traditional Light Source
void light lightcolor 0 0 3 10000 10000 10000
lightcolor source lightsource 0 0 4 1 1 1 2
```

In addition, the **genworm** program is invoked to create some curly iron rings around the spheres.

3. Add a traditional light source.

Next, let us add a traditional light source to the scene to get our first illuminated glimpse—without IBL—of what the scene looks like. Add the lines from Table 6.3 to **scene.rad** to specify a traditional light source.

4. Render the scene with traditional lighting

In this step, we create an image of the scene. First, we use the **oconv** program to process the scene file into an octree file for RADIANCE to render. Type the following command to the UNIX command prompt:

```
# oconv scene.rad > scene.oct
```

The **#** simply indicates the prompt and does not need to be typed. This creates an octree file `scene.oct` that can be rendered in RADIANCE's interactive renderer `rview`. Next, we specify a camera position. This can be done as command arguments to `rview`, but to make things simpler, let us store our camera parameters in a file. Use your text editor to create the file `camera.vp` with the following camera parameters as the first and only line of the file:

```
rview -vtv -vp 8 2.5 -1.5 -vd -8 -2.5 1.5 -vu 0 1 0 -vh 60
-vv 40
```

These parameters specify a perspective camera (`-vtv`) with a given viewing position (`-vp`), direction (`-vd`), up vector (`-vu`), and with horizontal (`-vh`) and vertical (`-vv`) fields of view of 60 and 40 degrees respectively. (The `rview` text at the beginning of the line is a standard placeholder in RADIANCE camera files, not an invocation of the `rview` executable.) Now let us render the scene in `rview`; type:

```
# rview -vf camera.vp scene.oct
```

In a few seconds, you should get an image window similar to what is shown in Figure 6.4. The image shows the spheres on the platform, surrounded by the curly rings and illuminated by the traditional light source. The image may or may not be pleasing, but it certainly looks very computer-generated. Let us now see if we can make it more realistic by lighting the scene with image-based lighting.

5. Download a light probe image.

Visit the Light Probe Image Gallery at `http://www.debevec.org/Probes/` and choose a light probe image to download. The light probe images without concentrated light sources tend to produce good-quality renders more quickly, so using one of the "beach," "uffizi," or "kitchen" probes to begin is recommended. We'll choose the "beach" probe described earlier in our first example. Download the `beach_probe.hdr` file to your computer by shift-clicking or right-clicking "Save Target As..." or "Save Link As...," and then view it using the RADIANCE image viewer `ximage`:

Figure 6.4. The RADIANCE `rview` interactive renderer viewing the scene as illuminated by a traditional light source.

```
# ximage beach_probe.hdr
```

If the probe downloaded properly, a window should pop up displaying the beach probe image. While the window is up, you can click-drag the mouse pointer over a region of the image and then press "=" to re-expose the image to properly display the region of interest. If the image did not download properly, try downloading and expanding the `all_probes.zip` or `all_probes.tar.gz` archive from the same web page, which will download all the light probes images and ensure that their binary format is preserved. When you are done examining the light probe image, press the "q" key in the `ximage` window to dismiss the window.

6. Map the light probe image onto the environment.

   Let us now add the light probe image to our scene by mapping it onto an environment surrounding our objects. First, we need to create a new file that will specify the mathematical formula for mapping the light probe image onto the environment. Use your text editor to create the file `angmap.cal` containing the lines from Table 6.4.

   This file will tell RADIANCE how to map direction vectors in the world (Dx, Dy, Dz) into light probe image coordinates (u,v). Fortunately,

Table 6.4. The angular map .cal file

```
{ angmap.cal
Convert from directions in the world (Dx, Dy, Dz)
into (u,v) coordinates on the light probe image
-z is forward (outer edge of sphere) +z is backward
(center of sphere) +y is up (toward top of sphere) }
d = sqrt(Dx*Dx + Dy*Dy);
r = if(d, 0.159154943*acos(Dz)/d, 0);
u = 0.5 + Dx * r; v = 0.5 + Dy * r;
```

Table 6.5. Commenting out the traditional light source in scene.rad

```
#lightcolor source lightsource 0 0 4 1 1 1 2
```

RADIANCE accepts these coordinates in the range of zero to one no matter the image size, making it easy to try out light probe images of different resolutions. The formula converts from the angular map version of the light probe images in the Light Probe Image Gallery, which is different than the mapping a mirrored ball produces. If you need to convert a mirrored ball image to this format, HDR Shop [27] has a panoramic transformations function for this purpose.

Next, comment out (by adding #'s at the beginning of the lines) the traditional light source in scene.rad that we added in step 3, as in Table 6.5.

Note that this is not a new line to add to the file, but a modification to a line already entered. Now, add the lines from Table 6.6 to the end of scene.rad to include the image-based lighting environment.

The colorpict sequence indicates the name of the light probe image and the calculations file to use to map directions to image coordinates. The glow sequence specifies a material property comprising the light probe image treated as an emissive glow. Finally, the source specifies the geometry of an infinite sphere mapped with the emissive glow of the light probe. When RADIANCE's rays hit this surface, their illumination contribution are taken to be the light specified for the

Table 6.6. The image-based lighting environment for `scene.rad`

```
# Image-Based Lighting Environment
void colorpict hdr_probe_image 7 red green blue
beach_probe.hdr angmap.cal u v 0 0
hdr_probe_image glow light_probe 0 0 4 1 1 1 0
light_probe source ibl_environment 0 0 4 0 1 0 360
```

corresponding direction in the light probe image.

Finally, since we changed the scene file, we need to update the octree file. Run `oconv` once more to do this:

```
# oconv scene.rad > scene.oct
```

7. Render the scene with image-based lighting.

Let us now render the scene using image-based lighting. Enter the following command to bring up a rendering in `rview`:

```
# rview -ab 1 -ar 5000 -aa 0.08 -ad 128 -as 64 -st 0 -sj 1
-lw 0 -lr 8 -vf camera.vp scene.oct
```

In a few moments, you should see the image in Figure 6.5 begin to take shape. What RADIANCE is doing is tracing rays from the camera into the scene, illustrated in Figure 6.6. When a ray hits the environment, it takes as its radiance value the corresponding value in the light probe image. When a ray hits a particular point on an object, RADIANCE calculates the color and intensity of the incident illumination (known as irradiance) on that point by sending out a multitude of rays in random directions to quantify the light arriving at that point on the object. The number of rays are specified by -ad and -as parameters.[2] Some of these rays will hit the environment, and others will hit other parts of the object, causing RADIANCE to recurse into computing the light coming from this new part of the object. The number of levels

---

[2]The -ad parameter specifies how many rays will be cast directly into the environment, and the -as parameter specifies how many additional rays will be sent to provide greater sampling density in regions of high variance.

Figure 6.5. The spheres on the pedestal illuminated by the "beach" light probe image from Figure 6.2.

of recursion is specified by the -ab parameter; usually an -ab value of 1 or 2 is sufficient. Once the illumination on the object point has been computed, RADIANCE calculates the light reflected toward the camera based on the object's material properties, and this becomes the pixel value of that point of the object. Images calculated in this way can take a while to render but produce a faithful rendition of how the captured illumination would illuminate the objects.

The command-line arguments to rview above tell RADIANCE the manner in which to perform the lighting calculations. -ab 1 indicates that RADIANCE should produce only one "ambient bounce" recursion in computing the object's illumination; more accurate simulations could be produced with a value of 2 or higher. The -ar and -aa sets the resolution and accuracy of the surface illumination calculations, and -ad and -as set the number of rays traced out from a surface point to compute its illumination. The -st, -sj, -lw, and -lr specify how the rays should be traced for glossy and shiny reflections. For more information on these and other RADIANCE parameters, see the reference guide on the RADIANCE web site.

When the render completes, you should see an image of the objects as illuminated by the beach lighting environment. The synthetic steel ball reflects the environment and the other objects directly; the glass ball both reflects and refracts the environment, and the diffuse white ball

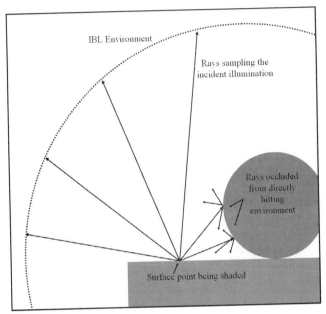

Figure 6.6. How RADIANCE traces rays to determine the incident illumination on a surface from an image-based lighting environment.

shows subtle shading, which is lighter toward the sunset and darkest where the ball contacts the pedestal. The rough specular reflections in the red and gold balls appear somewhat speckled in this medium-resolution rendering; the reason is that RADIANCE sends out just one ray for each specular sample (regardless of surface roughness) rather than the much greater number it sends out to compute the diffuse illumination. Rendering at a higher resolution and filtering the image down can alleviate this effect.

We might want to create a particularly high-quality rendering using the command-line `rpict` renderer, which outputs the rendered image to a file. Run the following `rpict` command:

```
# rpict -x 800 -y 800 -t 30 -ab 1 -ar 5000 -aa 0.08
-ad 128 -as 64 -st 0 -sj 1 -lw 0 -lr 8 -vf camera.vp
scene.oct > render.hdr
```

The command line arguments to `rpict` are identical to `rview` except that one also specifies the maximum x- and y-resolutions for the image (here, 800 by 800 pixels) as well as how often to report back on the rendering progress (here, every 30 seconds.) On a 2-GHz computer, this should take approximately five minutes. When complete, the rendered output image can by viewed with the `ximage` program. To produce very high-quality renderings, you can increase the x- and y-resolutions to very high numbers, such as 3000 by 3000 pixels, and then filter the image down to produce an antialiased rendering. This filtering down can be performed with either RADIANCE's `pfilt` command or using HDR Shop. To filter a 3000 by 3000 pixel image down to 1000 by 1000 pixels using `pfilt`, enter

```
# pfilt -1 -x /3 -y /3 -r 1 render.hdr > filtered.hdr
```

The high-quality renderings in this paper were produced using this method. To render the scene with different lighting environments, download a new probe image, change the `beach_probe.hdr` reference in the `scene.rad` file, and call `rview` or `rpict` once again. Light probe images with concentrated light sources, such as "grace" and "stpeters," will require increasing the `-ad` and `-as` sampling parameters to the renderer in order to avoid mottled renderings. Renderings of our objects illuminated by the light probes of Figure 6.2 are shown in Figure 6.7. Each rendering shows different effects of the lighting, from the particularly soft shadows under the spheres in the overcast Uffizi environment to the focused pools of light from the stained glass windows under the glass ball in the Grace Cathedral environment.

6.1.3   Lighting entire environments with IBL

So far, these techniques have shown how to illuminate synthetic objects with measurements of real light, which can help the objects appear as if they are actually in a real-world scene. The technique can also be used to light large-scale environments with captured illumination from real-world skies; Figure 6.8 shows a computer model of the Parthenon on the Athenian Acropolis illuminated by several real-world sky environments captured with high dynamic range photography using a fisheye lens.

6.2   Advanced Image-Based Lighting

The technique of image-based lighting can be used not only to illuminate synthetic objects with real-world light, but to place synthetic objects into a real-world scene. To do this realistically, the objects need to photometrically

Figure 6.7. The objects illuminated by the four light probe images from Figure 6.2.

affect their environment; that is, they need to cast shadows and reflect indirect illumination onto their nearby surroundings. Without these effects, the synthetic object will most likely look "pasted in" to the scene rather than as if it had actually been there.

A limitation of the techniques presented so far is that they require that the series of digital images taken span the complete range of intensities in the scene. For most scenes, the range of shutter speeds available on a digital camera can cover this range effectively. However, for sunlit scenes, the directly visible sun is often too bright to capture, even using the shortest exposure times available. This is a problem, since failing to record the color and intensity of the sun will cause significant inaccuracies in the captured lighting environment.

This section describes extensions to the image-based lighting techniques that solve both the problem of capturing the intensity of the sun and the problem of having the synthetic objects cast shadows onto the environment.

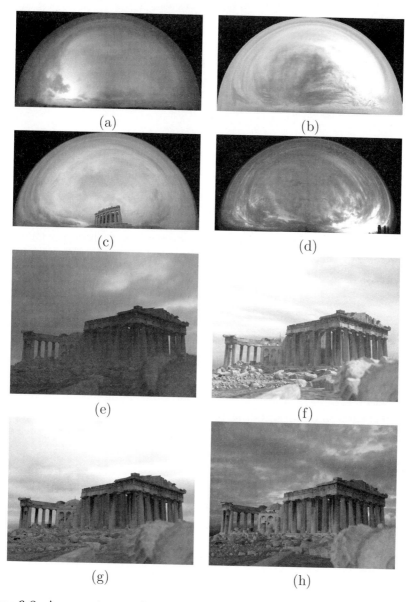

Figure 6.8. A computer graphics model of the Parthenon seen circa 1800 in (e–h) as illuminated by the image-based real-world lighting environments seen in (a–d).

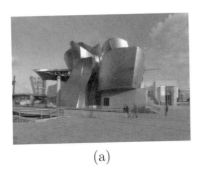
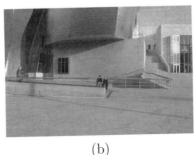

(a)   (b)

Figure 6.9. (a) The Guggenheim Museum in Bilbao. (b) A background plate image taken in front of the museum.

Although presented together, they are independent techniques that can be used separately as needed. These techniques are presented in the form of a new image-based lighting example, also using the RADIANCE rendering system.

6.2.1   Capturing a light probe in direct sunlight

For our example, we place a new sculpture and some spheres in front of the Guggenheim Museum in Bilbao, Spain, seen in Figure 6.9a. To create this example, I first photographed a background plate image using a Canon EOS-D60 digital camera on a tripod, as in Figure 6.9b. The image was taken with a 24mm lens at ASA 100 sensitivity, f/11 aperture, and 1/60 second shutter speed. A shadow from a nearby lamppost can be seen being cast along the ground.

To capture the lighting, I placed a three-inch mirrored sphere on a second tripod in front of the camera, as in Figure 6.10. I left the first camera in the same place, pointing in the same direction, since that would make it easier to align the probe and background images later. In order to keep the reflection of the camera small in the sphere, I placed the sphere approximately one meter away from the camera, and then changed to a 200mm telephoto lens in order for the sphere to appear relatively large in the frame. I used a remote release for the camera shutter to avoid moving the camera while capturing the lighting, and I took advantage of the camera's automatic exposure bracketing (AEB) feature to quickly shoot the mirrored sphere at 1/60, 1/250, and 1/1000 second. These images are shown in Figure 6.11a–c. The images were recorded using the camera's "RAW" mode, which saves uncompressed image data to the camera's memory card. This ensures capturing the best possible image quality and for this camera has the desirable effect that the image pixel values exhibit a linear response to incident light in this mode. Knowing that the response is linear is helpful for assembling high dynamic range images, since there is no need to derive the response curve of the camera.

Figure 6.10. Capturing a light probe image in front of the museum with a mirrored ball and a digital camera.

Since the sun was shining, it appears in the reflection in the mirrored sphere as seen by the camera. Because the sun is so bright, even the shortest 1/1000 second exposure failed to capture the sun's intensity without saturating the image sensor. After assembling the three images into a high dynamic range image, the area near the sun was not recorded properly, as seen in Figure 6.11d. As a result, using this image as a lighting environment would produce mismatched lighting, since a major source of the illumination would not have been accounted for.

I could have tried using several techniques to record an image of the mirrored sphere such that the sun did not saturate. I could have taken an even shorter exposure at 1/4000 second (the shortest shutter speed available on the camera), but it's almost certain the sun would still be too bright. I could have made the aperture smaller, taking the picture at f/22 or f/32 instead of f/11, but this would change the optical properties of the lens, and it would be unlikely to be a precise ratio of exposure to the previous images. Also, camera lenses tend to produce their best images at f/8 or f/11. I could have placed a neutral density filter in front of the lens, but this would also change the optical properties of the lens (the resulting image might not line up with the others due to refraction or from bumping the

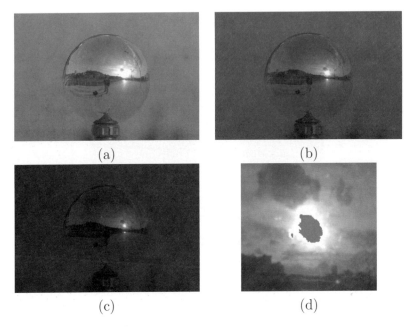

Figure 6.11. (a–c) A series of images of a mirrored sphere in the Bilbao scene taken at 1/60, 1/250, and 1/1000 second. (d) Detail of the sun area of (c), showing the overexposed areas in a dark color.

camera). Instead, I used a technique that allows us to indirectly measure the intensity of the sun, which is to photograph a diffuse gray sphere in the same location (Figure 6.12).

The image of the diffuse gray sphere allows us to reconstruct the intensity of the sun in the following way. Suppose that we had actually been able to capture the correct color and intensity of the sun in the light probe image. Then, if we were to use that light probe image to illuminate a synthetic diffuse gray sphere, it should have the same appearance as the real gray sphere actually illuminated by the lighting environment. However, since the light probe failed to record the complete intensity of the sun, the rendered diffuse sphere will not match the real diffuse sphere, and in fact we would expect it to be significantly darker. As we will see, the amount by which the rendered sphere is darker in each of the three color channels tells us the correct color and intensity for the sun.

The first part of this process is to make sure that all of the images that have been taken are radiometrically consistent. So far, we've recorded light

Figure 6.12. Photographing a diffuse gray sphere to determine the intensity of the sun.

in three different ways: seen directly with the 24mm lens, seen reflected in the mirrored sphere through the 200mm lens, and as reflected off of the diffuse sphere with the 200mm lens. Although the 24mm and 200mm lenses were both set to an f/11 aperture, they did not necessarily transmit the same amount of light to the image sensor. To measure this difference, we can photograph a white spot of light in our laboratory with the cameras to determine the transmission ratio between the two lenses at these two f/stops, as seen in Figure 6.13. The measurement recorded that the spot in the 24mm lens image appeared 1.42 times as bright in the red channel, 1.40 times as bright in green, and 1.38 times as bright in blue compared to the 200mm lens image. Using these results, we can multiply the pixel values of the images acquired with the 200mm lens by $(1.42, 1.40, 1.38)$ so that the images represent places in the scene with equal colors and intensities equally. I could have been even more precise by measuring the relative amount of light transmitted as a spatially varying function across each lens, perhaps by photographing the spot in different places in the field of view of each lens. At wider f/stops, many lenses, particularly wide-angle ones, transmit significantly less light to the corners of the image than to the center. Since the f/11 f/stop was reasonably small, I made the assumption that these effects would not be significant.

   Continuing toward the goal of radiometric consistency, we should note that neither the mirrored sphere nor the diffuse sphere are 100% reflective. The diffuse sphere is intentionally painted gray instead of white so that it is less likely to saturate the image sensor if an automatic light meter is employed. I measured the paint's reflectivity by painting a flat surface with

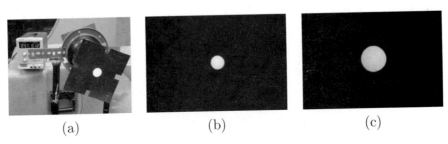

(a)   (b)   (c)

Figure 6.13. (a) A constant intensity illuminant created by placing a white LED inside a tube painted white on the inside, with a diffuser placed across the hole in the front. (b–c) Images taken of the illuminant using the 24mm and 200mm lenses at f/11 to measure the transmission ratio between the two lenses.

the same color, and then photographing the surface in diffuse natural illumination. In the scene, I also placed a similarly oriented *reflectance standard*, a white disk of material specially designed to reflect nearly 100% of all wavelengths of light across the visible spectrum. By dividing the average RGB pixel value of the painted sample to average pixel value of the reflectance standard, I determined that the reflectivity of the gray paint was $(0.032, 0.333, 0.346)$ in the red, green, and blue channels. By dividing the pixel values of the gray sphere image by these values, I obtained an image of how a perfectly white sphere would have appeared in the lighting environment. Reflectance standards such as the one that I used are available from companies such as Coastal Optics and LabSphere. A less expensive solution is to use the white patch of a MacBeth ColorChecker chart. We have measured the white patch of the chart to be 86% reflective, so any measurement taken with respect to this patch should be multiplied by 0.86 to scale it to be proportional to absolute reflectance.

The mirrored sphere is also not 100% reflective; even specially made first-surface mirrors are rarely greater than 90% reflective. To measure the reflectivity of the sphere to first order, I photographed the sphere on a tripod with the reflectance standard facing the sphere in a manner such that both the standard and its reflection were visible in the same frame, as in Figure 6.14. Dividing the reflected color by the original color revealed that the sphere's reflectivity was $(0.632, 0.647, 0.653)$ in the red, green, and blue channels. Using this measurement, I further scaled the pixel values of the image of the mirrored sphere in Figure 6.11 by $(\frac{1}{0.632}, \frac{1}{0.647}, \frac{1}{0.653})$ to approximate the image that a perfectly reflective sphere would have produced.

(a)   (b)

Figure 6.14. (a) An experimental setup to measure the reflectivity of the mirrored sphere. The reflectance standard at the left can be observed directly as well as in the reflection on the mirrored sphere seen in (b).

To be more precise, I could have measured and compensated for the sphere's variance in reflectivity with the angle of incident illumination. Due to the Fresnel reflection effect, the sphere will exhibit greater reflection toward grazing angles. However, my group has measured that this effect is less than a few percent for all measurable angles of incidence of the spheres we have used so far.

Now that the images of the background plate, the mirrored sphere, and the diffuse sphere have been mapped into the same radiometric space, we need to model the missing element, which is the sun. The sun is 1,390,000 km in diameter and on average 149,600,000 km away, making it appear as a disk in our sky with a subtended angle of 0.53 degrees. The direction of the sun could be calculated from standard formulas involving time and geographic coordinates, but it can also be estimated reasonably well from the position of the saturated region in the mirrored sphere. If we consider the mirrored sphere to have image coordinates (u, v) each ranging from -1 to 1, a unit vector (D_x, D_y, D_z) pointing toward the faraway point reflected toward the camera at (u, v) can be computed as

$$\theta = \tan^{-1} \frac{-v}{u}, \qquad (6.1)$$

$$\phi = 2\sin^{-1}(\sqrt{u^2 + v^2}), \qquad (6.2)$$

$$(D_x, D_y, D_z) = (\sin\phi\cos\theta, \sin\phi\sin\theta, -\cos\phi). \qquad (6.3)$$

In this case, the center of the saturated image was observed at coordinate $(0.414, 0.110)$ yielding a direction vector of $(0.748, 0.199, 0.633)$. Knowing the sun's size and direction, we can now model the sun in RADIANCE as an infinite `source` light:

Table 6.7. A unit intensity sun `sun.rad`

```
# sun.rad
void light suncolor 0 0 3 46700 46700 46700
suncolor source sun 0 0 4 0.748 0.199 0.633 0.53
```

We have not yet determined the radiance of the sun disk, so in this file `sun.rad`, I chose its radiance L to be $(46700, 46700, 46700)$. This radiance value happens to be such that the brightest point of a diffuse white sphere lit by this environment will reflect a radiance value of $(1, 1, 1)$. To derive this number, I computed the radiance L_r of the point of the sphere pointing toward the sun based on the rendering equation for Lambertian diffuse reflection [6]:

$$L_r = \int_0^{2\pi} \int_0^r L \, \frac{\rho_d}{\pi} \sin\theta \, \cos\theta \, \delta\theta \, \delta\phi$$

Since the radius of our 0.53° diameter sun is $r = 0.00465$ radians, the range of θ is small enough to approximate $\sin\theta \approx \theta$ and $\cos\theta \approx 1$, yielding

$$L_r \approx L \, \frac{\rho_d}{\pi} \int_0^{2\pi} \int_0^r \theta \, \delta\theta \, \delta\phi = \rho_d \, L \, r^2$$

Letting our desired radiance from the sphere be $L_r = 1$ and its diffuse reflectance to be a perfect white $\rho_d = 1$, we obtain that the required radiance value L for the sun is $1/r^2 = 46700$. A moderately interesting fact is that this tells us that on a sunny day, the sun is generally over 46,700 times as bright as the sunlit surfaces around you. Knowing this, it is not surprising that it is difficult to capture the sun's intensity with a camera designed to photograph people and places.

Now we are set to compute the correct intensity of the sun. Let us call our incomplete light probe image (the one that failed to record the intensity of the sun) $P_{incomplete}$. If we render an image of a virtual mirrored sphere illuminated by only the sun, we obtain a synthetic light probe image of the synthetic sun. Let us call this probe image P_{sun}. The desired complete light probe image, $P_{complete}$, can thus be written as the incomplete light probe plus some color $\alpha = (\alpha_r, \alpha_g, \alpha_b)$ times the sun light probe image, as in Figure 6.15.

$$P_{complete} \qquad = \qquad P_{incomplete} \qquad + \alpha \qquad P_{sun}$$

Figure 6.15. Modeling the complete image-based lighting environment, the sum of the incomplete light probe image, plus a scalar times a unit sun disk.

To determine $P_{complete}$, we somehow need to solve for the unknown α. Let us consider, for each of the three light probe images, the image of a diffuse white sphere as it would appear under each of these lighting environments, which we can call $D_{complete}$, $D_{incomplete}$, and D_{sun}. We can compute $D_{incomplete}$ by illuminating a diffuse white sphere with the light probe image $P_{incomplete}$ using the image-based lighting technique from the previous section. We can compute D_{sun} by illuminating a diffuse white sphere with the unit sun specified in sun.rad from Table 6.7. Finally, we can take $D_{complete}$ to be the image of the real diffuse sphere under the complete lighting environment. Since lighting objects is a linear operation, the same mathematical relationship for the light probe images must hold for the images of the diffuse spheres they are illuminating. This gives us the equation illustrated in Figure 6.16.

We can now subtract $D_{incomplete}$ from both sides to obtain the equation in Figure 6.17.

Since D_{sun} and $D_{complete} - D_{incomplete}$ are images consisting of thousands of pixel values, the equation in Figure 6.17 is actually a heavily overdetermined set of linear equations in α, with one equation resulting from each red, green, and blue pixel value in the images. Since the shadowed pixels in D_{sun} are zero, we cannot just divide both sides by D_{sun} to obtain an image with an average value of α. Instead, we can solve for α in a least squares sense to minimize $\alpha \, D_{sun} - (D_{complete} - D_{incomplete})$. In practice, however, we can more simply choose an illuminated region of D_{sun}, take this region's average

$D_{complete}$ $=$ $D_{incomplete}$ $+ \alpha$ D_{sun}

Figure 6.16. The equation of Figure 6.15 rewritten in terms of diffuse spheres illuminated by the corresponding lighting environments.

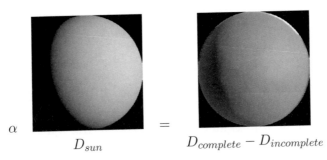

α D_{sun} $=$ $D_{complete} - D_{incomplete}$

Figure 6.17. Subtracting $D_{incomplete}$ from both sides of the equation of Figure 6.16 yields an equation in which it is straightforward to solve for the unknown sun intensity multiplier α.

value μ_{sun}, then take the same region of $D_{complete} - D_{incomplete}$, compute its average pixel value $\mu diff$, and then compute $\alpha = \mu diff / \mu_{sun}$.

Applying this procedure to the diffuse and mirrored sphere images from Bilbao, I determined $\alpha = (1.166, 0.973, 0.701)$. Multiplying the original sun radiance $(46700, 46700, 46700)$ by this number, we obtain $(54500, 45400, 32700)$ for the sun radiance, and we can update the `sun.rad` from Table 6.7 accordingly.

We can now verify that the recovered sun radiance has been modeled accurately by lighting a diffuse sphere with the sum of the recovered sun and the incomplete light probe $P_{incomplete}$. The easiest way to do this is to include both the `sun.rad` environment and the light probe image together as the RADIANCE lighting environment. Figure 6.18a and b show a comparison between such a rendering and the actual image of the diffuse sphere.

Subtracting the two images allows us to visually verify the accuracy of the lighting reconstruction. The difference image in Figure 6.18c is nearly black, which indicates a close match.

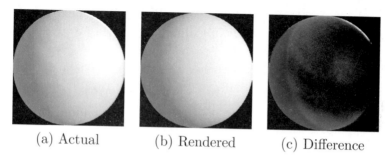

(a) Actual   (b) Rendered   (c) Difference

Figure 6.18. (a) The actual diffuse sphere in the lighting environment, scaled to show the reflection it would exhibit if it were 100% reflective. (b) A rendered diffuse sphere, illuminated by the incomplete probe and the recovered sun model. (c) The difference between the two, exhibiting an RMS error of less than 2% over all surface normals, indicating a good match.

The rendered sphere in Figure 6.18b is actually illuminated with a combination of the incomplete light probe image and the modeled sun light source rather than with a complete image-based lighting environment. To obtain a complete light probe, we can compute $P_{complete} = P_{incomplete} + \alpha \, P_{sun}$ as above and use this image on its own as the light probe image. However, it is actually advantageous to keep the sun separate as a computer-generated light source rather than as part of the image-based lighting environment. The reason for this has to do with how efficiently the Monte Carlo rendering system can generate the renderings. As seen in Figure 6.6, when a camera ray hits a point on a surface in the scene, a multitude of randomized indirect illumination rays are sent into the image-based lighting environment to sample the incident illumination. Since the tiny sun disk occupies barely 1/100000 of the sky, it is unlikely that any of the rays will hit the sun even if 1000 or more rays are traced to sample the illumination. Any such point will be shaded incorrectly as if it is in shadow. What is worse is that for some of the points in the scene at least one of the rays fired out *will* hit the sun. If 1000 rays are fired and one hits the sun, the rendering algorithm will light the surface with the sun's light as if it were 1/1000 of the sky, which is 100 times as large as the sun actually is. The result is that every 100 or so pixels in the render there will be a pixel that is approximately 100 times as bright as it should be, and the resulting rendering will appear very "speckled."

Leaving the sun as a traditional computer-generated light source avoids this problem almost entirely. The sun is still in the right place with the correct size, color, and intensity, but the rendering algorithm knows explicitly that it is there. As a result, the sun's illumination will be computed as part of the direct illumination calculation: for every point of every surface, the renderer will always fire at least one ray toward the sun. The renderer will still send a multitude of indirect illumination rays to sample the sky, but since the sky is relatively uniform, its illumination contribution can be sampled sufficiently accurately with a few hundred rays. (If any of the indirect illumination rays do hit the sun, they will not add to the illumination of the pixel, since the sun's contribution is already being considered in the direct calculation.) The result is that renderings with low noise can be computed with a small fraction of the rays that would otherwise be necessary if the sun were represented as part of the image-based lighting environment.

The technique of representing a concentrated image-based area of illumination as a traditional CG light source can be extended to multiple light sources within more complex lighting environments. Section 6.2.3 describes how the windows and ceiling lamps within St. Peter's Basilica were modeled as local area lights to efficiently render the animation *Fiat Lux*. Techniques have also been developed to approximate image-based lighting environments *entirely* as clusters of traditional point light sources [5, 17, 2]. These algorithms let the user specify the desired number of light sources and then use clustering algorithms to choose the light source positions and intensities to best approximate the lighting environment. Such techniques can make it possible to simulate image-based lighting effects using more traditional rendering systems.

At this point, we have successfully recorded a light probe in a scene where the sun is directly visible; in the next section, we use this recorded illumination to realistically place some new objects in front of the museum.

6.2.2   Compositing objects into the scene including shadows

The next step of rendering the synthetic objects into the scene involves reconstructing the scene's viewpoint and geometry rather than its illumination. We first need to obtain a good estimate of the position, orientation, and focal length of the camera at the time that it photographed the background plate image. Then, we must create a basic 3D model of the scene that includes the surfaces with which the new objects will photometrically interact—in our case, this is the ground plane upon which the objects will sit.

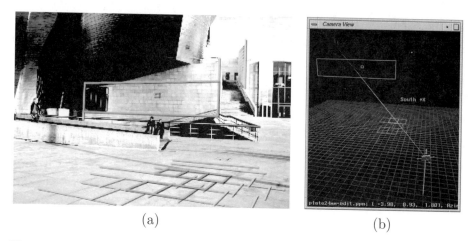

(a)   (b)

Figure 6.19. Reconstructing the camera pose and the local scene geometry using a photogrammetric modeling system. (a) Edges from the floor tiles and architecture are marked by the user. (b) The edges are corresponded to geometric primitives, which allows the computer to reconstruct the position of the local scene geometry and the pose of the camera.

Determining the camera pose and local scene geometry could be done through manual survey at the time of the shoot or through trial and error afterwards. For architectural scenes such as ours, it is usually possible to obtain these measurements through photogrammetry, often from just the one background photograph. Figure 6.19 shows the use of the Façade photogrammetric modeling system described in [12] to reconstruct the camera pose and the local scene geometry based on a set of user-marked edge features in the scene.

The objects we place in front of the museum are a statue of a Caryatid from the Athenian Acropolis and four spheres of differing materials: silver, glass, plaster, and onyx. By using the recovered camera pose, we can render these objects from the appropriate viewpoint, and using the light probe image, we can illuminate the objects with the lighting from the actual scene. However, overlaying such rendered objects onto the background plate image will not result in a realistic image, since the objects would not photometrically interact with the scene: the objects would not cast shadows on the ground, the scene would not reflect in the mirrored sphere or refract through the glass one, and there would be no light reflected between the objects and their local environment.

To have the scene and the objects appear to photometrically interact, we can model the scene in terms of its approximate geometry and reflectance and add this model of the scene to participate in the illumination calculations along with the objects. The ideal property for this approximate scene model to have is for it to precisely resemble the background plate image when viewed from the recovered camera position and illuminated by the captured illumination. We can achieve this property easily if we choose the model's geometry to be a rough geometric model of the scene, projectively texture-mapped with the original background plate image. Projective texture mapping is the image-based rendering process of assigning surface colors as if the image were being projected onto the scene as if from a virtual slide projector placed at the original camera position. Since we have recovered the camera position, performing this mapping is a straightforward process in many rendering systems. A useful property of projective texture mapping is that when an image is projected onto geometry, a rendering of the scene looks just like the original image when viewed from the same viewpoint as the projection. When viewed from different viewpoints, the similarity between the rendered viewpoint the original scene is a function of how far the camera has moved and the accuracy of the scene geometry. Figure 6.20 shows a bird's-eye view of the background plate projected onto a horizontal polygon for the ground plane and a vertical polygon representing the museum in the distance; the positions of these two polygons were determined from the photogrammetric modeling step in Figure 6.19.

When a texture map is applied to a surface, the renderer can be told to treat the texture map's pixel values either as measures of the surface's radiance, or as measures of the surface's reflectance. If we choose to map the values on as radiance, then the surface is treated as an emissive surface that appears precisely the same no matter how it is illuminated or what sort of objects are placed in its vicinity. This is in fact how we have been mapping the light probe images onto the sky dome as the "glow" material to become image-based lighting environments. However, for the ground beneath the objects, this is not the behavior we want. We instead need to texture-map the ground plane with reflectance values, which allows the ground plane to participate in the same illumination computation as the added objects, thus allowing the ground plane to receive shadows and interreflect light with the objects. In RADIANCE, we can do this by specifying the material as "plastic" with no specular component and then modulating the diffuse color values of the surface using the projective texture map.

Figure 6.20. The final scene, seen from above instead of from the original camera viewpoint. The background plate is mapped onto the local scene ground plane and a billboard for the museum behind. The light probe image is mapped onto the remainder of the ground plane, the billboard, and the sky. The synthetic objects are set on top of the local scene, and both the local scene and the objects are illuminated by the captured lighting environment.

The problem we now face is that the pixels of the background plate image are actually measurements of radiance—the amount light reflecting from the scene toward the camera—rather than reflectance, which is what we need to project onto the local scene. Reflectance values represent the proportion of incident light that a surface reflects back (also known as the material's *albedo*) and range from zero (0% of the light is reflected) to one (100% of the light is reflected). Considered in RGB for the three color channels, these reflectance values represent what we traditionally think of as the inherent "color" of a surface, since they are independent of the incident illumination. Radiance values are different in that they can range from zero to arbitrarily high values. Thus, it is generally inaccurate to use radiance values as if they were reflectance values.

As it turns out, we can convert the radiance values from the background plate image into reflectance values using the following relationship:

$$radiance = irradiance \times reflectance$$

or

$$reflectance = radiance \; / \; irradiance$$

Since the background plate image tells us the radiance for each point on the local scene, all we need to know is the irradiance of each point as well. These irradiance values can be determined using the illumination that we recorded within the environment in the following manner. We assign the local scene surfaces a diffuse material property of known reflectance, for example, 50% gray, or $(0.5, 0.5, 0.5)$ in RGB. We then render this local scene as illuminated by the image-based lighting environment to compute the appearance of the scene under the lighting, seen in Figure 6.21a. With this rendering, we can determine the irradiance of each surface point at $irradiance = radiance/reflectance$, where the reflectance is the 50% gray color. Finally, we can compute the reflectance, or *albedo map*, of the local scene as $reflectance = radiance/irradiance$, where the radiance values are the pixel values from the background plate. The result of this division is shown in Figure 6.21b.

(a)   (b)

Figure 6.21. Solving for the albedo map of the local scene. (a) A diffuse stand-in for the local scene is rendered as illuminated by the recovered lighting environment to compute the irradiance at each point of the local scene. (b) Dividing the background plate image from Figure 6.9b by the irradiance estimate image in (a) yields a per-pixel map of the albedo, or diffuse reflectance, of the local scene.

In the renderings in Figure 6.21, I kept the billboard polygon for the building mapped with the original scene radiance from the background plate, rather than attempting to estimate its reflectance values. The main reason I made this choice is that for the objects I would be placing in the scene, their photometric effect on the building in the background did not seem likely to be visually significant.

In this example, the local scene is a single polygon, and thus no point of the local scene is visible to any other point. As a result, it really makes no difference what reflectance value we assume for the scene when we compute its irradiance. However, if the local scene were a more complicated, nonconvex shape, such as we would see on a flight of stairs, the surfaces of the local scene would interreflect. Thus, the reflectance value chosen for computing the irradiance will affect the final reflectance values solved for by the algorithm. For this reason, it makes sense to try to choose an approximately correct value for the assumed reflectance; 50% gray is more likely to be accurate than 100% reflective. If better accuracy in determining the local scene reflectance is desired, then the reflectance estimation process could be iterated using the reflectance computed by each stage of the algorithm as the assumed reflectance in the next stage. A similar approach to the problem of estimating local scene reflectance is described in [7]. By construction, the local scene with its estimated reflectance properties produces precisely the appearance found in the background plate when illuminated by the recovered lighting environment.

At this point, it is a simple process to create the final rendering of the objects added in to the scene. We include the synthetic objects along with the local scene geometry and reflectance properties, and ask the renderer to produce a rendering of the scene as illuminated by the recovered lighting environment. This rendering for our example can be seen in Figure 6.22, using two bounces of indirect illumination. In this rendering, the objects are illuminated consistently with the original scene, they cast shadows in the appropriate direction, and their shadows are the appropriate color. The scene reflects realistically in the shiny objects and refracts realistically through the translucent objects. If all of the steps were followed for modeling the scene's geometry, photometry, and reflectance, the synthetic objects should look almost exactly as they would if they were really there in the scene.

6.2.3   Image-based lighting in *Fiat Lux*

The animation *Fiat Lux* shown at the SIGGRAPH 99 Electronic Theater used image-based lighting to place synthetic monoliths and spheres into the interior of St. Peter's Basilica. A frame from the film is shown in Figure 6.25. The project required making several extensions to the image-based

Figure 6.22. A final rendering of the synthetic objects placed in front of the museum using the recovered lighting environment.

lighting process to handle an interior environment with many concentrated light sources.

The source images for the St. Peter's sequences in the film were acquired on location with a Kodak DCS-520 camera. Each image was shot in high dynamic range at the following exposures: 2 sec, 1/4 sec, 1/30 sec, 1/250 sec, 1/1000 sec, and 1/8000 sec. The images consisted of 10 light probe images taken using a mirrored sphere along the nave and around the altar, and two partial panoramic image mosaics to be used as the background plate images (one is shown in Figure 6.24a).

The film not only required adding synthetic animated objects into the Basilica, but it also required rotating and translating the virtual camera. To accomplish this, we used the Façade system to create a basic 3D model of the interior of the Basilica from one of the panoramic photographs, shown in Figure 6.23. Projecting the panoramic image onto the 3D model allowed novel viewpoints of the scene to be rendered within several meters of the original camera position.

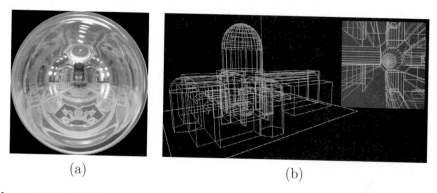

(a)   (b)

Figure 6.23. (a) One of the 10 light probe images acquired within the Basilica (b) A basic 3D model of the interior of St. Peter's obtained using the Façade system from the HDR panorama in 6.24a.

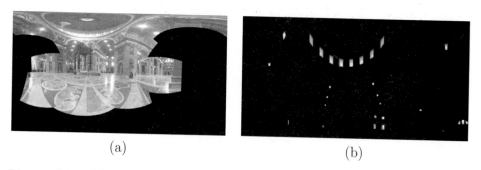

(a)   (b)

Figure 6.24. (a) One of the two HDR panoramas acquired to be a virtual background for *Fiat Lux*, assembled from 10 HDR images shot with a 14mm lens. (b) The modeled light sources from the St. Peter's environment.

Projecting the HDR panoramic image onto the Basilica's geometry caused the illumination of the environment to emanate from the local geometry rather than from infinitely far away, as seen in other IBL examples. As a result, the illumination from the lights in the vaulting and the daylight from the windows comes from different directions, depending on where a synthetic object in the model is placed. For each panoramic environment, the lighting was derived using a combination of the illumination within each HDR panorama as well as light probe images to fill in areas not seen in the partial panoramas.

As in the Bilbao example, the image-based lighting environment was used to solve for the reflectance properties of the floor of the scene so that it could participate in the lighting computation. A difference is that the marble floor of the Basilica is rather shiny, and significant reflections from the windows and light sources were present within the background plate. In theory, these shiny areas could have been computationally eliminated using the knowledge of the lighting environment. Instead, though, we decided to remove the specularities and recreate the texture of the floor using an image editing program. Then, after the diffuse albedo of the floor was estimated using the reflectometry technique from the last section, we synthetically added a specular component to the floor surface. For artistic reasons, we recreated the specular component as a perfectly sharp reflection with no roughness, virtually polishing the floor of the Basilica.

To achieve more efficient renderings, we employed a technique of converting concentrated sources in the incident illumination images into computer-generated area light sources. This technique is similar to how the sun was represented in our earlier Bilbao example, although there was no need to indirectly solve for the light source intensities, since even the brightest light sources fell within the range of the HDR images. The technique for modeling the light sources was semiautomatic. Working from the panoramic image, we clicked to outline a polygon around each concentrated source of illumination in the scene, including the Basilica's windows and lighting; these regions are indicated in Figure 6.24b. Then, a custom program projected the outline of each polygon onto the model of the Basilica. The program also calculated the average pixel radiance within each polygon and assigned each light source its corresponding color from the HDR image. RADIANCE includes a special kind of area light source called an `illum` that is invisible when directly viewed, but otherwise acts as a regular light source. Modeling the lights as `illum` sources made it so that the original details in the windows and lights were visible in the renders as though each was modeled as an area light of a single consistent color.

Even using the modeled light sources to make the illumination calculations more efficient, *Fiat Lux* was a computationally intensive film to render. The 2.5 minute animation required over 72 hours of computation on 100 dual 400MHz processor Linux computers. The full animation can be downloaded at `http://www.debevec.org/FiatLux/`.

Figure 6.25. A frame from the SIGGRAPH 99 film *Fiat Lux*, which combined image-based modeling, rendering, and lighting to place animated monoliths and spheres into a photorealistic reconstruction of St. Peter's Basilica.

6.2.4   Capturing and rendering spatially varying illumination

The techniques presented so far capture light at discrete points in space as 2D omnidirectional image data sets. In reality, light changes throughout space, and this changing illumination can be described as the 5D *plenoptic function* [1]. This function is essentially a collection of light probe images (θ, ϕ) taken at every possible (x, y, z) point in space. The most dramatic changes in lighting occur at shadow boundaries, where a light source may be visible in one light probe image and occluded in a neighboring one. More often, changes in light within space are more gradual, consisting of changes resulting from moving closer to and further from sources of direct and indirect illumination.

Two recent papers [19, 13] have shown that the plenoptic function can be reduced to a 4D function for free regions of space. Since light travels in straight lines and maintains its radiance along a ray, any ray recorded passing through a plane Π can be used to extrapolate the plenoptic function to any other point along that ray. These papers then showed that the appearance of a scene, recorded as a 2D array of 2D images taken from a planar grid of points in space, could be used to synthesize views of the scene from viewpoints both in front of and behind the plane.

A key observation in image-based lighting is that there is no fundamental difference between capturing images of a scene and capturing images of illumination. It is illumination that is being captured in both cases, and one need only pay attention to having sufficient dynamic range and linear sensor response when using cameras to capture light. It thus stands to reason that the light field capture technique would provide a method of capturing and reconstructing varying illumination within a volume of space.

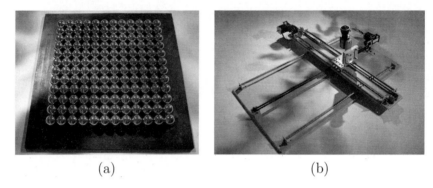

(a)   (b)

Figure 6.26. Two devices for capturing spatially varying incident lighting. (a) An array of mirrored spheres. (b) A 180° fisheye camera on a 2D translation stage. Both devices acquire 4D data sets of incident illumination.

To this end, our group has constructed two different devices for capturing *incident light fields* (ILFs) [28]. The first, shown in Figure 6.26a is a straightforward extrapolation of the mirrored ball capture technique, consisting of a 12 × 12 array of 1-inch mirrored spheres fixed to a board. Although the adjacent spheres reflect each other near the horizon, the array captures images with a nearly 160° unoccluded field of view. A single HDR image of the spheres provides a 4D data set of wide field-of-view images sampled across a planar surface. However, since each sphere occupied just a small part of the image, the angular resolution of the device was limited, and the 12 × 12 array of spheres provided even poorer sampling in the spatial dimensions.

To improve the resolution, we built a second device for capturing ILFs consisting of a video camera with a 185° fisheye lens mounted to an x-y translation stage, shown in 6.26b. While this device could not acquire an ILF in a single moment, it did allow for capturing 1024 × 1024 pixel light probes at arbitrary density across the ILF plane Π. We used this device to capture several spatially varying lighting environments that included spotlight sources and sharp shadows at spatial resolutions of up to 32×32 light probe images.

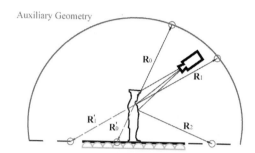

Figure 6.27. Lighting a 3D scene with an ILF captured at a plane beneath the objects. Illumination rays traced from the object, such as R_0 are traced back as R'_0 onto their intersection with the ILF plane. The closest light probe images to the point of intersection are used to estimate the color and intensity of light along the ray.

We saw in Figure 6.6 the simple process of how a rendering algorithm traces rays into a light probe image to compute the light incident upon a particular point on a surface. Figure 6.27 shows the corresponding process for rendering a scene as illuminated by an ILF. As before, the ILF is mapped on to a geometric shape covering the scene, referred to as the *auxiliary geometry*, which can be a finite-distance dome or a more accurate geometric model of the environment. However, the renderer also keeps track of the original plane Π from which the ILF was captured. When an illumination ray sent from a point on a surface hits the auxilary geometry, the ray's line is intersected with the plane Π to determine the location and direction of that ray as captured by the ILF. In general, the ray will not strike the precise center of a light probe sample, so bilinear interpolation can be used to sample the ray's color and intensity from the ray's four closest light probe samples. Rays that do not correspond to rays observed within the field of view of the ILF can be extrapolated from the nearest observed rays of the same direction or assumed to be black.

Figure 6.28a shows a real scene consisting of four spheres and a 3D print of a sculpture model illuminated by spatially varying lighting. The scene was illuminated by two spotlights, one orange and one blue, with a card placed in front of the blue light to create a shadow across the scene. We

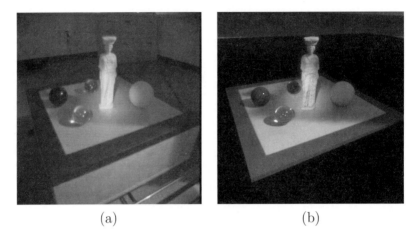

(a)   (b)

Figure 6.28. (a) A real scene, illuminated by two colored spotlights. (b) A computer model of the scene, illuminated with the captured, spatially varying illumination. The illumination was captured with the device in Figure 6.26b.

photographed this scene under the lighting, and then removed the objects and positioned the ILF capture device shown in Figure 6.26b to capture the illumination incident to the plane of the table top. We then captured this illumination as a 30×30 array of light probe images spaced 1 inch apart in the x and y dimensions. The video camera was electronically controlled to capture each image at 16 different exposures one stop apart, ranging from 1 sec. to $\frac{1}{32768}$ sec. To test the accuracy of the recovered lighting, we created a virtual version of the same scene and positioned a virtual camera to view the scene from the same viewpoint as the original camera. Then, we illuminated the virtual scene using the captured ILF to obtain the virtual rendering seen in Figure 6.28b. The virtual rendering shows the same spatially varying illumination properties on the objects, including the narrow orange spotlight and the half-shadowed blue light. Despite a slight misalignment between the virtual and real illumination, the rendering clearly shows the accuracy of using light captured on the plane of the table to extrapolate to areas above the table. For example, the base of the statue, where the ILF was captured, is in shadow from the blue light and fully illuminated by the orange light. Directly above, the head of the statue is illuminated by the blue light but does not fall within orange one. This effect is present in both the real photograph and the synthetic rendering.

6.3   Image-Based Relighting

The image-based lighting techniques we have seen so far are useful for lighting synthetic objects, lighting synthetic environments, and rendering synthetic objects into real-world scenes with consistent lighting. However, we have not yet seen techniques for creating renderings of *real-world* objects illuminated by captured illumination. Of particular interest would be to illuminate people with these techniques, since most images created for film and television involve people in some way. Certainly, if it were possible to create a very realistic computer model of a person, then this model could be illuminated using the image-based lighting techniques already presented. However, since creating photoreal models of people is still a difficult process, a more direct route is desirable.

There actually is a more direct route, and it requires nothing more than a special set of images of the person in question. The technique is based on the fact that light is *additive*, which can be explained by thinking of a person, two light sources, and three images. Suppose that the person sits still and is photographed lit by both light sources, one to each side. Then, two more pictures are taken, each with only one of the lights on. If the pixel values in the images are proportional to the light levels, then the additivity of light dictates that the sum of the two one-light images will look the same as the two-light image. More usefully, the color channels of the two images can be scaled before they are added, allowing us to create an image of a person illuminated with a bright orange light to one side and a dim blue light to the other. I first learned about this property from Paul Haeberli's Grafica Obscura web site [15].

In a complete lighting environment, the illumination comes not from just two directions, but from a whole sphere of incident illumination. If there were a way to light a person from a dense sampling of directions distributed across the whole sphere of incident illumination, it should be possible to recombine tinted and scaled versions of these images to show how the person would look in any lighting environment. The light stage device [8] shown in Figure 6.29 was designed to acquire precisely such a data set. The device's 250-watt halogen spotlight is mounted on a two-axis rotation mechanism such that the light can spiral from the top of the sphere to the bottom in approximately 1 minute. During this time, one or more digital video cameras can record the subject's appearance as illuminated by nearly 2000 lighting directions distributed throughout the sphere. A subsampled light stage data set of a person's face can be seen in Figure 6.30a.

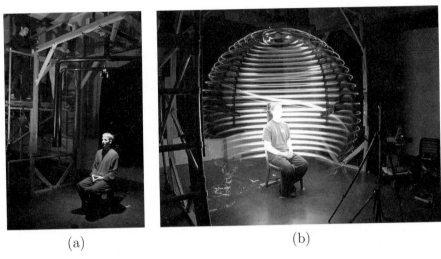

(a)   (b)

Figure 6.29. (a) Light Stage 1, a manually operated device for lighting a person's face from every direction in the sphere of incident illumination directions. (b) A long-exposure photograph of acquiring a Light Stage 1 data set. As the light spirals down, one or more video cameras record the different illumination directions on the subject in approximately 1 minute. The recorded data is shown in Figure 6.30.

Figure 6.30c shows the Grace Cathedral lighting environment remapped to be the same resolution and in the same longitude-latitude coordinate mapping as the light stage data set. For each image of the face in the data set, the resampled light probe indicates the color and intensity of the light from the environment in the corresponding direction. Thus, we can multiply the red, green, and blue color channels of each light stage image by the amount of red, green, and blue light in the corresponding direction in the lighting environment to obtain the modulated data set that appears in Figure 6.30d. Figure 6.31a shows the result of summing all of the scaled images in Figure 6.30d, producing an image of the subject as he would appear illuminated by the light of Grace Cathedral. Results obtained for three more lighting environments are shown in Figure 6.31b–d.

The process of computing the weighted sum of the face images is simple computation, but it requires accessing hundreds of megabytes of data for each rendering. The process can be accelerated by performing the computation on compressed versions of the original images. In particular, if the images are compressed using an orthonormal transform such as the discrete cosine transform, the linear combination of the images can be computed directly

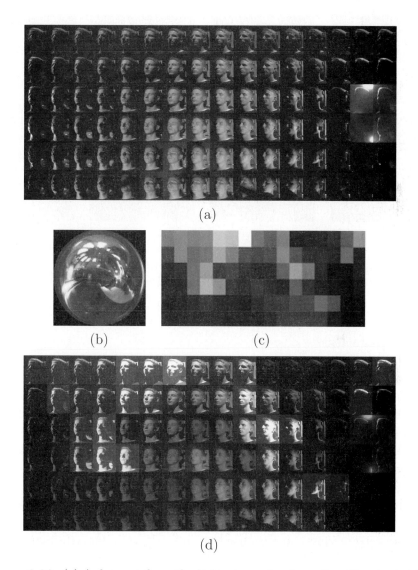

Figure 6.30. (a) A data set from the light stage, showing a face illuminated from the full sphere of lighting directions. Ninety-six of the 2,048 images in the full data set are shown. (b) The Grace Cathedral light probe image. (c) The light probe sampled into the same longitude-latitude space as the light stage data set. (d) The images in the data set scaled to the same color and intensity as the corresponding directions of incident illumination. Figure 6.31a shows a rendering of the face as illuminated by the Grace Cathedral environment obtained by summing these scaled images.

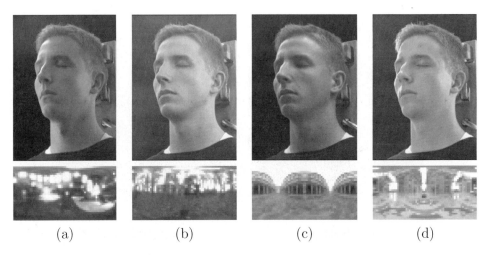

(a)   (b)   (c)   (d)

Figure 6.31. Renderings of the light stage data set from Figure 6.30 as illuminated by four image-based lighting environments. (a) Grace Cathedral. (b) Eucalyptus Grove. (c) Uffizi Gallery. (d) St. Peter's Basilica.

from the basis coefficients of the original images as shown in [25]. The Face Demo program written by Chris Tchou and Dan Maas (Figure 6.32) uses DCT-compressed versions of the face data sets to allow interactive face relighting with either light probe images or user-controlled light sources in real time. More recent work has used spherical harmonics [24, 23] and wavelets [22] to perform real-time image-based lighting on virtual light stage data sets of 3D computer graphics models.

The face renderings in Figure 6.31 are highly realistic and require no use of 3D models or recovered reflectance properties. However, the image summation technique that produced them works only for still images of the face. A project in our group's current work involves capturing light stage data sets of people in different expressions and from different angles, and blending between these expressions and viewpoints to create 3D animated faces. For this project, we have built a second light stage device that uses a rotating semicircular arm of strobe lights to capture a 500-image data set in less than 10 seconds. We have also used the device to record the reflectance properties of various cultural artifacts, described in [16].

Our most recent lighting apparatus, Light Stage 3, is a lighting reproduction system consisting of a full sphere of color LED light sources [10]. Each of the 156 light sources can be independently driven to any RGB color,

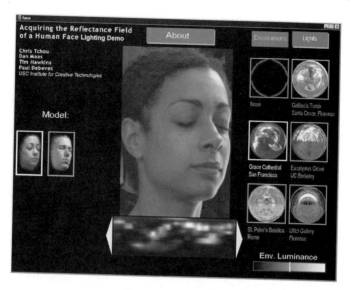

Figure 6.32. The Face Demo program written by Chris Tchou and Dan Maas allows light stage data sets to be interactively reilluminated using DCT-compressed data. The program is available at `http://www.debevec.org/FaceDemo/`.

which allows an approximate reproduction of a captured lighting environment to be recreated in a laboratory or studio. Figure 6.33 shows a subject standing within Light Stage 3 illuminated by a reproduction of the Grace Cathedral lighting environment. Figure 6.34 shows three subjects rendered into three different lighting environments. In these renderings, the subject has been composited onto an image of the background environment using a matte obtained from an infrared matting system.

6.4   Conclusion

Traditionally, the real and computer-generated worlds have been fundamentally separate spaces, each subject to their own laws of physics and aesthetics. The development of global illumination rendering techniques enabled light in computer-generated worlds to obey the same laws of physics as light in the real world, making it possible to render strikingly realistic images. Using the principles of global illumination, image-based lighting makes it possible for the real and computer-generated worlds to interact through lighting: real-world light can be captured and used to illuminate computer-generated objects, and light within the computer can be used to illuminate people and

Figure 6.33. Light Stage 3. A subject is illuminated by a real-time reproduction of the light of Grace Cathedral.

objects in the real world. This chapter has shown basic examples of how to perform image-based lighting using the freely available RADIANCE global illumination renderer. With some experimentation and your own camera and mirrored ball, you should be able to adapt these examples to your own models and applications. For more information on image-based lighting, please look for the latest developments at http://www.debevec.org/IBL/. Happy lighting!

Acknowledgments

The work described in this chapter represents the work of many individuals. Greg Ward wrote the RADIANCE rendering system and provided advice for using it optimally in an image-based lighting context. Marcos Fajardo wrote the Arnold renderer and provided helpful HDR and IBL support. Christine Waggoner and Son Chang helped carry out early light source identification experiments. Andreas Wenger wrote the Canon RAW-to-HDR converter, and Chris Tchou wrote the HDR Shop image editing program used to process the Bilbao images. Maya Martinez helped compose the Bilbao scene. Teddy Kim provided light probe assistance in Bilbao. Tim Hawkins helped acquire the HDR images used in *Fiat Lux* and helped lead the *Fiat Lux* team,

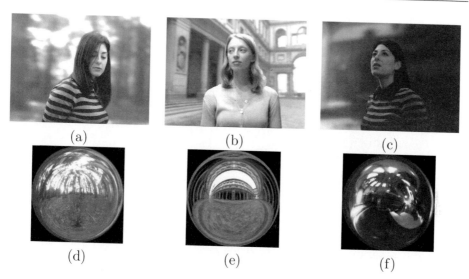

(a)   (b)   (c)

(d)   (e)   (f)

Figure 6.34. Light Stage 3 composite images. (a-c) Live-action subjects composited into three different environments using Light Stage 3 to reproduce the environment's illumination on the subject. (d–f) Corresponding lighting environments used in the composites. The subjects were composited over the backgrounds using an infrared compositing system [10].

including Westley Sarokin, H. P. Duiker, Tal Garfinkel, Jenny Huang, and Christine Cheng. The work presented here owes its support to the National Science Foundation, the California MICRO program, Interval Research Corporation, the Digital Media Innovation Program, Interactive Pictures Corporation, U.S. Army contract #DAAD19-99-D-0046, TOPPAN Printing Co. Ltd., and the University of Southern California Office of the Provost.

Bibliography

[1] E. H. Adelson and J. R. Bergen. *Computational Models of Visual Processing*. The Plenoptic Function and the Elements of Early Vision. MIT Press, 1991.

[2] S. Agarwal, R. Ramamoorthi, S. Belongie, and H. W. Jensen. Structured importance sampling of environment maps. *ACM Transactions on Graphics*, 22(3):605–612, July 2003.

[3] J. F. Blinn. Texture and reflection in computer generated images. *Communications of the ACM*, 19(10):542–547, October 1976.

[4] S. E. Chen. Quicktime VR—An image-based approach to virtual environment navigation. In *Proceedings of SIGGRAPH 95*, Computer Graphics Proceedings, Annual Conference Series, pages 29–38, August 1995.

[5] J. Cohen and P. Debevec. The LightGen HDRShop Plugin. Available at www.debevec.org/HDRShop/, 2001.

[6] M. F. Cohen and J. R. Wallace. *Radiosity and Realistic Image Synthesis*, chapter 2, page 32. Academic Press, 1993.

[7] P. Debevec. Rendering synthetic objects into real scenes: Bridging traditional and image-based graphics with global illumination and high dynamic range photography. In *SIGGRAPH 98*, July 1998.

[8] P. Debevec, T. Hawkins, C. Tchou, H.-P. Duiker, W. Sarokin, and M. Sagar. Acquiring the reflectance field of a human face. *Proceedings of SIGGRAPH 2000*, pages 145–156, July 2000.

[9] P. Debevec and D. Lemmon. Image based lighting. In *SIGGRAPH 2001 Notes for Course 14*, August 2001.

[10] P. Debevec, A. Wenger, C. Tchou, A. Gardner, J. Waese, and T. Hawkins. A lighting reproduction approach to live-action compositing. *ACM Transactions on Graphics*, 21(3):547–556, July 2002.

[11] P. E. Debevec and J. Malik. Recovering high dynamic range radiance maps from photographs. In *SIGGRAPH 97*, pages 369–378, August 1997.

[12] P. E. Debevec, C. J. Taylor, and J. Malik. Modeling and rendering architecture from photographs: A hybrid geometry- and image-based approach. In *SIGGRAPH 96*, pages 11–20, August 1996.

[13] S. J. Gortler, R. Grzeszczuk, R. Szeliski, and M. F. Cohen. The Lumigraph. In *SIGGRAPH 96*, pages 43–54, 1996.

[14] N. Greene. Environment mapping and other applications of world projections. *IEEE Computer Graphics and Applications*, 6(11):21–29, November 1986.

[15] P. Haeberli. Synthetic lighting for photography. Available at http://www.sgi.com/grafica/synth/index.html, January 1992.

[16] T. Hawkins, J. Cohen, and P. Debevec. A photometric approach to digitizing cultural artifacts. In *Proc. 2nd International Symposium on Virtual Reality, Archaeology, and Cultural Heritage (VAST 2001)*, pages 333–342, December 2001.

[17] T. Kollig and A. Keller. Efficient illumination by high dynamic range images. In *Eurographics Symposium on Rendering: 14th Eurographics Workshop on Rendering*, pages 45–51, June 2003.

[18] G. W. Larson. Overcoming gamut and dynamic range limitations in digital images. In *Proceedings of the Sixth Color Imaging Conference*, November 1998.

[19] M. Levoy and P. Hanrahan. Light field rendering. In *SIGGRAPH 96*, pages 31–42, 1996.

[20] G. S. Miller and C. R. Hoffman. Illumination and reflection maps: Simulated objects in simulated and real environments. In *SIGGRAPH 84 Course Notes for Advanced Computer Graphics Animation*, July 1984.

[21] T. Mitsunaga and S. K. Nayar. Radiometric self calibration. In *Proc. IEEE Conf. on Computer Vision and Pattern Recognition*, June 1999.

[22] R. Ng, R. Ramamoorthi, and P. Hanrahan. All-frequency shadows using non-linear wavelet lighting approximation. *ACM Transactions on Graphics*, 22(3):376–381, July 2003.

[23] R. Ramamoorthi and P. Hanrahan. Frequency space environment map rendering. *ACM Transactions on Graphics*, 21(3):517–526, July 2002.

[24] P.-P. Sloan, J. Kautz, and J. Snyder. Precomputed radiance transfer for real-time rendering in dynamic, low-frequency lighting environments. *ACM Transactions on Graphics*, 21(3):527–536, July 2002.

[25] B. Smith and L. Rowe. Compressed domain processing of JPEG-encoded images. *Real-Time Imaging*, 2(2):3–17, 1996.

[26] R. Szeliski and H.-Y. Shum. Creating full view panoramic mosaics and environment maps. In *Proceedings of SIGGRAPH 97*, Computer Graphics Proceedings, Annual Conference Series, pages 251–258, August 1997.

[27] C. Tchou and P. Debevec. HDR Shop. Available at http://www.debevec.org/HDRShop, 2001.

[28] J. Unger, A. Wenger, T. Hawkins, A. Gardner, and P. Debevec. Capturing and rendering with incident light fields. In *Eurographics Symposium on Rendering: 14th Eurographics Workshop on Rendering*, June 2003.

[29] G. Ward. Real pixels. *Graphics Gems II*, pages 80–83, 1991.

Chapter 7

COMPUTER VISION IN VISUAL EFFECTS

Doug Roble

7.1 Introduction

Computer vision has changed the way movies are made. Sure, computer graphics gets all the press, but computer vision techniques have made a significant impact on the way films with visual effects are envisioned, planned, and executed. This chapter examines the state of the practice of computer vision techniques in visual effects. We also examine the changes computer vision has brought to the industry, the new capabilities directors and visual effects creators have available, and what is desired in the future.

Let us examine the concept of visual effects in terms of computer vision. The main task of visual effects is to manipulate or add things to an image. Artists are very skilled, but the more information they have about the original image and the more tools the computer presents to them, the more effective they can be.

Computer vision has made a huge impact in the field of visual effects over the last decade. In the early 1990s, we saw the shift away from using physical models and photographic techniques to create and add fantastic things to an image. It has became commonplace to scan each frame of film into the computer and composite computer-generated elements with the filmed background image. Digitizing the filmed image is so common now that whole films are digitized, manipulated in the digital realm, and filmed out to a negative for printing. (*Oh Brother, Where Art Thou?* is a good

306

example of this. The color of each frame of the film was digitally adjusted to give the film its "look.") And, of course, digital cameras are starting to make inroads into Hollywood. *Star Wars, Attack of the Clones* was shot completely with a digital camera.

Once a sequence of frames is digitized, it becomes possible to apply standard computer vision algorithms to the digitized sequence in order to extract as much information as possible from the image. Structure from motion, feature tracking, and optical flow are just some of the techniques that we examine in relation to filmmaking.

7.2 Computer Vision Problems Unique to Film

7.2.1 Welcome to the set

A film set is a unique work environment. It is a place of high creativity, incredible tension, sometimes numbing boredom. The set is a place of conflict in the creation of art. The producers and their staff are responsible for keeping the show on budget and on time, and they fear and respect the set, for filming on a set is very expensive indeed. The director and crew are trying to get the best possible images and performances on film. These two forces are at odds with each other on the set, and often the computer vision people are caught in the middle!

The set is where it all begins for the digital artists responsible for what we call *data integration*. Data integration is acquiring, analyzing, and managing all the data recorded from the set. It is common for a visual effects facility to send a team of one or more people to the set with the standard crew. Before, during, and after filming, the data integration crew pop in and out of the set, recording everything they can about the makeup of the set. This is where the trouble starts: the producers are not crazy about the data integration crew; they are an added expense for a not so obvious result. The directors can get irritated with the data integration team because it is just one more disturbance on the set, slowing them down from getting one more take before the light fades or the producers start glaring again.

At least that is the way it was in the beginning. Movie makers are getting much more savvy and accustomed to the power that computer vision provides them in the filmmaking process. While a director might be somewhat annoyed that a vision person is slowing the process slightly, he or she is infinitely more annoyed when the visual effects supervisor tells him or her how to film a scene because of the inability of the effects people to deal with a particular camera move.

Watch some movies from the 1980s. Special effects were booming, and there were elaborate, effects-laden movies in that decade. But effects shots always telegraphed themselves to the audience because the camera stopped moving! If a stop-motion creature effect or some elaborate set extension was planned, the director could either lock the camera down or move it in very limited ways (a rotation about the lens' nodal axis or a simple zoom could usually be handled, but nothing more). Now directors can film however they want. The camera can be on a giant crane, a plane, a Steadicam, or even a handheld shot. If there is not enough information in the scene for the computer vision algorithms to work, we drop objects in the scene and paint them out later! (A couple of problem scenes for the ambitious researcher out there: a camera mounted in a plane flying over the ocean on a clear day—computing the camera location of each frame is tricky, because all your reference points are moving! And yet, if the camera track is not correct, inserting a boat on the surface of the ocean will look wrong. Or, a camera carried by a skier filming on a smooth glacier—this is where adding objects makes the problem go away, but without them, there is no hope!)

For a major production with many effects shots, a team of two to three people is sent to the set to record information. They typically stay with the film crew throughout the production, traveling to outdoor locations and working on the indoor sets that have been constructed. This data integration crew records everything it can to make the digital artists' lives easier. Here is a list of things that the crew tries to capture:

- **Environment maps:** Environment maps are spherical (or semispherical) photographs of the set, generally taken from near center of the set or near where the digital object is going to be inserted into the set. Digital artists began running into the center of sets with a chrome ball in the late 1980s and early 1990s. The quickest, cheapest way to acquire an environment map is to go to your garden supply center, buy a decorative chrome ball, and mount it on a tripod (Figure 7.1). Then, when the chance presents itself, run into the center of the set and snap two quick photographs of the chrome ball from both sides. It is relatively easy to map the pixels of the image to the area on the ball, and from that, it is possible to project the images onto a spherical environment map.

This environment map is then used in the renderer to render the digital objects as if they are in the environment of the set. Shiny objects

Figure 7.1. Chrome and diffuse balls used to capture environment maps.

reflect portions of the set, and even diffuse objects change their color depending on the incident light from the environment maps.

This ability to light digital objects with the light that existed on the set has caused a flurry of development over the last decade. Paul Debevec was a pioneer in the use of high dynamic range (HDR) imagery in conjunction with environment maps, showing just how powerful the combination can be.

A snapshot of a chrome ball is only so useful. It creates a picture of a 360-degree environment, but since film (be it digital or photochemical) has a limited dynamic range, much of the information in the image is inaccurate. All photographers have experienced this to some extent. Setting the exposure controls to capture detail in the dark areas of a scene means that the bright areas will be overexposed and clipped. Alternatively, setting the exposure for accurately capturing the bright areas of the scene means that the dark areas will have no information in them.

The concept behind HDR photography is simple: take multiple images of a scene each with a different exposure. Then, combine the images into one image by using only the nonclipped pixels of the images and an appropriate mapping function. The results are amazing. Instead of an image with 8-bit pixels and values that go from 0 to 256, HDR images store floating-point numbers that go from zero to a very large number indeed. It is even possible (and sometimes very useful) to have negative color values!

There are many advantages to HDR images, but in the context of environment maps, the main advantage is that they are an accurate way of capturing the light energy of the environment, not just a simple snapshot. These environment maps can be used as elaborate lights in a rendering package and to create stunningly convincing images. A good introduction to these concepts can be found in [11] and [3].

– **Camera and object motion:** Computing camera location and motion is where computer vision first made inroads into the film industry. Directors acknowledged, but did not like, the constraint that the camera had to stop moving or had to move in a very specific way if an effect was going to work.

The first method used to record camera motion did not rely on visual effects at all. Encoding devices were attached to the moving parts of a crane or a dolly. The measurements from the encoding devices were sent to a computer and, combined with an exact knowledge of the rig's geometry, the location and orientation of the camera were determined.

These measurements were then sent to a motion-control camera rig, and the motion of the camera was replicated. A motion-control camera rig is basically a large robotic camera mover. Very precisely manufactured and with very accurate stepper motors, the best motion-control rigs could move the camera starting at one end of a long set, perform a complicated move across the set, return to the starting point, and be only mere millimeters off from the original location.

Of course, the encoding process is rife with errors. It is well-nigh impossible to accurately encode the motion of the camera. The measurement and modeling of the rig is difficult to do exactly, it is impossible to encode any flex or looseness in the rig, and it is impossible to encode the inner workings of the camera. Given these problems, it is rare that a camera move is encoded on one camera system and replicated on another. More often, both the original camera move and the replication of the move was done on the same rig. This minimized the errors of the system and often worked well.

But, there is still a problem. Most motion-control rigs are big, heavy machines. They have to be, given the weight of the camera they are carrying and the need for stiffness in the rig. So, they could never move very quickly—certainly not as quickly as a handheld camera or a camera mounted on a dolly or a crane.

Given all these limitations, motion-control rigs were only used for shots with models or very limited shots with actors.

Recently, "real-time" motion-control rigs have been used on sets. The Bulldog Motion Control Camera rig from ImageG received a Technical Academy Award in 2001. It is extremely fast and can be used to reproduce very fast camera moves.

What does this all have to do with computer vision? Quite a bit, actually. Motion-control rigs are still used in visual effects because it is often much more cost effective to build a detailed model of a ship or location than to construct it all as a digital set. Industrial Light and Magic built a large model of a pirate ship and sailed it in a pond behind its facility for a large number of shots in the film *Pirates of the Caribbean* (2003). It was just more cost effective.

There are two problems associated with motion-control rigs and computer vision. The first problem is relatively easy. If a model is filmed with motion control, the images are often enhanced with digital effects. Unfortunately, even though the camera path is known from the motion-control move, it cannot be trusted. The path is very repeatable, but it is not subpixel accurate in the digital realm—the internal camera motion and flex of the devices is not recorded. However, the camera pose estimation is a very easy subset of the general problem. Once features in the image are identified, the motion of the features is easy to predict because of the known motion of the camera. In fact, while the overall path of the camera may have errors, the incremental motion from one frame to the next is quite accurate. So the camera motion is used to predict the feature motion, and a quick template match in the local area is performed to find the features to subpixel accuracy. Standard structure from motion techniques are used to refine the motion. Convergence is very fast, since the starting path is quite good.

The second problem with motion-control cameras is quite difficult and, to some extent, impossible to solve. Often it is required to replicate the motion of a handheld camera with the motion-control rig. Often, a director films the actors with a "wild" camera in front of a bluescreen. Then, the camera move is later reproduced on a motion-control rig filming an elaborate model. The actors are composited with the filmed model for the final shot. Tracking the camera on the stage is not terribly difficult. The data integration team places markers or known geometry on the bluescreen, and pose estimation techniques are used to recover the camera path. The problem comes when transferring the camera motion to the motion-control rig.

Motion-control rigs do not stop before a frame of film is shot. It is a common misconception, but they do not move, stop, shoot, move, stop.... Rather, they shoot the sequence of frames in one continuous motion. There are many advantages to continuous motion, but the two overriding benefits are speed of shooting and motion blur. Stopping before each frame slows down the motion-control rig terribly, and stage time is expensive. If the camera is moving while the shutter is open, one gets motion blur for free. Model shots look very unconvincing without motion blur.

This continuous motion of the rig leads to the problem. Because of the size of the motion-control rig, there are physical limits to the torque of the stepper motors and how fast they can move the rig. Therefore, at the speed that is needed for the shot, the camera may not be able to exactly reproduce the desired motion. When the foreground and background are merged, they may not line up, and high-frequency motions of the camera move will not be preserved in the background plate.

This problem is difficult to fix. If the problem is extreme enough, the motion-control camera motion is tracked (using methods described above) and compared with the original camera motion. Three-dimensional points in the scene are then projected onto the image plane of the real camera and the motion-control rig. The discrepancy of these points is then used to build a general warping field that is applied to the background images. Naturally, this is a 2D fix to what is a 3D problem, and it cannot correct all the problems, but it usually suffices. Of course, a generalized warp will filter the image and effectively reduce

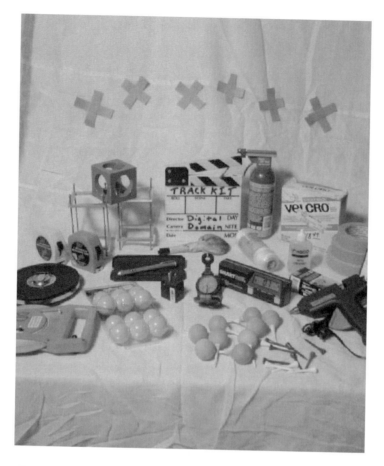

Figure 7.2. Typical gear in a data integration tracking kit.

the resolution of the background image so the images are typically digitized at a higher resolution.

- **Set and object geometry:** The data integration team is responsible for capturing geometric information about the set. This can range from some simple measurements of the height of a locked-off camera or where a light is in relation to the camera to full 3D reconstruction of the set with textures and BRDF (bidirectional reflectance distribution function) analysis of the surfaces.

Most data integration teams have assembled a "tracking kit" (Figure 7.2) that is brought to the set and (hopefully) has all the devices

needed to capture the required information. The following items are
included in most kits:

- **Tape measures:** Tape measures are still the fastest way to get
  at least some measurements from the set. Often, a data integration team member dashes onto the set and measures the distance
  between two notable landmarks. This data is used to give a scale
  to the geometry reconstructed using photogrammetry.

- **Tripods:** A good tripod is essential for the work the data integration team does. Being able to lock down a camera and have it
  stay locked down is very important for HDR image recording.

- **Chrome and diffuse spheres:** As discussed earlier, the data
  integration team is responsible for capturing environment data in
  the scene. These spheres are mounted on a tripod, set in the center of the scene, and photographed from various points of view.
  These images are then stitched together for a complete environment map.

- **Camera calibration charts:** High-quality checkerboard grids
  are taken on the set to determine the intrinsic parameters of the
  cameras being used. These charts are typically mounted on a rigid
  metal plate to insure accuracy. Sometimes, because of the lens
  being used or the situation, a larger calibration chart is needed.
  In this case, the grid is printed on large-format, plasticized paper.
  This large grid—typically 2 or 3 meters on a side—is attached to
  a flat wall.

- **Tracking geometry:** Visual effects companies typically have a
  machine shop to build necessary items for model shoots. The machine shop is also used to build highly accurate reference geometry
  for computer vision applications (Figure 7.3).

  Metal cubes, from a couple of centimeters to as large as a meter
  on a side, are placed in a set to provide some known geometry.
  The cubes help provide scale for photogrammetry and are also a
  way to measure and control error.

  These reference devices are not left in the set during actual filming, but are often filmed by the main camera and removed before
  main filming.

- **Brightly colored tape and Ping-Pong balls:** A huge problem, from a computer vision perspective, of shooting on a blue-

Figure 7.3. Large, rigid geometry used for photogrammetry and camera calibration.

screen set is that there is nothing to track! The background is a uniform blue, and the foreground is typically moving people. Quite challenging for camera tracking!

Hence the need for tracking objects. Pieces of tape or colored Ping-Pong balls are taped to the bluescreen background. These provide handy reference points for the tracking algorithms but do not cause too many problems for the matte extraction algorithms. As an aside, bluescreen shots can be quite challenging for a good camera track. While being able to place reference markers may seem ideal, there is a delicate balance between placing enough reference points and placing too many for an easy matte extraction. (The reason that the crew was filming in front of a blue screen was to make it easy!) But often, more of a problem is that the blue screen is usually quite a distance away from the actors and is typically a flat screen. Computing the pose of a camera with points that are all coplanar can be difficult sometimes, and the fact that the reference points are far away from a camera that probably does not move all that much makes accuracy even more challenging. Finally, the actors are always a factor. Depending on the number of actors, all the tracking markers may be obscured. What happens in the worst case, if the camera is moving but the actors have obscured all the reference points? We rely on the data integration artist to solve the problem by hand. Camera positions can be interpolated from known points, but in worst-case sections, there is no alternative to solving it using the artist's skill at knowing "what looks good."

- **A total station survey device:** Back in the early 1990s, when computer vision techniques were starting to make an impact on filming and the set, it was quite common to hire a survey crew to create a survey of reference points on the set. These points were used in a camera pose estimation algorithm to compute the location of a moving camera for each frame of film.

As computer vision techniques became commonplace, visual effects studios purchased their own survey devices and trained people in their use. Now, almost every studio owns a "total station." Total station is the generic name for the survey devices used by burly government workers often seen blocking traffic. A total station is a tripod-mounted, highly accurate, laser distance and angle measure. A total station has an accurate telescoping sight and a laser pointer. Once the station is set up, the survey person simply sights or points the laser at a spot in the terrain and presses a button. The distance to the spot is measured using phase measurement. The angle and distance provide the 3D location of the point. Key points in the set can be digitized quite quickly.

The problem with total stations is that they are time consuming to set up. And if the station is moved, there is a realignment process to compute the location of the total station in relation to the first setup point. Also, the only points that are digitized are the points that are selected during the session. If the data integration artist needs to compute a point that the person on the total station did not compute, different photogrammetry-based techniques must be used.

The huge advantage of a total station is that a highly accurate reference geometry of key points in the set is built. This can be enormously helpful for camera or object tracking later. Relying solely on structure from motion techniques to compute the location of the camera is unwise—those techniques are notoriously bad if the camera focal length changes during the shot. This certainly happens during a zoom and often happens during a simple change of focus. With reference geometry, the camera's intrinsic parameters can be calculated from frame to frame. This gives the director the ultimate freedom on a set.

– **Digital cameras with different kinds of lenses, including a fisheye:** Of course, the data integration team carries a digital camera on the set. This is used to record the layout of the set, reconstruct the geometry of the set, and build environment maps. The cameras are usually high resolution and have calibrated lenses.

   Data integration teams usually carry a high-quality fisheye lens to make environment maps, but there are issues with these lenses. Fisheye lenses can be used to quickly build a 360-degree panorama of the set, but they also typically have poor optic qualities, particularly large bloom around bright light sources or reflections. These can be difficult to deal with when assembling the HDR images—they are typically painted out by hand!

   To deal with this problem, there are special-purpose 2D slit cameras mounted on a motorized tripod head. These cameras slowly spin around, grabbing a slice of the environment at a time, usually changing the exposure as they go. They do not exhibit the bloom problem that fisheye lenses do, but they are much slower and much more expensive.

– **Video camera (progressive scan):** A video camera is also brought to some sets or location shoots. One or more can be used to augment what the main film (or digital video) camera is shooting. We have built stereo rigs that mount a video camera a fixed distance away from the main film rig so that a stereo correspondence can be built on every frame. This is useful for closeup shots of actors where the moving geometry of the face needs to be digitized.

   Of course, all the problems associated with stereo are present in this setup, including changing light sources, nonideal subjects, and changing lens parameters.

– **LIDAR (light detection and ranging):** Since the late 1990s, another option for building set geometry has been LIDAR. A LIDAR unit looks similar to a total station and accomplishes the same kind of task, just at a much finer detail and automatically. A LIDAR device is a time-of-flight laser mounted on a mechanical base that rotates the laser up and down and left and right. In a certain amount of time, a swath of the set can be scanned to high accuracy and great detail.

LIDAR produces a cloud of 3D points that represent the surfaces that the unit can see from a given position. By moving the device and scanning from different positions, the multiple clouds of points can be aligned and joined to provide a remarkably complete point-based representation of the set. The range of a professional LIDAR device is around 100 to 200 meters, so it is effective both inside and outside.

The density of points that LIDAR produces is both a strength and a weakness. Just a couple of scans produces hundreds of thousands of points which, though amazing to look at, are useless to a digital artist who is trying to recreate a set digitally. Some sort of surface representation is much more useful to the artist.

Computing the appropriate surface from a cloud of points is an ongoing research project. Some successful techniques have been developed based on the use of radial basis functions. And recently, point-based representations of geometry have started to make an impact on this problem [8, 9].

– **Textures and beyond:** Finally, the data integration team is responsible for recording the textures of the set. It is becoming quite popular to create completely digital representations of sets or locations. Director David Fincher has been a strong proponent of using photogrammetry techniques to completely recreate a set inside the computer. The visual effects company Buf Films recreated the interior of an entire house for him for the film *Panic Room*. But getting the geometry to mimic the real world is only part of the battle. A considerable amount of the "reality" of the real world is how light reflects off an object.

It is the data integration team's job to help the modelers who are building the digital set in any way they can. The most important thing to record is the diffuse textures of an object. Lighting the set with diffuse lights or using cross-polarizing filters on the light and the camera lens results in images with very few specular highlights. (Having a specular highlight baked into a texture is a bit of a pain. Part of the modeling process is cleaning up textures and painting out specular highlights. There are typically quite a few people working on this.)

Of course, as anyone who has played a video game can attest, textures are only one step on the way to reality. Next, shaders are written that use the textures and add things like bump mapping or displacement maps to give the surface of an object a feeling of reality. This is where the data integration team comes into play again. By photographing images of an object from different calibrated vantage points, the geometry of the object can be constructed, and the diffuse texture taken from one point of view can be compared with the diffuse texture from another point of view. Any disparity in the texture must be due to some displacement of the surface not modeled by the underlying geometry [4]. It becomes a simple stereo problem again. The disparity map produced by template matching along the epipolar lines intersecting the polygons of the surface can be used as a bump or displacement map in a shader. This helps the artists achieve a higher sense of reality.

Beyond displacement maps, changing the lighting model of the surface helps improve the reality of the digital model. Light rarely reflects off a surface in a purely diffuse model. Rather, it scatters according to a BRDF. This can be measured in the lab using a goniospectrometer, and a couple of big visual effects houses have purchased one of these devices. It consists of a moving camera that accurately measures the scattered light in a hemispherical area, producing an accurate distribution function for lighting models. Recently, there has been some exciting work on measuring approximate BRDF models without one of these cumbersome devices. It is now possible to measure a BRDF with a video camera and a flashlight moving in an approximate hemispherical arc [7]. Of course, the results of such measurements are quite crude, but they can be put to use by shader writers to make a surface look even better.

Finally, many objects do not simply reflect light using a diffuse and specular lighting model but rather allow the light to penetrate into the object where it scatters and bounces around before emerging again. This phenomena resulted in a modification to the BRDF model to produce the BSSRDF model in which subsurface scattering is taken into account. Marble exhibits this behavior, but even more notably, human skin is very translucent. Shine a laser pointer through your hand to see exactly how translucent it is. The data integration team often does just that. By shining a light into an object and photographing the glow produced by the subsurface scatter, a reasonable model of the scatter term can be built.

Obviously, the data integration team needs to be equipped to capture all aspects of geometry and light on the set. It is a fun job, the set is an exciting place to be, and there is a certain thrill with working alongside celebrities. But that thrill quickly fades as the reality of a high-pressure job with long hours and often a long schedule sets in. Once the film is exposed and the sets are struck, the team comes back to the office, and the processing of the images and data can begin in earnest.

7.3 Feature Tracking

Digital artists use one computer vision technique more often than all others: pattern tracking, or template matching. Digital artists who specialize in 2D effects (rotoscoping, wire removal, compositing) use pattern tracking nearly constantly. Being able to accurately follow a feature in an image is vital for quick, accurate effects.

As an example of how this is used, consider the film *Interview with the Vampire*, released in 1994. In that film, the vampire Lestat, played by Tom Cruise, is cut with scissors by another vampire. These cuts are on both cheeks. The director wanted the cuts to appear on the cheeks, bleed slightly, then fade as the vampire's healing power took hold. All within one continuous shot.

Of course, one really cannot cut Tom Cruise's cheek, and even if it was possible with makeup, getting it to heal convincingly on camera would be very difficult. Luckily, this kind of effect was fairly commonplace even back in the early 1990s. Digital artists modeled two pieces of "cut" geometry and animated it with standard modeling tools. These elements were rendered out as images, assuming no motion of Tom's head. They simply animated the cuts using the first frame of the sequence as a guide.

The challenge then was to track the cut elements to Tom's moving head throughout the shot. This is where computer vision techniques came in to play. At that time, the major digital image manipulation and compositing packages all had some method of pattern tracking. They are usually based on the minimization of the cross-correlation of the original image area in a search image area. J. P. Lewis wrote a seminal paper on template matching that many visual effects companies still use today [6].

In the case of *Interview with the Vampire*, two spots on Tom's right cheek were pattern tracked, and the resulting positions were used to scale and rotate the original images of the rendered cut to track along with the motion of the cheek. Because Tom did not move his head much, the artists

Figure 7.4. Brad Pitt's makeup was toned down using feature tracking in *Interview with the Vampire*.

were able to get away with only using a two-point technique. If the movement of the cheek was more complicated, the artists could use four points and a much more complicated warp of the image to track it exactly. (Of course, there are times where even a complicated deformation of a 2D image will not hold up; in that case, we need to do a more complicated 3D track of the camera and object move so that the effect can be rendered in the right position. This is discussed later.)

Pattern tracking has become such an integral part of a digital artist's toolbox that it is used all the time without audiences even noticing. In the same film, Brad Pitt's vampire makeup was too noticeable on screen. The director asked if it was possible to tone the makeup down. Feature tracking solved the problem again. The digital artist created a 2D digital makeup element—basically some pancake makeup—and this was tracked to the offending vampire veins. A quick composite later and the veins had been masked to a certain extent. See Figure 7.4.

The effect was quite successful, but it was not exactly easy. Pattern tracking in a film environment faces many challenges. First, it must be robust in the presence of noise. All vision techniques must deal with noise, and pattern tracking is no exception. A recurring theme in this chapter is the audience's ability to notice even the smallest error when it is magnified on the big screen. Some films are even being transferred to large-format film and projected on an Imax screen—sure to expose any flaws in the graphics or vision!

In dealing with noise, there are a couple of techniques that work with pattern tracking. One is relatively simple: the median filter. A small median filter does a nice job of removing the worst noise without destroying the underlying geometry of the image. More elaborate, the noise in the image can be characterized by analyzing a sequence of images over a still subject. By applying an adaptive low-pass filter over the image that follows the characteristics of the individual camera, the noise can be knocked down quite a bit.

Extreme lighting changes still stymie most pattern tracking algorithms. A high degree of specular reflection blows out most underlying information in the image and is quite difficult to deal with in a pattern track. Flashes of lightning or other extreme lighting changes also produce bloom on the image that change the geometry of the pattern being tracked. At this point, there is no recourse but to track the offending frames by hand.

7.4   Optical Flow

Recently, over the last five years, optical flow techniques have really started to make a big impact on the film industry. In fact, there are a couple of companies that do fairly well selling software based on optical flow techniques.

Optical flow is the 2D motion of the features in an image sequence, from one frame to the next [1, 10]. Consider a pattern track centered on every pixel. After an optical flow analysis of an image sequence, a secondary frame is generated for every original image in the sequence. This new image does not contain colors, but rather a 2D vector for each pixel. This vector represents the motion of the feature seen through the pixel from one frame to the next.

In an ideal case, if the pixels of the image n were moved all along their individual motion vectors, an exact replica of image $n + 1$ will be created. The pixels in image n would be pushed around to make image $n + 1$.

Of course, things never work out that perfectly, but it is possible to see the power of such information. The first application of optical flow follows naturally: change the timing of a sequence of images. If the director shoots a sequence using a standard film camera at normal speed, 24 frames are exposed per second. It happens occasionally that the director wants to change the speed of the film, either slow it down or speed it up. Optical flow helps in either case.

Slowing down an image sequence is relatively easy. If the director wanted to see the shot as if it were filmed at 48 frames per second, one simply duplicated each frame of film. This is what was done before the advent of digital image manipulation. A single film negative was printed onto two frames of print film. Playing the resulting film back resulted in slow motion. However, it looked a little jerky. The film pauses for 2/24 second, and it is a noticeable effect. The solution was to generate in-between frames that were blends of image n and image $n+1$. Again, this was done during the printing process: the in-between frames were generated by exposing the print film with the n and $n + 1$ negative frames for only half the exposure time each. This created an in-between frame that looked different than the original frames, but also looked pretty bad. It was a double exposure of two different images and exhibited quite a bit of strobing and streaking.

Optical flow came to the rescue. In the explanation above, it was shown that by moving the pixels along the optical flow motion vectors, a new image can be generated. By scaling the motion vectors by half, the new image is exactly between the two original images. This new image typically looks quite a bit better than a blend of the neighboring images, and it does not exhibit any of the blurred or double exposed edges. It also has the advantage that it can be used to slow the original sequence down by any factor. New images can be generated by scaling the motion vectors by any value from 0.0 to 1.0.

But pause a moment. It was stated earlier that if the pixels are pushed along their motion vectors, the image at $n+1$ can be created from the pixels that make up image n. Unfortunately, it is not that easy. The motion vectors typically go in all directions, and sometimes neighboring motion vectors diverge or converge. There is no guarantee that the pixels in image $n + 1$ will be all filled by simply following the motion of the pixels. Instead, the operation of creating the in-between image is a "pixel pull" operation. For each pixel in image $n + 1$, a search is done to find the motion vector that will land in the center of the pixel. This motion vector will probably be the motion vector located at a subpixel location that is the result of an interpolation between the four neighboring pixels. A search is made in the local neighborhood of the destination pixel to find a likely starting pixel location in image n. Then, a gradient descent is used to compute the exact subpixel location that, when added to the interpolated motion vector, lands on the center of the desired pixel.

But there is more! Interpolating the color of the subpixel leads to sampling and aliasing artifacts. In fact, if the motion vectors converge—that is, if the colors of a group of pixels all end up at one pixel in image $n + 1$—the original pixels should be filtered to generate an accurate pixel color. Typically, the previous backwards tracing is done from the four corners of the pixel. This creates a four-sided polygonal area that is sampled and filtered. These results create a superior in-between image.

Speeding up a sequence of images is just as straightforward, with a slight twist. Consider the problem of speeding up the action by a factor of two. In that case, it is desired to make it look like the action was filmed with a camera that ran at 12 fps. It should be easy, right? Simply throw away every other frame from the original sequence! This works fine, with one problem: motion blur. The objects in the images are now supposed to be moving twice as fast, and the motion blur recorded on the images looks like it was filmed with a 24-fps camera. This produces the strobing, harsh effect normally associated with sped-up film.

Optical flow comes to the rescue in this case as well. The motion vectors associated with each pixel can be used to streak the images so that the motion blur appears correct. Like the sampling issue before, blurring an image according to optical flow motion has some hidden issues. The easy, but incorrect, way to add motion blur to the image is to use the motion vector to create a convolution kernel. The entries in the convolution kernel that are within a certain distance from the motion vector are given a weight based on a Gaussian falloff. The final kernel is normalized and convolved with the image to produce the resultant pixel. This is done for each pixel in the image. This technique certainly blurs the image in a complex directional sense, but it does not quite produce correct motion blur. (It is reasonably fast—it does require a different kernel computation at every pixel—and quite acceptable in some situations.)

This method, as mentioned, is incorrect. Consider this example: a model of a spaceship has been filmed on the set with a black background behind it. After the shoot, it is decided to add more motion blur to the shot—maybe it makes the shot more dynamic or something. So, somehow, motion vectors for each pixel are computed. It need not be optical flow in this case. Since the model is rigid, it is possible to use the motion of the motion-control rig or track the motion of the model to generate a reproduction of the model shot in the computer. This can then be rendered so that instead of writing out the color of every pixel, the pixels contain the motion vector. Whether created by optical flow or some rendering technique, assume the pixels contain the 2D motion vector of the feature under the pixel. Now, consider what is happening at the edge of the model. A pixel covering the edge of the model will have the appropriate motion vector, but the neighboring pixel that only covers background contains no motion at all! The background is black, and even if the camera is moving, it is impossible to detect the motion. The problem becomes evident at this point. Using the previous technique to add motion blur if the spaceship is moving fast, the pixels covering the spaceship will be heavily blurred. The pixels right next to the spaceship will have no blurring at all. This causes a massive visual discontinuity that audiences (and visual effects supervisors) are sure to notice.

There is a (mostly) correct solution to this: instead of blurring the pixels, distribute the energy of the pixels along the motion vector. This gets a little more complicated in that the motion vectors are converging and diverging in places. This caused problems in interpolating images, and it causes problems in adding motion blur. Basically, the color of a pixel is distributed along the motion vector path from the pixel. Doing the naive approach of simply adding a fraction of the pixel's color to the pixels that intersect the path produces bright and dark streaks in the motion blur where the vectors converge and diverge. Rather, the interpolated motion vectors must be used at each corner to create a motion polygon of converge. Then, the color is added to each pixel under the motion polygon, using a weighting that is the area of coverage of the individual covered pixel divided by the total area of the motion polygon. This produces lovely motion blur at quite a cost in computation time. In the reality of visual effects, we try to get away with the cheap solution for motion blur, knowing that we can always pay the price for the more accurate version of motion blur.

The previous paragraphs seem to indicate that optical flow is a robust, trustworthy algorithm. This is not the case. Optical flow gets terribly confused with pixels that contain features that disappear (or appear) in the space of one frame. This is commonly referred to as foldover. Also, specular highlights and reflections in objects can fool even the best optical flow algorithm. And most optical flow algorithms assume a certain spatial continuity—one pixel moves in pretty much the same direction as its neighbor—which makes computing optical flow on complicated moving objects like water or blowing tree leaves quite difficult.

There are many solutions to these problems. If the foldover areas can be found, they can be dealt with by predicting the motion of the pixels from previous or following frames. Sequences of images are often analyzed both forwards and backwards; where the pixel motion vectors do not match up is usually a good indication of foldover. Most optical flow algorithms also compute an indication of confidence in the flow for a pixel or an area. Where foldover areas are indicated, the discontinuity in the vectors is encouraged rather than avoided. And interpolating the pixels for in-between frames is done by recognizing that the pixel will either be disappearing or appearing and interpolating in the appropriate direction.

Optical flow is not just for image manipulation. It is increasingly being used to extract information about the image. In 2002, Yung-Yu Chung et al. published an influential paper titled "Video Matting of Complex Scenes" [2]. The techniques in this paper have made an impact on the film industry: it is now quite a bit easier to extract a matte from an object that was not filmed in front of a bluescreen. The paper describes a technique for computing the matte based on Bayesian likelihood estimates. Optical flow comes into play by pushing the initial estimate of the matte (or in this case, the trimap—a segmenting of the image into foreground, background, and unknown elements) around as the contents of the image move. It uses the same technique as described above for computing an in-between image, but this time applied to the trimap image. Many visual effects companies have implemented compositing techniques based on this paper.

Additionally, ESC Entertainment used optical flow to help reconstruct the facial motion of the actors in the movies *Matrix Reloaded* and *Matrix Revolutions*. They filmed an actor in front of a bluescreen with three (or more) synced cameras. Instead of using traditional stereo techniques, which have problems with stereo correspondence, the artists placed a 3D facial mask on the face of the actor in each film sequence. This was relatively easy to do for the first frame of the film—the actor's face had not deformed from the digitized mask. The sequences of film were analyzed for optical flow. Then the vertices of the mask were moved in 2D according to the optical flow. Since each vertex could be seen in two or more camera views, the new 3D position of the vertex was calculated by triangulation. Doing this for each vertex on every frame produced quite clean-moving, deforming masks of face's of the actors. This avoided any stereo correspondence problem!

7.5 Camera Tracking and Structure from Motion

With the advent of computer vision techniques in visual effects, directors found that they could move the camera all over the place and the effects crew could still insert digital effects into the image. So, naturally, film directors ran with the ability and never looked back. Now all the large effects shops have dedicated teams of computer vision artists. Almost every shot is tracked in one way or another.

In the computer vision community, the world space location of the camera is typically called the camera's extrinsic parameters. Computing these parameters from information in an image or images is called *camera tracking* in the visual effects industry. In computer vision, this task has different names depending on how you approach the solution to the problem: camera calibration, pose estimation, structure from motion, and more. All the latest techniques are available to the artists computing the camera motion.

The overriding concern of the artists is accuracy. Once a shot gets assigned to an artist, the artist produces versions of the camera track that are rendered as wireframe over the background image. Until the wireframe objects line up exactly, the artist must continually revisit the sequence, adjusting and tweaking the motion curves of the camera or the objects.

A camera track starts its life on the set. As mentioned before, the data integration team gathers as much information as is practical. At the minimum, a camera calibration chart is photographed, but often a survey of key points is created or the set is photographed with a reference object so that photogrammetry can be used to determine the 3D location of points on the set. This data is used to create a polygonal model of the points on the set. (This model is often quite hard to decipher. It typically has only key points in the set, and they are connected in a rough manner. But it is usually enough to guide the tracking artist.)

The images from the shot are "brought online"—either scanned from the original film or transferred from the high-definition digital camera. Also, any notes from the set are transcribed to a text file and associated with the images.

Before the tracking begins, the lens distortion must be dealt with. While the lenses used in filmmaking are quite extraordinary, they still contain a certain amount of radial distortion on the image. Fixed lenses are quite good, but zoom lenses and especially anamorphic lens setups are notorious for the distortion they contain. In fact, for an anamorphic zoom lens, the distortion is quite hard to characterize. For the most part, the intrinsic parameters of the lens are easily computed from the calibration charts on set. For more complicated lenses, where the standard radial lens distortion equations do not accurately model the behavior of the lens, the distortion is modeled by hand. A correspondence between the grid shot through the lens and an ideal grid is built and used to warp the image into the ideal state.

However the lens distortion is computed, the images are warped to produce "straightened images," which approximate what would be seen through a pinhole camera. These images are used online throughout the tracking, modeling, and animation process. The original images are used only when the final results are composited with the rendered components of the shot. This seems like an unnecessary step; surely the intrinsic parameters of the lens can be incorporated into the camera model. The problem is that these images are not just used for tracking but also as background images in the animation packages that the artists use. And in those packages, a simple OpenGL pinhole camera model is the only one available. So, if the original images were used to track the camera motion and the tracking package uses a complicated camera model that includes radial (or arbitrary) lens distortion, the lineup would look incorrect when viewed through a standard animation package that does not support a more complicated camera model.

The implications of this are twofold. First, two different sets of images must be stored on disk, the original and the straightened version. Second, the artists must render their CG elements at a higher resolution because they will eventually be warped to fit the original images and this will soften (blur) the rendered image. Surprisingly, this does not impact the pipeline as much as one would expect; CG elements are typically blurred slightly to blend better with the original filmed plate.

The artist assigned to the shot uses either an inhouse computer vision program or one of many commercially available packages. For the most part, these programs have similar capabilities. They can perform pattern tracking and feature detection, and they can set up a constraint system and then solve for the location of the camera or moving rigid objects in the scene. Recently, most of these program have added the ability to compute structure from motion without the aid of any surveyed information.

In the most likely scenario, the artists begin with the sequence of images and the surveyed geometry. Correspondences between the points on the geometry and the points in the scene are made, and an initial camera location is determined. Typically, the data integration team has computed the intrinsic parameters of the camera based on the reference grid shot with the lens.

Once the initial pose is determined, the points associated with each point on the reference geometry are tracked in 2D over time. Pattern trackers are used to do this, though when the pattern tracks fail, the artist often tracks the images by hand.

Why would pattern tracks fail? One of the most common reasons is that feature points in the scene—the ones that are easy to digitize with a total station—are often not the best points to camera-track: corners of objects. Corners can be difficult to pattern-track because of the large amount of

Figure 7.5. Snapshot of a boat/car chase in the movie *xXx*. Left: Original. Right: Final composite image with digital hydrofoils inserted.

change that the pattern undergoes as the camera rotates around the corner. And often, the background behind the corner contributes to confusion of the pattern tracking algorithm. Corner detection algorithms can pick them out, but the subpixel precision of these algorithms is not as high as those based on cross-correlation template matching.

Motion blur poses many problems for the tracking artist. Pattern trackers often fail in the presence of motion blur. And even if the artist is forced to track the pattern by hand, it can be difficult to determine where a feature lies on the image if it is blurred over many pixels. Recently, the artists at Digital Domain were tracking a shot for a movie called *xXx*. In the shot, the camera was mounted on a speeding car traveling down a road parallel to a river. On the river, a hydrofoil boat was traveling alongside the car. The effect was to replace the top of the boat with a digitally created doomsday device and make it look like it was screaming down the river on some nefarious mission (Figure 7.5). The first task was to track the camera motion of the car with respect to the background landscape. Then the boat was tracked as a rigid, moving object with respect to the motion of the camera. Naturally, the motion blur for both cases was extreme. The automatic tracking program had a very difficult time due to the extreme motion of the camera. Feature points would travel nearly halfway across the image in the space of one or two frames. Optical flow algorithms failed for the same reason. Tracking the points proved quite difficult because of the extreme streaking of the information in the image. The artists placed the point in the center of the streak and hoped for the best.

This worked to some extent. In this extreme example, the motion blur of the boat was at times so extreme that the streaks caused by the motion blur were not linear at all. The motion of tracked camera and boat were represented as a linear interpolation between keyframe positions at every frame. The motion blur generated by the renderer did not match the motion blur of the background frame. It is surprising how much small detail the eye can pick up. When the artists used spline interpolation to interpolate the camera motion and added subframe keyframes to correctly account for sharp camera/model motion, the digital imagery fit with the background noticeably better.

Beyond tracking points, digital tracking artists have many tools available for a good track. The points can be weighted based on the certainty of the points—points that are obscured by objects can still be used; even though they cannot be seen, their positions are interpolated (or guessed) and their uncertainty set very high. Camera pose estimation is still an optimization problem, and often the position computed is the position that minimizes the error in the correspondence of points. That being the case, removing a point from the equation can have quite an effect on the computed location of the camera. This results in the dreaded "pop" in motion after a point leaves the image. This pop can be managed by ramping the point's uncertainty as the point approaches leaving the image. Alternatively, the point can be used after it leaves the image: the artist pretends that the point's location is known, and this can minimize any popping.

Other constraints beyond points are available in many packages. Linear constraints are often useful: an edge on the surveyed geometry is constrained to stay on the edge of an object in the image. Pattern trackers can follow the edge of the object as easily as a point on the object.

Of course, it is not necessary to have any surveyed information. Structure from motion techniques can solve for the location of both the camera and the 3D points in the scene without any surveyed information at all. If there is no absolute measurement of anything on the set, there is no way for structure from motion algorithms to determine the absolute scale of the 3D points and camera translation. Scale is important to digital artists; animated characters and physical effects are built at a specific scale, and having to play with the scale to get things to look right is not something that artists like to do. So, even when relying on structure from motion, some information—like the measurement of a distance between two points on the set—is quite useful for establishing the scale of the scene.

The convergence of optical flow techniques, following good features for tracks, and structure from motion solvers has caused large ripples through the film and video production facilities. Now, with no a priori knowledge

of the set, a reasonable track and reconstruction of the set can be created–
as long as the camera moves enough and enough points can be followed.
Some small visual effects facilities rely on almost all structure from motion
solutions. Larger facilities are finding it to be an enormous help, but rely on
more traditional camera tracking techniques for quite a few shots. Automatic
tracking is automatic when it is easy.

Consider a shot where the camera is chasing the hero of the film down a
street. At some point, the camera points up to the sky and then back down
to center on the main character. Typical effects might be to add some digital
buildings to the set (or at least enhance the buildings that are already in the
scene) and perhaps add some flying machines or creatures chasing our hero.
This makes the camera tracking particularly challenging: the camera, when
it points to the sky, will not be able to view any trackable points. But, if
buildings and flying objects are being added to the scene, the camera move
during that time must at least be "plausible" and certainly be smooth and
not exhibit any popping motion as tracking points are let go. Automatic
tracking techniques are generally not used on this kind of shot because the
artist will want complete control of the points used to compute the camera
motion. The camera will be tracked for the beginning of the shot and the
end of the shot, and then the camera motion for the area where the camera
is only pointing at sky or nonvalid tracking points (like the top of the actor's
head) is created by clever interpolation and gut feeling for what the camera
was doing. Often, something can be tracked, even if it is a far-off cloud, and
that can at least be used to determine the rotation of the camera.

7.6   The Future

Camera tracking and photogrammetry are the main applications of computer
vision in visual effects. But new techniques are making it into visual effects
all the time.

Space carving techniques are starting to show up in visual effects. In the
movie *Minority Report*, a 3D image of a person was projected in Tom Cruise's
apartment. The camera roamed around the projection, and it needed to look
like a true 3D display. The projected actor was filmed on a special greenscreen stage, surrounded by many synchronized video cameras. Each camera
was calibrated and fixed. Since the actor was surrounded by greenscreen, it
was relatively simple to determine what part of each image was actor and
what was background. From this, a view volume was formed. The intersection of all the view volumes of all the cameras produced a convex hull

of the actor. Projecting the images onto the resulting geometry produced a remarkably realistic model of the person. This is just the beginning for this technique. At recent conferences, similar techniques are being developed to extract amazingly detailed deforming models of humans with concavities and everything.

It is a terribly exciting time to be involved in computer vision and visual effects. Over the years, the advantages of robust computer vision tools have become evident to the artists and the management. Computer vision techniques allow the artists to do things that previously were not possible and to do them faster and more efficiently, which makes management happy.

Bibliography

[1] M. J. Black and P. Anandan. The robust estimation of multiple motions: Parametric and piecewise-smooth flow fields. *Computer Vision and Image Understanding*, 63(1):75–104, 1996.

[2] Y.-Y. Chuang, A. Agarwala, B. Curless, D. H. Salesin, and R. Szeliski. Video matting of complex scenes. *ACM Transactions on Graphics*, 21(3):243–248, July 2002.

[3] P. E. Debevec. Rendering synthetic objects into real scenes: Bridging traditional and image-based graphics with global illumination and high dynamic range photography. *Siggraph98, Annual Conference Series*, pages 189–198, 1998.

[4] P. E. Debevec, C. J. Taylor, and J. Malik. Modeling and rendering architecture from photographs: A hybrid geometry- and image-based approach. *Computer Graphics, Annual Conference Series*, pages 11–20, 1996.

[5] H. W. Jensen, S. R. Marschner, M. Levoy, and P. Hanrahan. A practical model for subsurface light transport. *Siggraph01, Annual Conference Series*, pages 511–518, 2001.

[6] J. Lewis. Fast normalized cross-correlation. *Vision Interface*, 1995.

[7] V. Masselus, P. Dutre, F. Anrys. The free-form light stage. *Proceedings of the 13th Eurographics Workshop on Rendering*, pages 247–256, June 2002.

[8] M. Pauly, R. Keiser, L. P. Kobbelt, and M. Gross. Shape modeling with point-sampled geometry. *ACM Transactions on Graphics*, 22(3):641–650, July 2003.

[9] H. Pfister, M. Zwicker, J. van Baar, and M. Gross. Surfels: Surface elements as rendering primitives. *Siggraph00, Annual Conference Series*, pages 335–342, 2000.

[10] R. Szeliski and J. Coughlin. Spline-based image registration. *Technical Report No. CRL-94-1*, Cambridge Research Lab., DEC, April 1994.

[11] Y. Yu, P. Debevec, J. Malik, and T. Hawkins. Inverse global illumination: Recovering reflectance models of real scenes from photographs. *Siggraph99, Annual Conference Series*, pages 215–224, Los Angeles, 1999.

Chapter 8

CONTENT-BASED IMAGE RETRIEVAL: AN OVERVIEW

Theo Gevers
and Arnold W. M. Smeulders

In this chapter, we present an overview on the theory, techniques, and applications of content-based image retrieval. We choose patterns of use, image domains, and computation as the pivotal building blocks of our survey. A graphical overview of the content-based image retrieval scheme is given in Figure 8.1. Derived from this scheme, we follow the data as they flow through the computational process (see Figure 8.3), with the conventions indicated in Figure 8.2. In all of this chapter, we follow the review in [155] closely.

We focus on still images and leave video retrieval as a separate topic. Video retrieval could be considered a broader topic than image retrieval, as video is more than a set of isolated images. However, video retrieval could also be considered simpler than image retrieval, since, in addition to pictorial information, video contains supplementary information such as motion, spatial constraints, and time constraints (e.g., video discloses its objects more easily, as many points corresponding to one object move together and are spatially coherent in time). In still pictures, the user's narrative expression of intention is in image selection, object description, and composition. Video, in addition, has the linear timeline as an important information cue to assist the narrative structure.

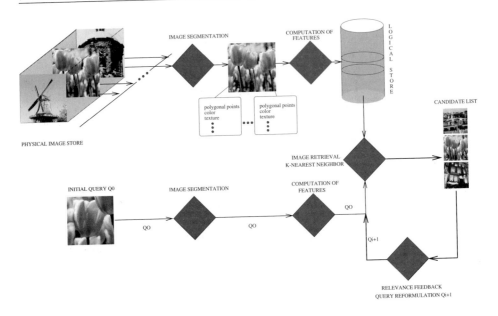

Figure 8.1. Overview of the basic concepts of the content-based image retrieval scheme considered in this chapter. First, features are extracted from the images in the database, then are stored and indexed. This is done offline. The online image retrieval process consists of a query example image from which image features are extracted. These image features are used to find the images in the database that are most similar. Then, a candidate list of most similar images is shown to the user. From the user feedback, the query is optimized and used as a new query in an iterative manner.

8.1 Overview of Chapter

The overview of the basic components of the content-based image retrieval scheme is given in Figure 8.1, and the corresponding dataflow process is shown in Figure 8.3. The sections in this chapter harmonize with the data as they flow from one computational component to another as follows:

- **Interactive query formulation:** Interactive query formulation is offered either by querying images and subimages or by offering a pattern of feature values and weights. To achieve interactive query formulation, an image is sketched, recorded, or selected from an image repository. With the query formulation, the aim is to search for particular images in the database. The mode of search might be one of the following three categories: *search by association, target search, category search.* For search by association, the intention of the user is to browse through a large collection of images

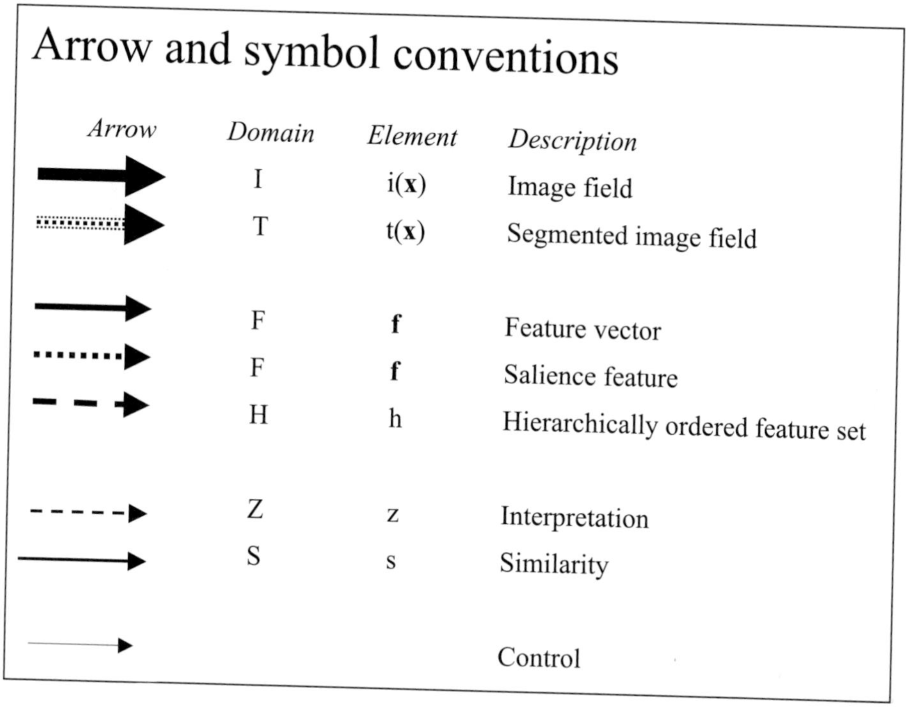

Arrow and symbol conventions

Arrow	Domain	Element	Description
	I	i($\mathbf{x}$)	Image field
	T	t($\mathbf{x}$)	Segmented image field
	F	$\mathbf{f}$	Feature vector
	F	$\mathbf{f}$	Salience feature
	H	h	Hierarchically ordered feature set
	Z	z	Interpretation
	S	s	Similarity
			Control

Figure 8.2. Data flow and symbol conventions used in this chapter. Different styles of arrows indicate different data structures.

without a specific aim. Search by association tries to find interesting images and is often applied in an iterative way by means of relevance feedback. Target search is to find similar (target) images in the image database. Note that *similar image* may imply a (partially) identical image or a (partially) identical object in the image. The third class is category search, where the aim is to retrieve an arbitrary image that is typical for a specific class or genre (e.g., indoor images, portraits, city views). Because many image retrieval systems are assembled around one of these three search modes, it is important to have insight into these categories and their structures. Search modes are discussed in Section 8.2.1.

• **Image domains:** The definition of image features depends on the repertoire of images under consideration. This repertoire can be ordered along the complexity of variations imposed by the imaging conditions, such as illumination and viewing geometry, going from *narrow domains* to *broad domains*. For images from a narrow domain, there is a restricted variability

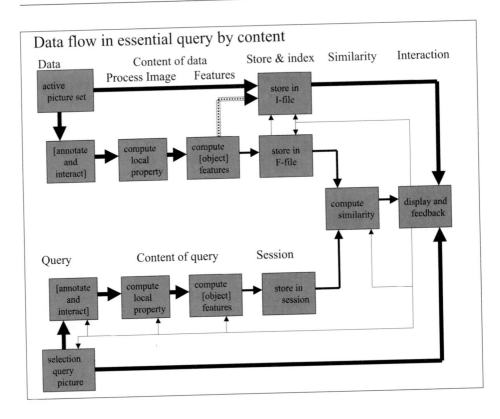

Figure 8.3. Basic algorithmic components of query by pictorial example captured in a dataflow scheme while using the conventions of Figure 8.2.

of their pictorial content. Examples of narrow domains are stamp collections and face databases. For broad domains, images may be taken from objects from unknown viewpoints and illumination. For example, two recordings taken from the same object from different viewpoints will yield different shadowing, shading, and highlighting cues, changing the intensity data fields considerably. Moreover, large differences in the illumination color will drastically change the photometric content of images even when they are taken from the same scene. Hence, images from broad domains have a large pictorial variety called the sensory gap, discussed in Section 8.2.2. Furthermore, low-level image features are often too restricted to describe images on a conceptual or semantic level. This semantic gap is a well-known problem in content-based image retrieval and is discussed in Section 8.2.3.

- **Image features:** Image feature extraction is an important step for image indexing and search. Image feature extraction modules should take into account whether the image domain is narrow or broad. In fact, they should consider to which of the imaging conditions they should be invariant to such a change in viewpoint, object pose, and illumination. Further, image features should be concise and complete, and at the same time, have high discriminative power. In general, a tradeoff exists between the amount of invariance and selectivity. In Section 8.3, a taxonomy on feature extraction modules is given from an image processing perspective. The taxonomy can be used to select the proper feature extraction method for a specific application based on whether images come from broad domains and which search goals are at hand (target, category, associate search). In Section 8.3.1, we first focus on color content descriptors derived from image processing technology. Various color-based image search methods are discussed based on different representation schemes, such as color histograms, color moments, color edge orientation, and color correlograms. These image representation schemes are created on the basis of RGB and other color systems such as HSI and CIE $L^*a^*b^*$. For example, the $L^*a^*b^*$ space has been designed to conform to the human perception of color similarity. If the appreciation of a human observer of an object is based on the perception of certain conspicuous items in the image, it is natural to direct the computation of broad domain features to these points and regions. Similarly, a biologically plausible architecture [84] of center-surround processing units is likely to select regions that humans would also focus on first. Further, we discuss color models which are robust to a change in viewing direction, object geometry, and illumination. Image processing for shape is outlined in Section 8.3.2. We focus on *local shapes*, which are image descriptors capturing salient details in images. Finally, in Section 8.3.3, we discuss texture, and a review is given on texture features describing local color characteristics and their spatial layout.

- **Representation and indexing:** Representation and indexing are discussed in Section 8.4. In general, the image feature set is represented by vector space, probabilistic, or logical models. For example, for the vector space model, weights can be assigned corresponding to the feature frequency giving the well-known histogram form. Further, for accurate image search, it is often desirable to assign weights in accordance to the importance of the image features. The image feature weights used for both images and queries can be computed as the product of the features frequency multiplied by the inverse collection frequency factor. In this way, features are emphasized having high feature frequencies but low overall collection frequencies. Feature accumulation and representation is further discussed in Section 8.4.2. In addition to feature representation, indexing is required to speed up the search process. Indexing techniques include adaptive histogram binning, signature files, and hashing. Further, tree-based indexing schemes have been developed for indexing the stored images so that similar images can be identified efficiently at some additional costs in memory, such as a k-d tree, R*-tree, or SS-tree [69].

Throughout the chapter, a distinction is made between *weak* and *strong* segmentation. Weak segmentation is a local grouping approach usually focusing on conspicuous regions such as edges, corners, and higher-order junctions. In Section 8.4.4, various methods are discussed to achieve weak segmentation. Strong segmentation is the extraction of the complete contour of an object in an image. Obviously, strong segmentation is far more difficult than weak segmentation and is hard to achieve, if not impossible, for broad domains.

- **Similarity and search:** The actual matching process can be seen as a search for images in the stored image set closest to the query specification. As both the query and the image data set is captured in feature form, the similarity function operates between the weighted feature sets. To make the query effective, close attention has to be paid to the selection of the similarity function. A proper similarity function should be robust to object fragmentation, occlusion, and clutter by the presence of other objects in the view. For example, it is known that the mean square and the Euclidean similarity measure provide accurate retrieval without any object clutter [59, 162]. A detailed overview on similarity and search is given in Section 8.5.

- **Interaction and Learning:** Visualization of the feature matching results gives the user insight into the importance of the different features. Windowing and information display techniques can be used to establish communications between system and user. In particular, new visualization techniques such as 3D virtual image clouds can used to designate certain images as relevant to the user's requirements. These relevant images are then further used by the system to construct subsequent (improved) queries. Relevance feedback is an automatic process designed to produce improved query formulations following an initial retrieval operation. Relevance feedback is needed for image retrieval where users find it difficult to formulate pictorial queries. For example, without any specific query image example, the user might find it difficult to formulate a query (e.g., to retrieve an image of a car) by image sketch or by offering a pattern of feature values and weights. This suggests that the first search is performed by an initial query formulation and a (new) improved query formulation is constructed based on the search results with the goal to retrieve more relevant images in the next search operations. Hence, from the user feedback giving negative/positive answers, the method can automatically learn which image features are most important. The system uses the feature weighting given by the user to find the images in the image database that are optimal with respect to the feature weighting. For example, search by association allows users to refine iteratively the query definition, the similarity, or the examples with which the search was started. Therefore, systems in this category are highly interactive. Interaction, relevance feedback, and learning are discussed in Section 8.6.

• **Testing** In general, image search systems are assessed in terms of precision, recall, query-processing time, and reliability of a negative answer. Further, the relevance feedback method is assessed in terms of the number of iterations to approach to the ground truth. Today, more and more images are archived, yielding a very large range of complex pictorial information. In fact, the average number of images used for experimentation, as reported in the literature, augmented from a few in 1995 to over a hundred thousand today. It is important that the data set should have ground truths, that is, images that are relevant and images that are nonrelevant to a given query. In general, it is hard to get these ground truths, especially for very large data sets. A discussion on system performance is given in Section 8.6.

8.2 Image Domains

In this section, we discuss patterns in image search applications, the repertoire of images, the influence of the image formation process, and the semantic gap between image descriptors and the user.

8.2.1 Search modes

We distinguish three broad categories of search modes when using a content-based image retrieval system; see Figure 8.4.

• *Search by association* encompasses a broad variety of methods and systems designed to browse through a large set of images from unspecified sources. At the start, users of search by association have no specific aims other than to find interesting images. Search by association often implies iterative refinement of the search, the similarity, or the examples with which the search was initiated. Systems in this category are highly interactive, where the query specification may be defined by sketch [28] or by example images. The oldest realistic example of such a system is probably [91]. The

Target, category, and association search in image retrieval

	Target	*Category*	*Association*
Object goal	1 specific object	an arbitrary object from 1 specific class	not defined at start
Query by example	1 ... N objects	1 ... N objects with class labels	N objects plus association
Similarity	feature-based	class driven	session-specific
Events in F-space	proximity to query	class membership	clusters
Feedback	rank ordered on proximity	likelihood on class membership	relevance feedback on association value
Interactive update:			
of images of query	-	expand query	refine on the way
of features of query	refine on the way	refine on the way	alter on the way
of similarity measure	-	adapt to group	reshape to goal

Figure 8.4. Three patterns in the purpose of content-based retrieval systems.

result of the search can be manipulated interactively by relevance feedback [76]. To support the quest for relevant results, sources other than images are also employed, for example [163].

- The purpose of *target search* is to find a specific image. The search may be for a precise copy of the image in mind, as in searching art catalogs, for example [47]. Target search may also be for another image of the same object the user has an image of. This is target search by example. Target search may also be applied when the user has a specific image in mind and the target is interactively specified as similar to a group of given examples, for instance [29]. These systems are suited to search for stamps, paintings, industrial components, textile patterns, and catalogs in general.

- *Category search* is aimed at retrieving an arbitrary image representative of a specific class. This is the case when the user has an example and the search is for other elements of the same class or genre. Categories may be derived from labels or may emerge from the database [164, 105]. In category search, the user may have available a group of images and the search is for additional images of the same class [25]. A typical application of category search is catalogs of varieties. In [82, 88], systems are designed for classifying trademarks. Systems in this category are usually interactive with a domain-specific definition of similarity.

8.2.2 The sensory gap

In the repertoire of images under consideration (the image domain), there is a gradual distinction between narrow and broad domains [154]. At one end of the spectrum, we have the narrow domain:

> *A narrow domain has a limited and predictable variability in all relevant aspects of its appearance.*

Hence, in a narrow domain, we find images with a reduced diversity in their pictorial content. Usually, the image formation process is similar for all recordings. When the object's appearance has limited variability, the semantic description of the image is generally well-defined and largely unique. An example of a narrow domain is a set of frontal views of faces recorded against a clear background. Although each face is unique and has large variability in the visual details, there are obvious geometrical, physical, and illumination constraints governing the pictorial domain. The domain would be wider if the faces had been photographed from a crowd or from an outdoor scene. In that case, variations in illumination, clutter in the scene, occlusion, and viewpoint will have a major impact on the analysis.

On the other end of the spectrum, we have the broad domain:

> *A broad domain has an unlimited and unpredictable variability in its appearance even for the same semantic meaning.*

In broad domains, images are polysemic and their semantics are described only partially. It might be the case that there are conspicuous objects in the scene for which the object class is unknown or even that the interpretation of the scene is not unique. The broadest class available today is the set of images available on the Internet.

Many problems of practical interest have an image domain in between these extreme ends of the spectrum. The notions of broad and narrow are helpful in characterizing patterns of use, in selecting features, and in designing systems. In a broad image domain, the gap between the feature description and the semantic interpretation is generally wide. For narrow, specialized image domains, the gap between features and their semantic interpretation is usually smaller, so domain-specific models may be of help.

For broad image domains in particular, we must resort to generally valid principles. Is the illumination of the domain white or colored? Does it assume fully visible objects, or may the scene contain clutter and occluded objects as well? Is it a 2D recording of a 2D scene or a 2D recording of a 3D scene? The given characteristics of illumination, presence or absence of occlusion, clutter, and differences in camera viewpoint determine the demands on the methods of retrieval.

> *The sensory gap is the gap between the object in the world and the information in a (computational) description derived from a recording of that scene.*

The sensory gap makes the description of objects an ill-posed problem: it yields uncertainty in what is known about the state of the object. The sensory gap is particularly poignant when a precise knowledge of the recording conditions is missing. The 2D records of different 3D objects can be identical. Without further knowledge, one has to decide that they *might* represent the same object. Also, a 2D recording of a 3D scene contains information accidental for that scene and that sensing but one does not know what part of the information is scene-related. The uncertainty due to the sensory gap holds not only for the viewpoint, but also for occlusion (where essential parts telling two objects apart may be out of sight), clutter, and illumination.

8.2.3　The semantic gap

As stated in the previous sections, content-based image retrieval relies on multiple low-level features (e.g., color, shape, and texture) describing the image content. To cope with the sensory gap, these low-level features should be consistent and invariant to remain representative of the repertoire of images in the database. For image retrieval by query by example, the online image retrieval process consists of a query example image, given by the user on input, from which low-level image features are extracted. These image features are used to find images in the database that are most similar to the query image. A drawback, however, is that these low-level image features are often too restricted to describe images on a conceptual or semantic level. It is our opinion that ignoring the existence of the semantic gap is the cause of many disappointments on the performance of early image retrieval systems.

> *The semantic gap is the lack of coincidence between the information that one can extract from the visual data and the interpretation that the same data have for a user in a given situation.*

A user wants to search for images on a conceptual level (e.g., images containing particular objects [target search] or conveying a certain message or genre [category search]). Image descriptions, on the other hand, are derived by low-level, data-driven methods. The *semantic* search by the user and the low-level *syntactic* image descriptors may be disconnected. Association of a complete semantic system to image data would entail, at least, solving the general object recognition problem. Since this problem is yet unsolved and is likely to stay unsolved in its entirety, research is focused on different methods to associate higher level semantics to data-driven observables.

Indeed, the most reasonable tool for semantic image characterization entails annotation by keywords or captions. This converts content-based image access to (textual) information retrieval [134]. Common objections to the practice of labeling are cost and coverage. On the cost side, labeling thousands of images is a cumbersome and expensive job to the degree that the deployment of the economic balance behind the database is likely to decrease. To solve the problem, systems presented in [140, 139] use a program that explores the Internet, collecting images and inserting them in a predefined taxonomy on the basis of the text surrounding them. A similar approach for digital libraries is taken by [19]. On the coverage side, labeling is seldom complete and context-sensitive. In any case, there is a significant fraction of requests whose semantics cannot be captured by labeling alone [7, 72]. Both methods cover the semantic gap only in isolated cases.

8.2.4 Discussion

We have discussed three broad types of search categories: target search, category search, and search by association. Target search is related to the classical methods in the field of pattern matching and computer vision such as object recognition and image matching. However, image retrieval differs from traditional pattern matching by considering more and more images in the database. Therefore, new challenges in content-based retrieval are in the huge number of images to search among, the query specification by multiple images, and the variability of imaging conditions and object states. Category search connects to statistical pattern recognition methods. However, compared to traditional pattern recognition, new challenges are in the interactive manipulation of results, the usually very large number of object classes, and the absence of an explicit training phase for feature and classifier tuning (active learning). Search by association is the most distant from the classical field of computer vision. It is severely hampered by the semantic gap. As long as the gap is there, use of content-based retrieval for browsing will not be within the grasp of the general public, as humans are accustomed to rely on the immediate semantic imprint the moment they see an image.

An important distinction we have discussed is that between broad and narrow domains. The broader the domain, the more browsing or search by association should be considered during system setup. The narrower the domain, the more target search should be taken as search mode.

The major discrepancy in content-based retrieval is that the user wants to retrieve images on a semantic level, but the image characterizations can provide similarity only on a low-level syntactic level. This is called the semantic gap. Furthermore, another discrepancy is that between the properties in an image and the properties of the object. This is called the sensory gap. Both the semantic and sensory gap play a serious limiting role in the retrieval of images based on their content.

8.3   Image Features

Before discussing image features, it is important to keep in mind that content-based retrieval does not depend on a complete description of the pictorial content of the image. It is sufficient that a retrieval system presents similar images (i.e., similar in some user-defined sense). The description of the content by image features should serve that goal primarily.

One such goal can be met by using invariance as a tool to deal with the accidental distortions in the image content introduced by the sensory gap. From Section 8.2.2, it is clear that invariant features may carry more object-specific information than other features, as they are insensitive to the accidental imaging conditions such as illumination, object pose, and camera viewpoint. The aim of invariant image features is to identify objects no matter how or from where they are observed, at the loss of some of the information content.

Therefore, the degree of invariance, should be tailored to the recording circumstances. In general, a feature with a very wide class of invariance loses the power to discriminate among object differences. The aim is to select the tightest set of invariants suited for the expected set of nonconstant conditions. What is needed in image search is a specification of the minimal invariant conditions in the specification of the query. The minimal set of invariant conditions can only be specified by the user, as it is part of his or her intention. For each image retrieval query, a proper definition of the desired invariance is in order. Does the applicant wish to search for the object in rotation and scale invariance? illumination invariance? viewpoint invariance? occlusion invariance? The oldest work on invariance in computer vision has been done in object recognition, as reported among others in

[119] for shape and [181] for color. Invariant description in image retrieval is relatively new but quickly gaining ground; for a good introduction, see [15, 30, 57].

8.3.1   Color

Color has been an active area of research in image retrieval, more than in any other branch of computer vision. Color makes the image take values in a color vector space. The choice of a color system is of great importance for the purpose of proper image retrieval. It induces the equivalent classes to the actual retrieval algorithm. However, no color system can be considered universal, because color can be interpreted and modeled in different ways. Each color system has its own set of color models, which are the parameters of the color system. Color systems have been developed for different purposes:

1. display and printing processes: RGB, CMY

2. television and video transmittion efficiency: YIQ, YUV

3. color standardization: XYZ

4. color uncorrelation: $I_1 I_2 I_3$

5. color normalization and representation: rgb, xyz

6. perceptual uniformity: $U^* V^* W^*$, $L^* a^* b^*$, $L^* u^* v^*$

7. intuitive description: HSI, HSV

With this large variety of color systems, the inevitable question is which color system to use for which kind of image retrieval application. To this end, criteria are required to classify the various color systems for the purpose of content-based image retrieval.

First, an important criterion is that the color system is independent of the underlying imaging device. This is required when images in the image database are recorded by different imaging devices, such as scanners, cameras, and camrecorders (e.g., images on the Internet). Another prerequisite might be that the color system should exhibit perceptual uniformity, meaning that numerical distances within the color space can be related to human perceptual differences. This is important when images that should be visually similar (e.g., stamps, trademarks, and paintings databases) are to be retrieved. Also, the transformation needed to compute the color system should be linear. A nonlinear transformation may introduce instabilities with respect to noise, causing poor retrieval accuracy. Further, the color system should be composed of color models that are understandable and intuitive to the user. Moreover, to achieve robust image retrieval, color invariance is an important criterion. In general, images and videos are taken from objects from different viewpoints. Two recordings made of the same object from different viewpoints will yield different shadowing, shading, and highlighting cues.

When there is no variation in the recording or in the perception, then the RGB color representation is a good choice. RGB-representations are widely used today. They describe the image in its literal color properties. An image expressed by RGB makes most sense when recordings are made in the absence of variance, as is the case, for instance, for art paintings [72], the color composition of photographs [47], and trademarks [88, 39], where 2D images are recorded in frontal view under standard illumination conditions.

A significant improvement over the RGB-color space (at least for retrieval applications) comes from the use of normalized color representations [162]. This representation has the advantage of suppressing the intensity information and hence is invariant to changes in illumination intensity and object geometry.

Others approaches use the Munsell or the $L^*a^*b^*$-spaces because of their relative perceptual uniformity. The $L^*a^*b^*$ color system has the property that the closer a point (representing a color) is to another point, the more visually similar the colors are. In other words, the magnitude of the perceived color difference of two colors corresponds to the Euclidean distance between the two colors in the color system. The $L^*a^*b^*$ system is based on the 3D coordinate system based on the opponent theory using black-white L^*, red-green a^*, and yellow-blue b^* components. The L^* axis corresponds to the lightness, where $L^* = 100$ is white and $L^* = 0$ is black. Further, a^* ranges from red $+a^*$ to green $-a^*$, while b^* ranges from yellow $+b^*$ to blue $-b^*$. The chromaticity coordinates a^* and b^* are insensitive to intensity and have the same invariant properties as normalized color. Care should be taken when digitizing the nonlinear conversion to $L^*a^*b^*$-space [117].

The HSV-representation is often selected for its invariant properties. Further, the human color perception is conveniently represented by these color models, where V is an attribute in terms of which a light or surface color may be ordered on a scale from dim to bright. S denotes the relative white content of a color, and H is the color aspect of a visual impression. The problem of H is that it becomes unstable when S is near zero due to the nonremovable singularities in the nonlinear transformation, in which a small perturbation of the input can cause a large jump in the transformed values [62]. H is invariant under the orientation of the object with respect to the illumination intensity and camera direction and hence is more suited for object retrieval. However, H is still dependent on the color of the illumination [57].

A wide variety of tight photometric color invariants for object retrieval were derived in [59] from the analysis of the dichromatic reflection model. They derive for matte patches under white light the invariant color space $(\frac{R-G}{R+G}, -\frac{B-R}{B+R}, \frac{G-B}{G+B})$, only dependent on sensor and surface albedo. For a shiny surface and white illumination, they derive the invariant representation as $\frac{|R-G|}{|R-G|+|B-R|+|G-B|}$ and two more permutations. The color models are robust against major viewpoint distortions.

Color constancy is the capability of humans to perceive the same color in the presence of variations in illumination that change the physical spectrum of the perceived light. The problem of color constancy has been the topic of much research in psychology and computer vision. Existing color constancy methods require specific a priori information about the observed scene (e.g., the placement of calibration patches of known spectral reflectance in the scene), which is not feasible in practical situations, [48, 52, 97] for example. In contrast, without any a priori information, [73, 45] use illumination-invariant moments of color distributions for object recognition. However, these methods are sensitive to object occlusion and cluttering, as the moments are defined as an integral property on the object as one. In global methods in general, occluded parts disturb recognition. [153] circumvents this problem by computing the color features from small object regions instead of the entire object. Further, to avoid sensitivity on object occlusion and cluttering, simple and effective illumination-independent color ratios have been proposed by [53, 121, 60]. These color constant models are based on the ratio of surface albedos rather than on recovering the actual surface albedo itself. However, these color models assume that the variation in spectral power distribution of the illumination can be modeled by the coefficient rule or von Kries model, where the change in the illumination color is approximated by a 3×3 diagonal matrix among the sensor bands and is equal to the multiplication of each RGB-color band by an independent scalar factor. The diagonal model of illumination change holds exactly in the case of narrow-band sensors. Although standard video cameras are not equipped with narrow-band filters, spectral sharpening could be applied [46] to achieve this to a large extent.

The color ratios proposed by [121] are given by $N(C^{\vec{x}_1}, C^{\vec{x}_2}) = \frac{C^{\vec{x}_1} - C^{\vec{x}_2}}{C^{\vec{x}_2} + C^{\vec{x}_1}}$ and those proposed by [53] are defined by $F(C^{\vec{x}_1}, C^{\vec{x}_2}) = \frac{C^{\vec{x}_1}}{C^{\vec{x}_2}}$, expressing color ratios between two neighboring image locations, for $C \in \{R, G, B\}$, where $\vec{x}_1$ and $\vec{x}_2$ denote the image locations of the two neighboring pixels.

The color ratios of [60] are given by $M(C_1^{\vec{x}_1}, C_1^{\vec{x}_2}, C_2^{\vec{x}_1}, C_2^{\vec{x}_2}) = \frac{C_1^{\vec{x}_1} C_2^{\vec{x}_2}}{C_1^{\vec{x}_2} C_2^{\vec{x}_1}}$, expressing the color ratio between two neighboring image locations, for $C_1, C_2 \in \{R, G, B\}$, where $\vec{x}_1$ and $\vec{x}_2$ denote the image locations of the two neighboring pixels. All these color ratios are device-dependent, not perceptually uniform, and they become unstable when intensity is near zero. Further, N and F are dependent on the object geometry. M has no viewing and lighting dependencies. In [55], a thorough overview is given on color models for the purpose of image retrieval. Figure 8.5 shows the taxonomy of color models with respect to their characteristics. For more information, refer to [55].

Color system	Device indep.	Perc. Uniform	Linear	Intuitive	View point	Object shape	Highlights	Illum. Intensity	Illum. SPD
RGB	-	-	+	-	-	-	-	-	-
XYZ	+	-	+	-	-	-	-	-	-
Norm. rgb	-	-	-	-	+	+	-	+	-
Norm. xyz	+	-	-	-	+	+	-	+	-
L*a*b*	+	+	-	-	-	-	-	-	-
U*V*W*	+	+	-	-	-	-	-	-	-
I1I2I3	-	-	+	-	-	-	-	-	-
YIQ	-	-	+	-	-	-	-	-	-
YUV	-	-	+	-	-	-	-	-	-
Intensity	-	-	+	+	-	-	-	-	-
Hue	-	-	-	+	+	+	+	+	-
Saturation	-	-	-	+	+	+	-	+	-
F, N	-	-	-	-	+	-	-	+	+
M	-	-	-	-	+	+	-	+	+

Figure 8.5. Overview of the dependencies differentiated for the various color systems. + denotes that the condition is satisfied; − denotes that the condition is not satisfied.

Rather than invariant descriptions, another approach to cope with the inequalities in observation due to surface reflection is to search for clusters in a color histogram of the image. In the RGB histogram, clusters of pixels reflected off an object form elongated streaks. Hence, in [126], a nonparametric cluster algorithm in RGB space is used to identify which pixels in the image originate from one uniformly colored object.

8.3.2   Shape

Under the name *local shape*, we collect all properties that capture conspicuous geometric details in the image. We prefer the name local shape over other characterizations, such as differential geometrical properties, to denote the result rather than the method.

Local shape characteristics derived from directional color derivatives are used in [117] to derive perceptually conspicuous details in highly textured patches of diverse materials. A wide, rather unstructured variety of image detectors can be found in [159].

In [61], a scheme is proposed to automatically detect and classify the physical nature of edges in images using reflectance information. To achieve this, a framework is given to compute edges by automatic gradient thresholding. Then, a taxonomy is given on edge types based upon the sensitivity of edges with respect to different imaging variables. A parameter-free edge classifier is provided, labeling color transitions into one of the following types: shadow-geometry edges, highlight edges, or material edges. In Figure 8.6.a, six frames are shown from a standard video often used as a test sequence in the literature. It shows a person against a textured background playing Ping-Pong. The size of the image is 260×135. The images are of low quality. The frames are clearly contaminated by shadows, shading, and interreflections. Note that each individual object-part (i.e., T-shirt, wall, and table) is painted homogeneously with a distinct color. Further, the wall is highly textured. The results of the proposed reflectance based edge classifier are shown in Figure 8.6. For more details, see [61].

Combining shape and color both in invariant fashion is a powerful combination, as described by [58], where the colors inside and outside affine curvature maximums in color edges are stored to identify objects.

Scale space theory was devised as the complete and unique primary step in preattentive vision, capturing all conspicuous information [178]. It provides the theoretical basis for the detection of conspicuous details on any scale. In [109], a series of Gabor filters of different directions and scale are used to enhance image properties [136]. Conspicuous shape geometric invari-

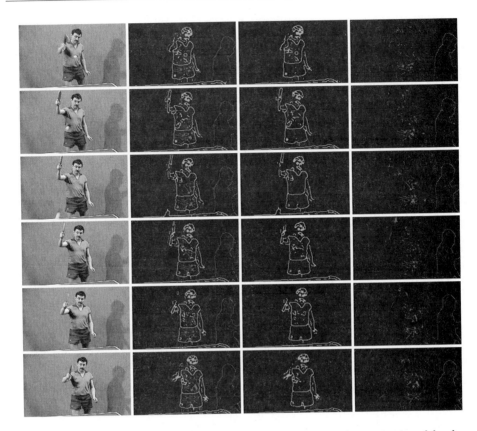

Figure 8.6. Frames from a video showing a person against a textured background playing Ping-Pong. From left to right column: original color frame; classified edges; material edges; shadow and geometry edges.

ants are presented in [135]. A method employing local shape and intensity information for viewpoint and occlusion invariant object retrieval is given in [143]. The method relies on voting among a complete family of differential geometric invariants. Also, [170] searches for differential affine-invariant descriptors. From surface reflection, in [5], the local sign of the Gaussian curvature is computed while making no assumptions on the albedo or the model of diffuse reflectance.

8.3.3 Texture

In computer vision, texture is considered all that is left after color and local shape have been considered, or it is given in terms of structure and randomness. Many common textures are composed of small textons, usually too large in number to be perceived as isolated objects. The elements can be placed more or less regularly or randomly. They can be almost identical or subject to large variations in their appearance and pose. In the context of image retrieval, research is mostly directed toward statistical or generative methods for the characterization of patches.

Basic texture properties include the Markovian analysis dating back to Haralick in 1973 and generalized versions thereof [95, 64]. In retrieval, the property is computed in a sliding mask for localization [102, 66].

Another important texture analysis technique uses multiscale, autoregressive MRSAR-models, which consider texture as the outcome of a deterministic dynamic system subject to state and observation noise [168, 110]. Other models exploit statistical regularities in the texture field [9].

Wavelets [33] have received wide attention. They have often been considered for their locality and their compression efficiency. Many wavelet transforms are generated by groups of dilations or dilations and rotations that have been said to have some semantic correspondent. The lowest levels of the wavelet transforms [33, 22] have been applied to texture representation [96, 156], sometimes in conjunction with Markovian analysis [21]. Other transforms have also been explored, most notably fractals [41]. A solid comparative study on texture classification from mostly transform-based properties can be found in [133].

When the goal is to retrieve images containing objects having irregular texture organization, the spatial organization of these texture primitives is, in worst case, random. It has been demonstrated that for irregular texture, the comparison of gradient distributions achieves satisfactory accuracy [122, 130], as opposed to fractal or wavelet features. Therefore, most of the work on texture image retrieval is stochastic in nature [12, 124, 190]. However, these methods rely on gray-value information, which is very sensitive to the imaging conditions. In [56], the aim is to achieve content-based image retrieval of textured objects in natural scenes under varying illumination and viewing conditions. To achieve this, image retrieval is based on matching feature distributions derived from color-invariant gradients. To cope with object cluttering, region-based texture segmentation is applied on the target images prior to the actual image retrieval process. In Figure 8.7, results are shown of color-invariant texture segmentation for image retrieval. From the results, we can observe that RGB and normalized color $\theta_1\theta_2$ is highly sensitive to a change in illumination color. Only M is insensitive to a change in illumination color. For more information, refer to [56].

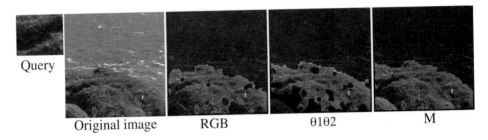

Query   Original image   RGB   θ1θ2   M

Figure 8.7. From left to right: query texture under different illumination; target image; segmentation result based on RGB; segmentation result based on variant of rgb; segmentation result based on color ratio gradient M.

Texture search proved also to be useful in satellite images [100] and images of documents [31]. Textures also served as a support feature for segmentation-based recognition [106], but the texture properties discussed so far offer little semantic referent. They are therefore ill-suited for retrieval applications in which the user wants to use verbal descriptions of the image. Therefore, in retrieval research, in [104] the Wold features of periodicity, directionality, and randomness are used, which agree reasonably well with linguistic descriptions of textures, as implemented in [127].

8.3.4 Discussion

First of all, image processing in content-based retrieval should primarily be engaged in enhancing the image information of the query, not in describing the content of the image in its entirety.

To enhance the image information, retrieval has set the spotlights on color, as color has a high discriminatory power among objects in a scene, much higher than gray levels. The purpose of most image color processing is to reduce the influence of the accidental conditions of the scene and sensing (i.e., the sensory gap). Progress has been made in tailored color space representation for well-described classes of variant conditions. Also, the application of geometrical description derived from scale space theory will reveal viewpoint- and scene-independent salient point sets, thus opening the way to similarity of images on a few most informative regions or points.

In this chapter, we have made a separation between color, local geometry, and texture. At this point, it is safe to conclude that the division is an artificial labeling. For example, wavelets say something about the local shape as well as the texture, and so may scale space and local filter strategies.

For the purposes of content-based retrieval, an integrated view on color, texture, and local geometry is urgently needed, as only an integrated view on local properties can provide the means to distinguish among hundreds of thousands of different images. A recent advancement in that direction is the fusion of illumination and scale invariant color and texture information into a consistent set of localized properties [74]. Also, in [16], homogeneous regions are represented as collections of ellipsoids of uniform color or texture, but invariant texture properties deserve more attention [167, 177]. Further research is needed in the design of complete sets of image properties with well-described variant conditions that they are capable of handling.

8.4   Representation and Indexing

We discussed the ultimate form of spatial data by grouping the data into object silhouettes, clusters of points, or point sets. In the next section, we leave the spatial domain to condense the pictorial information into feature values.

8.4.1   Grouping data

In content-based image retrieval, the image is often divided in parts before features are computed from each part. Partitionings of the image aim at obtaining more selective features by selecting pixels in a tradeoff against having more information in features when no subdivision of the image is used at all. We distinguish the following partitionings:

• When searching for an object, it would be most advantageous to do a complete object segmentation first:

> *Strong segmentation is a division of the image data into regions in such a way that region T contains the pixels of the silhouette of object O in the real world and nothing else specified by T = O.*

It should be noted immediately that object segmentation for broad domains of general images is not likely to succeed, with a possible exception for sophisticated techniques in very narrow domains.

• The difficulty of achieving strong segmentation may be circumvented by weak segmentation where grouping is based on data-driven properties:

> *Weak segmentation is a grouping of the image data in conspicuous regions T internally homogeneous according to some criterion, hopefully with T ⊂ O.*

The criterion is satisfied if region T is within the bounds of object O, but there is no guarantee that the region covers all of the object's area. When the image contains two nearly identical objects close to each other, the weak segmentation algorithm may falsely observe just one patch. Fortunately, in content-based retrieval, this type of error is rarely obstructive for the goal. In [125], the homogeneity criterion is implemented by requesting that colors be spatially coherent vectors in a region. Color is the criterion in [49, 126]. In [16, 114], the homogeneity criterion is based on color and texture. The limit case of weak segmentation is a set of isolated points [143, 59]. No homogeneity criterion is needed then, but the effectiveness of the isolated points rest on the quality of their selection. When occlusion is present in the image, weak segmentation is the best one can hope for. Weak segmentation is used in many retrieval systems either as a purpose of its own or as a preprocessing stage for data-driven, model-based object segmentation.

• When the object has a (nearly) fixed shape, like a traffic light or an eye, we call it a sign:

> *Localizing signs is finding an object with a fixed shape and semantic meaning, with* $T = \mathbf{x_{center}}$.

Signs are helpful in content-based retrieval, as they deliver an immediate and unique semantic interpretation.

• The weakest form of grouping is partitioning:

> *A partitioning is a division of the data array regardless of the data, symbolized by* $T \neq O$.

The area T may be the entire image or a conventional partitioning as the central part of the image against the upper, right, left, and lower parts [75]. The feasibility of fixed partitioning comes from the fact that images are created in accordance with certain canons or normative rules, such as placing the horizon about two-thirds up in the picture or keeping the main subject in the central area. This rule is often violated, but this violation in itself has semantic significance. Another possibility of partitioning is to divide the image in tiles of equal size and summarize the dominant feature values in each tile [129].

8.4.2   Features accumulation

In the computational process given in Figure 8.3, features are calculated next. The general class of accumulating features aggregate the spatial information of a partitioning irrespective of the image data. A special type of accumulative features are the global features, which are calculated from the entire image. F_j (see Figure 8.2) is the set of accumulative features or a set of accumulative features ranked in a histogram. F_j is part of feature space $\mathcal{F}$. T_j is the partitioning over which the value of F_j is computed. In the case of global features, $T_{j=void}$ represents the image; otherwise, T_j represents a fixed tiling of the image. The operator h may hold relative weights, for example, to compute transform coefficients.

A simple but very effective approach to accumulating features is to use the histogram: the set of features $\mathbf{F}(m)$ ordered by histogram index m.

One of the earlier approaches to color-based image matching using the color at pixels directly as indices, was proposed by Swain and Ballard [162]. If the RGB or normalized color distributions of two images are globally similar, the matching rate is high. The work by Swain and Ballard has had an enormous impact on color-based histogram matching, resulting in many histogram variations.

For example, the QBIC system [42] allows for a user-defined computation of the histogram by the introduction of variable k, denoting the number of bins of the histogram. Then, for each $3 \times k$ cells, the average modified Munsell color is computed together with the five most frequently occurring colors. Using a standard clustering algorithm, they obtain k super cells, resulting in the partitioning of the color system.

In [58], various color-invariant features are selected to construct color pattern-cards. First, histograms are created in a standard way. Because the color distributions of histograms depend on the scale of the recorded object (e.g., distance object-camera), they define color pattern-cards as thresholded histograms. In this way, color pattern-cards are scale-independent by indicating whether or not a particular color model value is substantially present in an image. Matching measures are defined, expressing similarity between color pattern-cards, robust to a substantial amount of object occlusion and cluttering. Based on the color pattern-cards and matching functions, a hashing scheme is presented offering runtime image retrieval independent of the number of images in the image database.

In the ImageRover system, proposed by [147], the $L^*u^*v^*$ color space is used; each color axis is split into four equally sized bins, resulting in a total of 64 bins. Further, [37] uses the $L^*a^*b^*$ system to compute the average color and covariance matrix of each of the color channels. [158] uses the HSV color space to obtain a partition into 144 bins, giving more emphasis on hue H than on value V and saturation I. Further, [4] also focuses on the HSV color space to extract regions of dominant colors. To obtain colors that are perceptually the same but still distinctive, [165] proposes to partition the RGB color space into 220 subspaces. [36] computes the average color describing a cell of a 4×4 grid, which is superimposed on the image. [149] uses the $L^*a^*b^*$ color space because the color space consists of perceptually uniform colors, which better matches the human perception of color. [65] roughly partitions the Munsell color space into 11 color zones. Similar partitioning is proposed by [29] and [24].

Another approach, proposed by [161], is the introduction of the cumulative color histogram, which generates more dense vectors. This enables coping with coarsely quantized color spaces. [186] proposes a variation of the cumulative histograms by applying cumulative histograms to each subspace.

Other approaches are based on the computation of moments of each color channel. For example, [6] represents color regions by the first three moments of the color models in the HSV-space. Instead of constructing histograms from color invariants, [73, 45] propose the computation of illumination-invariant moments from color histograms. In a similar way, [153] computes the color features from small object regions instead of the entire object.

[85] proposes to use integrated wavelet decomposition. In fact, the color features generate wavelet coefficients together with their energy distribution among channels and quantization layers. Similar approaches based on wavelets are proposed by [175, 101].

All of this is in favor of the use of histograms. When very large data sets are at stake, plain histogram comparison will saturate the discrimination. For a 64-bin histogram, experiments show that for reasonable conditions, the discriminatory power among images is limited to 25,000 images [160]. To keep up performance, in [125], a joint histogram is used, providing discrimination among 250,000 images in their database, rendering 80% recall among the best 10 for two shots from the same scene using simple features. Other joint histograms add local texture or local shape [68], directed edges [87], and local higher order structures [47].

Another alternative is to add a dimension representing the local distance. This is the correlogram [80], defined as a 3D histogram where the colors of any pair are along the first and second dimension, and the spatial distance between them is along the third. The autocorrelogram defining the distances between pixels of identical colors is found on the diagonal of the correlogram. A more general version is the geometric histogram [1] with the normal histogram, the correlogram, and several alternatives as special cases. This also includes the histogram of the triangular pixel values reported to outperform all of the above, as it contains more information.

A different view on accumulative features is to demand that all information (or all relevant information) in the image is preserved in the feature values. When the bit content of the features is less than the original image, this boils down to compression transforms. Many compression transforms are known, but the quest is for transforms simultaneously suited as retrieval features. As proper querying for similarity is based on a suitable distance function between images, the transform has to be applied on a metric space. In addition, the components of the transform have to correspond to semantically meaningful characteristics of the image. Finally, the transform should admit indexing in compressed form, yielding a big computational advantage over having the image be untransformed first. [144] is just one of many approaches in which the cosine-based JPEG-coding scheme is used for image retrieval. The JPEG transform fulfills the first and third requirements but fails on a lack of semantics. In the MPEG standard, the possibility to include semantic descriptors in the compression transform is introduced [27]. For an overview of feature indexes in the compressed domain, see [108]. In [96], a wavelet packet is applied to texture images and, for each packet, entropy and energy measures are determined and collected in a feature vector. In [83], vector quantization is applied in the space of coefficients to reduce its dimensionality. This approach was extended to incorporate the metric of the color space in [141]. In [86], a wavelet transform is applied independently to the three channels of a color image, and only the sign of the most significant coefficients is retained. In [3], a scheme is offered for a broad spectrum of invariant descriptors suitable for application on Fourier, wavelets, and splines and for geometry and color alike.

8.4.3   Feature accumulation and image partitioning

The lack of spatial information for methods based on feature accumulation might yield lower retrieval accuracy. As for general image databases, a manual segmentation is not feasible due to the sensory gap. A simple approach is to divide images into smaller subimages and then index them. This is known as fixed partitioning. Other systems use a segmentation scheme prior to the actual image search to partition each image into regions. Nearly all region-based partitioning schemes use some kind of weak segmentation, decomposing the image into coherent regions rather than complete objects (strong segmentation).

Fixed Partitioning

The simplest way is to use a fixed image decomposition in which an image is partitioned into equally sized segments. The disadvantage of a fixed partitioning is that blocks usually do not correspond with the visual content of an image. For example, [65] splits an image into nine equally sized subimages, where each subregion is represented by a color histogram. [67] segments the image by a quadtree, and [99] uses a multiresolution representation of each image. [36] also uses a 4×4 grid to segment the image. [148] partitions images into three layers, where the first layer is the whole image, the second layer is a 3×3 grid, and the third layer is a 5×5 grid. A similar approach is proposed by [107], where three levels of a quadtree are used to segment the images. [37] proposes the use of interhierarchical distances measuring the difference between color vectors of a region and its subsegments. [20] uses an augmented color histogram capturing the spatial information between pixels together with the color distribution. In [59], the aim is to combine color and shape invariants for indexing and retrieving images. Color-invariant edges are derived from which shape-invariant features are computed. Then, computational methods are described to combine the color and shape invariants into a unified, high-dimensional histogram for discriminatory object retrieval. [81] proposes the use of color correlograms for image retrieval. Color correlograms integrate the spatial information of colors by expressing the probability that a pixel of color c_i lies at a certain distance from a pixel of color c_j. It is shown that color correlograms are robust to a change in background, occlusion, and scale (camera zoom). [23] introduces the spatial chromatic histograms, where for every pixel the percentage of pixels having the same color is computed. Further, the spatial information is encoded by baricenter of the spatial distribution and the corresponding deviation.

Region-Based Partitioning

Segmentation is a computational method to assess the set of points in an image which represent one object in the scene. As discussed before, many different computational techniques exist, none of which is capable of handling any reasonable set of real-world images. However, in this case, *weak segmentation* may be sufficient to recognize an object in a scene. Therefore, in [12], an image representation is proposed providing a transformation from the raw pixel data to a small set of image regions that are coherent in color and texture space. This so-called Blobworld representation is based on segmentation using the expectation-maximization algorithm on combined color and texture features. In the Picasso system [13], a competitive learning clustering algorithm is used to obtain a multiresolution representation of color regions. In this way, colors are represented in the $L^*u^*v^*$ space through a set of 128 reference colors as obtained by the clustering algorithm. [63] proposes a method based on matching feature distributions derived from color ratio gradients. To cope with object cluttering, region-based texture segmentation is applied on the target images prior to the actual image retrieval process. [26] segments the image first into homogeneous regions by split and merge using a color distribution homogeneity condition. Then, histogram intersection is used to express the degree of similarity between pairs of image regions.

8.4.4   Salient features

As the information of the image is condensed into just a limited number of feature values, the information should be selected with precision for greatest saliency and proven robustness. That is why saliency in [103] is defined as the special points, which survive longest when gradually blurring the image in scale space. Also in [137], lifetime is an important selection criterion for salient points in addition to wiggliness, spatial width, and phase congruency. To enhance the quality of salient descriptions, in [170], invariant and salient features of local patches are considered. In each case, the image is summarized in a list of conspicuous points. In [143], salient and invariant transitions in gray-value images are recorded. Similarly, in [59, 54], photometric invariance is the leading principle in summarizing the image in salient transitions in the image. Salient feature calculations lead to sets of regions or points with known location and feature values capturing their salience.

In [16], first the most conspicuous homogeneous regions in the image are derived and mapped into feature space. Then, expectation-maximization [35] is used to determine the parameters of a mixture of Gaussians to model the distribution of points into the feature space. The means and covariance matrices of these Gaussians, projected on the image plane, are represented as ellipsoids characterized by their center $\mathbf{x}$, their area, eccentricity, and direction. The average values of the color and texture descriptions inside the ellipse are also stored.

Various color image segmentation methods have been proposed that account for the image formation process; see for instance the work collected by Wolff, Shafer, and Healey [181]. [150] presents the dichromatic reflection model, a physical model of reflection which states that two distinct types of reflection—surface and body reflection—occur and that each type can be decomposed into a relative spectral distribution and a geometric scale factor. [93] developed a color segmentation algorithm based on the dichromatic reflection model. The method is based on evaluating characteristic shapes of clusters in red-green-blue (RGB) space followed by segmentation independent of the object's geometry, illumination, and highlights. To achieve robust image segmentation, however, surface patches of objects in view must have a rather broad distribution of surface normals, which may not hold for objects in general. [10] developed a similar image segmentation method using the $H\text{-}S$ color space instead of the RGB color space. [73] proposed a method to segment images on the basis of normalized color. However, [92] showed that normalized color and hue are singular at some RGB values and unstable at many others.

8.4.5 Shape and object features

The theoretically best way to enhance object-specific information contained in images is by segmenting the object in the image. But, as discussed above, the brittleness of segmentation algorithms prevents the use of automatic segmentation in broad domains. And, in many cases, it is not necessary to know exactly where an object is in the image as long as one can identify the presence of the object by its unique characteristics. When the domain is narrow, a tailored segmentation algorithm may be needed more, and fortunately, it may also be more feasible.

The object's internal features are largely identical to the accumulative features, now computed over the object area. They need no further discussion here.

An abundant comparison of shape for retrieval can be found in [113], evaluating many features on a 500-element trademark data set. Straightforward features of general applicability include Fourier features and moment invariants of the object this time, sets of consecutive boundary segments, or encoding of contour shapes [40].

For retrieval, we need a shape representation that allows a robust measurement of distances in the presence of considerable deformations. Many sophisticated models widely used in computer vision often prove too brittle for image retrieval. On the other hand, the (interactive) use of retrieval makes some mismatch acceptable and, therefore precision can be traded for robustness and computational efficiency.

More sophisticated methods include elastic matching and multiresolution representation of shapes. In elastic deformation of image portions [34, 123] or modal matching techniques [145], image patches are deformed to minimize a cost functional that depends on a weighed sum of the mismatch of the two patches and on the deformation energy. The complexity of the optimization problem depends on the number of points on the contour. Hence, the optimization is computationally expensive and this, in spite of the greater precision of these methods, has limited their diffusion in image databases.

Multiscale models of contours have been studied as a representation for image databases in [118]. Contours were extracted from images and progressively smoothed by dividing them into regions of constant sign of the second derivative and progressively reducing the number of such regions. At the final step, every contour is reduced to an ellipsoid, which could be characterized by some of the features in [47]. A different view on multiresolution shape is offered in [98], where the contour is sampled by a polygon and then simplified by removing points from the contour until a polygon survives, selecting them on perceptual grounds. When computational efficiency is at stake, an approach for the description of the object boundaries is found in [189], where an ordered set of critical points on the boundary are found from curvature extremes. Such sets of selected and ordered contour points are stored in [112] relative to the basis spanned by an arbitrary pair of the points. All point pairs are used as a basis to make the redundant representation geometrically invariant, a technique similar to [182] for unordered point sets.

For retrieval of objects in 2D images of the 3D worlds, a viewpoint-invariant description of the contour is important. A good review of global shape invariants is given in [138].

8.4.6 Structure and layout

When feature calculations are available for different entities in the image, they may be stored with a relationship between them. Such a structural feature set may contain feature values plus spatial relationships, a hierarchically ordered set of feature values, or relationships between point sets or object sets. Structural and layout feature descriptions are captured in a graph, hierarchy, or any other ordered set of feature values and their relationships.

To that end, in [111, 49], layout descriptions of an object are discussed in the form of a graph of relations between blobs. A similar layout description of an image in terms of a graph representing the spatial relations between the objects of interest is used in [128] for the description of medical images. In [51], a graph is formed of topological relationships of homogeneous RGB regions. When selected features and the topological relationships are viewpoint-invariant, the description is viewpoint-invariant, but the selection of the RGB representation as used in the paper suit that purpose only to a limited degree. The systems in [78, 157] study spatial relationships between regions, each characterized by locations, size, and features. In the latter system, matching is based on the 2D string representation founded by Chang [17]. For a narrow domain, in [128, 132], the relevant elements of a medical X-ray image are characterized separately and joined together in a graph that encodes their spatial relations.

Starting from a shape description, the authors in [98] decompose an object into its main components, making the matching between images of the same object easier. Automatic identification of salient regions in the image based on nonparametric clustering followed by decomposition of the shapes found into limbs is explored in [50].

8.4.7   Discussion

General content-based retrieval systems have dealt with segmentation brittleness in a few ways. First, a weaker version of segmentation has been introduced in content-based retrieval. In weak segmentation, the result is a homogeneous region by some criterion, but not necessarily covering the complete object silhouette. It results in a fuzzy, blobby description of objects rather than a precise segmentation. Salient features of the weak segments capture the essential information of the object in a nutshell. The extreme form of the weak segmentation is the selection of a salient point set as the ultimately efficient data reduction in the representation of an object, very much like the focus-of-attention algorithms for an earlier age. Only points on the interior of the object can be used for identifying the object, and conspicuous points at the borders of objects have to be ignored. Little work has been done on how to make the selection. Weak segmentation and salient features are a typical innovation of content-based retrieval. It is expected that salience will receive much attention in the further expansion of the field, especially when computational considerations gain in importance.

The alternative is to do no segmentation at all. Content-based retrieval has gained from the use of accumulative features, computed on the global image or partitionings thereof, disregarding the content, the most notable being the histogram. Where most attention has gone to color histograms, histograms of local geometric properties and texture are following. To compensate for the complete loss of spatial information, recently the geometric histogram was defined with an additional dimension for the spatial layout of pixel properties. As it is a superset of the histogram, an improved discriminability for large data sets is anticipated. Accumulative features calculated from the central part of a photograph may be very effective in telling them apart by topic, but the center does not always reveal the purpose. Likewise, features calculated from the top part of a picture may be effective in telling indoor scenes from outdoor scenes, but again, this holds to a limited degree. A danger of accumulative features is their inability to discriminate among different entities and semantic meanings in the image. More work on semantic-driven groupings will increase the power of accumulative descriptors to capture the content of the image.

Structural descriptions match well with weak segmentation, salient regions, and weak semantics. We have to be certain that the structure is within one object and not an accidental combination of patches that have no meaning in the object world. The same brittleness of strong segmentation lurks here. We expect a sharp increase in the research of local, partial, or fuzzy structural descriptors for the purpose of content-based retrieval, especially of broad domains.

8.5   Similarity and Search

When the information from images is captured in a feature set, there are two possibilities for endowing them with meaning: one derives a unilateral interpretation from the feature set and one compares the feature set with the elements in a given data set on the basis of a similarity function.

8.5.1   Semantic interpretation

In content-based retrieval, it is useful to push the semantic interpretation of features derived from the image as far as possible.

> *Semantic features aim at encoding interpretations of the image that may be relevant to the application.*

Of course, such interpretations are a subset of the possible interpretations of an image. To that end, consider a feature vector $\mathbf{F}$ derived from an image i. For given semantic interpretations z from the set of all interpretations $\mathcal{Z}$, a strong semantic feature with interpretation z_j would generate a $P(z|\mathbf{F}) = \delta(z - z_j)$. If the feature carries no semantics, it would generate a distribution $P(z|\mathbf{F}) = P(z)$ independent of the value of the feature. In practice, many feature types will generate a probability distribution that is neither a pulse nor independent of the feature value. This means that the feature value skews the interpretation of the image, but does not determine it completely.

Under the umbrella *weak semantics*, we collect the approaches that try to combine features in some semantically meaningful interpretation. Weak semantics aims at encoding in a simple and approximate way a subset of the possible interpretations of an image that are of interest in a given application. As an example, the system in [28] uses color features derived from Itten's color theory to encode the semantics associated with color contrast and harmony in art application.

In the MAVIS2-system [90], data are considered at four semantic levels and embodied in four layers called the raw media, the selection, the selection expression, and conceptual layers. Each layer encodes information at an increasingly symbolic level. Agents are trained to create links between features, feature signatures at the selection layer, interrelated signatures at the selection expression layer, and concept (expressed as textual labels) at the conceptual layer. In addition to the vertical connections, the two top layers have intralayer connections that measure the similarity between concepts at that semantic level and contribute to the determination of the similarity between elements at the lower semantic level.

8.5.2 Similarity between features

A different road to assign a meaning to an observed feature set is to compare a pair of observations by a similarity function. While searching for a query image $i_q(\mathbf{x})$ among the elements of the data set of images, $i_d(\mathbf{x})$, knowledge of the domain is expressed by formulating a similarity measure $S_{q,d}$ between the images q and d on the basis of some feature set. The similarity measure depends on the type of features.

At its best use, the similarity measure can be manipulated to represent different semantic contents; images are then grouped by similarity in such a way that close images are similar with respect to use and purpose. A common assumption is that the similarity between two feature vectors $\mathbf{F}$ can be expressed by a positive, monotonically nonincreasing function. This assumption is consistent with a class of psychological models of human similarity perception [152, 142] and requires that the feature space be metric. If the feature space is a vector space, d often is a simple Euclidean distance, although there is indication that more complex distance measures might be necessary [142]. This similarity model was well suited for early query by example systems in which images were ordered by similarity with one example.

A different view sees similarity as an essentially probabilistic concept. This view is rooted in the psychological literature [8], and in the context of content-based retrieval, it has been proposed, for example, in [116].

Measuring the distance between histograms has been an active line of research since the early years of content-based retrieval, where histograms can be seen as a set of ordered features. In content-based retrieval, histograms have mostly been used in conjunction with color features, but there is nothing against being used in texture or local geometric properties.

Various distance functions have been proposed. Some of these are general functions such as Euclidean distance and cosine distance. Others are specially designed for image retrieval such as histogram intersection [162]. The Minkowski-form distance for two vectors or histograms $\vec{k}$ and $\vec{l}$ with dimension n is given by:

$$\mathcal{D}_M^k(\vec{k}, \vec{l}) = (\sum_{i=1}^{n} |k_i - l_i|^p)^{1/p}. \tag{8.1}$$

The Euclidean distance between two vectors $\vec{k}$ and $\vec{l}$ is defined as follows:

$$\mathcal{D}_E(\vec{k}, \vec{l}) = \sqrt{\sum_{i=1}^{n} (k_i - l_i)^2}. \tag{8.2}$$

The Euclidean distance is an instance of the Minkowski distance with $k = 2$.

The cosine distance compares the feature vectors of two images and returns the cosine of the angle between the two vectors:

$$\mathcal{D}_C(\vec{k}, \vec{l}) = 1 - \cos\phi, \tag{8.3}$$

where ϕ is the angle between the vectors $\vec{k}$ and $\vec{l}$. When the two vectors have equal directions, the cosine will add to one. The angle ϕ can also be described as a function of $\vec{k}$ and $\vec{l}$:

$$\cos\phi = \frac{\vec{k} \cdot \vec{l}}{||\vec{k}|| \; ||\vec{l}||}. \tag{8.4}$$

The cosine distance is well suited for features that are real vectors and not a collection of independent scalar features.

The histogram intersection distance compares two histograms $\vec{k}$ and $\vec{l}$ of n bins by taking the intersection of both histograms:

$$\mathcal{D}_H(\vec{k}, \vec{l}) = 1 - \frac{\sum_{i=1}^{n} \min(k_i, l_i)}{\sum_{i=1}^{n} k_i}. \tag{8.5}$$

When considering images of different sizes, this distance function is not a metric due to $\mathcal{D}_H(\vec{k}, \vec{l}) \neq \mathcal{D}_H(\vec{l}, \vec{k})$. In order to become a valid distance metric, histograms need to be normalized first:

$$\vec{k}^n = \frac{\vec{k}}{\sum_i^n k_i}.$$ (8.6)

For normalized histograms (total sum of 1), the histogram intersection is given by

$$\mathcal{D}_H^n(\vec{k}^n, \vec{l}^h) = 1 - \sum_i^n |k_i^n - l_i^n|.$$ (8.7)

This is again the Minkowski-form distance metric with $k = 1$. Histogram intersection has the property that it allows for occlusion; that is, when an object in one image is partly occluded, the visible part still contributes to the similarity [60, 59].

Alternatively, histogram matching is defined by normalized cross correlation:

$$\mathcal{D}_x(\vec{k}, \vec{l}) = \frac{\sum_{i=1}^n k_i l_i}{\sum_{i=1}^n k_i^2}.$$ (8.8)

The normalized cross correlation has a maximum of unity that occurs if and only if $\vec{k}$ exactly matches $\vec{l}$.

In the QBIC system [42], the weighted Euclidean distance has been used for the similarity of color histograms. In fact, the distance measure is based on the correlation between histograms $\vec{k}$ and $\vec{l}$:

$$\mathcal{D}_Q(\vec{k}, \vec{l}) = (k_i - l_i)^t A(k_i - l_j).$$ (8.9)

Further, A is a weight matrix with term a_{ij} expressing the perceptual distance between bin i and j.

The average color distance has been proposed by [70] to obtain a simpler, low-dimensional distance measure:

$$\mathcal{D}_{\text{Haf}}(\vec{k}, \vec{l}) = (k_{\text{avg}} - l_{\text{avg}})^t (k_{\text{avg}} - l_{\text{avg}}),$$ (8.10)

where k_{avg} and l_{avg} are 3×1 average color vectors of $\vec{k}$ and $\vec{l}$.

As stated before, for broad domains, a proper similarity measure should be robust to object fragmentation, occlusion, and clutter by the presence of other objects in the view. In [58], various similarity functions were compared for color-based histogram matching. From these results, it is concluded that retrieval accuracy of similarity functions depends on the presence of object clutter in the scene. The histogram cross correlation provides best retrieval accuracy without any object clutter (narrow domain). This is because this similarity function is symmetric and can be interpreted as the number of pixels with the same values in the query image that can be found present in the retrieved image, and vice versa. This is a desirable property when one object per image is recorded without any object clutter. In the presence of object clutter (broad domain), highest image retrieval accuracy is provided by the quadratic similarity function (e.g., histogram intersection). This is because this similarity measure counts the number of similar hits and hence is insensitive to false positives.

Finally, the natural measure to compare ordered sets of accumulative features is nonparametric test statistics. They can be applied to the distributions of the coefficients of transforms to determine the likelihood that two samples derive from the same distribution [14, 131]. They can also be applied to compare the equality of two histograms and all variations thereof.

8.5.3   Similarity of object outlines

In [176], a good review is given of methods to compare shapes directly after segmentation into a set of object points $t(\mathbf{x})$ without an intermediate description in terms of shape features.

For shape comparison, the authors distinguish between transforms, moments, deformation matching, scale space matching, and dissimilarity measurement. Difficulties for shape matching based on global transforms are the inexplicability of the result and the brittleness for small deviations. Moments, specifically their invariant combinations, have been frequently used in retrieval [94]. Matching a query and an object in the data file can be done along the ordered set of eigen shapes [145] or with elastic matching [34, 11]. Scale space matching is based on progressively simplifying the contour by smoothing [118]. By comparing the signature of annihilated zero crossings of the curvature, two shapes are matched in a scale- and rotation-invariant fashion. A discrete analogue can be found in [98], where points are removed from the digitized contour on the basis of perceptually motivated rules.

When based on a metric, dissimilarity measures render an ordered range of deviations suited for a predictable interpretation. In [176], an analysis is given for the Hausdorff and related metrics between two shapes on robustness and computational complexity. The directed Hausdorff metric is defined as the maximum distance between a point on query object q and its closest counterpart on d. The partial Hausdorff metric, defined as the k-th maximum rather than the absolute maximum, is used in [71] for affine-invariant retrieval.

8.5.4 Similarity of object arrangements

The result of a structural description is a hierarchically ordered set of feature values H. In this section, we consider the similarity of two structural or layout descriptions.

Many different techniques have been reported for the similarity of feature structures. In [180, 82], a Bayesian framework is developed for the matching of relational attributed graphs by discrete relaxation. This is applied to line patterns from aerial photographs.

A metric for the comparison of two topological arrangements of named parts, applied to medical images, is defined in [166]. The distance is derived from the number of editing steps needed to nullify the difference in the Voronoi diagrams of two images.

In [18], 2D strings describing spatial relationships between objects are discussed, and much later reviewed in [185]. From such topological relationships of image regions, in [79], a 2D indexing is built in trees of symbol strings, each representing the projection of a region on the coordinate axis. The distance between the H_q and H_d is the weighted number of editing operations required to transform one tree into the other. In [151], a graph is formed from the image on the basis of symmetry, as appears from the medial axis. Similarity is assessed in two stages via graph-based matching, followed by energy-deformation matching.

In [51], hierarchically ordered trees are compared for the purpose of retrieval by rewriting them into strings. A distance-based similarity measure establishes the similarity scores between corresponding leaves in the trees. At the level of trees, the total similarity score of corresponding branches is taken as the measure for tree and subtree similarity. From a small sized experiment, it is concluded that hierarchically ordered feature sets are more efficient than plain feature sets, with projected computational shortcuts for larger data sets.

8.5.5 Similarity of salient features

Salient features are used to capture the information in the image in a limited number of salient points. Similarity between images can then be checked in several ways.

In the first place, the color, texture, or local shape characteristics may be used to identify the salient points of the data as identical to the salient points of the query.

A measure of similarity between the feature values measured of the blobs resulting from weak segmentation consists of a Mahalanobis distance between the feature vector composed of the color, texture, position, area, eccentricity, and direction of the two ellipses [16].

In the second place, one can store all salient points from one image in a histogram on the basis of a few characteristics, such as color on the inside versus color on the outside. The similarity is then based on the groupwise presence of enough similar points [59]. The intersection model is used in image retrieval in [153] while keeping access to point location in the image by back-projection [162]. Further, a weight per dimension may favor the appearance of some salient features over another. See also [77] for a comparison with correlograms.

A third alternative for similarity of salient points is to concentrate only on the spatial relationships among the salient point sets. In point-by-point-based methods for shape comparison, shape similarity is studied in [89], where maximum curvature points on the contour and the length between them are used to characterize the object. To avoid the extensive computations, one can compute the algebraic invariants of point sets, known as the cross-ratio. Due to their invariant character, these measures tend to have only a limited discriminatory power among different objects. A more recent version for the similarity of nameless point sets is found in geometric hashing [182], where each triplet spans a base for the remaining points of the object. An unknown object is compared on each triplet to see whether enough similarly located points are found. Geometric hashing, though attractive in its concept, is too computationally expensive to be used on the very large data sets of image retrieval due to the anonymity of the points. Similarity of two points sets given in a row-wise matrix is discussed in [179].

8.5.6 Discussion

Whenever the image itself permits an obvious interpretation, the ideal content-based system should employ that information. A strong semantic interpretation occurs when a sign can be positively identified in the image. This is rarely the case due to the large variety of signs in a broad class of images and the enormity of the task to define a reliable detection algorithm for each of them. Weak semantics rely on inexact categorization induced by similarity measures, preferably online by interaction. The categorization may agree with semantic concepts of the user, but the agreement is in general imperfect. Therefore, the use of weak semantics is usually paired with the ability to gear the semantics of the user to his or her needs by interpretation. Tunable semantics is likely to receive more attention in the future, especially as data sets grow larger.

Similarity is an interpretation of the image based on the difference between it and another image. For each of the feature types, a different similarity measure is needed. For similarity between feature sets, special attention has gone to establishing similarity among histograms due to their computational efficiency and retrieval effectiveness.

Similarity of shape has received considerable attention in the context of object-based retrieval. Generally, global shape matching schemes break down when there is occlusion or clutter in the scene. Most global shape comparison methods implicitly require a frontal viewpoint against a clear enough background to achieve a sufficiently precise segmentation. With the recent inclusion of perceptually robust points in the shape of objects, an important step forward has been made.

Similarity of hierarchically ordered descriptions deserves considerable attention, as it is one mechanism to circumvent the problems with segmentation while maintaining some of the semantically meaningful relationships in the image. Part of the difficulty here is to provide matching of partial disturbances in the hierarchical order and the influence of sensor-related variances in the description.

8.6   Interaction and Learning

8.6.1   Interaction on a semantic level

In [78], knowledge-based type abstraction hierarchies are used to access image data based on context and a user profile, generated automatically from cluster analysis of the database. Also, in [19], the aim is to create a very large concept space inspired by the thesaurus-based search from the information retrieval community. In [117], a linguistic description of texture patch visual qualities is given, and ordered in a hierarchy of perceptual importance on the basis of extensive psychological experimentation.

A more general concept of similarity is needed for relevance feedback, in which similarity with respect to an ensemble of images is required. To that end, in [43], more complex relationships are presented between similarity and distance functions defining a weighted measure of two simpler similarities, $S(s, S_1, S_2) = w_1 \exp(-d(S_1, s)) + w_2 \exp(-d(S_2, s))$. The purpose of the bireferential measure is to find all regions that are similar to two specified query points, an idea that generalizes to similarity queries given multiple examples. The approach can be extended with the definition of a complete algebra of similarity measures with suitable composition operators [43, 38]. It is then possible to define operators corresponding to the disjunction, conjunction, and negation of similarity measures, much like traditional databases. The algebra is useful for the user to manipulate the similarity directly as a means to express characteristics in specific feature values.

8.6.2 Classification on a semantic level

To further enhance the performance of content-based retrieval systems, image classification has been proposed to group images into semantically meaningful classes [171, 172, 184, 188]. The advantage of these classification schemes is that simple, low-level image features can be used to express semantically meaningful classes. Image classification is based on unsupervised learning techniques such as clustering, self-organization maps (SOM) [188], and Markov models [184]. Further, supervised grouping can be applied. For example, vacation images have been classified based on a Bayesian framework into city vs. landscape by supervised learning techniques [171, 172]. However, these classification schemes are entirely based on pictorial information. Aside from image *retrieval* [44, 146], very little attention has been paid using both textual and pictorial information for *classifying* images on the Web. This is even more surprising given that images on Web pages are usually surrounded by text and discriminatory HTML tags such as IMG and the HTML fields SRC and ALT. Hence, WWW images have intrinsic annotation information induced by the HTML structure. Consequently, the set of images on the Web can be seen as an annotated image set.

8.6.3 Learning

As data sets grow and the processing power matches that growth, the opportunity arises to learn from experience. Rather than designing, implementing, and testing an algorithm to detect the visual characteristics for each different semantic term, the aim is to learn from the appearance of objects directly.

For a review on statistical pattern recognition, see [2]. In [174], a variety of techniques treating retrieval as a classification problem are discussed.

One approach is principal component analysis over a stack of images taken from the same class z of objects. This can be done in feature space [120] or at the level of the entire image; for example, faces as features in [115]. The analysis yields a set of eigenface images, capturing the common characteristics of a face without having a geometric model.

Effective ways to learn from partially labeled data have recently been introduced in [183, 32], both using the principle of transduction [173]. This saves the effort of labeling the entire data set, which is infeasible and unreliable as it grows.

In [169], a very large number of precomputed features is considered, of which a small subset is selected by boosting [2] to learn the image class.

An interesting technique to bridge the gap between textual and pictorial descriptions to exploit information at the level of documents is borrowed from information retrieval, called latent semantic indexing [146, 187]. First, a corpus of documents (in this case, images with a caption) is formed, from which features are computed. Then, by singular value decomposition, the dictionary covering the captions is correlated with the features derived from the pictures. The search is for hidden correlations of features and captions.

8.6.4 Discussion

Learning computational models for semantics is an interesting and relatively new approach. It gains attention quickly as the data sets and the machine power grow large. Learning opens up the possibility to an interpretation of the image without designing and testing a detector for each new notion. One such approach is appearance-based learning of the common characteristics of stacks of images from the same class. Appearance-based learning is suited for narrow domains. For the success of the learning approach, there is a tradeoff between standardizing the objects in the data set and the size of the data set. The more standardized the data are, the less data will be needed, but, on the other hand, the less broadly applicable the result will be. Interesting approaches to derive semantic classes from captions or a partially labeled or unlabeled data set have been presented recently.

8.7   Conclusion

In this chapter, we presented an overview on the theory, techniques, and applications of content-based image retrieval. We took patterns of use and computation as the pivotal building blocks of our survey.

From a scientific perspective, the following trends can be distinguished. First, large-scale image databases are being created. Obviously, large-scale data sets provide different image mining problems than do small, narrow-domain data sets. Second, research is directed toward the integration of different information modalities such as text, pictorial, and motion. Third, relevance feedback was and still is an important issue. Finally, invariance is necessary to get to general-purpose image retrieval.

From a societal/commercial perspective, it is obvious that there will be an enormous increase in the amount of digital images used in various communication frameworks, such as promotion, sports, education, and publishing. Further, digital images have become one of the major multimedia information sources on the Internet, where the amount of image/video on the Web is growing each day. Moreover, with the introduction of the new-generation cellphones, a tremendous market will be opened for the storage and management of pictorial data. Due to this tremendous amount of pictorial information, image mining and search tools are required, since indexing, searching, and assessing the content of large-scale image databases is inherently a time-consuming operation when done by human operators. Therefore, product suites for content-based video indexing and searching is not only necessary but essential for future content owners in the field of entertainment, news, education, video communication, and distribution.

We hope that from this review you get the picture in this new pictorial world.

Bibliography

[1] R. K. Srihari, A. Rao, and Z. Zhang. Geometric histogram: A distribution of geometric configurations of color subsets. In *Internet Imaging*, vol. 3,964, pages 91–101, 2000.

[2] R. P. W. Duin, A. K. Jain, and J. Mao. Statistical pattern recognition: A review. *IEEE Transactions on PAMI*, 22(1):4–37, 2000.

[3] R. Alferez and Y.-F. Wang. Geometric and illumination invariants for object recognition. *IEEE Transactions on PAMI*, 21(6):505–536, 1999.

[4] D. Androutsos, K. N. Plataniotis, and A. N. Venetsanopoulos. A novel vector-based approach to color image retrieval using a vector angular-based distance measure. *Image Understanding*, 75(1-2):46–58, 1999.

[5] E. Angelopoulou and L. B. Wolff. Sign of Gaussian curvature from curve orientation in photometric space. *IEEE Transactions on PAMI*, 20(10):1056–1066, 1998.

[6] A. R. Appas, A. M. Darwish, A. I. El-Desouki, and S. I. Shaheen. Image indexing using composite regional color channel features. In *IS&T/SPIE Symposium on Electronic Imaging: Storage and Retrieval for Image and Video Databases VII*, pages 492–500, 1999.

[7] L. Armitage and P. Enser. Analysis of user need in image archives. *Journal of Information Science*, 23(4):287–299, 1997.

[8] F. G. Ashby and N. A. Perrin. Toward a unified theory of similarity and recognition. *Psychological Review*, 95(1):124–150, 1988.

[9] D. Ashlock and J. Davidson. Texture synthesis with tandem genetic algorithms using nonparametric partially ordered Markov models. In *Proceedings of the Congress on Evolutionary Computation (CEC99)*, pages 1157–1163, 1999.

[10] R. Bajcsy, S. W. Lee, and A. Leonardis. Color image segmentation with detection of highlights and local illumination induced by inter-reflections. In *IEEE 10th ICPR'90*, pages 785–790, 1990.

[11] R. Basri, L. Costa, D. Geiger, and D. Jacobs. Determining the similarity of deformable shapes. *Vision Research*, 38(15-16):2365–2385, 1998.

[12] S. Belongie, C. Carson, H. Greenspan, and J. Malik. Color- and texture-based image segmentation using em and its application to content-based image retrieval. In *Sixth International Conference on Computer Vision*, 1998.

[13] A. del Bimbo, M. Mugnaini, P. Pala, and F. Turco. Visual querying by color perceptive regions. *Pattern Recognition*, 31(9):1241–1253, 1998.

[14] J. De Bonet and P. Viola. Texture recognition using a non-parametric multi-scale statistical model. In *Computer Vision and Pattern Recognition*, 1998.

[15] H. Burkhardt and S. Siggelkow. Invariant features for discriminating between equivalence classes. In I. Pitas et al. (Ed.), *Nonlinear model-based image video processing and analysis*. John Wiley and Sons, 2000.

[16] C. Carson, S. Belongie, H. Greenspan, and J. Malik. Region-based image querying. In *Proceedings of the IEEE International Workshop on Content-Based Access of Image and Video Databases*, 1997.

[17] S. K. Chang and A. D. Hsu. Image-information systems—where do we go from here? *IEEE Transactions on Knowledge and Data Engineering*, 4(5):431–442, 1992.

[18] S. K. Chang, Q. Y. Shi, and C. W. Yan. Iconic indexing by 2D strings. *IEEE Transactions on PAMI*, 9:413–428, 1987.

[19] H. Chen, B. Schatz, T. Ng, J. Martinez, A. Kirchhoff, and C. Lim. A parallel computing approach to creating engineering concept spaces for semantic retrieval: The Illinois digital library initiative project. *IEEE Transactions on PAMI*, 18(8):771–782, 1996.

[20] Y. Chen and E. K. Wong. Augmented image histogram for image and video similarity search. In *IS&T/SPIE Symposium on Electronic Imaging: Storage and Retrieval for Image and Video Databases VII*, pages 523–429, 1999.

[21] H. Choi and R. Baraniuk. Multiscale texture segmentation using wavelet-domain hidden Markov models. In *Conference Record of Thirty-Second Asilomar Conference on Signals, Systems and Computers*, vol. 2, pages 1692–1697, 1998.

[22] C. K. Chui, L. Montefusco, and L. Puccio. *Wavelets: Theory, algorithms, and applications*. Academic Press, 1994.

[23] L. Cinque, S. Levialdi, and A. Pellicano. Color-based image retrieval using spatial-chromatic histograms. In *IEEE Multimedia Systems*, vol. 2, pages 969–973, 1999.

[24] G. Ciocca and R. Schettini. A relevance feedback mechanism for content-based image retrieval. *Information Processing and Management*, 35:605–632, 1999.

[25] G. Ciocca and R. Schettini. Using a relevance feedback mechanism to improve content-based image retrieval. In *Proceedings of Visual Information and Information Systems*, pages 107–114, 1999.

[26] C. Colombo, A. Rizzi, and I. Genovesi. Histogram families for color-based retrieval in image databases. In *Proc. ICIAP'97*, 1997.

[27] P. Correira and F. Pereira. The role of analysis in content-based video coding and indexing. *Signal Processing*, 66(2):125–142, 1998.

[28] J. M. Corridoni, A. del Bimbo, and P. Pala. Image retrieval by color semantics. *Multimedia systems*, 7:175–183, 1999.

[29] I. J. Cox, M. L. Miller, T. P. Minka, and T. V. Papathomas. The Bayesian image retrieval system, PicHunter: Theory, implementation, and pychophysical experiments. *IEEE Transactions on Image Processing*, 9(1):20–37, 2000.

[30] G. Csurka and O. Faugeras. Algebraic and geometrical tools to compute projective and permutation invariants. *IEEE Transactions on PAMI*, 21(1):58–65, 1999.

[31] J. F. Cullen, J. J. Hull, and P. E. Hart. Document image database retrieval and browsing using texture analysis. In *Proceedings of the Fourth International Conference on Document Analysis and Recognition*, pages 718–721, 1997.

[32] M.-H. Yang, D. Roth, and N. Ahuja. Learning to recognize objects. In *Computer Vision and Pattern Recognition*, pages 724–731, 2000.

[33] I. Daubechies. *Ten lectures on wavelets*. Society for Industrial and Applied Mathematics, 1992.

[34] A. del Bimbo and P. Pala. Visual image retrieval by elastic matching of user sketches. *IEEE Transactions on PAMI*, 19(2):121–132, 1997.

[35] A. Dempster, N. Laird, and D. Rubin. Maximum likelihood from incomplete data via the em algorithm. *Journal of the Royal Statistical Society*, 39(1):1–38, 1977.

[36] E. Di Sciascio, G. Mingolla, and M. Mongielle. Content-based image retrieval over the web using query by sketch and relevance feedback. In *VISUAL99*, pages 123–30, 1999.

[37] A. Dimai. Spatial encoding using differences of global features. In *IS&T/SPIE Symposium on Electronic Imaging: Storage and Retrieval for Image and Video Databases IV*, pages 352–360, 1997.

[38] D. Dubois and H. Prade. A review of fuzzy set aggregation connectives. *Information Sciences*, 36:85–121, 1985.

[39] J. P. Eakins, J. M. Boardman, and M. E. Graham. Similarity retrieval of trademark images. *IEEE Multimedia*, 5(2):53–63, 1998.

[40] C. Esperanca and H. Samet. A differential code for shape representation in image database applications. In *Proceedings of the IEEE International Conference on Image Processing*, 1997.

[41] L. M. Kaplan et al. Fast texture database retrieval using extended fractal features. In I. Sethi and R. Jain (Eds.), *Proceedings of SPIE, vol. 3312, Storage and Retrieval for Image and Video Databases, VI*, pages 162–173, 1998.

[42] M. Flicker et al. Query by image and video content: The QBIC system. *IEEE Computer*, 28(9), 1995.

[43] R. Fagin. Combining fuzzy information from multiple systems. *J. Comput. Syst. Sci.*, 58(1):83–99, 1999.

[44] J. Favella and V. Meza. Image-retrieval agent: Integrating image content and text. 1999.

[45] G. D. Finlayson, S. S. Chatterjee, and B. V. Funt. Color angular indexing. In *ECCV96*, pages 16–27, 1996.

[46] G. D. Finlayson, M. S. Drew, and B. V. Funt. Spectral sharpening: Sensor transformation for improved color constancy. *JOSA*, 11:1553–1563, 1994.

[47] M. Flickner, H. Sawhney, W. Niblack, J. Ashley, Q. Huang, B. Dom, M. Gorkani, J. Hafner, D. Lee, D. Petkovic, D. Steele, and P. Yanker. Query by image and video content: The QBIC system. *IEEE Computer*, 1995.

[48] D. Forsyth. Novel algorithm for color constancy. *International Journal of Computer Vision*, 5:5–36, 1990.

[49] D. A. Forsyth and M. M. Fleck. Automatic detection of human nudes. *International Journal of Computer Vision*, 32(1):63–77, 1999.

[50] G. Frederix and E. J. Pauwels. Automatic interpretation based on robust segmentation and shape extraction. In D. P. Huijsmans and A. W. M. Smeulders (Eds.), *Proceedings of Visual 99, International Conference on Visual Information Systems*, vol. 1614 of *Lecture Notes in Computer Science*, pages 769–776, 1999.

[51] C.-S. Fuh, S.-W. Cho, and K. Essig. Hierarchical color image region segmentation for content-based image retrieval system. *IEEE Transactions on Image Processing*, 9(1):156–163, 2000.

[52] B. V. Funt and M. S. Drew. Color constancy computation in near-Mondrian scenes. In *Computer Vision and Pattern Recognition*, pages 544–549, 1988.

[53] B. V. Funt and G. D. Finlayson. Color constant color indexing. *IEEE Transactions on PAMI*, 17(5):522–529, 1995.

[54] J. M. Geusebroek, A. W. M. Smeulders, and R. van den Boomgaard. Measurement of color invariants. In *Computer Vision and Pattern Recognition*. IEEE Press, 2000.

[55] T. Gevers. Color-based image retrieval. In *Multimedia Search*. Springer Verlag, 2000.

[56] T. Gevers. Image segmentation and matching of color-texture objects. *IEEE Trans. on Multimedia*, 4(4), 2002.

[57] T. Gevers and A. W. M. Smeulders. Color-based object recognition. *Pattern Recognition*, 32(3):453–464, 1999.

[58] T. Gevers and A. W. M. Smeulders. Content-based image retrieval by viewpoint-invariant image indexing. *Image and Vision Computing*, 17(7):475–488, 1999.

[59] T. Gevers and A. W. M. Smeulders. Pictoseek: Combining color and shape invariant features for image retrieval. *IEEE Transactions on Image Processing*, 9(1):102–119, 2000.

[60] T. Gevers and A. W. M. Smeulders. Color-based object recognition. *Pattern Recognition*, 32:453–464, 1999.

[61] T. Gevers and H. M. G. Stokman. Classification of color edges in video into shadow-geometry, highlight, or material transitions. *IEEE Trans. on Multimedia*, 5(2), 2003.

[62] T. Gevers and H. M. G. Stokman. Robust histogram construction from color invariants for object recognition. *IEEE Transactions on PAMI*, 25(10), 2003.

[63] T. Gevers, P. Vreman, and J. van der Weijer. Color constant texture segmentation. In *IS&T/SPIE Symposium on Electronic Imaging: Internet Imaging I*, 2000.

[64] G. L. Gimel'farb and A. K. Jain. On retrieving textured images from an image database. *Pattern Recognition*, 29(9):1461–1483, 1996.

[65] Y. Gong, C. H. Chuan, and G. Xiaoyi. Image indexing and retrieval using color histograms. *Multimedia Tools and Applications*, 2:133–156, 1996.

[66] C. C. Gottlieb and H. E. Kreyszig. Texture descriptors based on co-occurrences matrices. *Computer Vision, Graphics, and Image Processing*, 51, 1990.

[67] L. J. Guibas, B. Rogoff, and C. Tomasi. Fixed-window image descriptors for image retrieval. In *IS&T/SPIE Symposium on Electronic Imaging: Storage and Retrieval for Image and Video Databases III*, pages 352–362, 1995.

[68] A. Gupta and R. Jain. Visual information retrieval. *Communications of the ACM*, 40(5):71–79, 1997.

[69] A. Guttman. R-trees: A dynamic index structure for spatial searching. In *ACM SIGMOD*, pages 47–57, 1984.

[70] J. Hafner, H. S. Sawhney, W. Equit, M. Flickner, and W. Niblack. Efficient color histogram indexing for quadratic form distance functions. *IEEE Transactions on PAMI*, 17(7):729–736, 1995.

[71] M. Hagendoorn and R. C. Veltkamp. Reliable and efficient pattern matching using an affine invariant metric. *International Journal of Computer Vision*, 35(3):203–225, 1999.

[72] S. Hastings. Query categories in a study of intellectual access to digitized art images. In *ASIS '95, Proceedings of the 58th Annual Meeting of the American Society for Information Science*, 1995.

[73] G. Healey. Segmenting images using normalized color. *IEEE Transactions on Systems, Man and Cybernetics*, 22(1):64–73, 1992.

[74] G. Healey and D. Slater. Computing illumination-invariant descriptors of spatially filtered color image regions. *IEEE Transactions on Image Processing*, 6(7):1002–1013, 1997.

[75] K. Hirata and T. Kato. Rough sketch-based image information retrieval. *NEC Res Dev*, 34(2):263–273, 1992.

[76] A. Hiroike, Y. Musha, A. Sugimoto, and Y. Mori. Visualization of information spaces to retrieve and browse image data. In D. P. Huijsmans and A. W. M. Smeulders (Eds.), *Proceedings of Visual 99, International Conference on Visual Information Systems*, vol. 1614 of *Lecture Notes in Computer Science*, pages 155–162, 1999.

[77] N. R. Howe and D. P. Huttenlocher. Integrating color, texture, and geometry for image retrieval. In *Computer Vision and Pattern Recognition*, pages 239–247, 2000.

[78] C. C. Hsu, W. W. Chu, and R. K. Taira. A knowledge-based approach for retrieving images by content. *IEEE Transactions on Knowledge and Data Engineering*, 8(4):522–532, 1996.

[79] F. J. Hsu, S. Y. Lee, and B. S. Lin. Similarity retrieval by 2D C-trees matching in image databases. *Journal of Visual Communication and Image Representation*, 9(1):87–100, 1998.

[80] J. Huang, S. R. Kumar, M. Mitra, W.-J. Zhu, and R. Zabih. Spatial color indexing and applications. *International Journal of Computer Vision*, 35(3):245–268, 1999.

[81] J. Huang, S. R. Kumar, M. Mitra, W.-J. Zhu, and R. Ramin. Image indexing using color correlograms. In *Computer Vision and Pattern Recognition*, pages 762–768, 1997.

[82] B. Huet and E. R. Hancock. Line pattern retrieval using relational histograms. *IEEE Transactions on PAMI*, 21(12):1363–1371, 1999.

[83] F. Idris and S. Panchanathan. Image indexing using wavelet vector quantization. In *Proceedings of the SPIE Vol. 2606–Digital Image Storage and Archiving Systems*, pages 269–275, 1995.

[84] L. Itti, C. Koch, and E. Niebur. A model for saliency-based visual attention for rapid scene analysis. *IEEE Transactions on PAMI*, 20(11):1254–1259, 1998.

[85] C. E. Jacobs, A. Finkelstein, and D. H. Salesin. Fast multiresolution image querying. In *Computer Graphics*, 1995.

[86] C. E. Jacobs, A. Finkelstein, and S. H. Salesin. Fast multiresolution image querying. In *Proceedings of SIGGRAPH 95*. ACM SIGGRAPH, 1995.

[87] A. K. Jain and A. Vailaya. Image retrieval using color and shape. *Pattern Recognition*, 29(8):1233–1244, 1996.

[88] A. K. Jain and A. Vailaya. Shape-based retrieval: A case study with trademark image databases. *Pattern Recognition*, 31(9):1369–1390, 1998.

[89] L. Jia and L. Kitchen. Object-based image similarity computation using inductive learning of contour-segment relations. *IEEE Transactions on Image Processing*, 9(1):80–87, 2000.

[90] D. W. Joyce, P. H. Lewis, R. H. Tansley, M. R. Dobie, and W. Hall. Semiotics and agents for integrating and navigating through multimedia representations. In M. M. Yeung, B.-L. Yeo, and C. Bouman (Eds.), *Proceedings of SPIE, Vol. 3972, Storage and Retrieval for Media Databases 2000*, pages 120–131, 2000.

[91] T. Kato, T. Kurita, N. Otsu, and K. Hirata. A sketch retrieval method for full color image database—Query by visual example. In *Proceedings of the ICPR, Computer Vision and Applications*, pages 530–533, 1992.

[92] J. R. Kender. Saturation, hue, and normalized colors: Calculation, digitization effects, and use. Technical report, Department of Computer Science, Carnegie-Mellon University, 1976.

[93] G. J. Klinker, S. A. Shafer, and T. Kanade. A physical approach to color image understanding. *International Journal of Computer Vision*, 4(1):7–38, 1990.

[94] A. Kontanzad and Y. H. Hong. Invariant image recognition by Zernike moments. *IEEE Transactions on PAMI*, 12(5):489–497, 1990.

[95] S. Krishnamachari and R. Chellappa. Multiresolution Gauss-Markov random field models for texture segmentation. *IEEE Transactions on Image Processing*, 6(2), 1997.

[96] A. Laine and J. Fan. Texture classification by wavelet packet signature. *IEEE Transactions on PAMI*, 15(11):1186–1191, 1993.

[97] E. H. Land. The retinex theory of color vision. *Scientific American*, 218(6):108–128, 1977.

[98] L. J. Latecki and R. Lakamper. Convexity rule for shape decomposition based on discrete contour evolution. *Image Understanding*, 73(3):441–454, 1999.

[99] K.-S. Leung and R. Ng. Multiresolution subimage similarity matching for large image databases. In *IS&T/SPIE Symposium on Electronic Imaging: Storage and Retrieval for Image and Video Databases VI*, pages 259–270, 1998.

[100] C.-S. Li and V. Castelli. Deriving texture feature set for content-based retrieval of satellite image database. In *Proceedings of the IEEE International Conference on Image Processing*, 1997.

[101] K. C. Liang and C. C. J. Kuo. Progressive image indexing and retrieval based on embedded wavelet coding. In *IEEE International Conference on Image Processing*, vol. 1, pages 572–575, 1997.

[102] H. C. Lin, L. L. Wang, and S. N. Yang. Color image retrieval based on hidden Markov models. *IEEE Transactions on Image Processing*, 6(2):332–339, 1997.

[103] T. Lindeberg and J. O. Eklundh. Scale space primal sketch construction and experiments. *Journ Image Vis Comp*, 10:3–18, 1992.

[104] F. Liu and R. Picard. Periodicity, directionality, and randomness: Wold features for image modelling and retrieval. *IEEE Transactions on PAMI*, 18(7):517–549, 1996.

[105] M. Welling, M. Weber, and P. Perona. Towards automatic discovery of object categories. In *Computer Vision and Pattern Recognition*, pages 101–108, 2000.

[106] W. Y. Ma and B. S. Manjunath. Edge flow: A framework of boundary detection and image segmentation. In *Proc. IEEE International Conference on Computer Vision and Pattern Recognition*, pages 744–749, 1997.

[107] J. Malki, N. Boujemaa, C. Nastar, and A. Winter. Region queries without segmentation for image retrieval content. In *Int. Conf. on Visual Information Systems, VISUAL99*, pages 115–122, 1999.

[108] M. K. Mandal, F. Idris, and S. Panchanathan. Image and video indexing in the compressed domain: A critical review. *Image and Vision Computing*, 2000.

[109] B. S. Manjunath and W. Y. Ma. Texture features for browsing and retrieval of image data. *IEEE Transactions on PAMI*, 18(8):837–842, 1996.

[110] J. Mao and A. K. Jain. Texture classification and segmentation using multiresolution simultaneous autoregressive models. *Pattern Recognition*, 25(2), 1992.

[111] J. Matas, R. Marik, and J. Kittler. On representation and matching of multi-coloured objects. In *Proc. 5th ICCV*, pages 726–732, 1995.

[112] R. Mehrotra and J. E. Gary. Similar-shape retrieval in shape data management. *IEEE Computer*, 28(9):57–62, 1995.

[113] B. M. Mehtre, M. S. Kankanhalli, and W. F. Lee. Shape measures for content-based image retrieval: A comparison. *Information Proc. Management*, 33(3):319–337, 1997.

[114] M. Mirmehdi and M. Petrou. Segmentation of color texture. *PAMI*, 22(2):142–159, 2000.

[115] B. Moghaddam and A. Pentland. Probabilistic visual learning for object representation. *IEEE Transactions on PAMI*, 19(7):696–710, 1997.

[116] B. Moghaddam, W. Wahid, and A. Pentland. Beyond eigenfaces: Probabilistic matching for face recognition. In *3rd IEEE International Conference on Automatic Face and Gesture Recognition*, 1998.

[117] A. Mojsilovic, J. Kovacevic, J. Hu, R. J. Safranek, and S. K. Ganapathy. Matching and retrieval based on the vocabulary and grammar of color patterns. *IEEE Transactions on Image Processing*, 9(1):38–54, 2000.

[118] F. Mokhtarian. Silhouette-based isolated object recognition through curvature scale-space. *IEEE Transactions on PAMI*, 17(5):539–544, 1995.

[119] J. L. Mundy, A. Zissermann, and D. Forsyth (Eds.). *Applications of invariance in computer vision*, vol. 825 of *Lecture Notes in Computer Science*. Springer Verlag GmbH, 1994.

[120] H. Murase and S. K. Nayar. Visual learning and recognition of 3D objects from appearance. *International Journal of Computer Vision*, 14(1):5–24, 1995.

[121] S. K. Nayar and R. M. Bolle. Reflectance-based object recognition. *International Journal of Computer Vision*, 17(3):219–240, 1996.

[122] T. Ojala, M. Pietikainen, and D. Harwood. A comparison study of texture measures with classification based on feature distributions. *Pattern Recognition*, 29:51–59, 1996.

[123] P. Pala and S. Santini. Image retrieval by shape and texture. *Pattern Recognition*, 32(3):517–527, 1999.

[124] D. K. Panjwani and G. Healey. Markov random field models for unsupervised segmentation of textured color images. *IEEE Transactions on PAMI*, 17(10):939–954, 1995.

[125] G. Pass and R. Zabith. Comparing images using joint histograms. *Multimedia systems*, 7:234–240, 1999.

[126] E. J. Pauwels and G. Frederix. Nonparametric clustering for image segmentation and grouping. *Image Understanding*, 75(1):73–85, 2000.

[127] A. Pentland, R. W. Picard, and S. Sclaroff. Photobook: Content-based manipulation of image databases. *International Journal of Computer Vision*, 18(3):233–254, 1996.

[128] E. Petrakis and C. Faloutsos. Similarity searching in medical image databases. *IEEE Transactions on Knowledge and Data Engineering*, 9(3):435–447, 1997.

[129] R. W. Picard and T. P. Minka. Vision texture for annotation. *Multimedia Systems*, 3, 1995.

[130] M. Pietikainen, S. Nieminen, E. Marszalec, and T. Ojala. Accurate color discrimination with classification based on feature distributions. In *Proc. Int'l Conf. Pattern Recognition*, pages 833–838, 1996.

[131] J. Puzicha, T. Hoffman, and J. M. Buhmann. Non-parametric similarity measures for unsupervised texture segmentation and image retrieval. In *Proceedings of the International Conference on Computer Vision and Pattern Recognition-CVPR*, 1997.

[132] W. Qian, M. Kallergi, L. P. Clarke, H. D. Li, D. Venugopal, D. S. Song, and R. A. Clark. Tree-structured wavelet transform segmentation of microcalcifications in digital mammography. *JI Med. Phys.*, 22(8):1247–1254, 1995.

[133] T. Randen and J. Hakon Husoy. Filtering for texture classification: A comparative study. *IEEE Transactions on PAMI*, 21(4):291–310, 1999.

[134] E. Riloff and L. Hollaar. Text databases and information retrieval. *ACM Computing Surveys*, 28(1):133–135, 1996.

[135] E. Rivlin and I. Weiss. Local invariants for recognition. *IEEE Transactions on PAMI*, 17(3):226–238, 1995.

[136] R. Rodriguez-Sanchez, J. A. Garcia, J. Fdez-Valdivia, and X. R. Fdez-Vidal. The RGFF representational model: A system for the automatically learned partitioning of "visual pattern" in digital images. *IEEE Transactions on PAMI*, 21(10):1044–1073, 1999.

[137] P. L. Rosin. Edges: Saliency measures and automatic thresholding. *Machine Vision and Appl.*, 9(7):139–159, 1997.

[138] I. Rothe, H. Suesse, and K. Voss. The method of normalization of determine invariants. *IEEE Transactions on PAMI*, 18(4):366–376, 1996.

[139] Y. Rui, T. S. Huang, M. Ortega, and S. Mehrotra. Relevance feedback: A power tool for interactive content-based image retrieval. *IEEE Transactions on Circuits and Video Technology*, 1998.

[140] M. Beigi, S.-F. Chang, J. R. Smith, and A. Benitez. Visual information retrieval from large distributed online repositories. *Comm. ACM*, 40(12):63–71, 1997.

[141] S. Santini, A. Gupta, and R. Jain. User interfaces for emergent semantics in image databases. In *Proceedings of the 8th IFIP Working Conference on Database Semantics (DS-8)*, 1999.

[142] S. Santini and R. Jain. Similarity measures. *IEEE Transactions on PAMI*, 21(9):871–883, 1999.

[143] C. Schmid and R. Mohr. Local grayvalue invariants for image retrieval. *IEEE Transactions on PAMI*, 19(5):530–535, 1997.

[144] M. Schneier and M. Abdel-Mottaleb. Exploiting the JPEG compression scheme for image retrieval. *IEEE Transactions on PAMI*, 18(8):849–853, 1996.

[145] S. Sclaroff. Deformable prototypes for encoding shape categories in image databases. *Pattern Recognition*, 30(4):627–641, 1997.

[146] S. Sclaroff, M. LaCascia, and S. Sethi. Using textual and visual cues for content-based image retrieval from the World Wide Web. *Image Understanding*, 75(2):86–98, 1999.

[147] S. Sclaroff, L. Taycher, and M. La Cascia. Imagerover: A content-based image browser for the World Wide Web. In *IEEE Workshop on Content-Based Access and Video Libraries*, 1997.

[148] N. Sebe, M.S. Lew, and D.P. Huijsmands. Multiscale subimage search. In *ACM Int. Conf. on Multimedia*, 1999.

[149] S. Servetto, Y. Rui, K. Ramchandran, and T. S. Huang. A region-based representation of images in Mars. *Journal on VLSI Signal Processing Systems*, 20(2):137–150, 1998.

[150] S. A. Shafer. Using color to separate reflection components. *COLOR Res. Appl.*, 10(4):210–218, 1985.

[151] D. Sharvit, J. Chan, H. Tek, and B. B. Kimia. Symmetry-based indexing of image databases. *Journal of Visual Communication and Image Representation*, 9(4):366–380, 1998.

[152] R. N. Shepard. Toward a universal law of generalization for physical science. *Science*, 237:1317–1323, 1987.

[153] D. Slater and G. Healey. The illumination-invariant recognition of 3D objects using local color invariants. *IEEE Transactions on PAMI*, 18(2):206–210, 1996.

[154] A. W. M. Smeulders, M. L. Kersten, and T. Gevers. Crossing the divide between computer vision and databases in search of image databases. In *Fourth Working Conference on Visual Database Systems*, pages 223–239, 1998.

[155] A. W. M. Smeulders, M. Worring, S. Santini, A. Gupta, and R. Jain. Content-based image retrieval at the end of the early years. *IEEE Transactions on PAMI*, 22(12):1349–1380, 2000.

[156] J. R. Smith and S. F. Chang. Automated binary feature sets for image retrieval. In C. Faloutsos (Ed.), *Proceedings of ICASSP*. Kluwer Academic, 1996.

[157] J. R. Smith and S.-F. Chang. Integrated spatial and feature image query. *Multimedia systems*, 7(2):129–140, 1999.

[158] J. R. Smith and S.-F. Chang. Visualseek: A fully automated content-based image query system. In *ACM Multimedia*, 1996.

[159] S. M. Smith and J. M. Brady. SUSAN—A new approach to low level image processing. *International Journal of Computer Vision*, 23(1):45–78, 1997.

[160] M. Stricker and M. Swain. The capacity of color histogram indexing. In *Computer Vision and Pattern Recognition*, pages 704–708. IEEE Press, 1994.

[161] M. A. Stricker and M. Orengo. Similarity of color images. In *IS&T/SPIE Symposium on Electronic Imaging: Storage and Retrieval for Image and Video Databases IV*, 1996.

[162] M. J. Swain and B. H. Ballard. Color indexing. *International Journal of Computer Vision*, 7(1):11–32, 1991.

[163] M. J. Swain. Searching for multimedia on the World Wide Web. In *IEEE International Conference on Multimedia Computing and Systems*, pages 33–37, 1999.

[164] D. J. Swets and J. Weng. Hierarchical discriminant analysis for image retrieval. *IEEE Transactions on PAMI*, 21(5):386–401, 1999.

[165] T. F. Syeda-Mahmood. Data and model-driven selection using color regions. *International Journal of Computer Vision*, 21(1):9–36, 1997.

[166] H. D. Tagare, F. M. Vos, C. C. Jaffe, and J. S. Duncan. Arrangement—A spatial relation between parts for evaluating similarity of tomographic section. *IEEE Transactions on PAMI*, 17(9):880–893, 1995.

[167] T. Tan. Rotation invariant texture features and their use in automatic script identification. *IEEE Transactions on PAMI*, 20(7):751–756, 1998.

[168] P. M. Tardif and A. Zaccarin. Multiscale autoregressive image representation for texture segmentation. In *Proceedings of SPIE Vol. 3026, Nonlinear Image Processing VIII*, pages 327–337, 1997.

[169] K. Tieu and P. Viola. Boosting image retrieval. In *Computer Vision and Pattern Recognition*, pages 228–235, 2000.

[170] T. Tuytelaars and L. van Gool. Content-based image retrieval based on local affinely invariant regions. In *Proceedings of Visual Information and Information Systems*, pages 493–500, 1999.

[171] A. Vailaya, M. Figueiredo, A. Jain, and H. Zhang. A Bayesian framework for semantic classification of outdoor vacation images. In C. A. Bouman, M. M. Yeung, and B. Yeo (Eds.), *Storage and Retrieval for Image and Video Databases VII, SPIE*, pages 415–426, 1999.

[172] A. Vailaya, M. Figueiredo, A. Jain, and H. Zhang. Content-based hierarchical classification of vacation images. In *IEEE International Conference on Multimedia Computing and Systems*, 1999.

[173] V. N. Vapnik. *The Nature of Statistical Learning Theory*. Springer-Verlag, 1995.

[174] N. Vasconcelos and A. Lippman. A probabilistic architecture for content-based image retrieval. In *Computer Vision and Pattern Recognition*, pages 216–221, 2000.

[175] A. Vellaikal and C. C. J. Kuo. Content-based retrieval using multiresolution histogram representation. *Digital Image Storage Archiving Systems*, pages 312–323, 1995.

[176] R. C. Veltkamp and M. Hagendoorn. State-of-the-art in shape matching. In *Multimedia Search: State of the Art*. Springer Verlag GmbH, 2000.

[177] L. Z. Wang and G. Healey. Using Zernike moments for the illumination and geometry invariant classification of multispectral texture. *IEEE Transactions on Image Processing*, 7(2):196–203, 1991.

[178] J. Weickert, S. Ishikawa, and A. Imiya. Linear scale space has first been proposed in Japan. *Journal of Mathematical Imaging and Vision*, 10:237–252, 1999.

[179] M. Werman and D. Weinshall. Similarity and affine invariant distances between 2D point sets. *IEEE Transactions on PAMI*, 17(8):810–814, 1995.

[180] R. C. Wilson and E. R. Hancock. Structural matching by discrete relaxation. *IEEE Transactions on PAMI*, 19(6):634–648, 1997.

[181] L. Wolff, S. A. Shafer, and G. E. Healey (Eds.). *Physics-based Vision: Principles and Practice*, vol. 2. Jones and Bartlett, 1992.

[182] H. J. Wolfson and I. Rigoutsos. Geometric hashing: An overview. *IEEE Computational Science and Engineering*, 4(4):10–21, 1997.

[183] Q. Tian, Y. Wu, and T.S. Huang. Discriminant-em algorithm with applications to image retrieval. In *Computer Vision and Pattern Recognition*, pages 222–227, 2000.

[184] H. H. Yu and W. Wolf. Scene classification methods for image and video databases. In *Proc. SPIE on Digital Image Storage and Archiving Systems*, pages 363–371, 1995.

[185] Q. L. Zhang, S. K. Chang, and S. S. T. Yau. A unified approach to iconic indexing, retrieval and maintenance of spatial relationships in image databases. *Journal of Visual Communication and Image Representation*, 7(4):307–324, 1996.

[186] Y. J. Zhang, Z. W. Liu, and Y. He. Comparison and improvement of color-based image retrieval. In *IS&T/SPIE Symposium on Electronic Imaging: Storage and Retrieval for Image and Video Databases IV*, pages 371–382, 1996.

[187] R. Zhao and W. Grosky. Locating text in complex color images. In *IEEE International Conference on Multimedia Computing and Systems*, 2000.

[188] Y. Zhong, K. Karu, and A. K. Jain. Locating text in complex color images. *Pattern Recognition*, 28(10):1523–1535, 1995.

[189] P. Zhu and P. M. Chirlian. On critical point detection of digital shapes. *IEEE Transactions on PAMI*, 17(8):737–748, 1995.

[190] S. C. Zhu and A. Yuille. Region competition: Unifying snakes, region growing, and bayes/mdl for multiband image segmentation. *IEEE Transactions on PAMI*, 18(9):884–900, 1996.

Chapter 9

FACE DETECTION, ALIGNMENT, AND RECOGNITION

Stan Z. Li

and Juwei Lu

9.1 Introduction

A face recognition system identifies faces in images and videos automatically using computers. It has a wide range of applications, such as biometric authentication and surveillance, human-computer interaction, and multimedia management. A face recognition system generally consists of four processing parts, as depicted in Figure 9.1: face detection, face alignment, facial feature extraction, and face classification. *Face detection* provides information about the location and scale of each detected face. In the case of video, the found faces may be tracked. In *face alignment*, facial components, such as eyes, nose, and mouth, and facial outline are located, and thereby the input face image is normalized in geometry and photometry. In *feature extraction*, features useful for distinguishing between different persons are extracted from the normalized face. In *face classification*, the extracted feature vector of the input face is matched against those of enrolled faces in the database, outputting the identity of the face when a match is found with a sufficient confidence or as an unknown face otherwise.

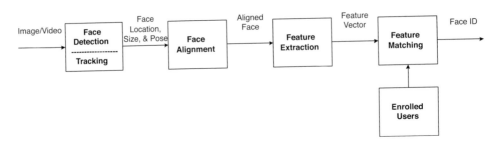

Figure 9.1. Structure of a face recognition system.

The underlying problems can be treated using pattern recognition and machine-learning techniques. There two central issues: (1) what features to use to represent a face, and (2) how to classify a new face image based on the chosen representation. A capable face recognition system should be able to deal with variations of face images in viewpoint, illumination, expression, and so on.

The geometric feature-based approach [44, 57, 101, 17] is based on the traditional computer vision framework [81]. Facial features such as eyes, nose, mouth, and chin are detected. Properties and relations (e.g., areas, distances, angles) between the features are used as descriptors of faces for recognition. Using this approach, Kanade built the first face recognition system in the world [57]. Advantages include economy and efficiency in achieving data reduction and insensitivity to variations in illumination and viewpoint. Disadvantages are that facial feature detection and measurement techniques developed to date have not been reliable enough [25], and geometric information only is insufficient for face recognition.

Great progress has been made in the past 15 or so years due to advances in the template-matching or appearance-based approach [122]. Such an approach generally operates directly on an image-based representation (i.e., pixel intensity array). It extracts features (instead of overabstract features) in a face subspace from images. A face subspace is constructed to best represent the face object only. Although it is much less general, it is more efficient and effective for face recognition. In the eigenface [122] or principal component analysis (PCA) method, the face space is spanned by a number of eigenfaces [111] derived from a set of training face images by using Karhunen-Loeve transform or the PCA [42]. A face image is efficiently represented by a feature point (i.e., a vector of weights) of low (e.g., 40 or lower) dimensionality. Such subspace features are more salient and richer for recognition.

Face recognition performance has been much improved as compared to that of the first automatic face recognition system of Kanade [57]. Nowadays, face detection, facial feature location, and recognition can be performed for image data of reasonable conditions, which was unachievable by the pioneer systems.

Although the progress has been encouraging, the task has also turned out to be a very difficult endeavor [121]. Face recognition evaluation reports such as [95, 1] and other independent studies indicate that the performance of many state-of-the-art face recognition methods deteriorates with changes in lighting, pose, and other factors [123, 18, 140]. The key technical challenges are summarized in the following:

- **Immense variability of facial appearance:** Whereas shape and reflectance are intrinsic properties of a face object, the appearance (i.e., the texture look) of a face is also subject to several other factors, including the facial pose (or, equivalently, camera viewpoint), the illumination, facial expression, and various imaging parameters such as aperture, exposure time, lens aberrations, and sensor spectral response. All these factors are confounded in the image data, so that "the variations between the images of the same face due to illumination and viewing direction are almost always larger than the image variation due to change in face identity" [89]. The complexity makes it very difficult to extract the intrinsic information of the face objects from their respective images.

- **Highly complex and nonlinear manifolds:** As illustrated above, the face manifold, as opposed to the manifold of nonface patterns, is highly nonconvex, and so are face manifolds of any individual under changes due to pose, lighting, facial wear, and so on. As linear subspace methods, such as principal component analysis (PCA) [111, 122], independent component analysis (ICA) [10], or linear discriminant analysis (LDA) [12] project the data in a high-dimensional space, such as the image space, to a low-dimensional subspace in an optimal direction in a linear way, they are unable to preserve the nonconvex variations of face manifolds necessary to differentiate between different individuals. In a linear subspace, Euclidean distance and more generally M-distance, which are normally used in template matching, do not apply well to the problem of classification between manifolds of face/nonface manifolds and between manifolds of different individuals. This is a crucial fact that limits the power of the linear methods to achieve highly accurate face detection and recognition.

• **High dimensionality and small sample size:** Another challenge is the ability to generalize. A canonical face example used in face recognition is an image of size 112×92 and resides in a $10,304$-dimensional real space. Nevertheless, the number of examples per person available for learning is usually much smaller than the dimensionality of the image space, e.g., less than 10 in most cases; a system trained on so few examples may not generalize well to unseen face instances. Besides, the computational cost caused by high dimensionality is also a concern for real-time systems.

The above problems can be handled in two ways. One is to normalize face images in geometry and photometry. This way, the face manifolds become simpler (less nonlinear and less nonconvex), so that the complex problems become easier to tackle. The other way is to devise powerful engines able to perform difficult nonlinear classification and regression and to generalize better. This relies on advances in pattern recognition and learning, and clever applications of them.

Both prior and observation constraints are needed for such powerful engines. Most successful approaches for tackling the above difficulties are based on subspace modeling of facial appearance and statistical learning. Constraints about the face include facial shape, texture, head pose, and illumination effect. Recent advances allow these to be effectively encoded into the system by learning from training data.

This chapter presents advanced techniques for face detection, face alignment, and face recognition (feature extraction and matching). The presentation is focused on appearance- and learning-based approaches.

9.2 Face Detection

Face detection is the first step in automated face recognition. Its reliability has a major influence on the performance and usability of the whole system. Given a single image or a video, an ideal face detection system should be able to identify and locate all faces regardless of their positions, scales, orientations, lighting conditions, expressions, and so on.

Face detection can be performed based on several different cues: skin color (for faces in color images), motion (for faces in videos), facial/head shape, and facial appearance, or a combination of them. Prior knowledge about them can be embedded into the system by learning from training data.

Appearance- and learning-based approaches have so far been the most effective ones for face detection, and this section focuses on such approaches. Refer to recent surveys [32, 136] for other methods. Great resources such as publications, databases, and codes can be found from face detection Web sites [41, 133].

9.2.1   Appearance-based and learning-based approaches

In appearance-based approach, face detection is treated as an intrinsically 2D problem. It is done in three steps: first, scan $\mathcal{I}$ exhaustively at all possible locations (u, v) and scales s, resulting in a large number of subwindows $x = x(u, v, s \mid \mathcal{I})$. Second, test for each x if it is a face:

$$H(x) \begin{array}{l} \geq 0 \quad \Rightarrow x \text{ is a face pattern} \\ < 0 \quad \Rightarrow x \text{ is a nonface pattern} \end{array} \tag{9.1}$$

Third, postprocess to merge multiple detects.

The key issue is the construction of a face detector that classifies a subwindow into either face or nonface. Face and nonface examples are given as the training set (Figure 9.2 shows a random sample of 10 face and 10 nonface examples). Taking advantage of the fact that faces are highly correlated, it is assumed that human faces can be described by some low-dimensional features that may be derived from a set of example face images. Large variation and complexity brought about by changes in facial appearance, lighting, and expression makes the face manifold highly complex [14, 110, 121]. Changes in facial view (head pose) further complicate the situation. Nonlinear classifiers are training to classify each subwindow into face or nonface. The following gives a brief review of exciting work.

Figure 9.2. Face (top) and nonface (bottom) examples.

Turk and Pentland [122] describe a detection system based on a PCA subspace or eigenface representation. Moghaddam and Pentland [87] present an improved method for Bayesian density estimation, where the high-dimensional image space is decomposed into a PCA subspace for prior distribution and the nullspace for the likelihood distribution. Sung and Poggio [114] first partition the image space into several clusters for face and nonface clusters. Each cluster is then further decomposed into the PCA and null subspaces.

The Bayesian estimation is then applied to obtain useful statistical features. Rowley and colleagues' system [100] uses retinally connected neural networks. Through a sliding window, the input images are examined after going through an extensive preprocessing stage. Osuna et al. [91] train support vector machines (SVMs) to classify face and nonface patterns. Roth et al. [99] use SNoW learning structure for face detection. In these systems, a bootstrap algorithm is used to iteratively collect meaningful nonface examples into the training set.

Viola and Jones [128, 127] build a successful face detection system in which AdaBoost learning is used to construct a nonlinear classifier (earlier work in application of AdaBoost for image classification and face detection can be found in [117] and [104]). AdaBoost is adapted to solving the following three fundamental problems in one boosting procedure: (1) learning effective features from a large feature set, (2) constructing weak classifiers, each based on one of the selected features, and (3) boosting the weak classifiers into stronger classifiers. Also, that work makes ingenius use of several techniques for effective computation of a large number of Haar wavelet-like features. Such features are steerable filters [92, 128]. Moreover, the simple-to-complex cascade of classifiers structure makes the computation even more efficient, which is in the principle of pattern rejection [8, 30] and coarse-to-fine search [5, 36]. Each such feature has a scalar value that can be computed very efficiently [109] from the summed-area table [26] or integral image [128]. Their system is the first real-time frontal-view face detector that runs at about 14 frames per second for a 320×240 image [128].

The ability to deal with nonfrontal faces is important for many real applications because statistics show that approximately 75% of the faces in home photos are nonfrontal [60]. A reasonable treatment for multiview is the view-based method [93] in which several face models are built, each describing faces in a certain view range. This way, explicit 3D modeling is avoided. Feraud et al. [34] adopt the view-based representation for face detection and use an array of five detectors, each responsible for one view. Wiskott et al. [130] build elastic bunch graph templates for multiview face detection and recognition. Gong and colleagues [45] study the trajectories of faces in linear PCA feature spaces as they rotate, and use kernel SVMs for multi-pose face detection and pose estimation [90, 70]. Huang et al. [51] use SVMs to estimate facial poses.

In the system of Schneiderman and Kanade [105], multiresolution information is used for different levels of wavelet transform. The algorithm consists of an array of five face detectors in the view-based framework. Each is constructed using statistics of products of histograms computed from examples of the respective view. It takes 1 min for a 320×240 image over only four octaves of candidate size, as reported in [105].

Li et al. [68, 67] present a multiview face detection system, extending the work of [128, 127] and [105]. A new boosting algorithm called FloatBoost is proposed to incorporate floating search [97] into AdaBoost. The backtrack mechanism therein allows deletions of weak classifiers that are ineffective in terms of the error rate, leading to a strong classifier consisting of fewer weak classifiers. An extended Haar feature set is proposed for dealing with out-of-plane rotations and a detector pyramid for improving the speed. A coarse-to-fine, simple-to-complex architecture called detector-pyramid is designed for the fast detection of multiview faces. This work leads to the first real-time multiview face detection system in the world. It runs at 200 ms per image of size 320×240 pixels on a Pentium-III CPU of 700 MHz.

Lienhart et al. [71] propose an extended set of rotated Haar features for dealing with in-plane rotations. Also, they use gentle AdaBoost [37] with small CART trees as base classifiers and show that this combination outperforms that of discrete AdaBoost and stumps.

In the following, face processing techniques are presented, including preprocessing and postprocessing, neural network-based methods, and boosting-based methods. Given that the boosting learning with Haar-like feature approach has achieved the best performance, the presentation focuses on such state-of-the-art methods.

9.2.2   Preprocessing

Skin color filtering

Human skin has its own distribution that differs from that of most of nonface objects. It can be used to filter the input image to obtain candidate regions of faces and as a standalone face detector (not an appearance-based detector). A simple color-based face detection algorithm can consists of two steps: (1) segmentation of likely face regions and (2) region merge.

Skin color likelihood model, $p(color \mid face)$, can be derived from skin color samples. This may be done in the H (of HSV) color space or in the normalized RGB color space (see a comparative evaluation in [138]). A Gaussian mixture model for $p(color \mid face)$ can give a better skin color modeling [132, 135]. Figure 9.3 shows a skin color segmentation map. A skin-colored pixel is found if $p(H \mid face)$, where H is the hue component of the pixel and is greater than a threshold (0.3), and its saturation (S) and value (V) are between some upper and lower bounds. A skin color map consists of a number of skin color regions that indicate potential candidate face areas. Refined face regions can be obtained by merging adjacent similar and homogeneous skin color pixels based on the color and spatial information. Heuristic postprocessing could be performed to remove false detection. For example, a human face contains eyes where the eyes correspond to darker regions inside the face region.

Figure 9.3. Skin color filtering: input image with multiple people (left); skin color filtered map (right).

While a color-based face detection system may work fast, the color constraint alone is insufficient for achieving high-performance face detection due to color variations for different lighting, shadow, and ethnic groups. Indeed, it is the appearance, albeit colored or gray leveled, rather than the color, that is essential for face detection. In fact, most successful systems need not use color to achieve good performance.

Image normalization

Preprocessing operations are usually performed to normalize the image pattern in a subwindow in its size, pixel value distribution, and lighting condition. These include resizing (say, 20×20 pixels), lighting correction, mean value normalization, and histogram equalization. A simple lighting correction operation is to subtract a best-fit plane $I'(x, y) = a \times x + b \times y + c$ from the subwindow $I(x, y)$, where the values of a, b, and c can be estimated using the least-squares method. Figure 9.4 gives an example of the effect. The mean value normalization operation subtracts the mean value of the window pattern from the window pixels so that the average intensity of all the pixels in the subwindow is zero. See Figure 9.5.

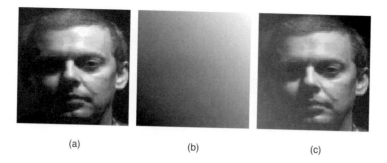

(a)   (b)   (c)

Figure 9.4. Effect of lighting correction: (a) before illumination correction; (b) best-fit linear function; (c) after illumination correction.

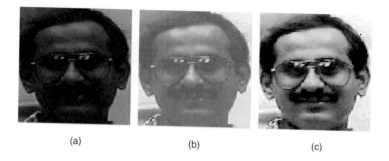

(a)   (b)   (c)

Figure 9.5. Mean value normalization and histogram equalization: (a) original image window; (b) linear stretch after mean value normalization; (c) histogram equalization after mean value normalization.

Multi-Gaussian clustering

The distribution of training patterns is very complex because of the variety of changes and high dimensionality. Therefore, it is hard to explain all such variations using a single distribution. Sung and Poggio [114] propose to deal with the complexity by dividing the face training data into $L_f = 6$ face clusters, and nonface training data into $L_n = 6$ nonface clusters, where cluster numbers 6 and 6 are empirically chosen. The clustering is performed by using a modified k-mean clustering algorithm based on the Mahalanobis distance [114]. Figure 9.6 shows the centroids of the obtained L_f face clusters and L_n nonface clusters. After the clustering, nonlinear classification is then based on the partition.

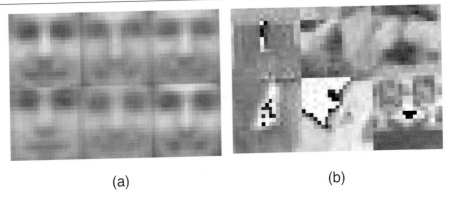

(a)   (b)

Figure 9.6. Centroids of six face clusters in (a) and six nonface clusters in (b).

9.2.3 Neural and kernel methods

Here we describe the methods of [100, 114]. In the training phase, a neural network (NN) classifier is trained by using normalized face and nonface training images. A bootstrap algorithm is used to collect meaningful nonface examples into the training set. In the test phase, the trained NN classifies each subwindow into either face or nonface. The main issue here is to train a nonlinear neural classifier by which highly nonlinear manifolds of face and nonface in the space of image subwindows are separated.

An NN is a fully connected, feedforward multilayer perceptron. The input feature vector can be simply the raw image in the subwindow or a feature vector extracted from it. For the latter case, it can be a preprocessed subimage [100] or vector of the distances from subspaces of face and nonface clusters [114]. A back-propagation (BP) algorithm is used for the training. Several copies of the same NN can be trained and their outputs combined by arbitration (ANDing) [100]—hopefully, this would give more reliable results than can be obtained by using a single NN.

Nonlinear classification for face detection can also be done using kernel SVMs [91, 90, 70]. While such methods can learn nonlinear boundaries, a large number of support vectors may result in order to capture high nonlinearity. This would create an issue unfavorable to real-time performance.

9.2.4 Boosting-based methods

In AdaBoost-based classification, a highly complex, nonlinear classifier is constructed as a linear combination of many simpler, easily constructible weak classifiers [37]. In AdaBoost face detection [128, 127, 68, 71], each weak classifier is built by thresholding on a scalar feature selected from an overcomplete set of Haar wavelet-like features [92, 117]. Such methods have so far been the most successful for face detection.

Haar-like features

Viola and Jones propose four basic types of scalar features for face detection [92, 128], as shown in Figure 9.7. Recently, such features have been extended for dealing with head rotations [68, 71]. Each such feature has a scalar value that can be computed very efficiently [109] from the summed-area table [26] or integral image [128]. Feature k, taking the value $z_k(x) \in \mathbb{R}$, can be considered as a transform from the n-dimensional (400D if a face example x is of size 20×20) data space to the real line. For a face example of size 20×20, there are tens of thousands of such features. These form an overcomplete feature set for the intrinsically low-dimensional face pattern.

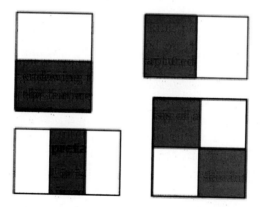

Figure 9.7. Rectangular Haar wavelet-like features. A feature takes a scalar value by summing up the white region and subtracting the dark region.

Learning feature selection

A weak classifier is associated with a single scalar feature; to find the best new weak classifier is to choose the best corresponding feature. AdaBoost learning is used to select most significant features from the feature set. More specifically, AdaBoost is adapted to solving the following three fundamental problems in one boosting procedure: (1) learning effective features from a large feature set; (2) constructing weak classifiers, each based on one of the selected features; and (3) boosting the weak classifiers into a stronger classifier.

The basic form of (discrete) AdaBoost [37] is for two class problems. A set of N labeled training examples is given as $(x_1, y_1), \ldots, (x_N, y_N)$, where $y_i \in \{+1, -1\}$ is the class label for the example $x_i \in \mathbb{R}^n$. AdaBoost assumes that a procedure is available for learning sequence of *weak classifiers* $h_m(x)$ ($m = 1, 2, \ldots, M$) from the training examples, with respect to the distributions $w_i^{(m)}$ of the examples. A *strong classifier* is a linear combination of the M weak classifiers:

$$H_M(x) = \frac{\sum_{m=1}^{M} \alpha_m h_m(x)}{\sum_{m=1}^{M} \alpha_m}, \tag{9.2}$$

where $\alpha_m \geq 0$ are the combining coefficients. The classification of x is obtained as $\widehat{y}(x) = sgn[H_M(x)]$ and the normalized confidence score is $|H_M(x)|$. The AdaBoost learning procedure is aimed to derive α_m and $h_m(x)$.

Learning weak classifiers

A weak classifier is constructed by thresholding one of those features according to the likelihoods (histograms) of the feature values for the target faces and the imposter faces:

$$h_k^{(M)}(x) = +1 \quad if \, z_k(x) > \tau_k^{(M)} \tag{9.3}$$
$$= -1 \quad otherwise, \tag{9.4}$$

where $z_k(x)$ is feature k extracted from x, and $\tau_k^{(M)}$ is the threshold for weak classifier k chosen to ensure a specified accuracy. The best weak classifier is the one for which the false alarm is minimized:

$$k^* = \arg\min_k FA(h_k^{(M)}(x)), \tag{9.5}$$

where FA is the false alarm caused by $h_k^{(M)}(x)$ (also w.r.t. $w^{(M-1)}$). This gives us the best weak classifier as

$$h_M(x) = h_{k^*}^{(M)}(x). \tag{9.6}$$

Boosted strong classifier

AdaBoost effectively learns to boost weak classifiers, h_m, into strong ones, H_M, by minimizing the upper bound on classification error achieved by H_M. The bound can be derived as the following exponential loss function [102]:

$$J(H_M) = \sum_i e^{-y_i H_M(x_i)} = \sum_i e^{-y_i \sum_{m=1}^{M} \alpha_m h_m(x)}. \tag{9.7}$$

AdaBoost constructs $h_m(x)$ by stagewise minimization of (9.7). Given the current $H_{M-1}(x) = \sum_{m=1}^{M-1} \alpha_m h_m(x)$, and the newly learned weak classifier h_M, the best combining coefficient α_M for the new strong classifier $H_M(x) = H_{M-1}(x) + \alpha_M h_M(x)$ minimizes the cost:

$$\alpha_M = \arg\min_\alpha J(H_{M-1}(x) + \alpha h_M(x)). \qquad (9.8)$$

The minimizer is

$$\alpha_M = \log \frac{1 - \epsilon_M}{\epsilon_M}, \qquad (9.9)$$

where ϵ_M is the weighted error

$$\epsilon_M = \sum_i w_i^{(M-1)} 1[\text{sign}(H_M(x_i)) \neq y_i], \qquad (9.10)$$

where $1[C]$ is one if C is true, or 0 otherwise.

Each example is reweighted after an iteration, that is, $w_i^{(M-1)}$ is updated according to the classification performance of H_M:

$$
\begin{aligned}
w^{(M)}(x, y) &= w^{(M-1)}(x, y) \exp\left(-\alpha_M y h_M(x)\right) \\
&= \exp\left(-y H_M(x)\right),
\end{aligned} \qquad (9.11)
$$

which is used for calculating the weighted error or another cost for training the weak classifier in the next round. This way, a more difficult example will be associated with a larger weight so that it will be more emphasized in the next round of learning. The algorithm is summarized in Figure 9.8.

FloatBoost learning

AdaBoost attempts to boost the accuracy of an ensemble of weak classifiers to a strong one. The AdaBoost algorithm [37] solved many of the practical difficulties of earlier boosting algorithms. Each weak classifier is trained one stagewise to minimize the empirical error in a given distribution reweighted according to classification errors of the previously trained classifier. It is shown that AdaBoost is a sequential forward search (SFS) procedure using the greedy selection strategy to minimize a certain margin on the training set [102].

A crucial heuristic assumption made in such a sequential forward search procedure is the monotonicity (i.e., when adding a new weak classifier to the current set, the value of the performance criterion does not decrease). The

0. (Input)
    (1) Training examples $\mathcal{Z} = \{(\mathbf{x}_1, y_1), \dots, (\mathbf{x}_N, y_N)\}$,
       where $N = a + b$; of which a examples have $y_i = +1$
       and b examples have $y_i = -1$;
    (2) The number M of weak classifiers to be combined;
1. (Initialization)
    $w_i^{(0)} = \frac{1}{2a}$ for those examples with $y_i = +1$ or
    $w_i^{(0)} = \frac{1}{2b}$ for those examples with $y_i = -1$.
2. (Forward Inclusion)
    For $m = 1, \dots, M$:
    (1) Choose optimal h_m to minimize weighted error;
    (2) Choose α_m according to (9.9);
    (3) Update $w_i^{(m)} \leftarrow w_i^{(m)} \exp[-y_i h_m(x_i)]$, and
       normalize to $\sum_i w_i^{(m)} = 1$;
3. (Output)
    $H(x) = \text{sign}[\sum_{m=1}^{M} h_m(x)]$.

Figure 9.8. AdaBoost learning algorithm.

premise offered by the sequential procedure can be broken down when the assumption is violated, that is, when the performance criterion function is nonmonotonic. This is the first topic to be dealt with in this chapter.

Floating search [97] is a sequential feature-selection procedure with backtracking, aimed to deal with nonmonotonic criterion functions for feature selection. A straight sequential selection method such as SFS or sequential backward search (SBS) adds or deletes one feature at a time. To make this work well, the monotonicity property has to be satisfied by the performance criterion function. Feature selection with a nonmonotonic criterion may be dealt with by using a more sophisticated technique, called plus-ℓ-minus-r, which adds or deletes ℓ features and then backtracks r steps [113, 59].

The sequential floating search method [97] allows the number of backtracking steps to be controlled instead of being fixed beforehand. Specifically, it adds or deletes $\ell = 1$ feature and then backtracks r steps, where r depends on the current situation. This helps handle the limitations associated with nonmonotonicity. Improvement on the quality of selected features is gained with the cost of increased computation due to the extended search. The sequential forward floating selection (SFFS) algorithm performs very well in several applications [97, 54]. The idea of floating search is further developed in [112] by allowing more flexibility for the determination of ℓ.

Let $\mathcal{H}_M = \{h_1, \ldots, h_M\}$ be the so-far-best set of M weak classifiers; $J(\mathcal{H}_M)$ be the criterion that measures the overall cost of the classification function $H_M(x) = \sum_{m=1}^{M} h_m(x)$ build on $\mathcal{H}_M$; $J_m^{\min}$ be the minimum cost achieved so far with a linear combination of m weak classifiers for $m = 1, \ldots, M_{\max}$ (which are initially set to a large value before the iteration starts). As shown in Figure 9.9, the FloatBoost learning procedure involves

0. (Input)
    (1) Training examples $\mathcal{Z} = \{(\mathbf{x}_1, y_1), \ldots, (\mathbf{x}_N, y_N)\}$,
        where $N = a + b$; of which a examples have
        $y_i = +1$ and b examples have $y_i = -1$;
    (2) The maximum number $M_{\max}$ of weak classifiers;
    (3) The cost function $J(\mathcal{H}_M)$ (e.g., error rate made by H_M), and
        the maximum acceptable cost J^*.
1. (Initialization)
    (1) $w_i^{(0)} = \frac{1}{2a}$ for those examples with $y_i = +1$ or
        $w_i^{(0)} = \frac{1}{2b}$ for those examples with $y_i = -1$;
    (2) $J_m^{\min}$ =max-value (for $m = 1, \ldots, M_{\max}$),
        $M = 0$, $\mathcal{H}_0 = \{\}$.
2. (Forward Inclusion)
    (1) $M \leftarrow M + 1$;
    (2) Choose h_M according to (9.8);
    (3) Update $w_i^{(M)} \leftarrow w_i^{(M-1)} \exp[-y_i h_M(x_i)]$,
        normalize to $\sum_i w_i^{(M)} = 1$;
    (4) $\mathcal{H}_M = \mathcal{H}_{M-1} \cup \{h_M\}$;
        If $J_M^{\min} > J(\mathcal{H}_M)$, then $J_M^{\min} = J(\mathcal{H}_M)$;
3. (Conditional Exclusion)
    (1) $h' = \arg\min_{h \in \mathcal{H}_M} J(\mathcal{H}_M - h)$;
    (2) If $J(\mathcal{H}_M - h') < J_{M-1}^{\min}$, then
        (a) $\mathcal{H}_{M-1} = \mathcal{H}_M - h'$;
            $J_{M-1}^{\min} = J(\mathcal{H}_M - h')$; $M = M - 1$;
        (b) if $h' = h_{m'}$, then
            re-calculate $w_i^{(j)}$ and h_j for $j = m', \ldots, M$;
        (c) goto 3.(1);
    (3) else
        (a) if $M = M_{\max}$ or $J(\mathcal{H}_M) < J^*$, then goto 4;
        (b) goto 2.(1);
4. (Output)
    $H(x) = \text{sign}[\sum_{m=1}^{M} h_m(x)]$.

Figure 9.9. FloatBoost algorithm.

training inputs, initialization, forward inclusion, conditional exclusion, and output.

In step 2 (forward inclusion), the currently most significant weak classifier is added one at a time, which is the same as in AdaBoost. In step 3 (conditional exclusion), FloatBoost removes the least significant weak classifier from $\mathcal{H}_M$, subject to the condition that the removal leads to a lower cost than $J_{M-1}^{\min}$. If the removed weak classifier was the m'-th in $\mathcal{H}_M$, then $h_{m'}, \ldots, h_M$ will be relearned. These are repeated until no more removals can be done.

For face detection, the acceptable cost J^* is the maximum allowable risk, which can be defined as a weighted sum of missing rate and false-alarm rate. The algorithm terminates when the cost is below J^* or the maximum number M of weak classifiers is reached.

FloatBoost usually needs fewer weak classifiers than AdaBoost to achieve a given objective J^*. We have two options with such a result: (1) Use the FloatBoost-trained strong classifier with its fewer weak classifiers to achieve similar performance, as can be done by an AdaBoost-trained classifier with more weak classifiers. (2) Continue FloatBoost learning to add more weak classifiers even if the performance on the training data does not increase. The reason for (2) is that even if the performance does not improve on the training data, adding more weak classifiers may lead to improvements on test data [102].

Cascade of strong classifiers

A boosted strong classifier effectively eliminates a large portion of nonface subwindows while maintaining a high detection rate. Nonetheless, such a single strong classifier may not meet the requirement of an extremely low false-alarm rate, such as 10^{-6} or even lower. A solution is to arbitrate between several detectors (strong classifier) [100], for instance, by using an AND operation.

Viola and Jones [128, 127] further extend this idea by training a serial cascade of a number of strong classifiers, as illustrated in Figure 9.10. A strong classifier is trained using bootstrapped nonface examples that pass through the previously trained cascade. Usually, 10 to 20 strong classifiers are cascaded. In detection, subwindows that fail to pass a strong classifier will not be further processed by the subsequent strong classifiers. This strategy can significantly speed up the detection and reduce false alarms, but these are achieved with a little sacrifice of the detection rate.

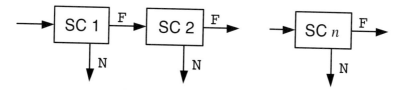

Figure 9.10. A cascade of n strong classifiers (SCs). The input is a subwindow x. It is sent to the next SC for further classification, only it has passed all the previous SCs as face (F) pattern; otherwise, it exits as nonface (N). x is finally considered to be a face when it passes all the n SCs.

9.2.5 Postprocessing

A face may be detected several times at close locations or multiple scales. False alarms may also occur but usually with less consistency than multiple face detections. The number of the multiple detections at a close location can be used as an effective indication for the existence of a face at the location. The observation leads to a heuristic for resolving the ambiguity caused by multiple detections and eliminates many false detections. A detection is confirmed if the number of multiple detections is large enough, and multiple detections are merged into one consistent detect. This is practiced in most face detection systems, for example, [114, 100]. Figure 9.11 gives an illustration. The image on the left shows a typical output of initial detection, where the face is detected four times with four false alarms on the cloth. On the right is the final result after merging. We can see that multiple detections are merged and false alarms eliminated. Figures 9.12 and 9.13 show some typical detection examples.

9.2.6 Evaluation

AdaBoost learning-based face detection methods have been the most effective of all methods developed so far. In terms of detection and false alarm rates, it is comparable to the neural network method of Henry Rowley [100]. But it is several times faster [128, 127].

Regarding the AdaBoost approach, the following conclusions can be made in terms of different feature sets, boosting algorithms, weak classifiers, subwindow sizes, and training-set sizes according to studies in [128, 127, 68, 71]:

- A set of Haar-like features, which can be computed very efficiently using the integral image, achieves true scale invariance and reduces the initial image processing required for object detection significantly. For the extended set of Haar features introduced in [71], the in-plane

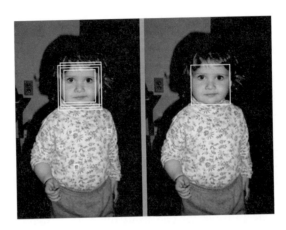

Figure 9.11. Merging multiple detections.

Figure 9.12. Results of face detection in gray images. Forty-two faces are detected, six faces are missing, and there are no false alarms.

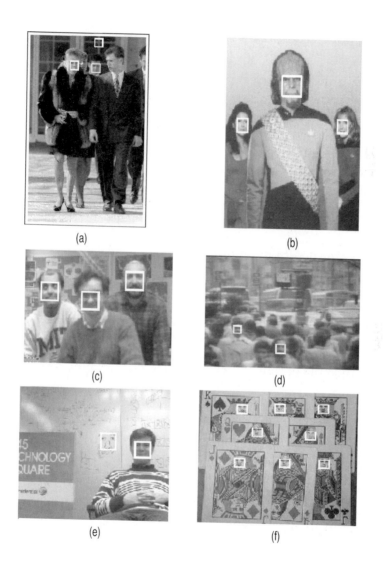

Figure 9.13. Results of face detection in gray images. Twenty-one faces are detected, with one face with large rotation missing and one false detect in the first image.

rotated features increased the accuracy, though frontal faces exhibited little diagonal structures.

– AdaBoost learning can select the best subset from a large feature set and construct a powerful, nonlinear classifier.

– The cascade structure significantly improves the speed and is effective in reducing false alarms, but with a little sacrifice of the detection rate [128, 127].

– FloatBoost effectively improves boosting-learning result [68]. It needs fewer weaker classifiers than AdaBoost to achieve a similar error rate or to achieve a lower error rate with the same number of weak classifiers. Such a performance improvement is achieved with the cost of longer training time, about five times longer.

– Gentle AdaBoost outperforms discrete and real AdaBoost [71].

– It is beneficial not just to use the simplest of all tree classifiers, or stumps, as the basis for the weak classifiers, but representationally more powerful classifiers such as small CART trees, which can model second-order and/or third-order dependencies.

– 20×20 is the optimal input pattern size for frontal face detection.

Face detection technologies are now mature enough to meet minimum needs of many practical applications. However, much work is needed before automated face detection can achieve performance comparable with the human. The Harr+AdaBoost approach is effective and efficient. However, the current form has almost reached its power limit. Within such a framework, possible improvements may be made by designing new sets of features that complement existing sets and adopting more advanced learning techniques that could lead to classifiers that have complex enough boundaries yet do not overfit.

9.3   Face Alignment

Both shape and texture (i.e., image pixels enclosed in the facial outline) provide important clues useful for characterizing the face [13]. Accurate extraction of features for the representation of faces in images offers advantages for many applications and is crucial for highly accurate face recognition and synthesis. The task of face alignment is to accurately locate facial features such as the eyes, nose, mouth, and outline, and normalize facial shape and texture.

The active shape model (ASM) [23] and active appearance model (AAM) [22, 29] are two popular models for the purpose of shape and appearance modeling and extraction. The standard ASM consists of two statistical models: (1) global shape model, which is derived from the landmarks in the object contour; (2) local appearance model, which is derived from the profiles perpendicular to the object contour around each landmark. ASM uses local models to find the candidate shape and the global model to constrain the searched shape. AAM uses subspace analysis techniques, PCA in particular, to model both shape variation and texture variation, and the correlations between them. In searching for a solution, it assumes linear relationships between appearance variation and texture variation and between texture variation and position variation, and it learns the two linear regression models from training data. This strategy is also developed in the active blob model of Sclaroff and Isidoro [108].

ASM and AAM can be expanded in several ways. The concept, originally proposed for the standard frontal view, can be extended to multiview faces either by using piecewise linear modeling [24] or nonlinear modeling [98]. Cootes and Taylor show that imposing constraints such as fixing eye locations can improve AAM search results [21]. Blanz and Vetter extended morphable models and AAM to model the relationships of 3D head geometry and facial appearance [16]. Li et al. [69] present a method for learning the 3D face-shape modeling from 2D images based on a shape-and-pose-free texture model. In Duta et al. [28], the shapes are automatically aligned using Procrustes analysis and clustered to obtain cluster prototypes and statistical information about intracluster shape variation. In Ginneken et al. [124], a K-nearest-neighbors classifier is used, and a set of features is selected for each landmark to build local models. Baker and colleagues [7] propose an efficient method called *inverse compositional algorithm* for alignment. Ahlberg [4] extends AAM to a parametric method called active appearance algorithm to extract positions parameterized by 3D rotation, 2D translation, scale, and six action units (controlling the mouth and the eyebrows). In the direct appearance model (DAM) [50, 66], shape is modeled as a linear function of texture. Using such an assumption, Yan et al. [131] propose texture-constrained ASM (TC-ASM), which has the advantage of ASM in having good localization accuracy and that of AAM in having insensitivity to initialization.

The following sections describe ASM, AAM, DAM, and TC-ASM. A training set of shape-texture pairs is assumed to be available and denoted as $\Omega = \{(S_0, T_0)\}$, where a *shape* $S_0 = ((x_1, y_1), \ldots, (x_K, y_K)) \in \mathbb{R}^{2K}$ is a sequence of K points in the 2D image plane, and a *texture* T_0 is the patch of pixel intensities enclosed by S_0. Let $\overline{S}$ be the mean shape of all the training shapes, as illustrated in Figure 9.14. All the shapes are aligned or warped to the tangent space of the mean shape $\overline{S}$. After that, the texture T_0 is warped correspondingly to $T \in \mathbb{R}^L$, where L is the number of pixels in the mean shape $\overline{S}$. The warping may be done by pixel-value interpolation, for example, using a triangulation or thin plate spline method.

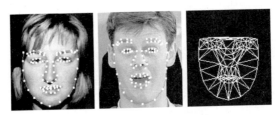

Figure 9.14. Two face instances labeled with 83 landmarks and the mesh of the mean shape.

9.3.1 Active shape model

In ASM, a shape is represented as a vector s in the low-dimensional shape subspace, denoted $\mathbb{S}_s$, in $\mathbb{R}^k$ spanned by the k $(< 2K)$ principal modes learned from the training shapes.

$$S = \overline{S} + \mathbf{U}s, \tag{9.12}$$

where $\mathbf{U}$ is the matrix consisting of k principal orthogonal modes of variation in $\{S_0\}$. Because the training shapes have been aligned to the tangent space of $\overline{S}$, the eigenvectors in $\mathbf{U}$ are orthogonal to the mean shape $\overline{S}$; that is, $\mathbf{U}^T \overline{S} = 0$, and the projection from S to s is

$$s = \mathbf{U}^T (S - \overline{S}) = \mathbf{U}^T S. \tag{9.13}$$

The local appearance models describing the typical images structure around each landmark are obtained from the sampled profiles perpendicular to the landmark contour. The first derivatives of the profiles are used to build these models. For a landmark, the mean profile and $\overline{g}$ and the covariance matrix $\mathbf{S}_g$ can be computed directly from the example profiles. The best candidate can by found by minimizing Mahalanobis distance between the candidate profile g and mean profile $\bar{g}$:

$$dist(g, \bar{g}) = (g - \bar{g})^T \mathbf{S}_g^{-1}(g - \bar{g}). \tag{9.14}$$

After relocating all the landmarks from local models, the resulting shape $\widehat{S}$ in the $\mathbb{S}_s$ can be derived from the likelihood distribution:

$$p(\widehat{S}|s) = \frac{\exp\{-E(\widehat{S}|s)\}}{Z}, \tag{9.15}$$

where Z is the normalizing constant, and the corresponding likelihood energy function can be defined as:

$$E(\widehat{S}|s) = \frac{\|\widehat{S} - S'\|}{2\sigma_1^2} + \sum_{i=1}^{k}\{\frac{\|s_i' - s_i\|^2}{2\epsilon_i}\}, \tag{9.16}$$

where σ_1 is a constant, ϵ_i is the ith largest eigenvalue of the covariance matrix of the training data $\{S_0\}$, $S' \in \mathbb{R}^{2K}$ is the projection of $\widehat{S}$ in $\mathbb{S}_s$, and s' is the corresponding shape parameters. The first term in (9.16) is the Euclidean distance of $\widehat{S}$ to the shape space $\mathbb{S}_s$, and the second term is the Mahalanobis distance between s' and s.

The maximum likelihood solution, $s_{ML} = \arg\max_{s \in S_s} p(\widehat{S}|s)$, is the projection of $\widehat{S}$ to the shape space $\mathbb{S}_s$:

$$s_{ML} = \mathbf{U}^T(\widehat{S} - \overline{S}). \tag{9.17}$$

Its corresponding shape in $\mathbb{R}^{2K}$ is

$$S = \overline{S} + \mathbf{U}s_{ML}. \tag{9.18}$$

9.3.2 Active appearance model

After aligning each training shape S_0 to the mean shape and warping the corresponding texture T_0 to T, the warped textures are aligned to the tangent space of the mean texture $\overline{T}$ by using an iterative approach [22]. The PCA model for the warped texture is obtained as

$$T = \overline{T} + \mathbf{V}t, \tag{9.19}$$

where $\mathbf{V}$ is the matrix consisting of ℓ principal orthogonal modes of variation in $\{T\}$, t is the vector of texture parameters. The projection from T to t is

$$t = \mathbf{V}^T(T - \overline{T}) = \mathbf{V}^T T. \tag{9.20}$$

By this, the L pixel values in the mean shape are represented as a point in the texture subspace $\mathbb{S}_t$ in $\mathbb{R}^\ell$.

The appearance of each example is a concatenated vector

$$A = \begin{pmatrix} \Lambda s \\ t \end{pmatrix}, \tag{9.21}$$

where Λ is a diagonal matrix of weights for the shape parameters, allowing for the difference in units between the shape and texture variation, typically defined as $r\mathbf{I}$. Again, by applying PCA on the set $\{A\}$, we get

$$A = \mathbf{W}a, \tag{9.22}$$

where $\mathbf{W}$ is the matrix consisting of principal orthogonal modes of the variation in $\{A \mid$ for all training samples$\}$. The appearance subspace $\mathbb{S}_a$ is modeled by

$$a = \mathbf{W}^T A. \tag{9.23}$$

The search for an AAM solution is guided by the following difference between the texture T_{im} in the image patch and the texture T_a reconstructed from the current appearance parameters:

$$\delta T = T_{im} - T_a. \tag{9.24}$$

More specifically, the search for a face in an image is guided by minimizing the norm $\|\delta T\|$. The AAM assumes that the appearance displacement δa and the position (including coordinates (x, y), scale s, and rotation parameter θ) displacement δp are linearly correlated to δT:

$$\delta a = \mathbf{A}_a \delta T, \tag{9.25}$$
$$\delta p = \mathbf{A}_p \delta T. \tag{9.26}$$

The prediction matrices $\mathbf{A}_a, \mathbf{A}_p$ are to be learned from the training data by using linear regression. In order to estimate $\mathbf{A}_a$, a is displaced systematically to induce $(\delta a, \delta T)$ pairs for each training image. Due to large consumption of memory required by the learning of $\mathbf{A}_a$ and $\mathbf{A}_p$, the learning has to be done with a small, limited set of $\{\delta a, \delta T\}$.

9.3.3   Modeling shape from texture

An analysis of mutual dependencies of shape, texture, and appearance parameters in the AAM subspace models shows that there exist admissible appearances that are not modeled and hence cannot be reached by AAM search processing [50]. Let us look into the relationship between shape and texture from an intuitive viewpoint. A texture (i.e., the patch of intensities) is enclosed by a shape (before aligning to the mean shape); the same shape can enclose different textures (i.e., configurations of pixel values). However, the reverse is not true: different shapes cannot enclose the same texture, so the mapping from the texture space to the shape space is many-to-one. The shape parameters should be determined completely by texture parameters but not vice versa.

Then, let us look further into the correlations or constraints between the linear subspaces $\mathbb{S}_s, \mathbb{S}_t$, and $\mathbb{S}_a$ in terms of their dimensionalities or ranks. We denote the rank of space $\mathbb{S}$ by $\dim(\mathbb{S})$. The following analysis is made in [50]:

1. When $\dim(\mathbb{S}_a) = \dim(\mathbb{S}_t) + \dim(\mathbb{S}_s)$, the shape and texture parameters are independent of each other and there exist no mutual constraints between the parameters s and t.

2. When $\dim(\mathbb{S}_t) < \dim(\mathbb{S}_a) < \dim(\mathbb{S}_t) + \dim(\mathbb{S}_s)$, not all the shape parameters are independent of the texture parameters. That is, one shape can correspond to more than one texture configuration, which seems intuitively correct.

3. We can also derive the relationship $\dim(\mathbb{S}_t) < \dim(\mathbb{S}_a)$ from (9.21) and (9.22) the formula

$$\mathbf{W}a = \begin{pmatrix} \Lambda s \\ t \end{pmatrix} \qquad (9.27)$$

   when that s contains some components that are independent of t.

4. However, in AAM, it is often the case that $\dim(\mathbb{S}_a) < \dim(\mathbb{S}_t)$ if the dimensionalities of $\mathbb{S}_a$ and $\mathbb{S}_t$ are chosen to retain, say, 98% of the total variations [22]. The consequence is that some admissible texture configurations cannot be seen in the appearance subspace because $\dim(\mathbb{S}_a) < \dim(\mathbb{S}_t)$, and therefore cannot be reached by the AAM search. This is a flaw of AAM's modeling of its appearance subspace.

From the above analysis, it is concluded in [50] that the ideal model should be such that $\dim(\mathbb{S}_a) = \dim(\mathbb{S}_t)$ and hence that s be completely linearly determinable by t. In other words, the shape should be linearly dependent on the texture so that $\dim(\mathbb{S}_t \cup \mathbb{S}_s) = \dim(\mathbb{S}_t)$.

Direct appearance models

A solution to this problem is made by assuming that the shape is linearly dependent on the texture [50, 66]:

$$s = \mathbf{R}t + \varepsilon, \tag{9.28}$$

where $\varepsilon = s - \mathbf{R}t$ is noise and $\mathbf{R}$ is a $k \times l$ projection matrix. Denoting the expectation by $E(\cdot)$, if all the elements in the variance matrix $E(\varepsilon\varepsilon^T)$ are small enough, the linear assumption made in (9.28) is approximately correct. This is true, as will be verified later by experiments. Define the objective cost function:

$$C(\mathbf{R}) = E(\varepsilon^T \varepsilon) = \text{trace}[E(\varepsilon\varepsilon^T)]. \tag{9.29}$$

$\mathbf{R}$ is learned from training example pairs $\{(s,t)\}$ by minimizing the above cost function. The optimal solution is

$$\mathbf{R}^* = E(st^T)[E(tt^T)]^{-1}. \tag{9.30}$$

The minimized cost is the trace of the following:

$$E(\varepsilon\varepsilon^T) = E(ss^T) - \mathbf{R}^* E(tt^T)\mathbf{R}^{*T}. \tag{9.31}$$

DAM [50, 66] assumes that the facial shape is a linear regression function of the facial texture and hence overcomes a defect in AAM; the texture information is used *directly* to predict the shape and to update the estimates of position and appearance (hence the name DAM). Also, DAM predicts the new face position and appearance based on the principal components of texture difference vectors instead of the raw vectors themselves, as in AAM. This cuts down the memory requirement to a large extent and further improves the convergence rate and accuracy.

Learning in DAM

The DAM consists of a shape model, two texture (original and residual) models, and two prediction (position and shape prediction) models. The shape and texture models and the position-prediction model (9.26) are built in the same way as in AAM. The residual texture model is built using the subspace analysis technique PCA. Abandoning AAM's crucial idea of combining shape and texture parameters into an appearance model, it predicts the shape parameters directly from the texture parameters. In the following, the last two models are demonstrated in detail.

Instead of using δT directly as in the AAM search (compare (9.26)), its principal components, denoted $\delta T'$, are used to predict the position displacement

$$\delta p = \mathbf{R}_p \delta T', \tag{9.32}$$

where $\mathbf{R}_p$ is the prediction matrix learned by using linear regression. To do this, samples of texture differences induced by small position displacements in each training image are collected, and PCA is performed to get the projection matrix $\mathbf{H}^T$. A texture difference is projected onto this subspace as

$$\delta T' = \mathbf{H}^T \delta T. \tag{9.33}$$

$\delta T'$ is normally about $1/4$ of δT in dimensionality. Results have shown that the use of δT instead of $\delta T'$, as in (9.32), makes the prediction more stable and more accurate.

Assume that a training set is given as $\mathbf{A} = \{(S_i, T_i^o)\}$, where a shape $S_i = ((x_1^i, y_1^i), \ldots, (x_K^i, y_K^i)) \in \mathbb{R}^{2K}$ is a sequence of K points in the 2D image plane, and a texture T_i^o is the patch of image pixels enclosed by S_i. The DAM learning consists of two parts: (1) learning $\mathbf{R}$, and (2) learning $\mathbf{H}$ and $\mathbf{R}_p$: (1) $\mathbf{R}$ is learned from the shape-texture pairs $\{s, t\}$ obtained from the landmarked images. (2) To learn $\mathbf{H}$ and $\mathbf{R}_p$, artificial training data is generated by perturbing the position parameters p around the landmark points to obtain $\{\delta p, \delta T\}$; then $\mathbf{H}$ is learned from $\{\delta T\}$ using PCA; $\delta T'$ is computed after that; and finally $\mathbf{R}_p$ is derived from $\{\delta p, \delta T'\}$.

The DAM regression in (9.32) requires much less memory than the AAM regression in (9.25); typically DAM needs only about $1/20$ of the memory required by AAM. For DAM, there are 200 training images, 4 parameters for the position $(x, y, \theta, scale)$, and 6 disturbances for each parameter to generate training data for the training $\mathbf{R}_p$. So, the size of training data for DAM is $200 \times 4 \times 6 = 4,800$. For AAM, there are 200 training images, 113 appearance parameters, and 4 disturbances for each parameter to generate training data for training $\mathbf{A}_a$. The size of training data for $\mathbf{A}_a$ is $200 \times 113 \times 4 = 90,400$. Therefore, the size of training data for AAM's prediction matrices is $90,400 + 4,800 = 95,200$, which is 19.83 times that for DAM. On a PC, for example, the memory capacity for AAM training with 200 images would allow DAM training with 3,966 images.

DAM Search

The DAM prediction models leads to the following search procedure: The DAM search starts with the mean shape and the texture of the input image enclosed by the mean shape, at a given initial position p_0. The texture difference δT is computed from the current shape patch at the current position, and its principal components are used to predict and update p and s using the DAM linear models described above. Note that the p can be computed from δT in one step as $\delta p = \mathbf{R}_T \delta T$, where $\mathbf{R}_T = \mathbf{R}_p \mathbf{H}^T$, instead of two steps as in (9.32) and (9.33). If $\|\delta T\|$ calculated using the new appearance at the position is smaller than the old one, the new appearance and position are accepted; otherwise, the position is updated by amount $\kappa \delta p$ with varied κ values. The search algorithm is summarized below:

1. Initialize the position parameters p_0 (with a given pose); set shape parameters $s_0 = 0$;

2. Get texture T_{im} from the current position, project it into the texture subspace $\mathbb{S}_t$ as t, reconstruct the texture T_a, and compute texture difference $\delta T_0 = T_{im} - T_a$ and the energy $E_0 = \|\delta T_0\|^2$;

3. Compute $\delta T' = \mathbf{H}^T \delta T$, get the position displacement $\delta p = \mathbf{R}_p \delta T'$;

4. Set step size $\kappa = 1$;

5. Update $p = p_0 - \kappa \delta p$, $s = \mathbf{R}t$;

6. Compute the difference texture δT using the new shape at the new position, and its energy $E = \|\delta T\|^2$;

7. If $|E - E_0| < \epsilon$, the algorithm is converged; exit;

8. If $E < E_0$, then let $p_0 = p$, $s_0 = s$, $\delta T_0 = \delta T$, $E_0 = E$, go to 3;

9. Change κ to the next number in $\{1.5, 0.5, 0.25, 0.125, \ldots, \}$, go to 5;

Texture-constrained active shape model

TC-ASM [131] imposes the linear relationship of DAM to improve ASM search. The motivation is the following: ASM has better accuracy in shape localization than AAM when the initial shape is placed close enough to the true shape, whereas the latter model incorporates information about texture enclosed in the shape and hence yields lower texture reconstruction error.

However, ASM uses constraints near the shape only, without a global optimality criterion, and therefore the solution is sensitive to the initial shape position. In AAM, the solution-finding process is based on the linear relationship between the variation of the position and the texture reconstruct error. The reconstruct error δT is influenced very much by the illumination. Since δT is orthogonal to $\mathbb{S}_t$ (projected back to $\mathbb{R}^L$) and $\dim(\mathbb{S}_t) \ll \dim(T)$, the dimension of the space $\{\delta T\}$ is very high, and it is hard to train the regression matrix $\mathbf{A}_a, \mathbf{A}_p$ and the prediction of the variance of position can be subject to significant errors. Also, it is time and memory consuming. TC-ASM is aimed to overcome the above problems.

TC-ASM consists of a shape model, a global texture model, K local appearance (likelihood) models, and a texture-shape model. The shape and texture model are built based on PCA in the same way as in AAM. The K local appearance models are built from the sampled profiles around each landmark, and the texture-shape model is learned from the pairs $\{(s,t)\}$ obtained from the landmarked images.

It assumes a linear relationship between shape and texture, as in DAM. The shape ($\in \mathbb{S}_s$) with a given texture ($\in \mathbb{S}_t$) can be approximately modeled by Gaussian distribution:

$$s \sim N(s_t, \Lambda'), \tag{9.34}$$

where

$$s_t = \mathbf{R}t + \varepsilon, \tag{9.35}$$

where ε is the noise, and Λ' can be defined as diagonal matrix for simplicity.

The matrix $\mathbf{R}$ can be precomputed from the training pairs $\{(s,t)\}$ obtained from the labeled images using the least-squares method. The matrix $\mathbf{R}$ will map the texture to the mean shape of all shapes with the same texture, that is, $E(s|$ shape S with parameter s encloses the given texture t), and it's right to be the expectation of the distribution, i.e., s_t. We denote

$$\mathbf{R}_{VT} = \mathbf{R}V^T. \tag{9.36}$$

From (9.20) and (9.35), one can get

$$s_t = \mathbf{R}_{VT}(T - \overline{T}). \tag{9.37}$$

Then, the prior conditional distribution of $s \in \mathbb{S}_s$ for a given s_t can be represented as

$$p(s|s_t) = \frac{\exp\{-E(s|s_t)\}}{Z'}, \tag{9.38}$$

where Z' is the normalizing constant, and the corresponding energy function is

$$E(s|s_t) = \sum_{i=1}^{k} \frac{\|s_i - s_{ti}\|^2}{2\epsilon_i'}. \tag{9.39}$$

TC-ASM search is formulated in the Bayesian framework. Intuitively, one could assume $\epsilon' = \beta\epsilon$ or $\epsilon' = \sigma^2 I$, where $\epsilon = (\epsilon_1, ..., \epsilon_k)^T$ is the k largest eigenvalues of the covariance matrix of the training data $\{S_0\}$ and β is a constant. The prior distribution of (9.38) and the likelihood distribution in (9.16) are used.

Let $\widehat{S}$ be the shape derived from local appearance models and s_t be the predicted shape from the global texture constrains. $\widehat{S}$ and s_t can be considered independent of each other; that is,

$$p(\widehat{S}, s_t) = p(\widehat{S})p(s_t) \text{ and } p(\widehat{S}|s_t) = p(\widehat{S}). \tag{9.40}$$

Shape extraction is posed as a problem of maximum a posterior (MAP) estimation. The posteriori distribution of $s \in \mathbb{S}_s$ is

$$p(s|\widehat{S}, s_t) = \frac{p(\widehat{S}|s, s_t)p(s, s_t)}{p(\widehat{S}, s_t)} \tag{9.41}$$

$$= \frac{p(\widehat{S}|s)p(s|s_t)}{p(\widehat{S})}. \tag{9.42}$$

The corresponding energy is

$$E(s|\widehat{S}, s_t) = E(\widehat{S}|s) + E(s|s_t), \tag{9.43}$$

and the MAP estimation can be defined as

$$s_{MAP} = arg \min_{s \in S_s} E(s|\widehat{S}, s_t). \tag{9.44}$$

The relationship between $\widehat{S}$, s_t and the MAP estimation s can be represented as in Figure 9.15, where the shape space, spanned by $\{e_1, e_2\}$, is assumed to be 2D.

Unlike ASM, TC-ASM uses the additional prior constrains between the shape and the texture, and this helps to avoid stagnating at the local minimum location in the shape search and tends to drive the shape to a more reasonable position. The shape is driven by the local appearance model instead of regression prediction as in AAM, so TC-ASM is more robust than

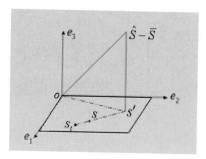

Figure 9.15. The relationship between $\widehat{S}$ (found from local appearance models), s_t (derived from the enclosed texture), and the MAP estimation $s \in \mathbb{R}^k$.

AAM to illumination variation. Unlike AAM, which must manually generate huge numbers of samples for training the regression matrix, TC-ASM has much lower memory consumption for its training.

The search in TC-ASM contains three main steps: (1) Search the candidate shape $\widehat{S}$ using local appearance models. (2) Warp the texture enclosed by the projection of $\widehat{S}$ in $\mathbb{S}_s$ and compute s_t using the texture-shape matrix **R**. (3) Make the MAP estimation from $\widehat{S}$ and s_t, and go to step (1) unless more than $K * \theta(\ 0 < \theta < 1)$ points in $\widehat{S}$ are converged or N_{max} iterations have been done. A multiresolution pyramid structure is used to improve speed and accuracy.

9.3.4   Dealing with head pose

We illustrate by using multiview DAM, a view-based approach with DAM as the base algorithm. The whole range of views from frontal to side views are partitioned into several subranges, and one DAM model is trained to represent the shape and texture for each subrange. Which view DAM model to use may be decided by using some pose estimate for static images. In the case of face alignment from video, the previous view plus the two neighboring view DAM models may be attempted, and the final result is chosen to be the one with the minimum texture residual error.

The full range of face poses is divided into five view subranges: $[-90°, 3-55°]$, $[-55°, -15°]$, $[-15°, 15°]$, $[15°, 55°]$, and $[55°, 90°]$, with $0°$ being the frontal view. The landmarks for frontal, half-side, and full-side view faces are illustrated in Figure 9.16. The dimensions of shape and texture vectors before and after the PCA dimension reductions are shown in Table 9.1, where the dimensions after PCA are chosen to be such that 98% of the corresponding total energies are retained. The texture appearances due to

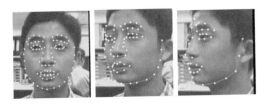

Figure 9.16. Frontal, half-side, and full-side view faces and the labeled landmark points.

respective variations in the first three principal components of texture are demonstrated in Figure 9.17.

The left-side models and right-side models are reflections of each other, so models for views on one side, plus the center view, need be trained—for example, $[-15°, 15°]$, $[15°, 55°]$, and $[55°, 90°]$ for the five models.

Table 9.1. Dimensionalities of shape and texture variations for face data. Numbered columns: (1) Number of landmark points. (2) Dimension of shape space $\mathbb{S}_s$. (3) Number of pixel points in the mean shape. (4) Dimension of texture space $\mathbb{S}_t$. (5) Dimension of texture variation space $(\delta T')$.

View	1	2	3	4	5
Frontal	87	69	3185	144	878
Half-side	65	42	3155	144	1108
Full-side	38	38	2589	109	266

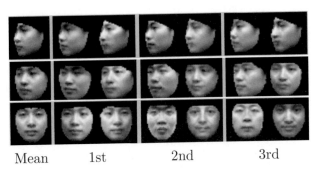

Mean   1st   2nd   3rd

Figure 9.17. Texture and shape variations due to variations in the first three principal components of the texture (the shapes change in accordance with $s = \mathbf{R}t$) for full-side $(\pm 1\sigma)$, half-side $(\pm 2\sigma)$, and frontal $(\pm 3\sigma)$ views.

Table 9.2. Comparisons of DAM, DAM', and AAM in terms of errors in estimated texture (appearance) parameters δT and position δp and convergence rates for the training images (first block of three rows) and test images (second block).

	$E(\|\delta T\|^2)$	std($\|\delta T\|^2$)	$E(\|\delta p\|)$	std($\|\delta p\|$)	cvg rate
DAM	0.156572	0.065024	0.986815	0.283375	100%
DAM'	0.155651	0.058994	0.963054	0.292493	100%
AAM	0.712095	0.642727	2.095902	1.221458	70%
DAM	1.114020	4.748753	2.942606	2.023033	85%
DAM'	1.180690	5.062784	3.034340	2.398411	80%
AAM	2.508195	5.841266	4.253023	5.118888	62%

9.3.5   Evaluation

DAM

Table 9.2 compares DAM and AAM in terms of the quality of position and texture parameter estimates [50] and the convergence rates. The effect of using $\delta T'$ instead of δT is demonstrated through DAM', which is DAM minus the PCA subspace modeling of δT. The initial position is a shift from the true position by $dx = 6, dy = 6$. The $\|\delta p\|$ is calculated for each image as the averaged distance between corresponding points in the two shapes, and therefore it is also a measure of difference in shape. The convergence is judged by the satisfaction of two conditions: $\|\delta T\|^2 < 0.5$ and $\|\delta p\| < 3$.

Figure 9.18 demonstrates scenarios of how DAM converges for faces of different pose [66]. By looking at $\|\delta T\|$ (compare (9.24)) as a function of iterate number, and by looking at the percentage of images whose texture reconstruction error δT is smaller than 0.2, it is concluded that DAM has

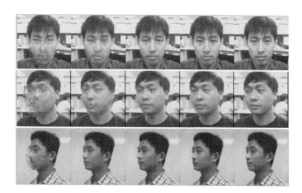

Figure 9.18. DAM-aligned faces (from left to right) at the 0th, 5th, 10th, and 15th iterations, and the original images for (top-bottom) frontal, half-side, and full-side view faces.

faster convergence rate and smaller error than AAM. When the face undergoes large variation due to stretch in either the x or y direction, the model fitting can be improved by allowing different scales in the two directions. This is done by splitting the scale parameter into two: s_x and s_y.

The DAM search is fairly fast. It takes on average 39 ms per iteration for frontal and half-side view faces, and 24 ms for full-side view faces in an image of size 320×240 pixels. Every view model takes about 10 iterations to converge. If three view models are searched with per face, as is done with image sequences from video, the algorithm takes about one second to find the best face alignment.

TC-ASM

Experiments are performed in both training data and testing data, and results are compared with ASM and AAM. A data set containing 700 face images with different illumination conditions and expressions selected from the AR database [82] is used in our experiments. They are all in frontal view with out-of-plane rotation within $[-10°, 10°]$. Eighty-three landmark points of the face are labeled manually. Six hundred images are randomly selected for the training set and the other 100 for testing.

TC-ASM is compared with ASM and AAM using the same data sets. The shape vector in the ASM shape space is 88-dimensional. The texture vector in the AAM texture space is 393-dimensional. The concatenated vector of $88 + 393$ dimensions is reduced with the parameter $r = 13.77$ to a 277-dimensional vector, which retains 98% of the total variation. Two types of experiments are presented: (1) comparison of the position accuracy and (2) comparison of the texture reconstruct error.

Position Accuracy

Consider the displacement D, i.e., the point-point Euclidean distance (in pixels) between the found shape and the manually labeled shape. One measure is the percentage of resulting shapes whose displacements are smaller than a given bound, given an initial displacement condition. Statistics calculated from 100 test images with different initial positions show that TC-ASM significantly improves the accuracy with all different initial conditions [131]. Stabilities of ASM and TC-ASM can be compared based on the average standard deviation of the shape results obtained with different initial positions deviated from the ground truth, say, by approximately 20 pixels. Results show TC-ASM is more stable than ASM to initialization.

Texture Reconstruction

The global texture constraints used in TC-ASM can improve the accuracy of texture matching in ASM. Results from [131] suggest that the texture accuracy of TC-ASM is close to that of AAM, while its position accuracy is much better than AAM's. Figure 9.19 compares the sensitivity of AAM and TC-ASM to illumination. While AAM is more likely to result in incorrect solution, TC-ASM is relatively robust to the illumination variation.

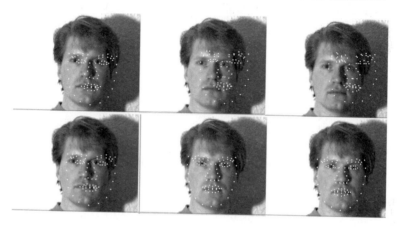

Figure 9.19. Sensitivity of AAM (upper) and TC-ASM (lower) to illumination condition not seen in the training set. From left to right are the results obtained at the 0th, 2nd, and 10th iterations.

In terms of speed, the standard ASM is a fast algorithm, and TC-ASM is computationally more expensive but still much faster than AAM. The training of TC-ASM is very fast, which takes only about one-tenth the time of AAM training. In our experiment (600 training images, 83 landmarks, and P-III 700 computer), TC-ASM takes on average 32 ms per iteration, which is twice of ASM (16 ms), while only one-fifth of AAM (172 ms). The training time of AAM is more than two hours, while TC-ASM is only about 12 minutes.

9.4   Face Recognition

To date, the appearance-based learning framework has been most influential in the face recognition research. Under this framework, the face recognition problem can be stated as follows: Given a training set, $\mathcal{Z} = \{\mathcal{Z}_i\}_{i=1}^{C}$, containing C classes with each class $\mathcal{Z}_i = \{\mathbf{z}_{ij}\}_{j=1}^{C_i}$ consisting of a number of localized face images $\mathbf{z}_{ij}$, a total of $N = \sum_{i=1}^{C} C_i$ face images are available in the set. For computational convenience, each image is represented as a column vector of length $J (= I_w \times I_h)$ by lexicographic ordering of the pixel elements, that is, $\mathbf{z}_{ij} \in \mathbb{R}^J$, where $(I_w \times I_h)$ is the image size, and $\mathbb{R}^J$ denotes the J-dimensional real space. Taking as input such a set $\mathcal{Z}$, the objectives of appearance-based learning are (1) to find based on optimization of certain separability criteria a low-dimensional feature representation $\mathbf{y}_{ij} = \varphi(\mathbf{z}_{ij})$, $\mathbf{y}_{ij} \in \mathbb{R}^M$, and $M \ll J$, with enhanced discriminatory power; (2) to design based on the chosen representation a classifier, $h : \mathbb{R}^M \to \mathbb{Y} = \{1, 2, \cdots, C\}$, such that h will correctly classify unseen examples $\varphi(\mathbf{z}, y)$, where $y \in \mathbb{Y}$ is the class label of $\mathbf{z}$.

9.4.1 Preprocessing

It has been shown that irrelevant facial portions such as hair, neck, shoulder, and background often provide misleading information to the face recognition systems [20]. A normalization procedure is recommended in [95], using geometric locations of facial features found in face detection and alignment. The normalization sequence consists of four steps: (1) the raw images are translated, rotated, and scaled so that the centers of the eyes are placed on specific pixels, (2) a standard mask, as shown in Figure 9.20 (middle) is applied to remove the nonface portions, (3) histogram equalization is performed in the nonmasked facial pixels, and (4) face data are further normalized to have zero mean and unit standard deviation. Figure 9.20 (right) depicts some samples obtained after the preprocessing sequence was applied.

9.4.2 Feature extraction

The goal of feature extraction is to generate a low-dimensional feature representation intrinsic to face objects with good discriminatory power for pattern classification.

Figure 9.20. Left: Original samples in the FERET database [3]. Middle: The standard mask. Right: The samples after the preprocessing sequence.

PCA subspace

In the statistical pattern recognition literature, PCA [55] is one of the most widely used tools for data reduction and feature extraction. The well-known face recognition method, eigenfaces [122] built on the PCA technique, has proven to be very successful. In the eigenfaces method [122], the PCA is applied to the training set $\mathcal{Z}$ to find the N eigenvectors (with nonzero eigenvalues) of the set's covariance matrix,

$$\mathbf{S}_{cov} = \frac{1}{N} \sum_{i=1}^{C} \sum_{j=1}^{C_i} (\mathbf{z}_{ij} - \bar{\mathbf{z}})(\mathbf{z}_{ij} - \bar{\mathbf{z}})^T, \qquad (9.45)$$

where $\bar{\mathbf{z}} = \frac{1}{N} \sum_{i=1}^{C} \sum_{j=1}^{C_i} \mathbf{z}_{ij}$ is the average of the ensemble. The eigenfaces are the first $M(\leq N)$ eigenvectors (denoted as Ψ_{ef}) corresponding to the largest eigenvalues, and they form a low-dimensional subspace, called *face space* where most energies of the face manifold are supposed to lie. Figure 9.21 (row 1) shows the first six eigenfaces, which appear, some re-

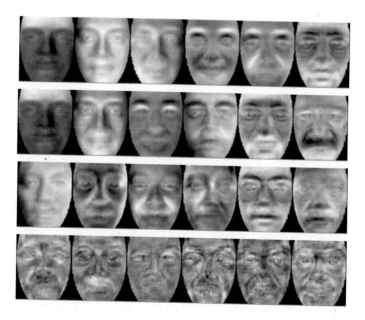

Figure 9.21. Visualization of four types of basis vectors obtained from a normalized subset of the FERET database. Row 1: The first six PCA bases. Row 2: The first six PCA bases for the extrapersonal variations. Row 3: The first six PCA bases for the intrapersonal variations. Row 4: The first six LDA bases.

searchers say, as ghostly faces. Transforming to the M-dimensional face space is a simple linear mapping: $\mathbf{y}_{ij} = \Psi_{ef}^T(\mathbf{z}_{ij} - \bar{\mathbf{z}})$, where the basis vectors Ψ_{ef} are orthonormal. The subsequent classification of face patterns can be performed in the face space using any classifier.

Dual PCA subspaces

The eigenfaces method is built on a single PCA, which suffers from a major drawback; that is, it cannot explicitly account for the difference between two types of face pattern variations key to the face recognition task: between-class variations and within-class variations. To this end, Moghaddam et al. [86] proposed a probabilistic subspace method, also known as dual eigenfaces. In the method, the distribution of face pattern is modeled by the intensity difference between two face images, $\Delta = \mathbf{z}_1 - \mathbf{z}_2$. The difference Δ can be contributed by two distinct and mutually exclusive classes of variations: intrapersonal variations Ω_I and extrapersonal variations Ω_E. In this way, the C-class face recognition task is translated to a binary pattern classification problem. Each class of variations can be modeled by a high-dimensional Gaussian distribution, $P(\Delta|\Omega)$. Since most energies of the distribution are assumed to exist in a low-dimensional PCA subspace, it is shown in [86] that $P(\Delta|\Omega)$ can be approximately estimated by

$$P(\Delta|\Omega) = P(\mathbf{z}_1 - \mathbf{z}_2|\Omega) = \exp(-\|\mathbf{y}_1 - \mathbf{y}_2\|^2/2)/\beta(\Omega), \qquad (9.46)$$

where $\mathbf{y}_1$ and $\mathbf{y}_2$ are the projections of $\mathbf{z}_1$ and $\mathbf{z}_2$ in the PCA subspace, and $\beta(\Omega)$ is a normalization constant for a given Ω. Any two images can be determined if they come from the same person by comparing the two likelihoods, $P(\Delta|\Omega_I)$ and $P(\Delta|\Omega_E)$, based on the *maximum-likelihood* (ML) classification rule. It is commonly believed that the extrapersonal PCA subspace, as shown in Figure 9.21 (row 2) represents more representative variations, such as those captured by the standard eigenfaces method, whereas the intrapersonal PCA subspace shown in Figure 9.21 (row 3) accounts for variations due mostly to changes in expression. Also, it is not difficult to see that the two Gaussian covariance matrices in $P(\Delta|\Omega_I)$ and $P(\Delta|\Omega_E)$ are equivalent to the within-class and between-class scatter matrices respectively of linear discriminant analysis (LDA), mentioned later. Thus, the dual eigenfaces method can be regarded as a quadratic extension of the so-called Fisherfaces method [12].

Other PCA extensions

The PCA-based methods are simple in theory, but they started the era of the appearance-based approach to visual object recognition [121]. In addition to the dual eigenfaces method, numerous extensions or variants of the eigenfaces method have been developed for almost every area of face research. For example, a multiple-observer eigenfaces technique is presented to deal with view-based face recognition in [93]. Moghaddam and Pentland derived two distance metrics, called *distance from face space* (DFFS) and *distance in face space* (DIFS), by performing density estimation in the original image space using Gaussian models for visual object representation [87]. Sung and Poggio built six face spaces and six nonface spaces, extracted the DFFS and DIFS of the input query image in each face/nonface space, and then fed them to a multilayer perceptron for face detection [114]. Tipping and Bishop [118] presented a *probabilistic PCA* (PPCA), which connects the conventional PCA to a probability density. This results in some additional practical advantages. For example, in classification, PPCA could be used to model class-conditional densities, thereby allowing the posterior probabilities of class membership to be computed. As another example, the single PPCA model could be extended to a mixture of such models. Due to its huge influences, the eigenfaces [122] was named the "most influential paper of the decade" by Machine Vision Applications in 2000.

ICA-based subspace methods

PCA is based on Gaussian models; that is, the found principal components are assumed to be subjected to independent Gaussian distributions. It is well known that the Gaussian distribution depends on only the first- and second-order statistical dependencies, such as the pairwise relationships between pixels. However, for complex objects such as human faces, much of the important discriminant information, such as phase spectrum, may be contained in the high-order relationships among pixels [9]. Independent component analysis (ICA) [61], a generalization of PCA, is one such method that can separate the high-order statistics of the input images in addition to the second-order statistic.

The goal of ICA is to search for a linear, nonorthogonal transformation $\mathbf{B} = [\mathbf{b}_1, \cdots, \mathbf{b}_M]$ to express a set of random variables $\mathbf{z}$ as linear combinations of statistically independent source random variables $\mathbf{y} = [\mathbf{y}_1, \cdots, \mathbf{y}_M]$,

$$\mathbf{z} \approx \sum_{m=1}^{M} \mathbf{b}_m \mathbf{y}_m = \mathbf{B}\mathbf{y}, \quad \mathbf{B}\mathbf{B}^T \neq \mathbf{I}, \quad p(\mathbf{y}) = \prod_{m=1}^{M} p(\mathbf{y}_m). \qquad (9.47)$$

These source random variables $\{\mathbf{y}_m\}_{m=1}^M$ are assumed to be subjected to non-Gaussian, such as heavy-tailed distributions. Compared to PCA, these characteristics of ICA often lead to a superior feature representation in terms of best fitting the input data, such as in the case shown in Figure 9.22 (left). There are several approaches for performing ICA, such as minimum mutual information, maximum neg-entropy, and maximum likelihood, and reviews can be found in [61, 43]. Recently, ICA has been widely attempted in face recognition studies such as [72, 9], where better performance than PCA-based methods was reported. Also, some ICA extensions like ISA [52] and TICA [53] have been shown to be particularly effective in view-based clustering of unlabeled face images [65].

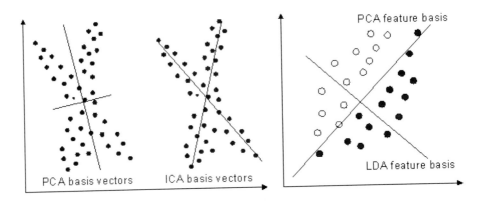

Figure 9.22. Left: PCA basis vs. ICA basis. Right: PCA basis vs. LDA basis.

LDA-based subspace methods

Linear discriminant analysis (LDA) [35] is also a representative technique for data reduction and feature extraction. In contrast with PCA, LDA is a class-specific one that utilizes supervised learning to find a set of $M (\ll J)$ feature basis vectors, denoted as $\{\psi_m\}_{m=1}^M$, in such a way that the ratio of the between-class and within-class scatters of the training image set is maximized. The maximization is equivalent to solve the following eigenvalue problem,

$$\Psi = \arg\max_{\Psi} \frac{|\Psi^T \mathbf{S}_b \Psi|}{|\Psi^T \mathbf{S}_w \Psi|}, \quad \Psi = [\psi_1, \dots, \psi_M], \quad \psi_m \in \mathbb{R}^J, \qquad (9.48)$$

where $\mathbf{S}_b$ and $\mathbf{S}_w$ are the between-class and within-class scatter matrices, having the following expressions,

$$\mathbf{S}_b = \frac{1}{N} \sum_{i=1}^{C} C_i (\bar{\mathbf{z}}_i - \bar{\mathbf{z}})(\bar{\mathbf{z}}_i - \bar{\mathbf{z}})^T = \sum_{i=1}^{C} \Phi_{b,i} \Phi_{b,i}^T = \Phi_b \Phi_b^T, \qquad (9.49)$$

$$\mathbf{S}_w = \frac{1}{N} \sum_{i=1}^{C} \sum_{j=1}^{C_i} (\mathbf{z}_{ij} - \bar{\mathbf{z}}_i)(\mathbf{z}_{ij} - \bar{\mathbf{z}}_i)^T, \qquad (9.50)$$

where $\bar{\mathbf{z}}_i$ is the mean of class $\mathcal{Z}_i$, $\Phi_{b,i} = (C_i/N)^{1/2}(\bar{\mathbf{z}}_i - \bar{\mathbf{z}})$ and $\Phi_b = [\Phi_{b,1}, \cdots, \Phi_{b,c}]$. The LDA-based feature representation of an input image $\mathbf{z}$ can be obtained simply by a linear projection, $\mathbf{y} = \Psi^T \mathbf{z}$.

Figure 9.21 (row 4) visualizes the first six basis vectors $\{\psi_i\}_{i=1}^6$ obtained by using the LDA version of [79]. Comparing Figure 9.21 (rows 1–3) to Figure 9.21 (row 4), it can be seen that the eigenfaces look more like a real human face than do LDA basis vectors. This can be explained by the different learning criteria used in the two techniques. LDA optimizes the low-dimensional representation of the objects with focus on the most discriminant feature extraction, while PCA achieves simple object reconstruction in a least-square sense. The difference may lead to significantly different orientations of feature bases as shown in Figure 9.22 (right). As a consequence, when it comes to solving problems of pattern classification such as face recognition, the LDA-based feature representation is usually superior to the PCA-based one (see, e.g., [12, 19, 137]).

When $\mathbf{S}_w$ is nonsingular, the basis vectors Ψ sought in (9.48) correspond to the first M most significant eigenvectors of $(\mathbf{S}_w^{-1}\mathbf{S}_b)$. However, in the particular tasks of face recognition, data are highly dimensional, while the number of available training samples per subject is usually much smaller than the dimensionality of the sample space. For example, a canonical example used for recognition is a (112×92) face image, which exists in a $10,304$-dimensional real space. Nevertheless, the number (C_i) of examples per class available for learning is not more than 10 in most cases. This results in the so-called *small sample size* (SSS) problem, which is known to have significant influences on the design and performance of a statistical pattern recognition system. In the application of LDA to face recognition tasks, the SSS problem often gives rise to high variance in the sample-based estimation for the two scatter matrices, which are either ill-posed or poorly-posed.

There are two methods for tackling the problem. One is to apply linear algebra techniques to solve the numerical problem of inverting the singular within-class scatter matrix $\mathbf{S}_w$. For example, the pseudoinverse is utilized to complete the task in [116]. Also, small perturbation may be added to $\mathbf{S}_w$ so that it becomes nonsingular [49, 139]. The other method is a subspace approach, such as the one followed in the development of the Fisherfaces method [12], where PCA is first used as a preprocessing step to remove the nullspace of $\mathbf{S}_w$, and then LDA is performed in the lower dimensional PCA subspace. However, it should be noted at this point that the discarded nullspace may contain significant discriminatory information. To prevent this from happening, solutions without a separate PCA step, called direct LDA (D-LDA) methods, have been presented recently in [19, 137, 79].

The basic premise behind the D-LDA approach is that the nullspace of $\mathbf{S}_w$ may contain significant discriminant information if the projection of $\mathbf{S}_b$ is not zero in that direction, while no significant information, in terms of the maximization in (9.48), will be lost if the nullspace of $\mathbf{S}_b$ is discarded. Based on the theory, it can be concluded that the optimal discriminant features exist in the complement space of the nullspace of $\mathbf{S}_b$, which has $M = C - 1$ nonzero eigenvalues denoted as $\mathbf{v} = [v_1, \cdots, v_M]$. Let $\mathbf{U} = [\mathbf{u}_1, \cdots, \mathbf{u}_M]$ be the M eigenvectors of $\mathbf{S}_b$ corresponding to the M eigenvalues $\mathbf{v}$. The complement space is spanned by $\mathbf{U}$, which is further scaled by $\mathbf{H} = \mathbf{U}\Lambda_b^{-1/2}$ so as to have $\mathbf{H}^T\mathbf{S}_b\mathbf{H} = \mathbf{I}$, where $\Lambda_b = \mathrm{diag}(\mathbf{v})$, and $\mathbf{I}$ is the $(M \times M)$ identity matrix. The projection of $\mathbf{S}_w$ in the subspace spanned by $\mathbf{H}$, $\mathbf{H}^T\mathbf{S}_w\mathbf{H}$ is then estimated using sample analogs. However, when the number of training samples per class is very small, even the projection $\mathbf{H}^T\mathbf{S}_w\mathbf{H}$ is ill-posed or poorly posed. To this end, a modified optimization criterion represented as $\Psi = \arg\max_{\Psi} \frac{|\Psi^T\mathbf{S}_b\Psi|}{|\Psi^T\mathbf{S}_b\Psi + \Psi^T\mathbf{S}_w\Psi|}$ was introduced in the D-LDA of [79] (hereafter LD-LDA) instead of (9.48) used in the D-LDA of [137] (hereafter YD-LDA). As will be seen later, the modified criterion introduces a considerable degree of regularization, which helps to reduce the variance of the sample-based estimates in ill-posed or poorly posed situations. The detailed process to implement the LD-LDA method is depicted in Figure 9.23.

The classification performance of traditional LDA may be degraded by the fact that their separability criteria are not directly related to their classification accuracy in the output space [74]. To this end, an extension of LD-LDA, called DF-LDA, is also introduced in [79]. In the DF-LDA method,

---

Input: A training set $\mathcal{Z}$ with C classes: $\mathcal{Z} = \{\mathcal{Z}_i\}_{i=1}^C$, each class contains $\mathcal{Z}_i = \{\mathbf{z}_{ij}\}_{j=1}^{C_i}$ face images, where $\mathbf{z}_{ij} \in \mathbb{R}^J$.

Output: An M-dimensional LDA subspace spanned by Ψ, an $M \times J$ matrix with $M \ll J$.

Algorithm:

  Step 1. Find the eigenvectors of $\Phi_b^T \Phi_b$ with nonzero eigenvalues, and denote them as $\mathbf{E}_m = [e_1, \ldots, e_m]$, $m \leq C - 1$.

  Step 2. Calculate the first m most significant eigenvectors ($\mathbf{U}$) of $\mathbf{S}_b$ and their corresponding eigenvalues (Λ_b) by $\mathbf{U} = \Phi_b \mathbf{E}_m$ and $\Lambda_b = \mathbf{U}^T \mathbf{S}_b \mathbf{U}$.

  Step 3. Let $\mathbf{H} = \mathbf{U}\Lambda_b^{-1/2}$. Find eigenvectors of $\mathbf{H}^T(\mathbf{S}_b + \mathbf{S}_w)\mathbf{H}$, $\mathbf{P}$.

  Step 4. Choose the $M(\leq m)$ eigenvectors in $\mathbf{P}$ with the smallest eigenvalues. Let $\mathbf{P}_M$ and Λ_w be the chosen eigenvectors and their corresponding eigenvalues respectively.

  Step 5. Return $\Psi = \mathbf{H}\mathbf{P}_M\Lambda_w^{-1/2}$.

Figure 9.23. The pseudocode implementation of the LD-LDA method.

the output LDA subspace is carefully rotated and reoriented by an iterative weighting mechanism introduced into the between-class scatter matrix,

$$\mathbf{S}_{b,t} = \sum_{i=1}^{C} \sum_{j=1}^{C} \varpi_{ij,t} (\bar{\mathbf{z}}_i - \bar{\mathbf{z}}_j)(\bar{\mathbf{z}}_i - \bar{\mathbf{z}}_j)^T. \tag{9.51}$$

In each iteration, object classes that are closer together in the preceding output space, and thus can potentially result in misclassification, are more heavily weighted (through $\varpi_{ij,t}$) in the current input space. In this way, the overall separability of the output subspace is gradually enhanced, and different classes tend to be equally spaced after a few iterations.

Gabor feature representation

The appearance images of a complex visual object are composed of many local structures. The Gabor wavelets are particularly aggressive at capturing the features of these local structures corresponding to spatial frequency (scale), spatial localization, and orientation selectively [103]. Consequently, it is reasonably believed that the Gabor feature representation of face images is robust against variations due to illumination and expression changes [73].

The kernels of the 2D Gabor wavelets, also known as Gabor filters, have the following expression at the spatial position $\vec{x} = (x_1, x_2)$ [27],

$$G_{\mu,\nu}(\vec{x}) = \frac{\|\kappa_{\mu,\nu}\|^2}{\sigma^2} \exp(-\frac{\|\kappa_{\mu,\nu}\|^2 \|\vec{x}\|^2}{2\sigma^2}) [\exp(ik_{\mu,\nu}\vec{x}) - \exp(-\sigma^2/2)], \quad (9.52)$$

where μ and ν define the orientation and scale of the Gabor filter, which is a product of a Gaussian envelope and a complex plane wave with wave vector $\kappa_{u,v}$. The family of Gabor filters is self-similar, since all of them are generated from one filter, the mother wavelet, by scaling and rotation via the wave vector $k_{\mu,\nu}$. Figure 9.24 depicts 40 commonly used Gabor kernels generalized at five scales $\nu \in \mathbb{U} = \{0, 1, 2, 3, 4\}$ and eight orientations $\mu \in \mathbb{V} = \{0, 1, 2, 3, 4, 5, 6, 7\}$ with $\sigma = 2\pi$. For an input image $\mathbf{z}$, its 2D Gabor feature images can be extracted by convoluting $\mathbf{z}$ with a Gabor filter, $\mathbf{g}_{\mu,\nu} = \mathbf{z} * G_{\mu,\nu}$, where $*$ denotes the convolution operator. Thus, a total of $(|\mathbb{U}| \times |\mathbb{V}|)$ Gabor feature images $\mathbf{g}_{\mu,\nu}$ can be obtained, where $|\mathbb{U}|$ and $|\mathbb{V}|$ denote the sizes of $\mathbb{U}$ and $\mathbb{V}$ respectively.

In [73], all the Gabor images $\mathbf{g}_{\mu,\nu}$ are down-sampled and then concatenated to form an augmented Gabor feature vector representing the input face image. In the sequence, PCA or LDA is applied to the augmented Gabor feature vector for dimensionality reduction and further feature extraction before it is fed to a classifier. Also, it can be seen that due to the similarity between the Gabor filters, there is a great deal of redundancy in the overcomplete Gabor feature set. To this end, Wiskott et al. [130] proposed to

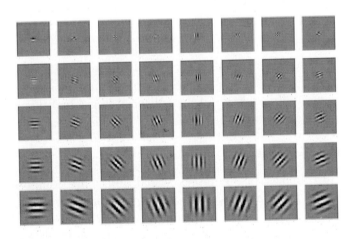

Figure 9.24. Gabor filters generalized at five scales $\nu \in \mathbb{U} = \{0, 1, 2, 3, 4\}$ and eight orientations $\mu \in \mathbb{V} = \{0, 1, 2, 3, 4, 5, 6, 7\}$ with $\sigma = 2\pi$.

utilize the Gabor features corresponding to some specific facial landmarks, called *Gabor jets*, instead of the holistic Gabor images. Based on the jet representation, the *elastic graph matching* (EGM) was then applied for landmark matching and face recognition. The Gabor-based EGM method was one of the top performers in the 1996/1997 FERET competition [95].

Mixture of linear subspaces

Although successful in many circumstances, linear appearance-based methods including the PCA- and LDA-based ones, may fail to deliver good performance when face patterns are subject to large variations in viewpoints, illumination or facial expression, which result in a highly nonconvex and complex distribution of face images. The limited success of these methods should be attributed to their linear nature. A cost-effective approach to address the nonconvex distribution is with a mixture of the linear models. The mixture- or ensemble-based approach embodies the principle of "divide and conquer," by which a complex face recognition problem is decomposed into a set of simpler ones, in each of which a locally linear pattern distribution can be generalized and dealt with by a relatively easy linear solution (see, for example, [93, 114, 38, 15, 75, 39, 77]).

From the designer's point of view, the central issue to the ensemble-based approach is to find an appropriate criterion to decompose the complex face manifold. Existing partition techniques, whether nonparametric clustering such as K-means [48] or model-based clustering such as EM [84], unanimously adopt *similarity criterion* based on which similar samples are within the same cluster and dissimilar samples are in different clusters. For example, in the view-based representation [93], every face pattern is manually assigned to one of several clusters according to its view angle with each cluster corresponding to a particular view. In the method considered in [114] and [38], the database partitions are automatically implemented using the K-means and EM clustering algorithms respectively. However, although such criterion may be optimal in the sense of approximating real face distribution for tasks such as face reconstruction, face pose estimation, and face detection, they may not be good for the recognition task considered in the section. It is not hard to see that from a classification point of view, the database partition criterion should be aimed to maximize the difference or separability between classes within each "divided" subset or cluster, which as a subproblem then can be relatively easy to "conquer" by a linear face recognition method.

With such a motivation, a novel method of *clustering* based on *separability criterion* (CSC) was introduced recently in [75]. Similar to LDA, the separability criterion is optimized in the CSC method by maximizing a widely used separability measure, the *between-class scatter* (BCS). Let $\mathcal{W}_k$ denote the kth cluster, $k = [1 \cdots K]$ with K: the number of clusters. Representing each class $\mathcal{Z}_i$ by its mean $\bar{\mathbf{z}}_i$, the total within-cluster BCS of the training set $\mathcal{Z}$ can be defined as

$$S_t = \sum_{k=1}^{K} \sum_{\bar{\mathbf{z}}_i \in \mathcal{W}_k} C_i \cdot (\bar{\mathbf{z}}_i - \mathbf{w}_k)^T (\bar{\mathbf{z}}_i - \mathbf{w}_k), \qquad (9.53)$$

where $\mathbf{w}_k = (\sum_{\bar{\mathbf{z}}_i \in \mathcal{W}_k} C_i \cdot \bar{\mathbf{z}}_i)/(\sum_{\bar{\mathbf{z}}_i \in \mathcal{W}_k} C_i)$ is the center of the cluster $\mathcal{W}_k$. (9.53) implies that a better class-separability intracluster is achieved if S_t has a larger value. The clustering algorithm maximizes S_t by iteratively reassigning those classes whose means have the minimal distances to their own cluster centers so that the separability between classes is enhanced gradually within each cluster. The maximization process can be implemented by the pseudocodes depicted in Figure 9.25.

Input: A training set $\mathcal{Z}$ with C classes: $\mathcal{Z} = \{\mathcal{Z}_i\}_{i=1}^{C}$, each class contains
$\mathcal{Z}_i = \{\mathbf{z}_{ij}\}_{j=1}^{C_i}$ face images.

Output: K maximally separable clusters $\{\mathcal{W}_k\}_{k=1}^{K}$, each class of images $\mathcal{Z}_i$
are assigned into one of K clusters.

Algorithm:
    Step 1. Calculate $\bar{\mathbf{z}}_i = \frac{1}{C_i} \sum_{j=1}^{C_i} \mathbf{z}_{ij}$ for class $\mathcal{Z}_i$, where $i = [1 \cdots C]$.
    Step 2. Randomly partition $\{\bar{\mathbf{z}}_i\}_{i=1}^{C}$ into K initial clusters $\{\mathcal{W}_k\}_{k=1}^{K}$,
        calculate their cluster center $\{\mathbf{w}_k\}_{k=1}^{K}$, and initial $\widehat{S}_t$ by (9.53).
    Step 3. Find $\widehat{\mathbf{z}}_k = \arg\min_{\bar{\mathbf{z}}_i \in \mathcal{W}_k} \left\{ (\bar{\mathbf{z}}_i - \mathbf{w}_k)^T (\bar{\mathbf{z}}_i - \mathbf{w}_k) \right\}$, $k = [1 \cdots K]$.
    Step 4. Compute distances of $\widehat{\mathbf{z}}_k$ to other cluster centers:
        $\mathbf{d}_{kh} = (\widehat{\mathbf{z}}_k - \mathbf{w}_h)^T (\widehat{\mathbf{z}}_k - \mathbf{w}_h)$, $h = [1 \cdots K]$.
    Step 5. Find the cluster $\widehat{h}$ so that $\widehat{h} = \arg\max_{h} \{\mathbf{d}_{kh}\}$, $h = [1 \cdots K]$,

        and reset $\widehat{\mathbf{z}}_k \in \mathcal{W}_{\widehat{h}}$.
    Step 6. Update the cluster centers $\mathbf{w}_k$ and recompute the total scatter S_t.
    Step 7. If $\widehat{S}_t < S_t$ then $\widehat{S}_t = S_t$; return to step 3;
        else proceed to Step 8; /* Maximal S_t has been found. */
    Step 8. Return current K clusters $\{\mathcal{W}_k\}_{k=1}^{K}$ and their centers $\{\mathbf{w}_k\}_{k=1}^{K}$.

Figure 9.25. The pseudocode implementation of the CSC method.

With the CSC process, the training set $\mathcal{Z}$ is partitioned into a set of subsets $\{\mathcal{W}_k\}_{k=1}^K$ called *maximally separable clusters* (MSCs). To take advantage of these MSCs, a two-stage *hierarchical classification framework* (HCF) was then proposed in [75]. The HCF consists of a group of face recognition subsystems, each one targeting a specific MSC. This is not a difficult task for most traditional face recognition methods such as the YD-LDA [137] used in [75] to work as such a subsystem in a single MSC with limited-sized subjects and high between-class separability.

Nonlinear subspace analysis methods

In addition to the approach using a mixture of locally linear models, another option to generate a representation for nonlinear face manifold is with a globally nonlinear approach. Recently, the so-called kernel machine technique has become one of the most popular tools for designing nonlinear algorithms in the communities of machine learning and pattern recognition [125, 106]. The idea behind the kernel-based learning methods is to construct a nonlinear mapping from the input space ($\mathbb{R}^J$) to an implicit high-dimensional feature space ($\mathbb{F}$) using a kernel function $\phi: \mathbf{z} \in \mathbb{R}^J \to \phi(\mathbf{z}) \in \mathbb{F}$. In the feature space, it is hoped that the distribution of the mapped data is linearized and simplified so that traditional linear methods could perform well. However, the dimensionality of the feature space could be arbitrarily large, possibly infinite. Fortunately, the exact $\phi(\mathbf{z})$ is not needed, and the nonlinear mapping can be performed implicitly in $\mathbb{R}^J$ by replacing dot products in $\mathbb{F}$ with a kernel function defined in the input space $\mathbb{R}^J$, $k(\mathbf{z}_i, \mathbf{z}_j) = \phi(\mathbf{z}_i) \cdot \phi(\mathbf{z}_j)$. Examples based on such a design include SVMs [125], kernel PCA (KPCA) [107], kernel ICA (KICA) [6], and generalized discriminant analysis (GDA, also known as kernel LDA) [11].

In the kernel PCA [107], the covariance matrix in $\mathbb{F}$ can be expressed as

$$\widetilde{\mathbf{S}}_{cov} = \frac{1}{N} \sum_{i=1}^{C} \sum_{j=1}^{C_i} (\phi(\mathbf{z}_{ij}) - \bar{\phi})(\phi(\mathbf{z}_{ij}) - \bar{\phi})^T, \qquad (9.54)$$

where $\bar{\phi} = \frac{1}{N} \sum_{i=1}^{C} \sum_{j=1}^{C_i} \phi(\mathbf{z}_{ij})$ is the average of the ensemble in $\mathbb{F}$. The KPCA is actually a classic PCA performed in the feature space $\mathbb{F}$. Let $\widetilde{\mathbf{g}}_m \in \mathbb{F}$ ($m = 1, 2, \ldots, M$) be the first M most significant eigenvectors of $\widetilde{\mathbf{S}}_{cov}$, and they form a low-dimensional subspace, called *KPCA subspace*, in $\mathbb{F}$. For any face pattern $\mathbf{z}$, its nonlinear principal components can be obtained by the dot product, $(\widetilde{\mathbf{g}}_m \cdot (\phi(\mathbf{z}) - \bar{\phi}))$, computed indirectly through the kernel function $k()$. When $\phi(\mathbf{z}) = \mathbf{z}$, KPCA reduces to PCA, and the KPCA subspace is equivalent to the eigenface space introduced in [122].

As such, GDA [11] is used to extract a nonlinear discriminant feature representation by performing a classic LDA in the high-dimensional feature space $\mathbb{F}$. However, GDA solves the SSS problem in $\mathbb{F}$ simply by removing the nullspace of the within-class scatter matrix $\widetilde{\mathbf{S}}_b$, although the null space may contain the most significant discriminant information, as mentioned earlier. To this end, a kernel version of LD-LDA [79], also called KDDA, was introduced recently in [78]. In the feature space, the between-class and within-class scatter matrices are given as follows:

$$\widetilde{\mathbf{S}}_b = \frac{1}{N} \sum_{i=1}^{C} C_i (\bar{\phi}_i - \bar{\phi})(\bar{\phi}_i - \bar{\phi})^T, \tag{9.55}$$

$$\widetilde{\mathbf{S}}_w = \frac{1}{N} \sum_{i=1}^{C} \sum_{j=1}^{C_i} (\phi(\mathbf{z}_{ij}) - \bar{\phi}_i)(\phi(\mathbf{z}_{ij}) - \bar{\phi}_i)^T, \tag{9.56}$$

where $\bar{\phi}_i = \frac{1}{C_i} \sum_{j=1}^{C_i} \phi(\mathbf{z}_{ij})$. Through eigen-analysis of $\widetilde{\mathbf{S}}_b$ and $\widetilde{\mathbf{S}}_w$ in the feature space $\mathbb{F}$, KDDA finds a low-dimensional discriminant subspace spanned by Θ, an $M \times N$ matrix. Any face image $\mathbf{z}$ is first nonlinearly transformed to an $N \times 1$ kernel vector,

$$\gamma(\phi(\mathbf{z})) = [k(\mathbf{z}_{11}, \mathbf{z}), k(\mathbf{z}_{12}, \mathbf{z}), \cdots, k(\mathbf{z}_{cc_c}, \mathbf{z})]^T. \tag{9.57}$$

Then, the KDDA-based feature representation $\mathbf{y}$ can be obtained by a linear projection: $\mathbf{y} = \Theta \cdot \gamma(\phi(\mathbf{z}))$.

9.4.3   Pattern classification

Given the feature representation of face objects, a classifier is required to learn a complex decision function to implement final classification. Whereas the feature representation optimized for the best discrimination would help reduce the complexity of the decision function, which is easy for the classifier design, a good classifier would be able to further learn the separability between subjects.

Nearest feature line classifier

The *nearest neighbor* (NN) is the simplest yet most popular method for template matching. In the NN-based classification, the error rate is determined by the representational capacity of a prototype face database. The representational capacity depends two issues: (1) how the prototypes are chosen to account for possible variations, and (2) how many prototypes are available. However, in practice, only a small number of them are available for a face class, typically from one to about a dozen. It is desirable to have a sufficiently large number of feature points stored to account for as many variations as possible. To this end, Li and Lu [64] proposed a method called the *nearest feature line* (NFL) to generalize the representational capacity of available prototype images.

The basic assumption behind the NFL method is based on an experimental finding, which revealed that although the images of appearances of the face patterns may vary significantly due to differences in imaging parameters such as lighting, scale, and orientation, these differences have an approximately linear effect when they are small [121]. Consequently, it is reasonable to use a linear model to interpolate and extrapolate the prototype feature points belonging to the same class in a feature space specific to face representation, such as the eigenfaces space. In the simplest case, the linear model is generalized by a *feature line* (FL) passing through a pair of prototype points $(\mathbf{y}_1, \mathbf{y}_2)$, as depicted in Figure 9.26. Denoting the FL as $\overline{\mathbf{y}_1\mathbf{y}_2}$, the *FL distance* between a query feature point $\mathbf{y}$ and $\overline{\mathbf{y}_1\mathbf{y}_2}$ is defined as

$$d(\mathbf{y}, \overline{\mathbf{y}_1\mathbf{y}_2}) = \|\mathbf{y} - b\|, \quad b = \mathbf{y}_1 + \varsigma(\mathbf{y}_2 - \mathbf{y}_1), \tag{9.58}$$

where b is $\mathbf{y}$'s projection point on the FL, and $\varsigma = \frac{(\mathbf{y}-\mathbf{y}_1)\cdot(\mathbf{y}_2-\mathbf{y}_1)}{(\mathbf{y}_2-\mathbf{y}_1)\cdot(\mathbf{y}_2-\mathbf{y}_1)}$ is a position parameter relative to $\mathbf{y}_1$ and $\mathbf{y}_2$. The FL approximates variants of the two prototypes under variations in illumination and expression (i.e., possible face images derived from the two). It provides a virtually infinite number of prototype feature points of the class. Also, assuming that there are $C_i > 1$

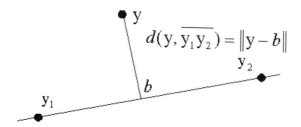

Figure 9.26. Generalizing two prototype feature points $\mathbf{y}_1$ and $\mathbf{y}_2$ by the feature line $\overline{\mathbf{y}_1\mathbf{y}_2}$. A query feature point $\mathbf{y}$ is projected onto the line as point b.

prototype feature points available for class i, a number of $K_i = C_i(C_i - 1)/2$ FLs can be constructed to represent the class (e.g., $K_i = 10$ when $C_i = 5$). As a result, the representational capacity of the prototype set is significantly enhanced in this way.

The class label (y) of the query feature point $\mathbf{y}$ can be inferred by the following NFL rule:

$$\text{Decide } y = i^* \text{ if } d(\mathbf{y}, \overline{\mathbf{y}_{i^*j^*}\mathbf{y}_{i^*k^*}}) = \min_{1 \le i \le C} \min_{1 \le j < k \le C_i} d(\mathbf{y}, \overline{\mathbf{y}_{ij}\mathbf{y}_{ik}}). \qquad (9.59)$$

The classification results not only determine the class label y, but also provide a quantitative position number ς^* as a by-product, which can be used to indicate the relative changes (in illumination and expression) between the query face and the two associated face images $\mathbf{y}_{i^*j^*}$ and $\mathbf{y}_{i^*k^*}$.

Regularized Bayesian classifier

LDA has its root in the optimal Bayesian classifier. Let $P(y = i)$ and $p(\mathbf{z}|y = i)$ be the prior probability of class i and the class-conditional probability density of $\mathbf{z}$ given the class label is i respectively. Based on the Bayes formula, we have the following *a posteriori* probability $P(y = i|\mathbf{z})$, that is, the probability of the class label being i given that $\mathbf{z}$ has been measured:

$$P(y = i|\mathbf{z}) = \frac{p(\mathbf{z}|y = i)P(y = i)}{\sum_{i=1}^{C} p(\mathbf{z}|y = i)P(y = i)}. \qquad (9.60)$$

The Bayesian decision rule to classify the unlabeled input $\mathbf{z}$ is then given as

$$\text{Decide } y = j \text{ if } j = \arg\max_{i \in \mathbb{Y}} P(y = i|\mathbf{z}). \qquad (9.61)$$

(9.60) is also known as the *maximum a posteriori* (MAP) rule, and it achieves minimal misclassification risk among all possible decision rules.

The class-conditional densities $p(\mathbf{z}|y = i)$ are seldom known. However, often it is reasonable to assume that $p(\mathbf{z}|y = i)$ is subjected to a Gaussian distribution. Let μ_i and Σ_i denote the mean and covariance matrix of the class i; we have

$$p(\mathbf{z}|y = i) = (2\pi)^{-J/2}|\Sigma_i|^{-1/2} \exp\left[-d_i(\mathbf{z})/2\right], \qquad (9.62)$$

where $d_i(\mathbf{z}) = (\mathbf{z} - \mu_i)^T \Sigma_i^{-1}(\mathbf{z} - \mu_i)$ is the squared Mahalanobis (quadratic) distance from $\mathbf{z}$ to the mean vector μ_i. With the Gaussian assumption, the classification rule of (9.61) turns to

$$\text{Decide } y = j \text{ if } j = \arg\min_{i \in \mathbb{Y}} \left(d_i(\mathbf{z}) + \ln|\Sigma_i| - 2\ln P(y = i)\right). \qquad (9.63)$$

The decision rule of (9.63) produces quadratic boundaries to separate different classes in the J-dimensional real space. Consequently, this is also called *quadratic discriminant analysis* (QDA). Often, the two statistics (μ_i, Σ_i) are estimated by their sample analogs,

$$\mu_i = \bar{\mathbf{z}}_i, \quad \Sigma_i = \frac{1}{C_i} \sum_{j=1}^{C_i} (\mathbf{z}_{ij} - \bar{\mathbf{z}}_i)(\mathbf{z}_{ij} - \bar{\mathbf{z}}_i)^T \tag{9.64}$$

LDA can be viewed as a special case of QDA when the covariance structure of all classes are identical (i.e., $\Sigma_i = \Sigma$). However, the estimation for either Σ_i or Σ is ill-posed in the small sample size (SSS) settings, giving rise to high variance. This problem becomes extremely severe due to $C_i \ll J$ in face recognition tasks, where Σ_i is singular with rank$\leq (C_i - 1)$. To deal with such a situation, a regularized QDA, built on the D-LDA idea and Friedman's regularization technique [40], called RD-QDA, was introduced recently in [80]. The purpose of the regularization is to reduce the variance related to the sample-based estimation for the class covariance matrices at the expense of potentially increased bias.

In the RD-QDA method, the face images $(\mathbf{z}_{ij})$ are first projected into the between-class scatter matrix $\mathbf{S}_b$'s complement null subspace spanned by $\mathbf{H}$, obtaining a representation $\mathbf{y}_{ij} = \mathbf{H}^T \mathbf{z}_{ij}$. The regularized sample covariance matrix estimate of class i in the subspace spanned by $\mathbf{H}$, denoted by $\widehat{\Sigma}_i(\lambda, \gamma)$, can be expressed as

$$\widehat{\Sigma}_i(\lambda, \gamma) = (1 - \gamma)\widehat{\Sigma}_i(\lambda) + \frac{\gamma}{M} tr[\widehat{\Sigma}_i(\lambda)]\mathbf{I} \tag{9.65}$$

where

$$\widehat{\Sigma}_i(\lambda) = \frac{1}{C_i(\lambda)} [(1 - \lambda)\mathbf{S}_i + \lambda\mathbf{S}], \; C_i(\lambda) = (1 - \lambda)C_i + \lambda N \tag{9.66}$$

$$\mathbf{S}_i = \sum_{j=1}^{C_i} (\mathbf{y}_{ij} - \bar{\mathbf{y}}_i)(\mathbf{y}_{ij} - \bar{\mathbf{y}}_i)^T, \; \mathbf{S} = \sum_{i=1}^{C} \mathbf{S}_i = N \cdot \mathbf{H}^T \mathbf{S}_w \mathbf{H} \tag{9.67}$$

$\bar{\mathbf{y}}_i = \mathbf{H}^T \bar{\mathbf{z}}_i$, and (λ, γ) is a pair of regularization parameters. The classification rule of (9.63) then turns to

$$\text{Decide } y = j \text{ if } j = \underset{i \in \mathbb{Y}}{\arg\min} \left(d_i(\mathbf{y}) + \ln |\widehat{\Sigma}_i(\lambda, \gamma)| - 2 \ln \pi_i \right) \tag{9.68}$$

where $d_i(\mathbf{y}) = (\mathbf{y} - \bar{\mathbf{y}}_i)^T \widehat{\Sigma}_i^{-1}(\lambda, \gamma)(\mathbf{y} - \bar{\mathbf{y}}_i)$ and $\pi_i = C_i/N$ is the estimate of the prior probability of class i. The regularization parameter λ $(0 \leq \lambda \leq 1)$ controls the amount that the $\mathbf{S}_i$ are shrunk toward $\mathbf{S}$. The other parameter γ $(0 \leq \gamma \leq 1)$ controls shrinkage of the class covariance matrix estimates toward a multiple of the identity matrix. Under the regularization scheme, the classification rule of (9.68) can be performed without experiencing high variance of the sample-based estimation even when the dimensionality of the subspace spanned by $\mathbf{H}$ is comparable to the number of available training samples, N.

RD-QDA has a close relationship with a series of traditional discriminant analysis classifiers, such as LDA, QDA, *nearest center* (NC), and *weighted nearest center* (WNC). First, the four corners defining the extremes of the (λ, γ) plane represent four well-known classification algorithms, as summarized in Table 9.3, where the prefix D- means that all these methods are developed in the subspace spanned by $\mathbf{H}$ derived from the D-LDA technique. Based on Fisher's criterion (9.48) used in YD-LDA [137], it is obvious that the YD-LDA feature extractor followed by an NC classifier is actually a standard LDA classification rule implemented in the subspace $\mathbf{H}$. Also, we have $\widehat{\Sigma}_i(\lambda, \gamma) = \alpha\left(\frac{\mathbf{S}}{N} + \mathbf{I}\right) = \alpha\left(\mathbf{H}^T\mathbf{S}_w\mathbf{H} + \mathbf{I}\right)$ when $(\lambda = 1, \gamma = \eta)$, where $\alpha = \left(\frac{tr[\mathbf{S}/N]}{tr[\mathbf{S}/N]+M}\right)$ and $\eta = \frac{M}{tr[\mathbf{S}/N]+M}$. In this situation, it is not difficult to see that RD-QDA is equivalent to LD-LDA followed by an NC classifier. In addition, a set of intermediate discriminant classifiers between the five traditional ones can be obtained when we smoothly slip the two regularization parameters in their domains. The purpose of RD-QDA is to find the optimal (λ^*, γ^*) that give the best correct recognition rate for a particular face recognition task.

Table 9.3. A series of discriminant analysis algorithms derived from RD-QDA.

Algs.	D-NC	D-WNC	D-QDA	YD-LDA	LD-LDA
λ	1	0	0	1	1
γ	1	1	0	0	η
$\widehat{\Sigma}_i(\lambda, \gamma)$	$\frac{1}{M}tr[\frac{\mathbf{S}}{N}]\mathbf{I}$	$\frac{1}{M}tr[\frac{\mathbf{S}_i}{C_i}]\mathbf{I}$	$\frac{\mathbf{S}_i}{C_i}$	$\frac{\mathbf{S}}{N}$	$\alpha\left(\frac{\mathbf{S}}{N} + \mathbf{I}\right)$

Neural networks classifiers

Linear or quadratic classifiers may fail to deliver good performance when the feature representation $\mathbf{y}$ of face images $\mathbf{z}$ is subject to a highly nonconvex

distribution—for example, in the case depicted in Figure 9.27 (right), where
a nonlinear decision boundary much more complex than the quadratic one
is required. One option to construct such a boundary is to utilize a neural
network classifier. Figure 9.27 (left) depicts the architecture of a general
multilayer *feedforward neural network* (FNN), which consists of one input
layer, one hidden layer and one output layer. The input layer has M units
to receive the M-dimensional input vector $\mathbf{y}$. The hidden layer is composed
of L units, each one operated by a nonlinear activation function $h_l(\cdot)$ to
nonlinearly map the input to an L-dimensional space $\mathbb{R}^L$, where the patterns
are hoped to become linearly separable, so that linear discriminants can be
implemented by the activation function $f(\cdot)$ in the output layer. The process
can be summarized as

$$t_k(\mathbf{y}) = f\left\{ \sum_{j=0}^{L} h\left(\sum_{i=0}^{M} w_{ji}^h \mathbf{y}_i \right) \cdot w_{kj}^o \right\} \qquad (9.69)$$

where w_{ji}^h and w_{kj}^o are the connecting weights between neighboring layers of
units.

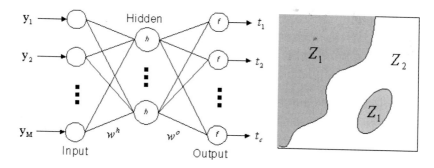

Figure 9.27. Left: A general multilayer feedforward neural network. Right: An
example requires complex decision boundaries.

The key to the neural networks is to learn the involved parameters. One
of the most popular methods is the *back-propagation* (BP) algorithm based
on error gradient descent. The most widely used activation function in both
hidden and output units of a BP network is a sigmoidal function given by
$f(\cdot) = h(\cdot) = 1/(1+e^{(\cdot)})$. Most BP-like algorithms utilize local optimization
techniques. As a result, the training results are very much dependent on the
choices of initial estimates. Recently, a *global* FNN learning algorithm was
proposed in [120, 119]. The global FNN method was developed by addressing two issues: (1) characterization of global optimality of an FNN learning

objective incorporating the weight decay regularizer, and (2) derivation of an efficient search algorithm based on results of (1). The face recognition simulations reported in [120] indicate that the global FNN classifier can perform well in conjunction with various feature extractors, including eigenfaces [122], Fisherfaces [12], and D-LDA [137].

In addition to the classic BP networks, the radial basis function (RBF) neural classifiers have recently attracted extensive interest in the community of pattern recognition. In the RBF networks, the Gaussian function is often preferred as the activation function in the hidden units, $h_l(\mathbf{y}) = \exp(-\|\mathbf{y} - u_i\| / \sigma_i^2)$, while the output activation function $f(\cdot)$ is usually a linear function. In this way, it can be seen that the output of the RBF networks is actually a mixture of Gaussians. Consequently, it is generally believed that the RBF networks possess the best approximation property. Also, the learning speed of the RBF networks is fast due to locally tuned neurons. Attempts to apply the RBF neural classifiers to solve face recognition problems have been reported recently. For example, in [31], a face recognition system built on an LDA feature extractor and an enhanced RBF neural network produced one of the lowest error rates reported on the ORL face database [2].

Support vector machine classifiers

Assuming that all the examples in the training set $\mathcal{Z}$ are drawn from a distribution $P(\mathbf{z}, y)$, where y is the label of the example $\mathbf{z}$, the goal of a classifier learning from $\mathcal{Z}$ is to find a function $f(\mathbf{z}, \alpha^*)$ to minimize the expected risk:

$$R(\alpha) = \int |f(\mathbf{z}, \alpha) - y| dP(\mathbf{z}, y) \qquad (9.70)$$

where α is a set of abstract parameters. Since $P(\mathbf{z}, y)$ is unknown, most traditional classifiers, such as nearest neighbor, Bayesian classifier, and neural network, solve the specific learning problem using the so-called empirical risk (i.e., training error) minimization (ERM) induction principle, where the expected risk function $R(\alpha)$ is replaced by the empirical risk function: $R_{emp}(\alpha) = \frac{1}{N} \sum_{i=1}^{N} |f(\mathbf{z}_i, \alpha) - y_i|$. As a result, the classifiers obtained may be entirely unsuitable for classification of unseen test patterns, although they may achieve the lowest training error. To this end, Vapnik and Chervonenkis [126] provide a bound on the deviation of the empirical risk from the expected risk. The bound, also called Vapnik-Chervonenkis (VC) bound, holding with probability $(1 - \eta)$, has the following form:

$$R(\alpha) \leq R_{emp}(\alpha) + \sqrt{\frac{1}{N}\left(h\left(\ln\frac{2N}{h} + 1\right) - \ln\frac{\eta}{4}\right)}, \qquad (9.71)$$

where h is the VC-dimension as a standard measure to the complexity of the function space that $f(\mathbf{z}_i, \alpha)$ is chosen from. It can be seen from the VC bound that both $R_{emp}(\alpha)$ and (h/N) have to be small to achieve good generalization performance.

Based on the VC theory, the SVMs embody the structural risk minimization principle, which aims to minimize the VC bound. However, often it is intractable to estimate the VC dimension of a function space. Fortunately, it has been shown in [125] that for the function class of hyperplanes, $f(\mathbf{z}) = \mathbf{w} \cdot \mathbf{z} + b$, its VC dimension can be controlled by increasing the *margin*, which is defined as the minimal distance of an example to the decision surface (see Figure 9.28). The main idea behind SVMs is to find a separating hyperplane with the largest margin, as shown in Figure 9.28, where the margin is equal to $2/\|\mathbf{w}\|$.

For a binary classification problem where $y_i \in \{1, -1\}$, the general optimal separating hyperplane sought by SVM is the one that

$$Minimizes: \quad P = \frac{1}{2}\|\mathbf{w}\|^2 + \zeta\sum_{i=1}^{n}\xi_i \qquad (9.72)$$

subject to $y_i(\mathbf{w}^T\mathbf{z}_i + b) \geq 1 - \xi_i$, $\xi_i \geq 0$, where ξ_i are slack variables, ζ is a regularization constant, and the hyperplane is defined by parameters

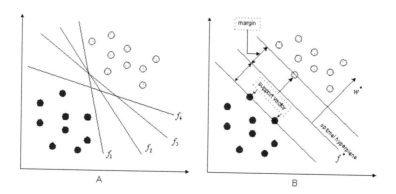

Figure 9.28. A binary classification problem solved by hyperplanes: (A) arbitrary separating hyperplanes; (B) the optimal separating hyperplanes with the largest margin.

$\mathbf{w}$ and b. After some transformations, the minimization in (9.72) can be reformulated as

$$Maximizing: \quad D = \sum_{i=1}^{N} \alpha_i - \frac{1}{2} \sum_{i=1}^{N} \sum_{j=1}^{N} \alpha_i \alpha_j y_i y_j \mathbf{z}_i \cdot \mathbf{z}_j \qquad (9.73)$$

subject to $0 \leq \alpha_i \leq \zeta$ and $\sum_{i=1}^{N} \alpha_i y_i = 0$, where α_i are positive Lagrange multipliers. Then, the solution, $\mathbf{w}^* = \sum_{i=1}^{N} \alpha_i^* y_i \mathbf{z}_i$ and $b^* = y_i - \mathbf{w}^* \cdot \mathbf{z}_i$ $(\alpha_i^* > 0)$, can be derived from (9.73) using quadratic programming. The support vectors are those examples $(\mathbf{z}_i, y_i)$ with $\alpha_i^* > 0$.

For a new data point $\mathbf{z}$, the classification is performed by a decision function,

$$f(\mathbf{z}) = sign\left(\mathbf{w}^* \cdot \mathbf{z} + b^*\right) = sign\left(\sum_{i=1}^{N} \alpha_i^* y_i (\mathbf{z}_i \cdot \mathbf{z}) + b^*\right) \qquad (9.74)$$

In the case where the decision function is not a linear function of the data, SVMs first map the input vector $\mathbf{z}$ into a high-dimensional feature space by a nonlinear function $\phi(\mathbf{z})$, and then construct an optimal separating hyperplane in the high-dimensional space with linear properties. The mapping $\phi(\mathbf{z})$ is implemented using kernel machine techniques, as it is done in KPCA and GDA. Examples of applying SVMs into the face recognition tasks can be found in [94, 56, 47, 76].

9.4.4   Evaluation

In this section, we introduce several experiments from our recent studies on subspace analysis methods. Following standard face recognition practices, any evaluation database ($\mathcal{G}$) used here is randomly partitioned into two subsets: the training set $\mathcal{Z}$ and the test set $\mathcal{Q}$. The training set consists of $N (= \sum_{i=1}^{C} C_i)$ images: C_i images per subject are randomly chosen. The remaining images are used to form the test set $\mathcal{Q} = \mathcal{G} - \mathcal{Z}$. Any face recognition method evaluated here is first trained with $\mathcal{Z}$, and the resulting face recognizer is then applied to $\mathcal{Q}$ to obtain a correct recognition rate (CRR), which is defined as the fraction of the test examples correctly classified. To enhance the accuracy of the assessment, all the CRRs reported here are averaged over ≥ 5 runs. Each run is executed on a random partition of the database $\mathcal{G}$ into $\mathcal{Z}$ and $\mathcal{Q}$.

Linear and quadratic subspace analysis

The experiment is designed to assess the performance of various linear and quadratic subspace analysis methods, including eigenfaces, YD-LDA, LD-

LDA, RD-QDA, and those listed in Table 9.3. The evaluation database used in the experiment is a midsized subset of the FERET database. The subset consists of 606 grayscale images of 49 subjects, each one having more than 10 samples. The performance is evaluated in terms of the CRR and the sensitivity of the CRR measure to the SSS problem, which depends on the number of training examples per subject C_i. To this end, six tests were performed with various values of C_i ranging from $C_i = 2$ to $C_i = 7$.

The CRRs obtained by RD-QDA in the (λ, γ) grid are depicted in Figure 9.29. Also, a quantitative comparison of the best found CRRs and their corresponding parameters among the seven methods is summarized in Table 9.4. The parameter λ controls the degree of shrinkage of the individual class covariance matrix estimates $\mathbf{S}_i$ toward the within-class scatter matrix, $(\mathbf{H}^T \mathbf{S}_w \mathbf{H})$. Varying the values of λ within $[0, 1]$ leads to a set of intermediate classifiers between D-QDA and YD-LDA. In theory, D-QDA should be the best performer among the methods evaluated here if sufficient training samples are available. It can be observed at this point from Figure 9.29 that the CRR peaks gradually moved from the central area toward the corner $(0, 0)$; that is the case of D-QDA as C_i increases. Small values of λ have been good enough for the regularization requirement in many cases ($C_i \geq 3$). However, both D-QDA and YD-LDA performed poorly when $C_i = 2$. This should be attributed to the high variance of the estimates of $\mathbf{S}_i$ and $\mathbf{S}$ due

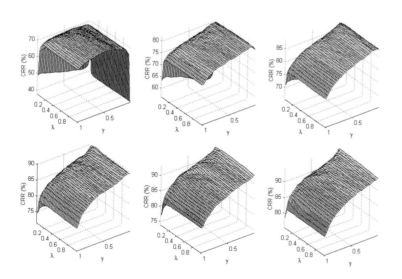

Figure 9.29. CRRs obtained by RD-QDA as functions of (λ, γ). Top: $C_i = 2, 3, 4$; Bottom: $C_i = 5, 6, 7$.

to insufficient training samples. In this case, $\mathbf{S}_i$ and even $\mathbf{S}$ are singular or close to singular, and the resulting effect is to dramatically exaggerate the importance associated with the eigenvectors corresponding to the smallest eigenvalues. Against the effect, the introduction of another parameter γ helps to decrease the larger eigenvalues and increase the smaller ones, thereby counteracting to some extent the bias. This is also why LD-LDA greatly outperformed YD-LDA when $C_i = 2$. Although LD-LDA seems to be a little overregularized compared to the optimal RD-QDA(λ^*, γ^*), the method almost guarantees a stable suboptimal solution. A CRR difference of 4.5% on average over the range $C_i \in [2, 7]$ has been observed between the top performer RD-QDA(λ^*, γ^*) and LD-LDA. It can be concluded, therefore, that LD-LDA should be preferred when insufficient prior information about the training samples is available and a cost-effective processing method is sought.

Table 9.4. Comparison of correct recognition rates (CRRs) (%).

$C_i =$	2	3	4	5	6	7
Eigenfaces	59.8	67.8	73.0	75.8	81.3	83.7
D-NC	67.8	72.3	75.3	77.3	80.2	80.5
D-WNC	46.9	61.7	68.7	72.1	73.9	75.6
D-QDA	57.0	79.3	87.2	89.2	92.4	93.8
YD-LDA	37.8	79.5	87.8	89.5	92.4	93.5
LD-LDA	70.7	77.4	82.8	85.7	88.1	89.4
(η)	0.84	0.75	0.69	0.65	0.61	0.59
RD-QDA	73.2	81.6	88.5	90.4	93.2	94.4
(λ^*)	0.93	0.93	0.35	0.11	0.26	0.07
(γ^*)	0.47	0.10	0.07	0.01	1e-4	1e-4

In addition to the regularization, it is found that the performance of the LDA-based methods can be further improved with an ensemble-based approach [75]. The approach uses cluster analysis techniques like the K-mean and the CSC method mentioned earlier to form a mixture of LDA subspaces. Experiments conducted on a compound database with 1654 face images of 157 subjects and a large FERET subset with 2400 face images of 1200 subjects indicate that the performance of both YD-LDA and LD-LDA can be greatly enhanced under the ensemble framework. The average CRR improvement that has been observed so far ranges from 6% to 22%.

Nonlinear subspace analysis

Applications of kernel-based methods in face research have been widely reported (e.g., [69, 63, 58, 134, 78, 62]). Here, we introduce two experiments to illustrate the effectiveness of the kernel-based discriminant analysis methods in face recognition tasks. The two experiments were conducted on a multi-view face database, the UMIST database [46], consisting of 575 images of 20 people, each covering a wide range of poses from profile to frontal views.

The first experiment provided insights on the distribution of face pattern in four types of subspaces generalized by utilizing the PCA [122], KPCA [107], LD-LDA [79], and KDDA [78] algorithms respectively. The projections of five face subjects in the first two most significant feature bases of each subspace are shown in Figure 9.30. Figure 9.30 (upper) shows the first two most significant principal components extracted by PCA and KPCA, and they provide a low-dimensional representation for the samples in a least-square sense. Thus we can roughly learn the original distribution of the samples

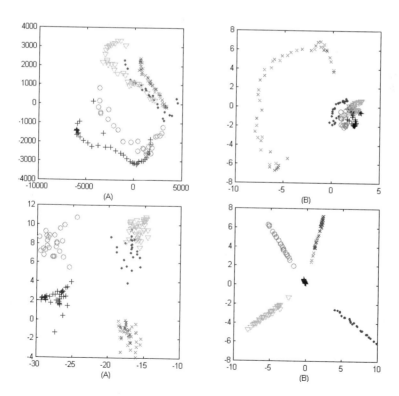

Figure 9.30. Distribution of five subjects in four subspaces; upper A: PCA-based; upper B: KPCA-based; lower A: LD-LDA-based; and lower B: KDDA-based.

from Figure 9.30 (upper A), where it can been seen that the distribution is as expected nonconvex and complex. Also, it is hard to find any useful improvement in Figure 9.30 (upper B) for the purpose of pattern classification. Figure 9.30 depicts the first two most discriminant features extracted by LD-LDA and KDDA respectively. Obviously, these features outperform, in terms of discriminant power, those obtained by using the PCA-like techniques. However, subject to limitation of linearity, some classes are still nonseparable in Figure 9.30 (lower A). In contrast to this, we can see the linearization property of the KDDA-based subspace, as shown in Figure 9.30 (lower B), where all of the classes are well linearly separable.

The second experiment compares the CRR performance among the three kernel-based methods, KPCA [107], GDA (also called KDA) [11], and KDDA [78]. The two most popular face recognition algorithms, eigenfaces [122] and Fisherfaces [12], were also implemented to provide performance baselines. $C_i = 6$ images per person were randomly chosen to form the training set $\mathcal{Z}$. The nearest neighbor was chosen as the classifier following these feature extractors. The obtained results with best found parameters are depicted in Figure 9.31, where KDDA is clearly the top performer among all the methods evaluated here. Also, it can be observed that the performance of KDDA is more stable and predictable than that of GDA. This should be attributed to the introduction of the regularization, which significantly reduces the variance of the estimates of the scatter matrices arising from the SSS problem.

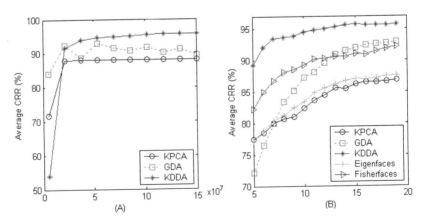

Figure 9.31. A comparison of CRR based on an RBF kernel, $k(\mathbf{z}_1, \mathbf{z}_2) = exp\left(-||\mathbf{z}_1 - \mathbf{z}_2||^2/\sigma^2\right)$. A: CRR as a function of the parameter σ^2 with best found M^*. B: CRR as a function of the feature number M with best found σ^*.

Other performance evaluations

Often, it is desired in the face recognition community to give the overall evaluation and benchmarking of various face recognition algorithms. Here, we introduce several recent face recognition evaluation reports in literature. Before we proceed to an evaluation, it should be noted that the performance of a learning-based pattern recognition system is very data- and application-dependent, and there is no theory that can accurately predict them for unknown distribution data/new applications. In other words, some methods that have reported almost perfect performance in certain scenarios may fail in other scenarios.

To date, the FERET program incepted in 1993 has made a significant contribution to the evaluation research of face recognition algorithms by building the FERET database and the evaluation protocol [3, 96, 95]. Their availability has made it possible to objectively assess the laboratory algorithms under close to real-world conditions. In the FERET protocol, an algorithm is given two sets of images: the target set and the query set. The target set is a set of known facial images, while the query set consists of unknown facial images to be identified. Furthermore, multiple galleries and probe sets can be constructed from the target and query sets respectively. For a pair of given gallery $\mathcal{G}$ and probe set $\mathcal{P}$, the CRR is computed by examining the similarity between the two sets of images. Table 9.5 depicts

Table 9.5. CRR performance rank of algorithms in the September 1996 FERET evaluation [95].

Algorithms	on FB Probes						on Duplicate I Probes				
	$\mathcal{G}_1$	$\mathcal{G}_2$	$\mathcal{G}_3$	$\mathcal{G}_4$	$\mathcal{G}_5$	$\mathcal{G}_6$	$\mathcal{G}_1$	$\mathcal{G}_2$	$\mathcal{G}_3$	$\mathcal{G}_4$	$\mathcal{G}_5$
B-PCA	9	10	8	8	10	8	6	10	5	5	9
B-Corr	9	9	9	6	9	10	10	7	6	6	8
X	6	7	7	5	7	6	3	5	4	4	3
Eigenfaces	4	2	1	1	3	3	2	1	2	2	3
D-Eigenfaces	7	5	4	4	5	7	7	4	7	8	10
LDA.M	3	4	5	8	4	4	9	6	8	10	6
GrayProj	7	8	9	6	7	9	5	7	10	7	6
LDA.U1	4	6	6	10	5	5	7	9	9	9	3
LDA.U2	1	1	3	2	2	1	4	2	3	3	1
EGM-GJ	2	3	2	2	1	1	1	3	1	1	1

some test results reported in the September 1996 FERET evaluation [95]. B-PCA and B-Corr are two baseline methods based on PCA and normalized correlation [122, 88]. D-eigenfaces is the dual eigenfaces method [86]. LDA.M [115] and LDA.U1/U2 [33, 141] are three LDA-based algorithms. GrayProj is a method using grayscale projection [129]. EGM-GJ is the elastic graph matching method with Gabor jet [130]. X denotes an unknown algorithm from Excalibur Corporation. It can be seen that among these evaluated methods, D-eigenfaces, LDA.U2, and EGM-GJ are the three top performers. Based on the FERET program, the face recognition vendor tests (FRVTs) that systematically measured commercial face recognition products were also developed, and the latest FRVT 2002 reports can be found in [1].

Recently, Moghaddam [85] evaluated several unsupervised subspace analysis methods and showed that dual PCA (dual eigenfaces) > KPCA > PCA $\approx$ ICA, where > denotes "outperform" in terms of the average CRR measure. Compared to these unsupervised methods, it is generally believed that algorithms based on LDA are superior in face recognition tasks. However, it is shown recently in [83] that this is not always the case. PCA may outperform LDA when the number of training samples per subject is small or when the training data nonuniformly sample the underlying distribution. Also, PCA is shown to be less sensitive to different training data sets. More recent evaluation results of subspace analysis methods in different scenarios can be found in Table 9.6, where LDA is based on the version of [33, 12], PPCA is the probabilistic PCA [118], and KICA is the kernel version of ICA [6]. The overall performance of linear subspace analysis methods was summarized as LDA > PPCA > PCA > ICA, and it was also observed that kernel-based methods are not necessarily better than linear methods [62].

Table 9.6. CRR performance rank of subspace methods in [62].

Rank	1	2	3	4	5	6	7
Pose	KDA	PCA	KPCA	LDA	KICA	ICA	PPCA
Expression	KDA	LDA	PPCA	PCA	KPCA	ICA	KICA
Illumination	LDA	PPCA	KDA	KPCA	KICA	PCA	ICA

Acknowledgments

Portions of the research in this chapter use the FERET database of facial images collected under the FERET program [96]. The authors would like to thank the FERET Technical Agent, the U.S. National Institute of Standards and Technology (NIST), for providing the FERET database.

Bibliography

[1] Face recognition vendor tests (FRVT) [Online]. Available: http://www.frvt.org.

[2] ORL face database [Online]. Available: http://www.cam-orl.co.uk/3facedatabase.html. Released by AT&T Laboratories Cambridge.

[3] FERET [Online]. Available: http://www.itl.nist.gov/iad/humanid/feret/. Released by Image Group, Information Access Division, ITL, NIST, USA, March 2001.

[4] J. Ahlberg. Using the active appearance algorithm for face and facial feature tracking. In *IEEE ICCV Workshop on Recognition, Analysis and Tracking of Faces and Gestures in Real-time Systems*, pages 68–72, July 2001.

[5] Y. Amit, D. Geman, and K. Wilder. Joint induction of shape features and tree classifiers. *IEEE Transactions on Pattern Analysis and Machine Intelligence*, 19:1300–1305, 1997.

[6] F. R. Bach and M. I. Jordan. Kernel independent component analysis. Computer Science Division, University of California Berkeley, Technical Report No. UCB/CSD-01-1166, November 2001.

[7] S. Baker and I. Matthews. Equivalence and efficiency of image alignment algorithms. In *Proc. of IEEE Computer Society Conference on Computer Vision and Pattern Recognition*, vol. 1, pages 1090–1097, December 2001.

[8] S. Baker and S. Nayar. Pattern rejection. In *Proc. of IEEE Computer Society Conference on Computer Vision and Pattern Recognition*, pages 544–549, June 1996.

[9] M. Bartlett, J. Movellan, and T. Sejnowski. Face recognition by independent component analysis. *IEEE Transactions on Neural Networks*, 13(6):1450–1464, November 2002.

[10] M. S. Bartlett, H. M. Lades, and T. J. Sejnowski. Independent component representations for face recognition. *Proc. of the SPIE, Conference on Human Vision and Electronic Imaging III*, 3299:528–539, 1998.

[11] G. Baudat and F. Anouar. Generalized discriminant analysis using a kernel approach. *Neural Computation*, 12:2385–2404, 2000.

[12] P. N. Belhumeur, J. P. Hespanha, and D. J. Kriegman. Eigenfaces vs. Fisherfaces: Recognition using class-specific linear projection. *IEEE Transactions on Pattern Analysis and Machine Intelligence*, 19(7):711–720, July 1997.

[13] D. Beymer. Vectorizing face images by interleaving shape and texture computations. A. I. Memo 1537, MIT, 1995.

[14] M. Bichsel and A. P. Pentland. Human face recognition and the face image set's topology. *CVGIP: Image Understanding*, 59:254–261, 1994.

[15] C. M. Bishop and J. M. Winn. Nonlinear Bayesian image modelling. In *ECCV (1)*, pages 3–17, 2000.

[16] V. Blanz and T.Vetter. A morphable model for the synthesis of 3d faces. In *SIGGRAPH'99 Conference Proc.*, pages 187–194, 1999.

[17] R. Brunelli and T. Poggio. Face recognition: Features versus templates. *IEEE Transactions on Pattern Analysis and Machine Intelligence*, 15(10):1042–1052, 1993.

[18] R. Chellappa, C. Wilson, and S. Sirohey. Human and machine recognition of faces: A survey. *Proc. of the IEEE*, 83:705–740, 1995.

[19] L.-F. Chen, H.-Y. M. Liao, M.-T. Ko, J.-C. Lin, and G.-J. Yu. A new LDA-based face recognition system which can solve the small sample size problem. *Pattern Recognition*, 33:1713–1726, 2000.

[20] L.-F. Chen, H.-Y. M. Liao, J.-C. Lin, and C.-C. Han. Why recognition in a statistics-based face recognition system should be based on the pure face portion: A probabilistic decision-based proof. *Pattern Recognition*, 34(7):1393–1403, 2001.

[21] T. Cootes and C. Taylor. Constrained active appearance models. In *Proc. of IEEE International Conference on Computer Vision*, 1:748–754, 2001.

[22] T. F. Cootes, G. J. Edwards, and C. J. Taylor. Active appearance models. In *ECCV98*, vol. 2, pages 484–498, 1998.

[23] T. F. Cootes, C. J. Taylor, D. H. Cooper, and J. Graham. Active shape models: Their training and application. *CVGIP: Image Understanding*, 61:38–59, 1995.

[24] T. F. Cootes, K. N. Walker, and C. J. Taylor. View-based active appearance models. In *Proc. International Conference on Face and Gesture Recognition*, pages 227–232, 2000.

[25] I. J. Cox, J. Ghosn, and P. Yianilos. Feature-based face recognition using mixture-distance. In *Proc. of IEEE Computer Society Conference on Computer Vision and Pattern Recognition*, pages 209–216, 1996.

[26] F. Crow. Summed-area tables for texture mapping. In *SIGGRAPH*, 18(3):207–212, 1984.

[27] J. G. Daugman. Complete discrete 2D Gabor transform by neural network for image analysis and compression. *IEEE Trans. On Acoustics, Speech and Signal Processing*, 36(7):1169–1179, July 1988.

[28] N. Duta, A. Jain, and M. Dubuisson-Jolly. Automatic construction of 2-d shape models. *IEEE Transactions on Pattern Analysis and Machine Intelligence*, 23(5):433–446, 2001.

[29] G. J. Edwards, T. F. Cootes, and C. J. Taylor. Face recognition using active appearance models. In *Proc. of the European Conference on Computer Vision*, vol. 2, pages 581–695, 1998.

[30] M. Elad, Y. Hel-Or, and R. Keshet. Pattern detection using a maximal rejection classifier. *Pattern Recognition Letters*, 23:1459–1471, October 2002.

[31] M. J. Er, S. Wu, J. Lu, and H. L. Toh. Face recognition with radial basis function (RBF) neural networks. *IEEE Transactions on Neural Networks*, 13(3):697–710, May 2002.

[32] B. K. L. Erik Hjelmas. Face detection: A survey. *Computer Vision and Image Understanding*, 3(3):236–274, September 2001.

[33] K. Etemad and R. Chellappa. Discriminant analysis for recognition of human face images. *J. Optical Soc. Am. A*, 14(8):1724–1733, August 1997.

[34] J. Feraud, O. Bernier, and M. Collobert. A fast and accurate face detector for indexation of face images. In *Proc. 4th IEEE International Conference on Automatic Face and Gesture Recognition*, 2000.

[35] R. Fisher. The use of multiple measures in taxonomic problems. *Ann. Eugenics*, 7:179–188, 1936.

[36] F. Fleuret and D. Geman. Coarse-to-fine face detection. *International Journal of Computer Vision*, 20:1157–1163, 2001.

[37] Y. Freund and R. Schapire. A decision-theoretic generalization of on-line learning and an application to boosting. *Journal of Computer and System Sciences*, 55(1):119–139, August 1997.

[38] B. J. Frey, A. Colmenarez, and T. S. Huang. Mixtures of local linear subspaces for face recognition. In *Proc. of The IEEE Conference on Computer Vision and Pattern Recognition*, June 1998.

[39] B. J. Frey and N. Jojic. Transformation-invariant clustering using the EM algorithm. *IEEE Transactions on Pattern Analysis and Machine Intelligence*, 25(1):1–17, January 2003.

[40] J. H. Friedman. Regularized discriminant analysis. *Journal of the American Statistical Association*, 84:165–175, 1989.

[41] R. Frischholz. The face detection homepage [Online]. Available: http://home.3t-online.de/home/Robert.Frischholz/face.htm.

[42] K. Fukunaga. *Introduction to statistical pattern recognition*, 2nd ed. Academic Press, 1990.

[43] M. Girolami. *Advances in Independent Component Analysis*. Springer-Verlag, 2000.

[44] A. J. Goldstein, L. D. Harmon, and A. B. Lesk. Identification of human faces. *Proc. of the IEEE*, 59(5):748–760, May 1971.

[45] S. Gong, S. McKenna, and J. Collins. An investigation into face pose distribution. In *Proc. IEEE International Conference on Face and Gesture Recognition*, Vermont, 1996.

[46] D. B. Graham and N. M. Allinson. Characterizing virtual eigensignatures for general purpose face recognition. In H. Wechsler, P. J. Phillips, V. Bruce, F. Fogelman-Soulie, and T. S. Huang (Eds.), *Face Recognition: From Theory to Applications, NATO ASI Series F, Computer and Systems Sciences*, vol. 163, pages 446–456. 1998.

[47] G. Guo, S. Li, and K. Chan. Face recognition by support vector machines. In *Proc. of 4th IEEE International Conference on Automatic Face and Gesture Recognition 2000*, March 2000.

[48] J. Hartigan. Statistical theory in clustering. *Journal of Classification*, 2:63–76, 1985.

[49] Z.-Q. Hong and J.-Y. Yang. Optimal discriminant plane for a small number of samples and design method of classifier on the plane. *Pattern Recognition*, 24(4):317–324, 1991.

[50] X. W. Hou, S. Z. Li, and H. J. Zhang. Direct appearance models. In *Proc. of IEEE Computer Society Conference on Computer Vision and Pattern Recognition*, vol. 1, pages 828–833, December 2001.

[51] J. Huang, X. Shao, and H. Wechsler. Face pose discrimination using support vector machines (SVM). In *Proc. of International Conference Pattern Recognition*, 1998.

[52] A. Hyvrinen and P. Hoyer. Emergence of phase and shift invariant features by decomposition of natural images into independent feature subspaces. *Neural Computation*, 12:1705–1720, 2000.

[53] A. Hyvrinen and P. Hoyer. Emergence of topography and complex cell properties from natural images using extensions of ICA. In *Proc. of Advances in Neural Information Processing Systems*, vol. 12, pages 827–833, 2000.

[54] A. Jain and D. Zongker. Feature selection: evaluation, application, and small sample performance. *IEEE Trans. on PAMI*, 19(2):153–158, 1997.

[55] L. Jolliffe. *Principle Component Analysis*. Springer-Verlag, 1986.

[56] K. Jonsson, J. Matas, J. Kittler, and Y. Li. Learning support vectors for face verification and recognition. In *Proc. of 4th IEEE International Conference on Automatic Face and Gesture Recognition*, March 2000.

[57] T. Kanade. *Picture Processing by Computer Complex and Recognition of Human Faces*. PhD thesis, Kyoto University, 1973.

[58] K. I. Kim, K. Jung, and H. J. Kim. Face recognition using kernel principal component analysis. *IEEE Signal Processing Letters*, 9(2):40–42, February 2002.

[59] J. Kittler. Feature set search algorithm. In C. H. Chen (Ed.), *Pattern Recognition in Practice*, pages 41–60. NorthHolland, Sijthoff, and Noordhoof, 1980.

[60] A. Kuchinsky, C. Pering, M. L. Creech, D. Freeze, B. Serra, and J. Gwizdka. FotoFile: A consumer multimedia organization and retrieval system. In *Proc. of ACM SIG CHI'99 Conference*, Pittsburgh, May 1999.

[61] T.-W. Lee. *Independent Component Analysis: Theory and Applications*. Kluwer Academic, 1998.

[62] J. Li, S. Zhou, and C. Shekhar. A comparison of subspace analysis for face recognition. In *Proc. of the 28th IEEE International Conference on Acoustics, Speech, and Signal Processing*, April 2003.

[63] S. Z. Li, Q. D. Fu, L. Gu, B. Scholkopf, Y. M. Cheng, and H. J. Zhang. Kernel machine-based learning for multiview face detection and pose estimation. In *Proc. of 8th IEEE International Conference on Computer Vision*, July 2001.

[64] S. Z. Li and J. Lu. Face recognition using the nearest feature line method. *IEEE Transactions on Neural Networks*, 10:439–443, March 1999.

[65] S. Z. Li, X. G. Lv, and H. .Zhang. View-based clustering of object appearances based on independent subspace analysis. In *Proc. of The 8th IEEE International Conference on Computer Vision*, vol. 2, pages 295–300, July 2001.

[66] S. Z. Li, S. C. Yan, H. J. Zhang, and Q. S. Cheng. Multiview face alignment using direct appearance models. In *Proc. of IEEE International Conference on Automatic Face and Gesture Recognition*, 20-21 May 2002.

[67] S. Z. Li, Z. Zhang, H.-Y. Shum, and H. Zhang. FloatBoost learning for classification. In *Proc. of Neural Information Processing Systems*, pages –, December 2002.

[68] S. Z. Li, L. Zhu, Z. Q. Zhang, A. Blake, H. Zhang, and H. Shum. Statistical learning of multiview face detection. In *Proc. of the European Conference on Computer Vision*, vol. 4, pages 67–81, May 2002.

[69] Y. Li, S. Gong, and H. Liddell. Constructing facial identity surfaces in a nonlinear discriminating space. In *Proc. of IEEE Computer Society Conference on Computer Vision and Pattern Recognition*, December 2001.

[70] Y. M. Li, S. G. Gong, and H. Liddell. Support vector regression and classification based multiview face detection and recognition. In *IEEE Int. Conf. Oo Face & Gesture Recognition*, pages 300–305, March 2000.

[71] R. Lienhart, A. Kuranov, and V. Pisarevsky. Empirical analysis of detection cascades of boosted classifiers for rapid object detection. MRL technical report, Intel Labs, December 2002.

[72] C. Liu and H. Wechsler. Comparative assessment of independent component analysis (ica) for face recognition. In *Proc. of the 2nd International Conference on Audioand Video-based Biometric Person Authentication*, March 1999.

[73] C. Liu and H. Wechsler. Gabor feature-based classification using the enhanced fisher linear discriminant model for face recognition. *IEEE Transactions on Image Processing*, 11(4):467–476, April 2002.

[74] R. Lotlikar and R. Kothari. Fractional-step dimensionality reduction. *IEEE Transactions on Pattern Analysis and Machine Intelligence*, 22(6):623–627, 2000.

[75] J. Lu and K. Plataniotis. Boosting face recognition on a large-scale database. In *Proc. of the IEEE International Conference on Image Processing*, pages II.109–II.112, September 2002.

[76] J. Lu, K. Plataniotis, and A. Venetsanopoulos. Face recognition using feature optimization and ν-support vector learning. In *Proc. of the IEEE International Workshop on Neural Networks for Signal Processing*, pages 373–382, September 2001.

[77] J. Lu, K. Plataniotis, and A. Venetsanopoulos. Boosting linear discriminant analysis for face recognition. In *Proc. of the IEEE International Conference on Image Processing*, September 2003.

[78] J. Lu, K. Plataniotis, and A. Venetsanopoulos. Face recognition using kernel direct discriminant analysis algorithms. *IEEE Transactions on Neural Networks*, 14(1), January 2003.

[79] J. Lu, K. Plataniotis, and A. Venetsanopoulos. Face recognition using LDA based algorithms. *IEEE Transactions on Neural Networks*, 14(1), January 2003.

[80] J. Lu, K. Plataniotis, and A. Venetsanopoulos. Regularized discriminant analysis for the small sample size problem in face recognition. *Pattern Recognition Letter*, July 2003.

[81] D. Marr. *Vision*. W. H. Freeman, 1982.

[82] A. Martinez and R. Benavente. The AR face database. Technical Report 24, CVC, June 1998.

[83] A. M. Martnez and A. C. Kak. PCA versus LDA. *IEEE Transactions on Pattern Analysis and Machine Intelligence*, 23(2):228–233, 2001.

[84] G. McLachlan and D. Peel. *Finite Mixture Models*. John Wiley and Sons, 2000.

[85] B. Moghaddam. Principal manifolds and probabilistic subspaces for visual recognition. *IEEE Transactions on Pattern Analysis and Machine Intelligence*, 24(6):780–788, June 2002.

[86] B. Moghaddam, T. Jebara, and A. Pentland. Bayesian face recognition. *Pattern Recognition*, 33:1771–1782, 2000.

[87] B. Moghaddam and A. Pentland. Probabilistic visual learning for object representation. *IEEE Transactions on Pattern Analysis and Machine Intelligence*, 7:696–710, July 1997.

[88] H. Moon and P. Phillips. Analysis of pca-based face recognition algorithms. In K. Bowyer and P. Phillips (Eds.), *Empirical Evaluation Techniques in Computer Vision*, pages 57–71. IEEE CS Press, 1998.

[89] Y. Moses, Y. Adini, and S. Ullman. Face recognition: The problem of compensating for changes in illumination direction. In *Proc. of the European Conference on Computer Vision*, vol. A, pages 286–296, 1994.

[90] J. Ng and S. Gong. Performing multiview face detection and pose estimation using a composite support vector machine across the view sphere. In *Proc. IEEE International Workshop on Recognition, Analysis, and Tracking of Faces and Gestures in Real-Time Systems*, pages 14–21, September 1999.

[91] E. Osuna, R. Freund, and F. Girosi. Training support vector machines: An application to face detection. In *CVPR*, pages 130–136, 1997.

[92] C. P. Papageorgiou, M. Oren, and T. Poggio. A general framework for object detection. In *Proc. of IEEE International Conference on Computer Vision*, pages 555–562, 1998.

[93] A. P. Pentland, B. Moghaddam, and T. Starner. View-based and modular eigenspaces for face recognition. In *Proc. of IEEE Computer Society Conference on Computer Vision and Pattern Recognition*, pages 84–91, 1994.

[94] P. Phillips. Support vector machines applied to face recognition. In M. Kearns, S. Solla, and D. Cohn (Eds.), *NIPS*, 1998.

[95] P. J. Phillips, H. Moon, S. A. Rizvi, and P. J. Rauss. The FERET evaluation methodology for face-recognition algorithms. *IEEE Transactions on Pattern Analysis and Machine Intelligence*, 22(10):1090–1104, 2000.

[96] P. J. Phillips, H. Wechsler, J. Huang, and P. Rauss. The FERET database and evaluation procedure for face recognition algorithms. *Image and Vision Computing J.*, 16(5):295–306, 1998.

[97] P. Pudil, J. Novovicova, and J. Kittler. Floating search methods in feature selection. *Pattern Recognition Letters*, 15(11):1119–1125, 1994.

[98] S. Romdhani, A. Psarrou, and S. Gong. Learning a single active face shape model across views. In *Proc. IEEE International Workshop on Recognition, Analysis, and Tracking of Faces and Gestures in Real-Time Systems*, September 1999.

[99] D. Roth, M. Yang, and N. Ahuja. A snow-based face detector. In *Proc. of Neural Information Processing Systems*, 2000.

[100] H. A. Rowley, S. Baluja, and T. Kanade. Neural network-based face detection. *IEEE Transactions on Pattern Analysis and Machine Intelligence*, 20(1):23–28, 1998.

[101] A. Samal and P. A. Iyengar. Automatic recognition and analysis of human faces and facial expressions: A survey. *Pattern Recognition*, 25:65–77, 1992.

[102] R. Schapire, Y. Freund, P. Bartlett, and W. S. Lee. Boosting the margin: A new explanation for the effectiveness of voting methods. *The Annals of Statistics*, 26(5):1651–1686, October 1998.

[103] B. Schiele and J. L. Crowley. Recognition without correspondence using multidimensional receptive field histograms. *International Journal of Computer Vision*, 36(1):31–52, 2000.

[104] H. Schneiderman. *A Statistical Approach to 3D Object Detection Applied to Faces and Cars*. PhD thesis (CMU-RI-TR-00-06), Robotics Institute, Carnegie Mellon University, 2000.

[105] H. Schneiderman and T. Kanade. A statistical method for 3d object detection applied to faces and cars. In *Proc. of IEEE Computer Society Conference on Computer Vision and Pattern Recognition*, 2000.

[106] B. Schölkopf, C. Burges, and A. J. Smola. *Advances in Kernel Methods: Support Vector Learning*. MIT Press, 1999.

[107] B. Schölkopf, A. Smola, and K. R. Müller. Nonlinear component analysis as a kernel eigenvalue problem. *Neural Computation*, 10:1299–1319, 1999.

[108] S. Sclaroff and J. Isidoro. Active blobs. In *Proc. of IEEE International Conference on Computer Vision*, 1998.

[109] P. Y. Simard, L. Bottou, P. Haffner, and Y. L. Cun. Boxlets: A fast convolution algorithm for signal processing and neural networks. In M. Kearns, S. Solla, and D. Cohn (Eds.), *Advances in Neural Information Processing Systems*, vol. 11, pages 571–577. MIT Press, 1998.

[110] P. Y. Simard, Y. A. L. Cun, J. S. Denker, and B. Victorri. Transformation invariance in pattern recognition: Tangent distance and tangent propagation. In G. B. Orr and K.-R. Muller (Eds.), *Neural Networks: Tricks of the Trade*, Springer, 1998.

[111] L. Sirovich and M. Kirby. Low-dimensional procedure for the characterization of human faces. *Journal of the Optical Society of America A*, 4(3):519–524, March 1987.

[112] P. Somol, P. Pudil, J. Novoviova, and P. Paclik. Adaptive floating search methods in feature selection. *Pattern Recognition Letters*, 20:1157–1163, 1999.

[113] S. D. Stearns. On selecting features for pattern classifiers. In *Proc. of International Conference Pattern Recognition*, pages 71–75, 1976.

[114] K.-K. Sung and T. Poggio. Example-based learning for view-based human face detection. *IEEE Transactions on Pattern Analysis and Machine Intelligence*, 20(1):39–51, 1998.

[115] D. L. Swets and J. Weng. Using discriminant eigenfeatures for image retrieval. *IEEE Transactions on Pattern Analysis and Machine Intelligence*, 18:831–836, 1996.

[116] Q. Tian, M. Barbero, Z. Gu, and S. Lee. Image classification by the Foley-Sammon transform. *Opt. Eng.*, 25(7):834–840, 1986.

[117] K. Tieu and P. Viola. Boosting image retrieval. In *Proc. of IEEE Computer Society Conference on Computer Vision and Pattern Recognition*, vol. 1, pages 228–235, 2000.

[118] M. Tipping and C. Bishop. Probabilistic principal component analysis. *Journal of the Royal Statistical Society, Series B.*, 61:611–622, 1999.

[119] K. A. Toh. Global optimization by monotonic transformation. *Computational Optimization and Applications*, 23:77–99, October 2002.

[120] K.-A. Toh, J. Lu, and W.-Y. Yau. Global feedforward neural network learning for classification and regression. In M. Figueiredo, J. Zerubia, and A. K. Jain (Eds.), *Proc. of the Energy Minimization Methods in Computer Vision and Pattern Recognition*, September 2001.

[121] M. Turk. A random walk through eigenspace. *IEICE Trans. Inf. & Syst.*, E84-D(12):1586–1695, December 2001.

[122] M. A. Turk and A. P. Pentland. Eigenfaces for recognition. *Journal of Cognitive Neuroscience*, 3(1):71–86, March 1991.

[123] D. Valentin, H. Abdi, A. J. O'Toole, and G. W. Cottrell. Connectionist models of face processing: A survey. *Pattern Recognition*, 27(9):1209–1230, 1994.

[124] B. van Ginneken, A. F. Frangi, J. J. Staal, B. M. ter Haar Romeny, and M. A. Viergever. A non-linear gray-level appearance model improves active shape model segmentation. *In Proc. of Mathematical Methods in Biomedical Image Analysis*, 2001.

[125] V. N. Vapnik. *The Nature of Statistical Learning Theory*. Springer-Verlag, 1995.

[126] V. N. Vapnik. An overview of statistical learning theory. *IEEE Transactions on Neural Networks*, 10(5):439–443, September 1999.

[127] P. Viola and M. Jones. Rapid object detection using a boosted cascade of simple features. In *Proc. of IEEE Computer Society Conference on Computer Vision and Pattern Recognition*, December 2001.

[128] P. Viola and M. Jones. Robust real time object detection. In *IEEE ICCV Workshop on Statistical and Computational Theories of Vision*, July 2001.

[129] J. Wilder. Face recognition using transform coding of grayscale projection projections and the neural tree network. In R. J. Mammone (Ed.), *Artificial Neural Networks with Applications in Speech and Vision*, pages 520–536, Chapman Hall, 1994.

[130] L. Wiskott, J. Fellous, N. Kruger, and C. V. malsburg. Face recognition by elastic bunch graph matching. *IEEE Transactions on Pattern Analysis and Machine Intelligence*, 19(7):775–779, 1997.

[131] S. C. Yan, C. Liu, S. Z. Li, L. Zhu, H. J. Zhang, H. Shum, and Q. Cheng. Texture-constrained active shape models. In *Proc. of the First International Workshop on Generative-Model-Based Vision*, May 2002.

[132] J. Yang, W. Lu, and A. Waibel. Skin-color modeling and adaptation. In *Proc. of the 1st Asian Conference on Computer Vision*, pages 687–694, 1998.

[133] M.-H. Yang. Resources for face detection [Online]. Available: http://vision.3ai.uiuc.edu/mhyang/face-detection-survey.html.

[134] M.-H. Yang. Kernel eigenfaces vs. kernel fisherfaces: Face recognition using kernel methods. In *Proc. of the 5th IEEE International Conference on Automatic Face and Gesture Recognition*, May 2002.

[135] M.-H. Yang and N. Ahuja. Gaussian mixture model for human skin color and its application in image and video databases. In *Proc. of the SPIE Conf. on Storage and Retrieval for Image and Video Databases*, vol. 3656, pages 458–466, January 1999.

[136] M.-H. Yang, D. Kriegman, and N. Ahuja. Detecting faces in images: A survey. *IEEE Transactions on Pattern Analysis and Machine Intelligence*, 24(1):34–58, 2002.

[137] H. Yu and J. Yang. A direct LDA algorithm for high-dimensional data with application to face recognition. *Pattern Recognition*, 34:2067–2070, 2001.

[138] B. D. Zarit, B. J. Super, and F. K. H. Quek. Comparison of five color models in skin pixel classification. In *IEEE ICCV Workshop on Recognition, Analysis and Tracking of Faces and Gestures in Real-time Systems*, pages 58–63, September 1999.

[139] W. Zhao, R. Chellappa, and J. Phillips. Subspace linear discriminant analysis for face recognition. Technical Report, CS-TR4009, University of Maryland, 1999.

[140] W. Zhao, R. Chellappa, A. Rosenfeld, and P. Phillips. Face recognition: A literature survey. Technical Report, CFAR-TR00-948, University of Maryland, 2000.

[141] W. Zhao, A. Krishnaswamy, R. Chellappa, D. Swets, and J. Weng. Discriminant analysis of principal components for face recognition. In H. Wechsler, P. Phillips, V. Bruce, F. Soulie, and T. Huang (Eds.), *Face Recognition: From Theory to Applications*, pages 73–85, Springer-Verlag, 1998.

Chapter 10

PERCEPTUAL INTERFACES

Matthew Turk

and Mathias Kölsch

A keyboard! How quaint.

— Scotty, in the film *Star Trek IV: The Voyage Home* (1986)

10.1 Introduction

Computer vision research has traditionally been motivated by a few main areas of application. The most prominent of these include biological vision modeling, robot navigation and manipulation, surveillance, medical imaging, and various inspection, detection, and recognition tasks. In recent years, a new area, often referred to as *perceptual interfaces*, has emerged to motivate an increasingly large amount of research within the machine vision community. The general focus of this effort is to integrate multiple perceptual modalities (such as computer vision, speech and sound processing, and haptic I/O) into the user interface. For computer vision technology in particular, the primary aim is to use vision as an effective input modality in human–computer interaction (HCI). Broadly defined, perceptual interfaces are highly interactive, multimodal interfaces that enable rich, natural, and efficient interaction with computers. More specifically, perceptual interfaces seek to leverage sensing (input) and rendering (output) technologies in order to provide interactions not feasible with standard interfaces and the common triumvirate of I/O devices: the keyboard, mouse, and monitor.

The motivation behind perceptual interfaces is twofold: (1) the changing nature of computers and (2) the desire for a more powerful, compelling user experience than what has been available with graphical user interfaces (GUI) and the associated WIMP (windows, icons, menus, pointing devices) implementations. As computers evolve away from their recent past—desktop machines used primarily for word processing, spreadsheet manipulation, and information browsing—and move toward new environments with a plethora of computing form factors, uses, and interaction scenarios, the desktop metaphor will become less relevant and more cumbersome. Keyboard-based alphanumeric input and mouse-based 2D pointing and selection can be very limiting, and in some cases awkward and inefficient, modes of interaction. Neither mouse nor keyboard, for example, is very appropriate for communicating 3D information or the subtleties of human emotions.

Moore's law has driven computer hardware over the decades, increasing performance (measured in various ways) exponentially. This observation predicts an improvement in chip density in 5 years by a factor of 10; in 10 years by a factor of 100; and in 20 years by a factor of 10,000. Unfortunately, human capacity does not grow at such a rate (if at all) so there is a serious problem in scaling HCI as machines evolve. It is unlikely that a user interface paradigm developed at an early point in the Moore's law curve will continue to be appropriate much later on.

New computing scenarios, such as in automobiles and other mobile environments, rule out many traditional approaches to user interaction. Computing is becoming something that permeates daily life rather than something people do only at distinct times and places. In order to accommodate a wider range of scenarios, tasks, users, and preferences, interfaces must become more natural, intuitive, adaptive, and unobtrusive. These are primary goals of research in perceptual interfaces.

We will certainly need new and different interaction techniques in a world of small, powerful, connected, ubiquitous computing. Since small, powerful, connected sensing and display technologies should be available, there has been increased interest in building interfaces that use these technologies to leverage the natural human capabilities to communicate via speech, gesture, expression, touch, and so on. While these are unlikely to completely replace traditional desktop and GUI-based interfaces, they will complement existing interaction styles and enable new functionality not otherwise possible or convenient.

In this chapter, we seek to communicate the motivations and goals of perceptual interfaces, to enumerate the relevant technologies, to discuss the integration of multiple modalities, and to describe in more detail the role of computer vision in HCI. We cover vision problems, constraints, and approaches that are apropos to the area, survey the state of the art in computer vision research applied to perceptual interfaces, describe several near-term applications, and suggest promising research directions.

10.2 Perceptual Interfaces and HCI

HCI is the study of people, computer technology, and the ways these influence each other. In practice, HCI involves the design, evaluation, and implementation of interactive computing systems for human use. The discipline has its early roots in studies of human performance in manufacturing industries. Human factors, or ergonomics, originally attempted to maximize worker productivity by designing equipment to reduce operator fatigue and discomfort. With the arrival of computers and their spread into the workforce, many human factors researchers began to specialize in the various issues surrounding the use of computers by people. HCI is now a very broad interdisciplinary field involving computer scientists, psychologists, cognitive scientists, human factors researchers, and many other disciplines, and it involves the design, implementation, and evaluation of interactive computer systems in the context of the work or tasks in which a user is engaged [30].

As one element of HCI, the user interface is typically considered as the portion of a computer program with which the user interacts; that is, the point of contact between the human and the computer. Shneiderman [124] describes five human factors objectives that should guide designers and evaluators of user interfaces:

1. Time to learn
2. Speed of performance
3. Rate of errors by users
4. Retention over time
5. Subjective satisfaction

Shneiderman also identifies the accommodation of human diversity as a major goal and challenge in the design of interactive systems, citing the remarkable diversity of human abilities, backgrounds, motivations, personalities, and work styles of users. People have a range of perceptual, cognitive, and motor abilities and limitations. In addition, different cultures produce different perspectives and styles of interaction, a significant issue in today's

Table 10.1. The evolution of user interface paradigms

When	**Implementation**	**Paradigm**
1950s	Switches, punch cards, lights	None
1970s	Command-line interface	Typewriter
1980s	WIMP-based GUI	Desktop
2000s	Perceptual interfaces	Natural interaction

international markets. Users with various kinds of disabilities, elderly users, and children all have distinct preferences or requirements to enable a positive user experience.

In addition to human factors considerations for HCI in the context of typical workplace and consumer uses of computers, the cutting-edge uses of computer technology in virtual and augmented reality systems, wearable computers, ubiquitous computing environments, and other such scenarios demand a fresh view of usability and user interface design. Theoretical and experimental advances (such as the concept of Fitts' law [39]) have to be translated into new arenas, which also require new analyses of usability. Despite the apparent ubiquity of GUIs, they are not the answer to all interactive system needs.

Historically, a few major user interface paradigms have dominated computing. Table 10.1 describes one view of the evolution of user interfaces. In the early days of computing, there was no real model of interaction—data was entered into the computer via switches or punched cards, and the output was produced (some time later) via punched cards or lights. The second phase began with the arrival of command-line interfaces in perhaps the early 1960s, first using teletype terminals and later electronic keyboards and text-based monitors. This "typewriter" model—where the user types a command (with appropriate parameters) to the computer, hits carriage return, and gets typed output from the computer—was spurred on by the development of timesharing systems and continued with the popular UNIX and DOS operating systems.

In the 1970s and 1980s, the GUI and its associated desktop metaphor arrived, often described by the acronym WIMP (windows, icons, menus, and a pointing device). For over two decades, graphical interfaces have dominated both the marketplace and HCI research, and for good reason: WIMP-based

GUIs have provided a standard set of direct manipulation techniques that largely rely on recognition rather than recall. That is, GUI-based commands can typically be easily found and do not have to be remembered or memorized. Direct manipulation is appealing to novice users, it is easy to remember for occasional users, and it can be fast and efficient for frequent users [124]. Direct manipulation interfaces, in general, allow easy learning and retention, encourage exploration (especially with Undo commands), and can give users a sense of accomplishment and responsibility for the sequence of actions leading to the completion of a task or subtask. The direct manipulation style of interaction with GUIs has been a good match with the office productivity and information access applications that have been the "killer apps" of computing to date.

However, as computing changes in terms of physical size and capacity, usage, and ubiquity, the obvious question arises: What is the next major generation in the evolution of user interfaces? Is there a paradigm (and its associated technology) that will displace GUI and become the dominant user interface model? Or will computer interfaces fragment into different models for different tasks and contexts? There is no shortage of HCI research areas billed as "advanced" or "future" interfaces—these include various flavors of immersive environments (virtual, augmented, and mixed reality), 3D interfaces, tangible interfaces, haptic interfaces, affective computing, ubiquitous computing, and multimodal interfaces. These are collectively called "post-WIMP" interfaces by van Dam [138], a general phrase describing interaction techniques not dependent on classical 2D widgets such as menus and icons.

The (admittedly grandiose) claim of this chapter is that the next dominant, long-lasting HCI paradigm is what many people call *perceptual interfaces*:

> Perceptual User Interfaces (PUIs) are characterized by interaction techniques that combine an understanding of natural human capabilities (particularly communication, motor, cognitive, and perceptual skills) with computer I/O devices and machine perception and reasoning. They seek to make the user interface more natural and compelling by taking advantage of the ways in which people naturally interact with each other and with the world—both verbal and non-verbal communications. Devices and sensors should be transparent and passive if possible, and machines should both perceive relevant human communication channels and generate output that is naturally understood. This is expected to require integration at multiple levels of technologies such as speech and sound recognition and generation, computer

vision, graphical animation and visualization, language understanding, touch-based sensing and feedback (haptics), learning, user modeling, and dialog management. (Turk and Robertson [136])

There are two key features of perceptual interfaces. First, they are highly interactive. Unlike traditional passive interfaces that wait for users to enter commands before taking any action, perceptual interfaces actively sense and perceive the world and take actions based on goals and knowledge at various levels. (Ideally, this is an "active" interface that uses "passive," or nonintrusive, sensing.) Second, they are multimodal, making use of multiple perceptual modalities (e.g., sight, hearing, touch) in both directions: from the computer to the user and from the user to the computer. Perceptual interfaces move beyond the limited modalities and channels available with a keyboard, mouse, and monitor to take advantage of a wider range of modalities, either sequentially or in parallel.

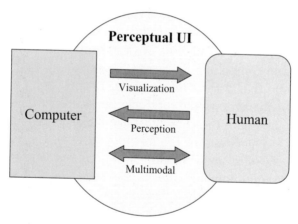

Figure 10.1. Information flow in perceptual interfaces.

The concept of perceptual interfaces is closely related to multimodal, multimedia and recognition-based interfaces, as depicted in Figure 10.1. Multimodal interfaces tend to focus on the input direction (input to the computer) and are most often extensions to current GUI- and language-based interfaces. Multimedia interfaces describe the presentation of various media to users along with some (again, GUI-based) interaction techniques to control and query the media sources. Interfaces based on individual recognition technologies (such as speech, vision, pen-gesture) focus on the individual recognition technologies, with little integration across modalities. Although each of these classes has significant overlap with the idea of perceptual interfaces, none of them provides a clear conceptual model, or an overarching interaction paradigm, to the user.

The general model for perceptual interfaces is that of human-to-human communication. While this is not universally accepted in the HCI community as the ultimate interface model (e.g., see [121, 122, 123]), there are several practical and intuitive reasons that it makes sense to pursue this goal. Human interaction is natural and in many ways effortless; beyond an early age, people do not need to learn special techniques or commands to communicate with one another. There is a richness in human communication via verbal, visual, and haptic modalities, underscored by shared social conventions, shared knowledge, and the ability to adapt and to model another person's point of view, that is very different from current computer interfaces, which essentially implement a precise command-and-control interaction style. Figure 10.2 depicts natural interaction between people and, similarly, between humans and computers. Perceptual interfaces can potentially effect improvements in the human factors objectives mentioned earlier in the section, as they can be easy to learn and efficient to use, they can reduce error rates by giving users multiple and redundant ways to communicate, and they can be very satisfying and compelling for users.

People are adaptive in their interactions. Despite an abundance of ambiguity in natural language, people routinely pursue directions in conversation intended to disambiguate the content of the message. We do the same task in multiple ways, depending on the circumstances of the moment. A computer system that can respond to different modalities or interaction methods depending on the context would allow someone to perform a given task with ease whether he or she is in the office, in the car, or walking along a noisy street. Systems that are aware of the user and his or her activities can make appropriate decisions on how and when to best present information.

A number of studies by Reeves and Nass and their colleagues [96, 109, 95] provide compelling evidence that human interaction with computers and other communication technologies is fundamentally social and natural. These studies have produced (though typically reduced) social effects in HCI similar to those found in person-to-person interactions. The general approach to this work has been to choose a social-science finding regarding people's behaviors or attitudes and to determine if the relevant social rule still applies (and to what magnitude) when one of the roles is filled by a computer rather than a human. For example, will users apply norms of politeness or gender stereotypes to computers? In general, which social rules will people apply to computers, and how powerful are these rules?

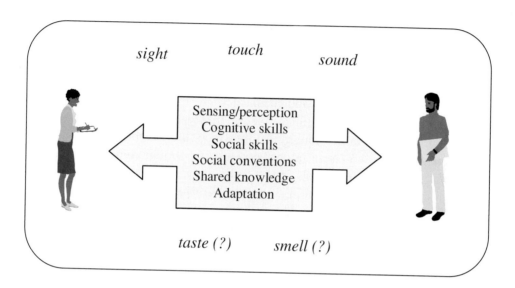

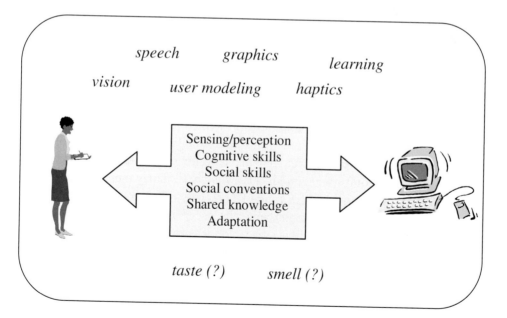

Figure 10.2. Models of interaction: human–human interaction and perceptual human–computer interaction.

Such studies have found that social responses are automatic and unconscious, and can be elicited by basic cues. People show social responses to cues regarding manners and politeness, personality, emotion, gender, trust, ethics, and other social concepts. Much of the research in social interfaces has focused on embodied conversational agents, or lifelike characters, that use speech and language processing to communicate with a human user [4, 3, 16, 1]. Although the application of these findings is not simple and straightforward, there are implications for future HCI technologies and for perceptual interfaces in particular. An interface that perceives, reasons, and acts in a social manner—even imperfectly—is not too far removed from people's current conceptual models of the technology.

Despite technical advances in speech recognition, speech synthesis, computer vision, and artificial intelligence, computers are still, by and large, deaf, dumb, and blind. Many have noted the irony of public restrooms that are "smarter" than computers because they can sense when people come and go, and can act appropriately, while a computer may wait all night for a user (who has gone home for the evening) to respond to the dialog that asks "Are you sure you want to do this?" While general-purpose machine intelligence is still a difficult and elusive goal, our belief is that much can be gained by pursuing an agenda of technologies to support the human-to-human model of interaction. Even if the Holy Grail of perceptual interfaces is far in the future, the near-term benefits may be transformational as component technologies and integration techniques mature and provide new tools to improve the ways in which people interact with technology.

In addition to the general goal of interfaces that better match human capabilities and provide a more natural and compelling user experience, there are immediate application areas that are ripe for early perceptual interface technologies. The computer game industry is particularly compelling, as it is large and its population of users tends to be early adopters of new technologies. Game interfaces that can perceive the user's identity, body movement, and speech, for example, are likely to become very popular. Another group of people who may have a lot to gain from early perceptual interfaces are users with physical disabilities. Interfaces that are more adaptable and flexible, and not limited to particular ways of moving a mouse or typing keys, will provide a significant benefit to this community.

Several other areas—such as entertainment, personal robotics, multimedia learning, and biometrics—would clearly seem to benefit from initial advances in perceptual interfaces. Eventually, the applications of technologies that can sense, perceive, understand, and respond appropriately to human behavior appear to be unlimited.

What is necessary in order to bring about this vision of perceptual interfaces? A better understanding of human capabilities, limitations, and preferences in interaction is important, including physical, social, and cognitive aspects. Advances in several technical areas, some quite old and some relatively new, are also vital. Speech understanding and generation (i.e., speech recognition, natural language processing, speech synthesis, discourse modeling, and dialogue management) are vital to leverage the natural modality of spoken language as well as other sound recognition and synthesis tools (e.g., in addition to words, systems should recognize a sneeze, a cough, a loud plane passing overhead, a general noisy environment, etc.). Computer graphics and information visualization are important to provide richer ways of communicating to users. Affective computing [105] may be vital to understand and generate natural interaction, especially subtle cues to aspects of humor and irony, and appropriate context-dependent displays of emotion (or lack thereof). Haptic and tangible interfaces (e.g., [7, 57]) that leverage physical aspects of interaction may also be important in building truly perceptual interfaces. User and task modeling (e.g., [52, 53]) is key to understanding the whole context of interaction.

Computer vision is also a vital element of perceptual interfaces. Whether alone or in conjunction with other perceptual modalities, visual information provides useful and important cues to interaction. The presence, location, and posture of a user may be important contextual information; a gesture or facial expression may be a key signal; the direction of gaze may disambiguate the object referred to in language as "this" or "that thing." The next section of the chapter describes the scope of computer vision research and development as it relates to this area of perceptual interfaces.

In addition to advances in individual component areas, the integration of multiple modalities is of fundamental importance in perceptual interfaces. Both lower-level fusion techniques and higher-level integration frameworks are necessary to build interactive, multimodal interfaces that provide a compelling, natural user experience.

10.3   Multimodal Interfaces

A multimodal interface is a system that combines two or more input modalities in a coordinated manner. Perceptual interfaces are inherently multimodal. In this section, we define more precisely what we mean by *modes* and *channels*, and discuss research in multimodal interfaces and how this relates to the more general concept of perceptual interfaces.

Humans interact with the world by way of information being sent and received, primarily through the five major senses of sight, hearing, touch, taste, and smell. A *modality* (informally, a mode) refers to a particular sense. A communication *channel* is a course or pathway through which information is transmitted. In typical HCI usage, a channel describes the interaction technique that utilizes a particular combination of user and computer communication—that is, the user output/computer input pair or the computer output/user input pair.[1] This can be based on a particular device, such as the keyboard channel or the mouse channel, or on a particular action, such as spoken language, written language, or dynamic gestures. In this view, the following are all channels: text (which may use multiple modalities when typing in text or reading text on a monitor), sound, speech recognition, images/video, and mouse pointing and clicking.

Unfortunately, there is some ambiguity in the use of the word *mode* in HCI circles, as sometimes it means "modality" and at other times it means "channel." So, are multimodal interfaces "multimodality" or "multichannel"? Certainly, every command-line interface uses multiple modalities, as sight and touch (and sometimes sound) are vital to these systems. The same is true for GUIs, which in addition use multiple channels of keyboard text entry, mouse pointing and clicking, sound, images, and so on.

What, then, distinguishes multimodal interfaces from other HCI technologies? As a research field, multimodal interfaces focus on integrating sensor recognition-based input technologies such as speech recognition, pen gesture recognition, and computer vision, into the user interface. The function of each technology is better thought of as a channel than as a sensing modality; hence, in our view, a multimodal interface is one that uses multiple modalities to implement multiple channels of communication. Using multiple modalities to produce a single interface channel (e.g., vision and sound to produce 3D user location) is multisensor fusion, not a multimodal interface. Similarly, using a single modality to produce multiple channels (e.g., a left-hand mouse to navigate and a right-hand mouse to select) is a multichannel (or multidevice) interface, not a multimodal interface.

An early prototypical multimodal interfaces was the "Put That There" prototype system demonstrated at MIT in the early 1980s [10]. In this system, the user communicated via speech and pointing gestures in a "media room." The gestures served to disambiguate the speech (Which object does

---

[1]Input means *to the computer*; output means *from the computer*.

the word "this" refer to? What location is meant by "there"?) and effected other direct interactions with the system. More recently, the QuickSet architecture [18] is a good example of a multimodal system using speech and pen-based gesture to interact with map-based and 3D visualization systems. QuickSet is a wireless, handheld, agent-based, collaborative multimodal system for interacting with distributed applications. The system analyzes continuous speech and pen gesture in real time and produces a joint semantic interpretation using a statistical unification-based approach. The system supports unimodal speech or gesture as well as multimodal input.

Multimodal systems and architectures vary along several key dimensions or characteristics, including

- Number and type of input modalities;
- Number and type of communication channels;
- Ability to use modes in parallel, serially, or both;
- Size and type of recognition vocabularies;
- Methods of sensor and channel integration;
- Kinds of applications supported.

There are many potential advantages of multimodal interfaces, including the following [101]:

- They permit the flexible use of input modes, including alternation and integrated use.
- They support improved efficiency, especially when manipulating graphical information.
- They can support shorter and simpler speech utterances than a speech-only interface, which results in fewer disfluencies and more robust speech recognition.
- They can support greater precision of spatial information than a speech-only interface, since pen input can be quite precise.
- They give users alternatives in their interaction techniques.
- They lead to enhanced error avoidance and ease of error resolution.
- They accommodate a wider range of users, tasks, and environmental situations.
- They are adaptable during continuously changing environmental conditions.
- They accommodate individual differences, such as permanent or temporary handicaps.

– They can help prevent overuse of any individual mode during extended computer usage.

Oviatt and Cohen and their colleagues at the Oregon Health and Science University (formerly Oregon Graduate Institute) have been at the forefront of multimodal interface research, building and analyzing multimodal systems over a number of years for a variety of applications. Oviatt's "Ten Myths of Multimodal Interaction" [100] are enlightening for anyone trying to understand the area. We list Oviatt's myths in italics, with our accompanying comments:

Myth 1. *If you build a multimodal system, users will interact multimodally.* In fact, users tend to intermix unimodal and multimodal interactions; multimodal interactions are often predictable based on the type of action being performed.

Myth 2. *Speech and pointing is the dominant multimodal integration pattern.* This is only one of many interaction combinations, comprising perhaps 14% of all spontaneous multimodal utterances.

Myth 3. *Multimodal input involves simultaneous signals.* Multimodal signals often do not co-occur temporally.

Myth 4. *Speech is the primary input mode in any multimodal system that includes it.* Speech is not the exclusive carrier of important content in multimodal systems, nor does it necessarily have temporal precedence over other input modes.

Myth 5. *Multimodal language does not differ linguistically from unimodal language.* Multimodal language is different, and often much simplified, compared with unimodal language.

Myth 6. *Multimodal integration involves redundancy of content between modes.* Complementarity of content is probably more significant in multimodal systems than is redundancy.

Myth 7. *Individual error-prone recognition technologies combine multimodally to produce even greater unreliability.* In a flexible multimodal interface, people figure out how to use the available input modes effectively; in addition, there can be mutual disambiguation of signals that also contributes to a higher level of robustness.

Myth 8. *All users' multimodal commands are integrated in a uniform way.* Different users may have different dominant integration patterns.

Myth 9. *Different input modes are capable of transmitting comparable content.* Different modes vary in the type and content of their information, their functionality, the ways they are integrated, and their suitability for multimodal integration.

Myth 10. *Enhanced efficiency is the main advantage of multimodal systems.* While multimodal systems may increase efficiency, this may not always be the case. The advantages may reside elsewhere, such as in decreased errors, increased flexibility, or increased user satisfaction.

A technical key to multimodal interfaces is the specific integration levels and technique(s) used. Integration of multiple sources of information is generally characterized as "early," "late," or somewhere in between. In early integration (or *feature fusion*), the raw data from multiple sources (or data that has been processed somewhat, perhaps into component features) are combined and recognition or classification proceeds in the multidimensional space. In late integration (or *semantic fusion*), individual sensor channels are processed through some level of classification before the results are integrated. Figure 10.3 shows a view of these alternatives. In practice, integration schemes may combine elements of early and late integration, or even do both in parallel.

There are advantages to using late, semantic integration of multiple modalities in multimodal systems. For example, the input types can be recognized independently and therefore do not have to occur simultaneously. The training requirements are smaller, $O(2N)$, for two separately trained modes as opposed to $O(N^2)$ for two modes trained together. The software development process is also simpler in the late integration case, as exemplified by the QuickSet architecture [148]. Quickset uses temporal and semantic filtering, unification as the fundamental integration technique, and a statistical ranking to decide among multiple consistent interpretations.

Multimodal interface systems have used a number of non-traditional modes and technologies. Some of the most common are the following:

- **Speech recognition.** Speech recognition has a long history of research and commercial deployment, and has been a popular component of multimodal systems for obvious reasons. Speech is a very important and flexible communication modality for humans and is much more

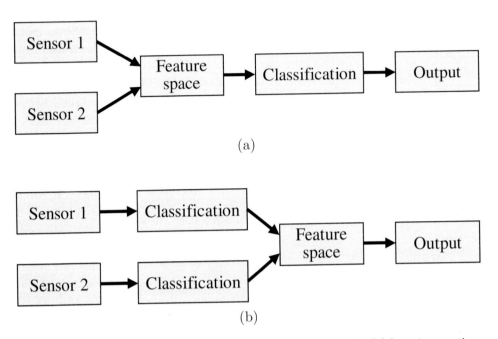

Figure 10.3. (a) Early integration, fusion at the feature level. (b) Late integration, fusion at the semantic level.

natural than typing or any other way of expressing particular words, phrases, and longer utterances. Despite the decades of research in speech recognition and over a decade of commercially available speech recognition products, the technology is still far from perfect, due to the size, complexity, and subtlety of language; the limitations of microphone technology; the plethora of disfluencies in natural speech; and problems of noisy environments. Systems using speech recognition have to be able to recover from the inevitable errors produced by the system.

– **Language understanding.** Natural language processing attempts to model and understand human language, whether spoken or written. In multimodal interfaces, language understanding may be hand-in-hand with speech recognition (together forming a "speech understanding" component), or it may be separate, processing the user's typed or handwritten input. Typically, the more a system incorporates natural language, the more users will expect sophisticated semantic understanding from the system. Current systems are unable to deal with completely unconstrained language, but can do quite well with limited vocabularies and subject matter. Allowing for user feedback to clarify and disambiguate language input can help language understanding systems significantly.

– **Pen-based gesture.** Pen-based gesture has been popular in part because of computer form factors (PDAs and tablet computers) that include a pen or stylus as a primary input device. Pen input is particularly useful for deictic (pointing) gestures, defining lines, contours, and areas, and specially defined gesture commands (e.g., minimizing a window by drawing a large M on the screen). Pen-based systems are quite useful in mobile computing, where a small computer can be carried, but a keyboard is impractical.

– **Sensors (such as magnetic and inertial) for body tracking.** Sturman's 1991 thesis [131] thoroughly documented the early use of sensors worn on the hand for input to interactive systems. Magnetic tracking sensors such as the Ascension Flock of Birds[2] product, various instrumented gloves, and sensor- or marker-based motion capture devices have been used in multimodal interfaces, particularly in immersive environments (e.g., see [50]).

– **Nonspeech sound.** Nonspeech sounds have traditionally been used in HCI to provide signals to the user: for example, warnings, alarms, and status information. (Ironically, one of the most useful sounds for computer users is rather serendipitous: the noise made by many hard drives that lets a user know that the machine is still computing rather than hung.) However, nonspeech sound can also be a useful input channel, as sound made by users can be meaningful events in human-to-human communication—utterances such as "uh-huh" used in *backchannel* communication (communication events that occur in the background of an interaction rather than being the main focus), a laugh, a sigh, or a clapping of hands.

– **Haptic input and force feedback.** Haptic, or touch-based, input devices measure pressure, velocity, location—essentially perceiving aspects of a user's manipulative and explorative manual actions. These can be integrated into existing devices (e.g., keyboards and mice that

[2]http://www.ascension-tech.com

Figure 10.4. SensAble Technologies, Inc. PHANTOM haptic input/output device (reprinted with permission).

know when they are being touched, and possibly by whom). Or they can exist as standalone devices, such as the well-known PHANTOM device by SensAble Technologies, Inc.[3] (see Figure 10.4), or the DELTA device by Force Dimension.[4] These and most other haptic devices integrate force feedback and allow the user to experience the "touch and feel" of simulated artifacts as if they were real. Through the mediator of a handheld stylus or probe, haptic exploration can now receive simulated feedback, including rigid boundaries of virtual objects, soft tissue, and surface texture properties. A tempting goal is to simulate all haptic experiences and to be able to recreate objects with all their physical properties in virtual worlds so they can be touched and handled in a natural way. The tremendous dexterousity of the human hand makes this very difficult. Yet, astonishing results can already be achieved, for example with the CyberForce device, which can produce forces on each finger and the entire arm. The same company, Immer-

---

[3]http://www.sensable.com
[4]http://www.forcedimension.com

sion Corp.,[5] also supplies the iDrive, a hybrid of a rotary knob and joystick input interface to board computers of BMW's flagship cars. This is the first attempt outside the gaming industry to bring haptic and force-feedback interfaces to the general consumer.

- **Computer vision.** Computer vision has many advantages as an input modality for multimodal or perceptual interfaces. Visual information is clearly important in human–human communication, as meaningful information is conveyed through identity, facial expression, posture, gestures, and other visually observable cues. Sensing and perceiving these visual cues from video cameras appropriately placed in the environment is the domain of computer vision. The following section describes relevant computer vision technologies in more detail.

10.4   Vision-Based Interfaces

Vision supports a wide range of human tasks, including recognition, navigation, balance, reading, and communication. In the context of perceptual interfaces, the primary task of computer vision is to detect and recognize meaningful visual cues to communication—that is, to "watch the users" and report on their locations, expressions, gestures, and so on. While vision is one of possibly several sources of information about the interaction to be combined multimodally in a perceptual interface, in this section we focus solely on the vision modality. Using computer vision to sense and perceive the user in an HCI context is often called vision-based interaction, or vision-based interfaces (VBI).

In order to accurately model human interaction, it is necessary to take every observable behavior into account [69, 68]. The analysis of human movement and gesture coordinated with speech conversations has a long history in areas such as sociology, communication, and therapy (e.g., Scheflen and Birdwhistell's Context Analysis [115]). These analyses, however, are often quite subjective and ill-suited for computational analysis. VBI aims to produce precise and real-time analysis that will be useful in a wide range of applications, from communication to games to automatic annotation of human–human interaction.

There is a range of human activity that has occupied VBI research over the past decade; Figure 10.5 shows some of these from the camera's view-

---

[5]http://www.immersion.com

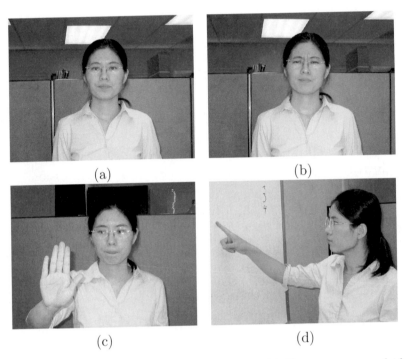

Figure 10.5. Some common visual cues for VBI. (a) User presence and identity. (b) Facial expression. (c) A simple gesture. (d) A pointing gesture and focus of attention.

point. Key aspects of VBI include the detection and recognition of the following elements:

- Presence and location: Is someone there? How many people? Where are they (in 2D or 3D)? (face detection, body detection, head and body tracking)

- Identity: Who are they? (face recognition, gait recognition)

- Expression: Is a person smiling, frowning, laughing, speaking, ... ? (facial feature tracking, expression modeling and analysis)

- Focus of attention: Where is a person looking? (head/face tracking, eye gaze tracking)

- Body posture and movement: What is the overall pose and motion of the person? (body modeling and tracking)

– Gesture: What are the semantically meaningful movements of the head, hands, body? (gesture recognition, hand tracking)

– Activity: What is the person doing? (analysis of body movement)

Surveillance and VBI are related areas with different emphases. Surveillance problems typically require less precise information and are intended not for direct interaction but to record general activity or to flag unusual activity. VBI demands more fine-grained analysis, where subtle facial expressions or hand gestures can be very important.

These computer vision problems of tracking, modeling, and analyzing human activities are quite difficult. In addition to the difficulties posed in typical computer vision problems by noise, changes in lighting and pose, and the general ill-posed nature of the problems, VBI problems add particular difficulties because the objects to be modeled, tracked, and recognized are people rather than simple, rigid, unchanging widgets. People change hairstyles, get sunburned, grow facial hair, wear baggy clothing, and in general make life difficult for computer vision algorithms. Robustness in the face of the variety of human appearances is a major issue in VBI research.

The main problem that computer vision faces is an overload of information. The human visual system effortlessly filters out unimportant visual information, attending to relevant details like fast-moving objects, even if they are in the periphery of the visible hemisphere. But this is a very complex computational task. At the low level in human vision, a great deal of preprocessing is done in the retina in order to decrease the bandwidth requirements of the nervous channel into the visual cortex. At the high level, humans leverage a priori knowledge of the world in ways that are not well understood computationally. For example, a computer does not simply know that objects under direct sunlight cast sharp shadows. The difficult question is how to extract only relevant observations from the visual information so that vision algorithms can concentrate on a manageable amount of work. Researchers frequently circumvent this problem by making simplifying assumptions about the environment, which makes it possible to develop working systems and investigate the suitability of computer vision as a user interface modality.

In recent years, there has been increased interest in developing practical vision-based interaction methods. The technology is readily available, inexpensive, and fast enough for most real-time interaction tasks. CPU speed has continually increased following Moore's law, allowing increasingly complex vision algorithms to run at frame rate (30 frames per second, *fps*). Fig-

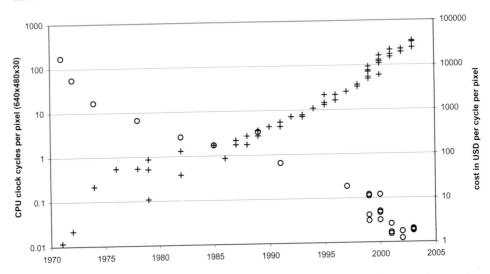

Figure 10.6. CPU processing power over the last 35 years. Each + data point denotes the release of the fastest CPU for the PC market from one of the major manufacturers. Multiple data points per year are shown if one manufacturer released multiple CPUs that year or competing manufacturers' CPUs were released that year. The available clock cycles per pixel per frame of a video stream with 30 full frames per second of size 640 × 480 pixels determine the y-value. The "o" data points describe the cost in U.S. dollars per MHz CPU speed.

ure 10.6 shows a history of available clock cycles per pixel of a VGA-sized video stream with a top-of-the-line CPU for the PC market over the last 35 years. Higher processing speeds, as well as the recent boom in digital imaging in the consumer market, could have far-reaching implications for VBI. It is becoming more and more feasible to process large, high-resolution images in near real-time, potentially opening the door for numerous new applications and vision algorithms.

Fast, high-resolution digital image acquisition devices and fast processing power are only as effective as the link between them. The PC market has recently seen a revolution in connector standards. Interface speeds to peripheral devices used to be orders of magnitude lower than the connection speed between the motherboard and internal devices. This was largely due to the parallel (32 bits or more) connector structure for internal boards and serial links to external devices. The introduction of Firewire (also called 1394 and i-Link) in 1995 and more recently USB 2.0 pushed interface speeds for peripheral devices into the same league as internal interfaces. While other high-speed interfaces for external devices exist (e.g., ChannelLink), they have not made inroads to the consumer market.

Using computer vision in HCI can enable interaction that is difficult or impossible to achieve with other interface modalities. As a picture is worth a thousand words to a person, a video stream may be worth a thousand words of explanation to a computer. In some situations, visual interaction is very important to human–human interaction—hence, people fly thousands of miles to meet face to face. Adding visual interaction to computer interfaces, if done well, may help to produce a similarly compelling user experience.

Entirely unobtrusive interfaces are possible with computer vision because no special devices must be worn or carried by the user (although special, easily tracked objects may be useful in some contexts). No infrastructure or sensors need to be placed in the interaction space because cameras can cover a large physical range. In particular, no wires or active transmitters are required by the user. A camera operates extremely quietly, allowing input to a computer without disturbing the environment. Also, modern cameras can be very lightweight and compact, well suited for mobile applications. Even in environments unsuitable for moving or exposed parts, cameras can be utilized since a camera can be completely encapsulated in its housing with a transparent window.

The cost of cameras and their supporting hardware and software has dropped dramatically in recent years, making it feasible to expect a large installed based in the near future. Software for image and video processing (e.g., for movie and DVD editing) has entered the consumer market and is frequently preinstalled, bundled with a computer's operating system.

The versatility of a camera makes it reasonable and compelling to use as an interface device. A camera may be used for several different purposes, sometimes simultaneously. For example, a single camera, affixed to a person's head or body, may function as a user interface device by observing the wearer's hands [73]; it can videotape important conversations or other visual memories at the user's request [58]; it can store and recall the faces of conversation partners and associate their names [112]; it can be used to track the user's head orientation [140, 2]; it can guide a museum visitor by identifying and explaining paintings [111].

As a particular example of VBI, hand gesture recognition offers many promising approaches for interaction. Hands can operate without obstructing high-level activities of the brain, such as sentence-forming, thus being a good tool for interface tasks while the user is thinking. Generating speech, on the other hand, is said to take up general-purpose brain resources, impeding the thought process [66]. Hands are very dextrous physical tools, and their capabilities have been quite successfully employed in the HCI context in conjunction with devices such as the keyboard and mouse. Human motor skills are, in many cases, easily trained to execute new tasks with high precision and incredible speeds. With the aid of computer vision, we have the chance to go beyond the range of activities that simple physical devices can capture and instead to let hands gesture with all their capabilities. The goal is to leverage the full range of both static hand postures and dynamic gestures in order to communicate (purposefully or otherwise) and perhaps to command and control. Data gloves accomplish some of this goal, yet they have an unnatural feel and are cumbersome to the user.

10.4.1 Terminology

This subsection reviews the essential terminology relevant to VBI. Current VBI tasks focus on modeling (and detecting, tracking, recognizing, etc.) one or more body parts. These parts could be a face, a hand, a facial feature such as an eye, or an entire body. We will interchangeably call this the object of focus, the feature of attention, or simply the "body part."

Determining the presence (or absence) of an object of focus is the problem of *detection* and has so far primarily been applied to people detection [43] and face detection [155, 49]. Strictly speaking, the output is binary ("person present" versus "no person present"), but typically the location of the object is also reported. Object *localization* is sometimes used to describe the special case that the presence of an object is assumed and its location is to be determined at a particular point in time. *Registration* refers to the problem of aligning an object model to the observation data, often both object position and orientation (or *pose*). Object *tracking* locates objects and reports their changing pose over time [43].

Although tracking can be considered as a repeated frame-by-frame detection or localization of a feature or object, it usually implies more than discontinuous processing. Various methods improve tracking by explicitly taking temporal continuity into account and using prediction to limit the space of possible solutions and speed up the processing. One general approach uses filters to model the object's temporal progression. This can be as simple as linear smoothing, which essentially models the object's inertia. A Kalman filter [63] assumes a Gaussian distribution of the motion process and can thus also model nonconstant movements. The frequently used extended Kalman filter (EKF) relieves the necessity of linearity. Particle filtering methods (or sequential Monte Carlo methods [34], and frequently called condensation [55] in the vision community) make no assumptions about the characteristics of underlying probability distributions but instead sample the probability and build an implicit representation. They can therefore deal with non-Gaussian processes and also with multimodal densities (caused by multiple, statistically independent sources) such as arise from object tracking in front of cluttered backgrounds. In addition to the filtering approach, different algorithms can be used for initialization (detection) and subsequent tracking. For example, some approaches detect faces with a learning-based approach [139] and then track with a shape- and color-based head tracker [8].

Recognition (or *identification*) involves comparing an input image to a set of models in a database. A recognition scheme usually determines confidence scores or probabilities that define how closely the image data fits each model. Detection is sometimes called recognition, which makes sense if there are very different classes of objects (faces, cars, and books) and one of them (faces) must be recognized. A special case of recognition is *verification* or *authentication*, which judges whether the input data belongs to one particular identity with very high confidence. An important application of verification is in biometrics, which has been applied to faces, fingerprints, and gait characteristics [17].

A *posture* is a static configuration of the human body—for example, sitting and thinking or holding a coffee cup or pen. *Gestures* are dynamic motions of the body or body parts and can be considered as temporally consecutive sequences of postures. Hand and arm gestures in particular are covered extensively in the social sciences literature, especially in conversational and behavioral psychology [36, 69, 86, 87, 108]. As a result, the term gesture often refers to the semantic interpretation associated with a particular movement of the body (e.g., happiness associated with a smile). We limit our attention in this chapter to a mostly syntactic view of gesture and gesture recognition, leaving the difficult problem of semantic interpretation and context [26, 86] to others.

Facial gestures are more commonly called *facial expressions*. Detecting and tracking facial features are typically the first steps of facial expression analysis, although holistic appearance-based approaches may also be feasible. Subsequent steps try to recognize known expressions (e.g., via FACS action units [37]) and to infer some meaning from them, such as the emotional state of the human [6, 9].

The parameter set for a rigid body consists of its *location* (x, y, z) in 3D space and its *orientation* (rx, ry, rz) with respect to a fixed coordinate frame. Deformable objects such as human faces require many parameters for an accurate description of their form, as do articulated objects such as human bodies. An object's *appearance* describes its color and brightness properties at every point on its surface. Appearance is caused by texture, surface structure, lighting, and view direction. Since these attributes are view-dependent, it only makes sense to talk about appearance from a given viewpoint.

The *view sphere* is an imaginary sphere around the object or scene of interest. Every surface point of the sphere defines a different view of the object. When taking perspective into account, the object's appearance changes even for different distances, despite a constant viewing angle. Vision techniques that can detect, track, or recognize an object regardless of the viewing angle are called *view-independent*; those that require a certain range of viewing angles for good performance, for example, frontal views for face tracking, are called *view-dependent*.

10.4.2 Elements of VBI

VBI techniques apply computer vision to specific body parts and objects of interest. Depending on the application and environmental factors, it may be most useful to search for or track the body as a whole, individual limbs, a single hand, or an artifact such as a colored stick held by the user. VBI techniques for different body parts and features can complement and aid each other in a number of ways.

1. Coarse-to-fine hierarchy: Here, in a sequence of trackers targeted to successively more detailed objects, each stage benefits from the results of the previous stage by drastically reducing the search space. For example, whole-body detection limits the search space for face detection. After the face is detected in a scene, the search for particular facial features such as eyes can be limited to small areas within the face. After the eyes are identified, one might estimate the eye gaze direction [129].

2. Fine-to-coarse hierarchy: This approach works in the opposite way. From a large number of cues of potential locations of small features, the most likely location of a larger object that contains these features is deduced. For example, feature-based face recognition methods use this approach. They first try to identify prominent features such as eyes, nose, and lips in the image. This may yield many false positives, but knowledge about the features' relative locations allows the face to be detected. Bodies are usually attached to heads at the same location:

thus, given the position of a face, the search for limbs, hands, and so on, becomes much more constrained and therefore simpler [91].

3. Assistance through more knowledge: The more elements in a scene that are modeled, the easier it is to deal with their interactions. For example, if a head tracking interface also tracks the hands in the image, although it does not need their locations directly, it can account for occlusions of the head by the hands (e.g., [146]). The event is within its modeled realm, whereas otherwise occlusions would constitute an unanticipated event, diminishing the robustness of the head tracker.

Face tracking in particular is often considered a good anchor point for other objects [135, 24], since faces can be fairly reliably detected based on skin color or frontal view appearance.

Person-level, whole body, and limb tracking

VBI at the person level has probably produced the most commercial applications to date. Basic motion sensing is a perfect example how effective VBI can be in a constrained setting: the scene is entirely stationary, so that frame-to-frame differencing is able to detect moving objects. Another successful application is in traffic surveillance and control. A number of manufacturers offer systems that automate or augment the push button for pedestrian crossings. The effectiveness of this technology was demonstrated in a study [137] that compared the number of people crossing an intersection during the "Don't Walk" signal with and without infrastructure that detected people in the waiting area on the curb or in the crossing. The study found that in all cases these systems can significantly decrease the number of dangerous encounters between people and cars.

Motion capture has frequently been used in the film industry to animate characters with technology from companies such as Adtech, eMotion, Motion Analysis, and Vicon Motion Systems. In optical motion capture, many infrared-reflecting markers are placed on an artist's or actor's body. Typically, more than five cameras observe the acting space from different angles, so at least two cameras can see each point at any time. This allows for precise reconstruction of the motion trajectories of the markers in 3D and eventually the exact motions of the human body. This information is used to drive animated models and can result in much more natural motions than those generated automatically. Other common applications of motion capture technology include medical analysis and sports training (e.g., to analyze golf swings or tennis serves).

Detecting and tracking people passively using computer vision, without the use of markers, has been applied to motion detection and other surveillance tasks. In combination with artificial intelligence, it is also possible to detect unusual behavior, for example in the context of parking lot activity analysis [47]. Some digital cameras installed in classrooms and auditoriums follow a manually selected person or head through pan-tilt-zoom image adjustments. Object-based image encoding such as defined in the MPEG-4 standard is an important application of body tracking technologies.

The difficulties of body tracking arise from the many degrees of freedom (DOF) of the human body. Adults have 206 bones, which are connected by over 230 joints. Ball and socket joints such as in the shoulder and hip have three DOFs: They can abduct and adduct, flex and extend, and rotate around the limb's longitudinal axis. Hinge joints have one DOF and are found in the elbow and between the phalanges of hands and feet. Pivot joints also allow for one DOF; they allow the head as well as radius and ulna to rotate. The joint type in the knee is called a condylar joint. It has two DOF because it allows for flexion and extension and for a small amount of rotation. Ellipsoid joints also have two DOF, one for flexion and extension, the other for abduction and adduction (e.g., the wrist's main joint and the metacarpophalangeal joint, depicted in Figure 10.8). The thumb's joint is a unique type, called a saddle joint; in addition to the two DOF of ellipsoid joints, it also permits a limited amount of rotation. The joints between the human vertebrae each allow limited three-DOF motion, and all together they are responsible for the trunk's flexibility.

Recovering the DOF for all these joints is an impossible feat for today's vision technology. Body models for computer vision purposes must therefore abstract from this complexity. They can be classified by their dimensionality and by the amount of knowledge versus learning needed to construct them.

The frequently used 3D kinematic models have between 20 and 30 DOF. Figure 10.7 shows an example in which the two single-DOF elbow and radioulnar joints have been combined into a 2-DOF joint at the elbow. In fact, the rotational DOF of the shoulder joint is often transferred to the elbow joint. This is because the humerus shows little evidence of its rotation, while the flexed lower arm indicates this much better. Similarly, the rotational DOF of the radioulnar joint can be attributed to the wrist. This transfer makes a hierarchical model parameter estimation easier.

Most of the vision-based efforts to date have concentrated on detecting and tracking people while walking, dancing, or performing other tasks in a mostly upright posture. Pedestrian detection, for example, has seen

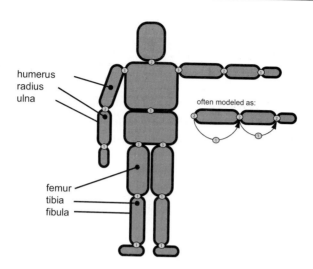

humerus
radius
ulna

often modeled as:

femur
tibia
fibula

Figure 10.7. The "sausage link man" shows the structure of a 3D body model. The links can have cylindrical shape, but especially the trunk is more accurately modeled with a shape with noncircular cross-section.

methods employed that had previously shown success in face detection, such as wavelets [99] and a combination of depth information and a learning method [162].

Two systems with comprehensive functionality, Pfinder [145] and W4 [48], both show well how computer vision must be tailored to the task and properties of the particular environment. First, they rely on a static camera mounting, which gives the opportunity to model the background and achieve fast and reliable segmentation of moving foreground objects. Second, they make assumptions of the body posture; namely, they expect a mostly upright person. This can be easily distinguished from other moving objects such as cars or windblown objects. Third, heuristics about the silhouettes enable classification of a few typical postures or actions such as carrying an object, making use of the fact that only a small number of scenarios are likely for a person entering the field of view.

Hands

Hands are our most dextrous body parts, and they are heavily used in both manipulation and communication. Estimation of the hands' configuration is extremely difficult due to the high DOF and the difficulties of occlusion.

Even obtrusive data gloves[6] cannot acquire the hand state perfectly. Compared with worn sensors, computer vision methods are at a disadvantage. With a monocular view source, it is impossible to know the full state of the hand unambiguously for all hand configurations, as several joints and finger parts may be hidden from the camera's view. Applications in VBI have to keep these limitations in mind and focus on obtaining information that is relevant to gestural communication, which may not require full hand pose information.

Generic hand detection is a largely unsolved problem for unconstrained settings. Systems often use color segmentation, motion flow, and background subtraction techniques, and especially a combination of these, to locate and track hands in images. In a second step and in settings where the hand is the prominent object in view, a shape recognition or appearance-based method is often applied for hand posture classification.

Anatomically, the hand is a connection of 18 elements: the five fingers with three elements each, the thumb-proximal part of the palm, and the two parts of the palm that extend from the pinky and ring fingers to the wrist (see Figure 10.8). The 17 joints that connect the elements have 1, 2, or 3 DOFs. There are a total of 23 DOFs, but for simplicity, the joints inside the palm are frequently ignored, as is the rotational DOF of the trapeziometacarpal joint, leaving 20 DOF. Each hand configuration is a point in this 20-dimensional configuration space. In addition, the hand reference frame has 6 DOF (location and orientation). See Braffort et al. [12] for an exemplary anthropomorphic hand model.

It is clearly difficult to automatically match a hand model to a point in such a high-dimensional space for posture recognition purposes. Lin et al. [80] suggest limiting the search to the interesting subspace of natural hand configurations and motions, and they define three types of constraints. Type I constraints limit the extent of the space by considering only anatomically possible joint angles for each joint (see also earlier work by Lee and Kunii [77]). Type II constraints reduce the dimensionality by assuming direct correlation between DIP and PIP flexion. Type III constraints limit the extent of the space again by eliminating generally impossible configurations and unlikely transitions between configurations. With a 7D space, they cover 95% of configurations observed in their experiments.

---

[6]Data gloves are gloves with sensors embedded in them that can read out the fingers' flexion and abduction. Their locations and orientations in 3D space are often tracked with supplemental means such as electromagnetic trackers.

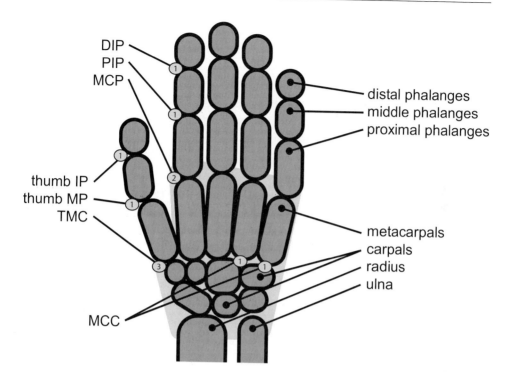

Figure 10.8. The structure of the hand. The joints and their degrees of freedom: distal interphalangeal joints (DIP, 1 DOF), proximal interphalangeal joints (PIP, 1 DOF), metacarpophalangeal joints (MCP, 2 DOF), metacarpocarpal joints (MCC, 1 DOF for pinky and ring fingers), thumb's interphalangeal joint (IP, 1 DOF), thumb's metacarpophalangeal joint (MP, 1 DOF), and thumb's trapeziometacarpal joint (TMC, 3 DOF).

An introduction to the state of the art in hand modeling and recognition can be found in a survey by Wu and Huang [152]. One of the early papers that described whole hand tracking and posture classification as real-time input was written by Fukumoto et al. [42]. They moved a cursor around on a projection screen by making a pointing hand posture and moving the hand within a space observed from two cameras. Two different postures can be distinguished (thumb up and down) with various interpretations to control a VCR and to draw. The paper also deals with the problem of estimating the pointing direction. Cutler and Turk [23] use a rule-based system for gesture recognition in which image features are extracted by optical flow. The location and trajectory of the hand(s) constitutes the input to various simple interfaces such as controlling musical instruments. Mysliwiec [94, 107] tracks the hands by detecting the hands anew in every frame based on a skin color model. Then, a simple, hand-specific heuristic is used to classify the posture and find the index fingertip. Freeman and Roth [41] use histograms of edge orientations in hand images to distinguish different gestures. Moghaddam and Pentland [90] apply a density estimation to the PCA-transformed edge images and obtain a scale-invariant shape similarity measure. GREFIT [98] uses fingertip locations in 2D images to deduce possible 3D configurations of view-dependent hand images with an underlying anatomical hand model with Type I constraints. Zhu et al. [163] combine color segmentation with frame differencing to find handlike objects, using higher-level dynamic models together with shape similarity based on image moments to distinguish gestures. They observed that it was easier and more reliable to classify gestures based on their motion trajectory than on finger configurations.

View independence is a significant problem for hand gesture interfaces. Wu and Huang [151] compared a number of classifiers for their suitability to view-independent hand posture classification.

The above approaches operate in the visible light spectrum and do not use any auxiliary aids such as marked gloves. Segmenting the hand from the background is much simplified by using infrared (IR) light. Since the skin reflects near-IR light well, active IR sources placed in proximity to the camera in combination with an IR pass filter on the lens make it easy to locate hands that are within range of the light source. Dorfmüller-Ulhaas and Schmalstieg [32] use special equipment: users must wear gloves with IR-reflecting markers, the scene is illuminated with IR light sources, and a pair of cameras is used for stereo vision. The system's accuracy and robustness are quite high even with cluttered backgrounds. It can deliver the accuracy necessary to grab and move virtual checkerboard figures.

Hand and arm gestures in 4D

The temporal progression of hand gestures, especially those that accompany speech, are generally composable into three stages: pre-stroke, stroke, and post-stroke [86]. The pre-stroke prepares the movement of the hand. The hand waits in this ready state until the speech arrives at the point in time when the stroke is to be delivered. The stroke is often characterized by a peak in the hand's velocity and distance from the body. The hand is retracted during the post-stroke, but this phase is frequently omitted or strongly influenced by the gesture that follows (similar to coarticulation issues in speech processing).

Automatic sign language recognition has long attracted vision researchers. It offers enhancement of communication capabilities for the speech-impaired and deaf, promising improved social opportunities and integration. For example, the signer could wear a head-mounted camera and hand a device to his or her conversation partner that displays the recognized and translated text. Alternatively, a text-to-speech module could be used to output the sign language interpretation. Sign languages exist for several dozen spoken languages, such as American English (ASL), British English, French, German, and Japanese. The semantic meanings of language components differ, but most of them share common syntactic concepts. The signs are combinations of hand motions and finger gestures, frequently augmented with mouth movements according to the spoken language. Hand motions are distinguished by the spatial motion pattern, the motion speed, and in particular by which body parts the signer touches at the beginning, during, or at the end of a sign. The finger configuration during the slower parts of the hand movements is significant for the meaning of the gesture. Usually, uncommon words can be spelled out as a concatenation of letter symbols and then be assigned to a context-dependent symbol for more efficient signing. Trained persons achieve speeds that equal that of conversational speech.

Most computer vision methods applicable to the task of sign language recognition have extracted feature vectors composed of hand location and contour. These feature vectors have their temporal evolution and variability in common with feature vectors stemming from audio data; thus, tools applied to the speech recognition domain may be suited to recognizing the visual counterpart of speech as well. An early system for ASL recognition [127] fed such a feature vector into a hidden Markov model (HMM) and achieved high recognition rates for a small vocabulary of mostly unambiguous signs from a constrained context. However, expansion into a wider semantic domain is difficult; the richness of syntactic signs is a big hurdle for a universal sign language recognition system. The mathematical methods that perform well for speech recognition need adaptation to the specifics of spatial data with temporal variability. What is more, vision-based recognition must achieve precision in two complementary domains: very fast tracking of the position of the hand in 3D space and exact estimation of the configuration of the fingers of the hand. To combine these requirements in one system is a major challenge. The theoretical capabilities of sign language recognition—assuming the computer vision methods are fast and precise enough—can be evaluated with glove-based methods in which research has a longer history (e.g., [93, 79]).

Wu and Huang [150] present a review of recognition methods for dynamic gestures up to 1999. Overall, vision-based hand gesture recognition has not yet advanced to a stage where it can be successfully deployed for user interfaces in consumer-grade applications. The big challenges are robustness, user independence, and some measure of view independence.

Head and Face

Head and face detection and tracking contributes an essential component to VBIs. Heads and faces can safely be presumed to be present and visible for almost all kinds of human tasks. Heads are rarely occluded entirely, and they convey a good deal of information about the human, such as identity and focus of attention. In addition, this is an attractive area of research in computer vision because the appearance variability of heads and faces is limited yet complex enough to touch on many fundamental problems of computer vision. Methods that perform well on head or face detection or tracking may also perform well on other objects. This area of VBI has therefore received the most attention, and its maturity can be observed by the existence of standard evaluation methods (test databases), the availability of software tools, and commercial developments and products. This progress raises the question of whether at least parts of the problem are solved to a degree that computers can satisfactorily perform these tasks.

Applications of the technology include face detection followed by face recognition for biometrics—for example, to spot criminals at airports or to verify access to restricted areas. The same technology can be useful for personalized user interfaces—for example, to recall stored car seat positions, car stereo volume, and car phone speed-dial lists depending on the driver. Head pose estimation and gaze direction have applications for video conferencing. A common problem occurs when watching the video of one's interlocutor on a screen while the camera is *next to* the monitor. This causes an apparent offset of gaze direction, which can be disturbing if eye contact is expected. Computer vision can be used to correct for this problem [44, 156]. Head tracking has been used for automatically following and focusing on a speaker with fixed-mounted pan/tilt/zoom cameras. Future applications could utilize face recognition to aid human memory, and attempts are already being made to use face detection and tracking for low bandwidth, object-based image coding. Face tracking is usually a prerequisite for efficient and accurate locating of facial features and expression analysis.

Face tracking methods can be characterized along two dimensions: whether they track a planar face or a 3D face and whether they assume a rigid or a deformable face model. The usual tradeoffs apply: a model with more DOFs is harder to register with the image, but it can be more robust. For example, it may explicitly handle rotations out of the image plane. Planar methods can deal with only limited shape and appearance variation caused by out-of-plane rotations, for instance, by applying learning methods. Fixed shape and appearance models such as polygons [130], ellipses [8], cylinders [74], and ellipsoids [84], are efficient for coarse head tracking, especially when combined with other image features such as color [8]. Models that can describe shape and/or appearance variations have the potential to yield more precise results and handle varying lighting conditions and even sideways views. Examples for 2D models are Snakes [147], Eigenfaces [134, 103], active shape models [21] and active appearance models (AAM) [20, 154], Gabor and other wavelets [75, 104, 160, 38], and methods based on independent component analysis (ICA). Three-dimensional model examples are 3D AAM [33], point-distribution models [45, 78], and meshes [27, 157].

The major difficulties for face detection arise from in-plane (tilted head, upside down) and out-of-plane (frontal view, side view) rotations of the head, facial expressions (see below), facial hair, glasses, and, as with all computer vision methods, lighting variation and cluttered backgrounds. There are several good surveys of head and face VBI,[7] including face detection [155, 49], face recognition [153, 46], and face tracking [54].

Facial expression analysis, eye tracking

Facial expressions are an often overlooked aspect of human–human communication. However, they make a rich contribution to our everyday life. Not only can they signal emotions and reactions to specific conversation topics, but on a more subtle level, they also regularly mark the end of a contextual piece of information and help in turn-taking during a conversation. In many situations, facial gestures reveal information about a person's true feelings, while other bodily signals can be artificially distorted more easily. Specific facial actions, such as mouth movements and eye gaze direction changes, have significant implications. Facial expressions serve as interface medium between the mental states of the participants in a conversation.

Eye tracking has long been an important facility for psychological experiments on visual attention. The quality of results of automatic systems that used to be possible only with expensive hardware and obtrusive head-mounted devices is now becoming feasible with off-the-shelf computers and

---

[7]See also Chapter 9.

cameras, without the need for head-mounted structures.[8] Application deployment in novel environments such as cars is now becoming feasible.

Face-driven animation has begun to make an impact in the movie industry, just as motion capture products did a few years earlier. Some systems still require facial markers, but others operate without markers. The generated data is accurate enough to animate a virtual character's face with ease and with a degree of smoothness and naturalness that is difficult to achieve (and quite labor-intensive) with conventional, scripted animation techniques.

Facial expressions in general are an important class of human motion that behavioral psychology has studied for many decades. Much of the computer vision research on facial expression analysis has made use of the Facial Action Coding System (FACS) [37] (see Figure 10.9), a fine-grained classification of facial expressions. It describes on the order of 50 individual action units (AU) such as raising the upper lip, stretching the lips, or parting the lips, most of which are oriented on a particular facial muscle and its movements. Some AUs are combinations of two muscles' movements, and some muscles have more than one AU associated with them if contracting different parts of the muscle results in very different facial expressions. Expressions such as a smile are composed of one or more AU. AU thus do not carry semantic meaning; they only describe physical deformations of skin and facial tissues. FACS was originally developed so that human observers could succinctly describe facial expressions for use in behavioral psychology. While many degrees of freedom allow for precise distinction of even the most subtle facial notions— even distinction between natural and purposefully forced smiles is said to be possible—the high expressiveness also poses additional challenges over less fine-grained classifications, such as the semantics-oriented classification into facial expressions caused by emotions (for example, happiness and anger).

For eye tracking, a coarse-to-fine approach is often employed. Robust and accurate performance usually necessitates brief user training and constant lighting conditions. Under these conditions, the eye gaze direction can be accurately estimated despite moderate head translations and rotations.

According to a survey by Donato et al. [31], most approaches to automatic analysis of facial expression are based on either optical flow, global appearance, or local features. The authors reimplemented some of the most promising methods and compared their performance on one data set. Optical flow methods (such as [85]) were used early on to analyze the short-term, dynamical evolution of the face. Naturally, these require an image sequence

---

[8]See, for example, the Arrington Research, Inc. ViewPoint EyeTracker remote camera system (http://www.tobii.se).

AU	Facial expression	FACS description	AU	Facial expression	FACS description
1		inner brow raiser	2		outer brow raiser
4	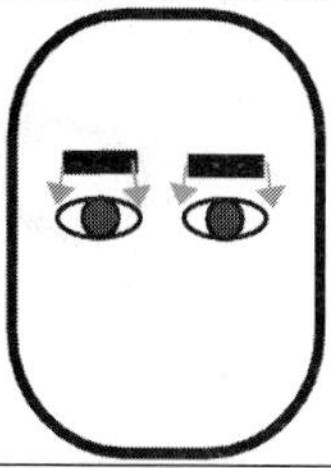	brow lower	5		upper lid raiser
6	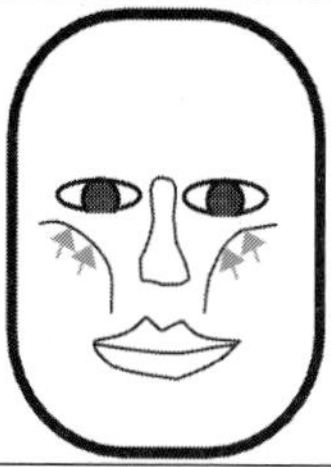	brow lower	10	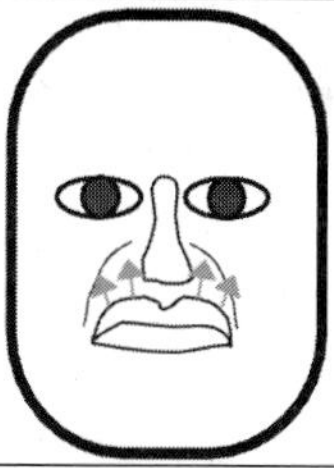	upper lip raiser

Figure 10.9. Some of the action units of the FACS [37].

of the course of a facial expression and do not apply to static images due to the lack of motion information. These methods were found to perform significantly better when the image is not smoothed in a preprocessing step. This leads to the conclusion that small image artifacts are important when it comes to tracking and/or recognizing facial gestures. Another hint that high spatial frequencies contribute positively stems from comparing the performance of Gabor wavelets. Using only high frequencies had a less detrimental effect than using only low frequencies, compared to the baseline of unrestricted Gabor filters.

Approaches using Gabor filters and ICA, two methods based on spatially constrained filters, outperformed other methods such as those using PCA or local feature analysis [102] in the investigation by Donato et al. [31].

The authors conclude that local feature extraction and matching is very important to good performance. This alone is insufficient, however, as the comparison between global and local PCA-based methods showed. Good performance with less precise feature estimation can be achieved with other methods (e.g., [161].

Global methods that do not attempt to separate the individual contributors to visual appearances seem to be ill-suited to model multimodal distributions. Refinement of mathematical methods for computer vision tasks, such as ICA, appears to be promising in order to achieve high accuracy in a variety of appearance-based applications such as detection, tracking, and recognition. Optical flow and methods that do not attempt to extract FACS action units are currently better suited to the real-time demands of VBI.

Handheld objects

Vision-based interfaces that detect and track objects other than human body parts—that is, handheld objects used in interaction—have a high potential for successful commercial development in the transition period from traditional to perceptual interfaces. Tracking such objects can be achieved more reliably than tracking high DOF body parts, and they may be easier to use than freeform gestures. Handheld artifacts can be used in much the same way as one's hands, for example to make pointing gestures, to perform rhythmic commands, or to signal the spatial information content of sign languages. Possible useful objects include passive wands [143], objects with active transmitters such as LEDs,[9] and specially colored objects—in fact, anything that is easily trackable with computer vision methods. An alternative approach to having fixed-mounted cameras and tracking moving objects is to embed the camera in the moving object and recognize stationary objects in the environment or egomotion to enable user interaction. Since detecting arbitrary objects is very hard, fiducials can make this reverse approach practical [72].

10.4.3 Computer vision methods for VBI

In order for a computer vision method to be suitable for VBI, its performance must meet certain requirements with respect to speed, precision, accuracy, and robustness. A system is said to experience real-time behavior if no delay is apparent between an event (occurrence) and its effect. Precision concerns

---

[9]Head trackers can utilize this technology to achieve high accuracy and short latencies. One commercial product is the Precision Position Tracker, available at http://www.worldviz.com.

the repeatability of observations for identical events. This is particularly important for recognition tasks and biometrics: only if the VBI consistently delivers the same result for a given view can this information be used to identify people or scenes. Accuracy describes the deviation of an observation from ground truth. Ground truth must be defined in a suitable format. This format can, for example, be in the image domain, in feature space, or it can be described by models such as a physical model of a human body. It is often impossible to acquire ground-truth data, especially if no straightforward translation from observation to the ground-truth domain is known. In that case, gauging the accuracy of a VBI method may be quite difficult.

Robustness, on the other hand, is easier to determine by exposing the VBI to different environmental conditions, including different lighting (fluorescent and incandescent lighting and sunlight), different users, cluttered backgrounds, occlusions (by other objects or self-occlusion), and nontrivial motion. Generally, the robustness of a vision technique is inversely proportional to the amount of information that must be extracted from the scene.

With currently available hardware, only a very specific set of fairly fast computer vision techniques can be used for truly interactive VBI. One of the most important steps is to identify constraints on the problem (regarding the environment or the user) in order to make simplifying assumptions for the computer vision algorithms. These constraints can be described by various means. Prior probabilities are a simple way to take advantage of likely properties of the object in question, both in image space and in feature space. When these properties vary significantly, but the variance is not random, PCA, neural networks, and other learning methods frequently do a good job in extracting these patterns from training data. Higher-level models can also be employed to limit the search space in the image or feature domain to physically or semantically possible areas.

Frequently, interface-quality performance can be achieved with multimodal or multicue approaches. For example, combining the results from a stereo-based method with those from optical flow may overcome the restrictions of either method used in isolation. Depending on the desired tradeoff between false positives and false negatives, early or late integration (see Figure 10.3) lends itself to this task. Application- and interface-oriented systems must also address issues such as calibration or adaptation to a particular user, possibly at runtime, and reinitialization after loss of tracking. Systems tend to become very complex and fragile if many hierarchical stages rely on each other. Alternatively, flat approaches (those that extract high-level information straight from the image domain) do not have to deal with scheduling many components, feedback loops from higher levels to lower levels, and performance estimation at each of the levels. Robustness in computer vision systems can be improved by devising systems that do not make irrevocable decisions in the lower layers but instead model uncertainties explicitly. This requires modeling of the relevant processes at all stages, from template matching to physical object descriptions and dynamics.

All computer vision methods need to specify two things. First, they need to specify the mathematical or algorithmic tool used to achieve the result. This can be PCA, HMM, a neural network, and so on. Second, the domain to which this tool is applied must be made explicit. Sometimes this will be the raw image space with grayscale or color pixel information, and sometimes this will be a feature space that was extracted from the image by other tools in a preprocessing stage. One example would be using wavelets to find particular regions of a face. The feature vector, composed of the image-coordinate locations of these regions, is embedded in the domain of all possible region locations. This can serve as the input to an HMM-based facial expression analysis.

Edge- and shape-based methods

Shape properties of objects can be used in three different ways. Fixed shape models, such as an ellipse for head detection or rectangles for body limb tracking, minimize the summative energy function from probe points along the shape. At each probe, the energy is lower for sharper edges (in the intensity or color image). The shape parameters (size, ellipse foci, rectangular size ratio) are adjusted with efficient, iterative algorithms until a local minimum is reached. On the other end of the spectrum are edge methods that yield unconstrained shapes. Snakes [67] operate by connecting local edges to global paths. From these sets, paths are selected as candidates for recognition that resemble a desired shape as much as possible. In between these extremes lie the popular statistical shape models, such as the active shape model (ASM) [21]. Statistical shape models learn typical deformations from a set of training shapes. This information is used in the recognition phase to register the shape to deformable objects. Geometric moments can be computed over entire images or alternatively over select points such as a contour.

These methods require sufficient contrast between the foreground object and the background, which may be unknown and cluttered. Gradients in color space [119] can alleviate some of the problems. Even with perfect segmentation, nonconvex objects are not well suited for recognition with shape-based methods, since the contour of a concave object can translate into a landscape with many, deep local minima in which gradient descent methods get stuck. Only near-perfect initialization allows the iterations to descend into the global minimum.

Color-based methods

Compelling results have been achieved merely by using skin color properties, for example, to estimate gaze direction [116] or for interface-quality hand gesture recognition [71]. This is because the appearance of skin color varies mostly in intensity while the chrominance remains fairly consistent [114]. According to Zarit et al. [159], color spaces that separate intensity from chrominance (such as the HSV color space) are better suited to skin segmentation when simple threshold-based segmentation approaches are used. However, some of these results are based on a few images only, while more recent work examined a huge number of images and found an excellent classification performance with a histogram-based method in RGB color space as well [61]. It seems that simple threshold methods or other linear filters achieve better results in HSV space, while more complex methods, particularly learning-based, nonlinear models, do well in any color space. Jones et al. [61] also state that Gaussian mixture models are inferior to histogram-based approaches, which makes sense given the multimodality of random image scenes and the fixed amount of Gaussians available to model this. The CAMShift algorithm (continuously adaptive mean shift) [11, 19] is a fast method to dynamically parameterize a thresholding segmentation that can deal with a certain amount of lighting and background changes. Together with other image features such as motion, patches, or blobs of uniform color, this makes for a fast and easy way to segment skin-colored objects from backgrounds.

Infrared Light: One "color" is particularly well suited to segment human body parts from most backgrounds, and that is energy from the IR portion of the EM spectrum. All objects constantly emit heat as a function of their temperature in form of infrared radiation, which are electromagnetic waves in the spectrum from about 700 nm (visible red light) to about 1 mm (microwaves). The human body emits the strongest signal at about 10 μm, which is called long wave IR or thermal infrared. Not many common background objects emit strongly at this frequency in modest climates; therefore, it is easy to segment body parts given a camera that operates in this spectrum. Unfortunately, this requires very sensitive sensors that often need active cooling to reduce noise. While the technology is improving rapidly in this field, the currently easier path is to actively illuminate the body part with short-wave IR. The body reflects it just like visible light, so the illuminated body part appears much brighter than background scenery to a camera that filters out all other light. This is easily done for short-wave IR because most digital imaging sensors are sensitive this part of the spectrum. In fact, consumer digital cameras require a filter that limits the sensitivity to the visible spectrum to avoid unwanted effects. Several groups have used IR in VBI-related projects (e.g., [113, 32, 133]).

Color information can be used on its own for body part localization, or it can create attention areas to direct other methods, or it can serve as a validation and "second opinion" about the results from other methods (multicue approach). Statistical color as well as location information is thus often used in the context of Bayesian probabilities.

Motion flow and flow fields

Motion flow is usually computed by matching a region from one frame to a region of the same size in the following frame. The motion vector for the region center is defined as the best match in terms of some distance measure (e.g., least-squares difference of the intensity values). Note that this approach is parameterized by both the size of the region ("feature") as well as the size of the search neighborhood. Other approaches use pyramids for faster, hierarchical flow computation; this is especially more efficient for large between-frame motions. The most widely used feature tracking method is the KLT tracker. The tracker and the selection of good features to track (usually corners or other areas with high image gradients) are described by Shi and Tomasi [120]. KLT trackers have limitations due to the constancy assumption (no change in appearance from frame to frame), match window size (aperture problem), and search window size (speed of moving object, computational effort).

A flow field describes the apparent movement of entire scene components in the image plane over time. Within these fields, motion blobs are defined as pixel areas of (mostly) uniform motion—that is, motion with similar speed and direction. Especially in VBI setups with static camera positions, motion blobs can be very helpful for object detection and tracking. Regions in the image that are likely locations for a moving body part direct other computer vision methods and the VBI in general to these "attention areas."

The computational effort for tracking image features between frames increases dramatically with lower frame rates, since the search window size has to scale according to the tracked object's estimated maximal velocity.

Since motion flow computation can be implemented as a local image operation that does not require a complex algorithm or extrinsic state information (only the previous image patch and a few parameters), it is suited for on-chip computation. Comprehensive reviews of optical flow methods can be found in Barron et al. [5] and Mitiche and Bouthemy [89].

Texture- and appearance-based methods

Information in the image domain plays an important role in every object recognition or tracking method. This information is extracted to form image features: higher-level descriptions of what was observed. The degree of abstraction of the features and the scale of what they describe (small, local image artifacts or large, global impressions) have a big impact on the method's characteristics. Features built from local image information such as steep gray-level gradients are more sensitive to noise; they need a good spatial initialization and frequently a large collection of those features is required. Once these features are found, they need to be brought into context with each other, often involving an iterative and computationally expensive method with multiple, interdependent, and thus more fragile, stages.

If instead the features are composed of many more pixels, cover a larger region in the image, and abstract to more complex visuals, the methods are usually better able to deal with clutter and might flatten the hierarchy of processing levels (since they already contain much more information than smaller-scale features). The benefits do not come without a price—in this case, increased computational requirements.

Appearance-based methods attempt to identify the patterns that an object frequently produces in images. The simplest approach to comparing one appearance to another is to use metrics such as least-squared difference on a pixel-by-pixel basis (i.e., the lowest-level feature vector). This is not very efficient and is too slow for object localization or tracking. The key is to encode as much information as possible in as small as possible a feature vector—to maximize the entropy.

One of the most influential procedures uses a set of training images and the Karhunen-Loeve transform [65, 81]. This transformation is an orthogonal basis rotation of the training space that maximizes sample variance along the new basis vectors. It is frequently known in the computer vision literature as PCA [59] and is directly related to singular value decomposition (SVD). In the well-known eigenfaces approach, Turk and Pentland [134] applied this method to perform face detection and recognition, extending the work by Kirby and Sirovich for image representation and compression [70].

AAMs [20] encode shape and appearance information in one model, built in a two-step process with PCA. Active blobs [118] are similar to AAM. In these approaches, the observed object appearance steers object tracking by guiding initialization for subsequent frames, similar to the concept of the Kalman filter. During training, a parameterization is learned that correlates observation-estimation error with translational offsets.

A Gabor wavelet is a sine wave enveloped by a Gaussian, modeled after the function of the human visual cortical cell [106, 60, 29, 25]. Wavelets are well suited to encode both spatial and frequency information in a compact, parameterized representation. This alleviates problems of FFT approaches where all local spatial information is lost. Feris et al. [117] showed good face tracking performance with a hierarchical wavelet network, a collection of collections of wavelets. Each feature is represented by a set of wavelets, enabling tracking in the manner of KLT trackers, but comparatively more robust to lighting and deformation changes.

Another approach to learn and then test for common appearances of objects is to use neural networks; however, in some cases their performance (in terms of speed and accuracy) has been surpassed by other methods. One extremely fast detection procedure proposed by Viola and Jones [139] has attracted much attention. In this work, very simple features based on intensity comparisons between rectangular image areas are combined by Ada-boosting into a number of strong classifiers. These classifiers are arranged in sequence and achieve excellent detection rates on face images with a low false-positive rate. The primary advantage of their method lies in the constant-time computation of features that have true spatial extent as opposed to other techniques that require time proportional to the area of the extent. This allows for very high-speed detection of complex patterns at different scales. However, the method is rotation-dependent.

A more complete review of appearance-based methods for detection and recognition of patterns (faces) can be found in Yang et al.'s survey on face detection [155].

Background modeling

Background modeling is often used in VBI to account for (or subtract away) the nonessential elements of a scene. There are essentially two techniques: segmentation by thresholding and dynamic background modeling.

Thresholding requires that the foreground object has some unique property that distinguishes it from all or most background pixels. For example, this property can be foreground brightness, so that pixels with values above a particular grayscale intensity threshold are classified as foreground, and values below as belonging to the background. Color restrictions on the background are also an effective means for simple object segmentation. There, it is assumed that the foreground object's color does not appear very frequently or in large blobs in the background scene. Of course, artificial coloring of the foreground object avoids problems induced by wildly varying or unknown object colors—for example, using a colored glove for hand detection and tracking.

Dynamic background modeling requires a static camera position. The values of each pixel are modeled over time with the expectation to find a pattern of values that this pixel assumes. The values may be described by a single contiguous range, or it may be multimodal (two or more contiguous ranges). If the value suddenly escapes these boundaries that the model describes as typical, the pixel is considered part of a foreground object that temporarily occludes the background. Foreground objects that are stationary for a long time will usually be integrated into the background model over time and segmentation will be lost. The mathematics to describe the background often use statistical models with one or more Gaussians [48].

Temporal filtering and modeling

Temporal filtering typically involves methods that go beyond motion flow to track on the object or feature level rather than at the pixel or pattern level. For example, hand and arm gesture recognition (see subsection "Hand and arm gestures in 4D" on page 486) requires temporal filtering.

Once the visual information has been extracted and a feature vector has been built, general physics-based motion models are often used. For example, Kalman filtering in combination with a skeletal model can deal with resolving simple occlusion ambiguities [146]. Other readily available mathematical tools can be applied to the extracted feature vectors, independently of the preceding computer vision computation. Kalman filtering takes advantage of smooth movements of the object of interest. At every frame, the filter predicts the object's location based on the previous motion history. The image matching is initialized with this prediction, and once the object is found, the Kalman parameters are adjusted according to the prediction error.

One of the limitations of Kalman filtering is the underlying assumption of a Gaussian probability. If this is not the case, and the probability function is essentially multimodal, as is the case for scenes with cluttered backgrounds, Kalman filtering cannot cope with the non-Gaussian observations. The particle filtering or factored sampling method, often called condensation (conditional density propagation) tracking, has no implicit assumption of a particular probability function but rather represents it with a set of sample points of the function. Thus, irregular functions with multiple "peaks"—corresponding to multiple hypotheses for object states—can be handled without violating the method's assumptions. Factored sampling methods [55] have been applied with great success to tracking of fast-moving, fixed-shape objects in very complex scenes [76, 28, 14]. Various models, one for each typical motion pattern, can improve tracking, as shown by Isard and Blake [56]. Partitioned sampling reduces the computational complexity of particle filters [82]. The modeled domain is usually a feature vector, combined from shape-describing elements (such as the coefficients of B-splines) and temporal elements (such as the object velocities).

Dynamic gesture recognition (i.e., recognition and semantic interpretation of continuous gestures and body motions) is an essential part of perceptual interfaces. Temporal filters exploit motion information only to improve tracking, while the following methods aim at detecting meaningful actions such as waving a hand for goodbye.

Discrete approaches can perform well at detecting spatio-temporal patterns [51]. Most methods in use, however, are borrowed from the more evolved field of speech recognition due to the similarity of the domains: multidimensional, temporal, and noisy data. Hidden Markov Models (HMMs) [15, 158] are frequently employed to dissect and recognize gestures due to their suitability to processing temporal data. However, the learning methods of traditional HMMs cannot model some structural properties of moving limbs very well.[10] Brand [13] uses another learning procedure to train HMMs that overcomes these problems. This allows for estimation of 3D model configurations from a series of 2D silhouettes and achieves excellent results. The advantage of this approach is that no knowledge has to be hardcoded but instead everything is learned from training data. This has its drawbacks when it comes to recognizing previously unseen motion.

Higher-level models

To model a human body or its parts very precisely, as is required for computer graphics applications, at least two models are necessary. One component describes the kinematic structure of the skeleton, the bone connections and

---

[10]Brand [13] states that the traditional learning methods are not well suited to model state transitions, since they do not improve much on the initial guess about connectivity. Estimating the structure of a manifold with these methods thus is extremely suboptimal in its results.

the joint characteristics. Even complex objects such as the entire human body or the hand can thus—with reasonable accuracy—be thought of as a kinematic chain of rigid objects. The second component describes the properties of the flesh around the bones, either as a surface model or as a volumetric model. Very complex models that even describe the behavior of skin are commonplace in the animation industry. Additional models are frequently used to achieve even greater rendering accuracy, such as models of cloth or hair.

Computer vision can benefit from these models to a certain degree. A structural, anatomical, 3D model of the human body has advantages over a 2D model because it has explicit knowledge of the limbs and the fact that they can occlude each other from a particular view [110]. Given a kinematic model of the human body (e.g., [77]), the task to estimate the limb locations becomes easier compared to the case when the knowledge of interdependency constraints is lacking. Since the locations and orientations of the individual rigid objects (limbs, phalanges) are constrained by their neighboring chain links, the effort to find them in the image decreases dramatically.

It is important for VBI applications to exploit known characteristics of the object of interest as much as possible. A kinematic model does exactly that, as do statistical models that capture the variation in appearance from a given viewpoint or the variation of shape (2D) or form (3D). Often, the model itself is ad hoc; that is, it is manually crafted by a human based on prior knowledge of the object. The model's parameter range is frequently learned from training data.

Higher-level models describe properties of the tracked objects that are not immediately visible in the image domain. Therefore, a translation between image features and model parameters needs to occur. A frequently used technique is *analysis by synthesis* in which the model parameters are estimated by an iterative process. An initial configuration of the model is back-projected into the image domain. The difference between projection and observation then drives adjustment of the parameters, following another back-projection into the image domain and so forth until the error is sufficiently small (e.g., see Kameda et al. [64] and Ström et al. [130]). These methods lack the capability to deal with singularities that arise from ambiguous views [92]. When using more complex models that allow methods from projective geometry to be used to generate the synthesis, self-occlusion is modeled again and thus it can be dealt with. Stenger et al. [128] use a recently proposed modification of the Kalman filter, the "unscented Kalman filter" [62], to align the model with the observations. Despite speed and accuracy advantages over more traditional approaches, the main drawbacks of all Kalman-based filters is that they assume a unimodal distribution. This assumption is most likely violated by complex, articulated objects such as the human body or the hand.

Particle filtering methods for model adjustment are probably better suited for more general applications because they allow for any kind of underlying distribution, even multimodal distributions. Currently, their runtime requirements are not immediately suitable for real-time operation, but more efficient modifications of the original algorithm are available.

Real-time systems, frame rate, latency, processing pipelines

Most user interfaces require real-time responses from the computer: for feedback to the user, to execute commands immediately, or both. But what exactly does real-time mean for a computer vision application?

There is no universal definition for real-time in computer terms. However, a system that responds to user input is said to operate in real time if the user of the system perceives no delay between action command and action. Hence, real time pertains to a system's ability to respond to an event without noticeable delay. The opposite would be a delayed response, a response after a noticeable processing time. Mouse input, for example, is usually processed in real time; that is, a mouse motion is immediately visible as a mouse pointer movement on the screen. Research has shown that delays as low as 50ms are noticeable for visual output [142, 83]. However, people are able to notice audible delays of just a few milliseconds, since this ability is essential to sound-source localization.

The terms *frame rate* and *latency* are well suited to describe a computer vision system's performance. The (minimum, maximum, average) frame rate determines how many events can be processed per time unit. About five frames per second is the minimum for a typical interactive system. The system latency, on the other hand, describes the time between the event occurrence and the availability of the processed information. As mentioned above, about 50ms is tolerable for a system's performance to be perceived as real time.

To illustrate the difference between frame rate and latency, imagine a pipeline of four processors, as shown in Figure 10.10. The first one grabs a frame from the camera and performs histogram normalization. The normalized image frame is input to the second processor, which segments the image into regions based on motion flow and color information, and then outputs the frame to the third processor. This processor applies an AAM

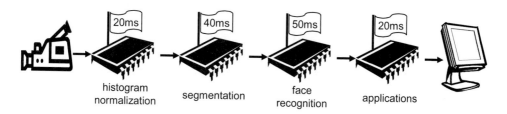

Figure 10.10. A four-stage pipeline processing of a video stream.

for face recognition. The recognized face is input to the last processor, which performs a saliency test and then uses the information to drive various applications. Altogether, this process takes 20ms + 40ms + 50ms + 20ms = 130ms per frame. This is the latency: input is available after 130ms as output. The frame rate is determined by the pipeline's bottleneck, the third processor. A processed frame is available at the system output every 50ms, that is, the frame rate is maximal 20fps.

If the input occurs at a rate higher than 20fps, there are two options for pipelined systems:

The first option is to process every frame. A 10-second input sequence with 30fps, for example, has 300 frames. Thus, it requires 300f/20fps = 15s to process them all. The first frame is available 130ms after its arrival, so the last frame is available as output 5s 130ms after its input. It also means that there must be sufficient buffer space to hold the images; in our example, for 5s * 30fps = 150 frames. It is obvious that a longer input sequence increases the latency of the last frame and the buffer requirements. This model is therefore not suitable to real-time processing.

With the second option, frames are dropped somewhere in the system. In our pipeline example, a 30fps input stream is converted into at most a 20fps output stream. Valid frames (those that are not dropped) are processed and available in at most a constant time. This model is suitable to real-time processing.

Dropping frames, however, brings about other problems. There are two cases: First, there is no buffer available anywhere in the system, not even for partial frames. This is shown on the left-hand side in Figure 10.11. In our example, the first processor has no problems keeping up with the input pace, but the second processor will still be working on the first frame when it receives the second frame 33.3ms after arrival of the first. The second frame needs to be dropped. Then the second processor would idle for 2 * 33.3ms

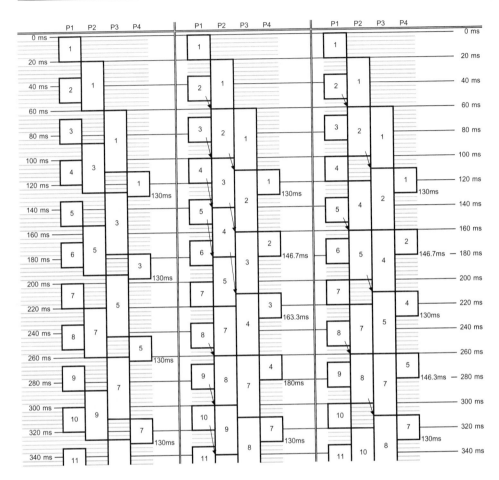

Figure 10.11. An example of pipelined processing of frames, with and without buffering inside the pipeline. Buffering is indicated with arrows. The example on the right uses feedback from stage 3 to avoid unnecessary processing and increased latency.

— 40ms = 26.7ms. Similar examples can be constructed for the subsequent pipeline stages.

Second, frames can be buffered at each pipeline stage output or input. This is shown in the center drawing of Figure 10.11. It assumes no extra communication between the pipeline stages. Each subsequent stage requests a frame as soon as it completed processing one frame, and the preceding stage keeps the latest processed frame around in a buffer until it can replace it with the next processed frame. In the example, the second frame would be buffered for 6.7ms at the transition to the second processor, the third frame for $2 * 40\text{ms} - 2 * 33.3\text{ms} = 13.3\text{ms}$, the fourth for $3 * 40\text{ms} - 3 * 33.3\text{ms} = 20\text{ms}$, and so on. The highest latency after the second stage would be for every fifth frame, which is buffered for a total of $5 * 40\text{ms} - 5 * 33.3\text{ms} = 33.3\text{ms}$ before processing. Then, in fact, the next processed frame (frame number 7) is finished being processed already and is sent to stage 3 instead of the buffered frame, which is dropped. These latencies can accumulate throughout the pipeline. The frame rate is maximal (20fps) in this example, but some of the frames' latencies have increased dramatically.

The most efficient method in terms of system latency and buffer usage facilitates pipeline feedback from the later stages to the system input, as shown on the right in Figure 10.11. The feedback is used to adjust every stages' behavior in order to maximize utilization of the most time-critical stage in the pipeline. Alternatively to the scenario shown, it is also possible to buffer frames only buffered before the first stage and feed them into the system with a speed that the pipeline's bottleneck stage can keep up with. This is not shown, but it completely avoids the need for in-pipeline buffering. Note that even in this efficient scenario with pipeline feedback, the increased frame rates in comparison to the leftmost case are bought with an increase in average latency.

Examples of architectures that implement component communication and performance optimizing scheduling are the Quality-Control Management in the DirectShow subsystem of Microsoft's DirectX technology and the Modular Flow Scheduling Middleware [40].

10.4.4 VBI summary

Vision-based interfaces have numerous applications; the potential of VBI has only begun to be explored. But computational power is getting to a stage where it can handle the vast amounts of data of live video streams. Progress has been made in many relevant areas of computer vision; many methods have been demonstrated that begin to provide HCI quality translation of body actions into computer commands. While a large amount of work is still required to improve the robustness of these methods, especially in modeling and tracking highly articulated objects, the community has begun to take steps toward standardizing interfaces of popular methods and providing toolkits for increasingly higher level tasks. These are important steps in bringing the benefits of VBI to a wider audience.

The number of consumer-grade commercial applications of computer vision has significantly increased in recent years, and this trend will continue, driven by ongoing hardware progress. To advance the state of the art of VBI—at the intersection of the disciplines of computer vision and HCI—it is vital to establish evaluation criteria, such as benchmarks for the quality and speed of the underlying methods and the resulting interfaces. Evaluation databases must be made accessible for all components of VBI (such as those already available for faces), both for static images and increasingly dynamic data for real-time video processing.

10.5   Brain-Computer Interfaces

Perhaps the ultimate interface to computers would be a direct link to the thoughts and intentions of the user, a "Your wish is my command" model of interaction, involving no physical action or interpretation of any kind. While this kind of mind-reading technology is not likely to be developed in the foreseeable future, the nascent research area of brain–computer interfaces (BCI) is perhaps a step in this direction. BCI technology attempts to perceive commands or control parameters by sensing relevant brain activity of the user. While not fitting completely within the perceptual interface model of natural human–human interaction, BCI may eventually be an integral component of perceptual interfaces. The computer vision community's extensive experience with learning, statistical, and other pattern recognition methods and techniques can be of tremendous value to this new field.

A BCI does not depend on the brain's normal output channels of peripheral nerves and muscles, but instead measures electrical activity either at the scalp or in the cortex. By measuring the electroencephalographic (EEG) activity at the scalp, certain features of the EEG signal can be used to produce a control signal. Alternatively, implanted electrodes can be used to measure the activity of individual cortical neurons or an array of neurons. These technologies have primarily been targeted to be used by people with neuromuscular impairments that prevent them from communicating via conventional methods. In recent years, researchers have begun to consider more general uses of the technologies. A review by Wolpaw et al. [144] notes that while the rising interest in BCI technologies in recent years has produced exciting developments with considerable promise, they are currently low-bandwidth devices with a maximum information transfer rate of 10 to 25 bits per minute, and this rate is likely to improve only gradually.

Wolpaw et al. argue that in order to make progress in BCIs, researchers must understand that BCI is not simply mind-reading or "wire-tapping" the brain, determining a person's thoughts and intentions by listening in on brain activity. Rather, BCI should be considered as a new output channel for the brain, one that is likely to require training and skill to master.

Brain–machine interface [97] is the traditional term as it grew out of initial uses of the technology: to interface to prosthetic devices. Sensors were implanted primarily in motoric nerves in the extremities and a one-to-one function was typically used to map the sensor outputs to actuator control signals. Brain–computer interface more accurately captures the necessity for computational power between the neurosensors and the controlled devices or application-specific software. As the sensors increasingly move into the brain (intracortical electrodes) and target not only motoric nerves but generic neurons, the mapping from neuron activity to (desired, normal, or pathologic) output becomes less direct. Complex mathematical models translate the activity of many neurons into a few commands—computational neuroscience focuses on such models and their parameterizations. Mathematical models that have proven to be well suited to the task of replicating human capabilities, in particular the visual sense, seem to perform well for BCIs as well—for example, particle and Kalman filters [149]. Two feature extraction and classification methods frequently used for BCIs are reviewed in [35].

Figure 10.12 schematically explains the principles of a BCI for prosthetic control. The independent variables, signals from one or many neural sensors, are processed with a mathematical method and translated into the dependent variables, spatial data that drives the actuators of a prosthetic device. Wolpaw et al. [144] stress that BCI should eventually comprise three levels of adaptation. In the first level, the computational methods (depicted in the right upper corner of Figure 10.12) are trained to learn the correlation between the observed neural signals and the user's intention for arm movement. Once trained, the BCI then must translate new observations into actions. We quote from [144]: "However, EEG and other electro-physiological signals typically display short- and long-term variations linked to time of day, hormonal levels, immediate environment, recent events, fatigue, illness, and other factors. Thus, effective BCIs need a second level of adaptation: periodic online adjustments to reduce the impact of such spontaneous variations."

Since the human brain is a very effective and highly adaptive controller, adaptation on the third level means to benefit from the combined resources of the two adaptive entities brain and BCI. As the brain adapts to the demands and characteristics of the BCI by modifying its neural activity, the BCI should detect and exploit these artifacts and communicate back to the brain that it appreciates the effort, for example through more responsive, more

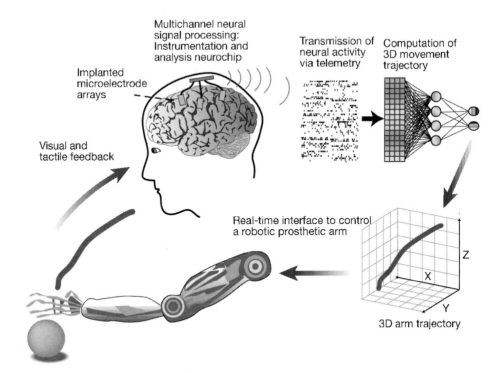

Multichannel neural
signal processing:
Instrumentation and
analysis neurochip

Transmission of   Computation of
neural activity   3D movement
via telemetry   trajectory

Implanted
microelectrode
arrays

Visual and
tactile feedback

Real-time interface to control
a robotic prosthetic arm

Z

X

Y

3D arm trajectory

Figure 10.12. The control path of a closed-loop BCI. Figure reprinted with permission from Nicolelis et al. [97].

precise, or more expressive command execution. This level of adaptation is difficult to achieve, but promises to yield vastly improved performance.

The number of monitored neurons necessary to accurately predict a task such as 3D arm movement is open to debate. Early reports employed open-loop (no visual feedback to the study subject) experiments with offline model building and parameterization. Those studies suggest by extrapolation that between 400 and 1350 neurons are necessary, depending on the brain area in which the sensors are implanted [141]. A more recent study by Taylor et al. provided real-time visual feedback and repeatedly updated the mathematical model underlying the translation function from neurons to the controlled object [132, 97]. They used only 18 neurons to achieve sufficient performance for a 2D cursor task, with the closed-loop method being significantly superior to the open-loop method. Currently, up to about 100 neurons can be recorded simultaneously. All currently used electro-physiological artifacts can be detected with a temporal resolution of 10ms to 100ms, but some develop only over the course of many seconds.

In addition to the "input" aspect of BCIs, there are several examples of the reverse technology: computers connecting into the sensorimotor system providing motor output to the human (see [22]). Well-known examples include heart pace makers and cochlear implants, which directly stimulate auditory nerves, obviating the need for a mechanical hearing mechanism. Another device is able to prevent tremors caused by Parkinson's disease or "essential tremor" by blocking erroneous nervous signals from reaching the thalamus, where they would trigger involuntary muscle contractions [88].

10.6 Summary

The topic of perceptual interfaces is very broad, covering many technologies and their applications in advanced HCI. In this chapter, we gave an overview of perceptual interfaces and went a bit deeper into how the field of computer vision can contribute to the larger goal of natural, adaptive, multimodal, interactive interfaces. Vision-based interaction (VBI) is useful in itself, providing information about human identity, location, movement, and expression through noninvasive and nonintrusive methods. VBI has many near-term application areas, including computer games, accessibility, intelligent environments, biometrics, movement analysis, and social robots.

If the technical goal of building perceptual interfaces can be achieved to any reasonable degree, the ways in which people interact with computers— and with technology in general—will be transformed significantly. In addition to computer vision, this will require advances in many areas, including speech and sound recognition and synthesis, natural language processing, user modeling, haptic and tangible interfaces, and dialogue modeling. More difficult yet, it will require collaboration and integration among these various research areas. In recent years, several workshops and conferences have begun to focus on these issues, including the Workshop on Perceptual/Perceptive User Interfaces (PUI), the International Conference on Multimodal Interfaces (ICMI), and International Conference on Intelligent User Interfaces (IUI). In addition, large major conferences that attract a wide variety of participants—such as CHI and SIGGRAPH—now frequently showcase perceptual interface research or demonstrations.

As the separate technical communities continue to interact and work together on these common goals, there will be a great need for multimodal data sets for training and testing perceptual interfaces, with task data, video, sound, and so on, and associated ground truth. Building such a database is not an easy task. The communities will also need standard benchmark suites for objective performance evaluation, similar to those that exist for individual modalities of speech, fingerprint, and face recognition. Students need to be trained to be conversant with multiple disciplines, and courses must be developed to cover various aspects of perceptual interfaces.

The fact that perceptual interfaces have great promise but will require herculean efforts to reach technical maturity leads to the question of short- and medium-term viability. One possible way to move incrementally toward the long-term goal is to to "piggyback" on the current paradigm of GUIs. Such a "strawman perceptual interface" could start by adding just a few new events in the standard event stream that is part of typical GUI-based architectures. The event stream receives and dispatches events of various kinds: mouse movement, mouse button click and release, keyboard key press and release, window resize, and so on. A new type of event—a "perceptual event"—could be added to this infrastructure that would, for example, be generated when a person enters the visual scene in front of the computer; or when a person begins to speak; or when the machine (or object of interest) is touched; or when some other simple perceptual event takes place. The benefit of adding to the existing GUI event-based architecture is that thousands upon thousands of developers already know how to deal with this architecture and how to write event handlers that implement various functionality. Adding even a small number of perceptual events to this structure would allow developers to come up with creative, novel uses for them and help lead to their acceptance in the marketplace.

This proposed development framework raises several questions. Which perceptual events would be most useful and feasible to implement? Is the event-based model the best way to bootstrap perceptual interfaces? Can we create perceptual events that are reliable enough to be useful? How should developers think about nondeterministic events (as opposed to current events, which are for all practical purposes deterministic)? For example, will visual events work when the lights are turned off, or if the camera lens is obstructed?

There are numerous issues, both conceptual and practical, surrounding the idea of perceptual interfaces. Privacy is one of the utmost importance. What are the implications of having microphones, cameras, and other sensors in computing environments? Where does the data go? What behavioral parameters are stored or sent elsewhere? For perceptual interfaces, to have any chance of success, these issues must be dealt with directly, and it must be made clear to users exactly where the data goes (and does not go). Acceptance of perceptual interfaces depends on instilling confidence that one's privacy is not violated in any way.

Some argue against the idea of interface technologies that attempt to be intelligent or anthropomorphic, claiming that HCI should be characterized by direct manipulation, providing the user with predictable interactions that are accompanied by a sense of responsiblity and accomplishment [121, 122, 123, 126, 125]. While these arguments seem quite appropriate for some uses of computers—particularly when a computer is used as a tool for calculations, word processing, and the like—it appears that future computing environments and uses will be well suited for adaptive, intelligent, agent-based perceptual interfaces.

Another objection to perceptual interfaces is that they just won't work, that the problems are too difficult to be solved well enough to be useful. This is a serious objection—the problems are, indeed, very difficult. It would not be so interesting otherwise. In general, we subscribe to the "If you build it, they will come" school of thought. Building it is a huge and exciting endeavor, a grand challenge for a generation of researchers in multiple disciplines.

Bibliography

[1] E. André and T. Rist. Presenting through performing: On the use of multiple lifelike characters in knowledge-based presentation systems. In *Proc. of the 5th International Conference on Intelligent User Interfaces*, pages 1–8. ACM Press, 2000.

[2] R. Azuma, J. W. Lee, B. Jiang, J. Park, S. You, and U. Neumann. Tracking in unprepared environments for augmented reality systems. *ACM Computers and Graphics*, 23(6):787–793, December 1999.

[3] G. Ball and J. Breese. Emotion and personality in a conversational character. In *Workshop on Embodied Conversational Characters*, pages 83–84, October 1998.

[4] G. Ball, D. Ling, D. Kurlander, J. Miller, D. Pugh, T. Skelly, A. Stankosky, D. Thiel, M. van Dantzich, and T. Wax. Lifelike computer characters: The persona project at Microsoft. In J. Bradshaw (Ed.), *Software Agents*. AAAI/MIT Press, 1997.

[5] J. L. Barron, D. J. Fleet, and S. S. Beauchemin. Performance of optical flow techniques. *Int. Journal of Computer Vision*, 12(1):43–77, 1994.

[6] J. N. Bassili. Emotion recognition: The role of facial movement and the relative importance of upper and lower areas of the face. *Journal of Personality and Social Psychology*, 37:2049–2059, 1979.

[7] J. Biggs and M. A. Srinivasan. Haptic interfaces. In K. Stanney (Ed.), *Handbook of Virtual Environments*. Lawrence Earlbaum, Inc., 2002.

[8] S. Birchfield. Elliptical head tracking using intensity gradients and color histograms. In *Proc. of the IEEE Conference on Computer Vision and Pattern Recognition*, pages 232–237, June 1998.

[9] M. J. Black and Y. Yacoob. Recognizing facial expressions in image sequences, using local parameterized models of image motion. *Int. Journal of Computer Vision*, 25(1):23–48, 1997.

[10] R. A. Bolt. Put-that-there: Voice and gesture in the graphics interface. *Computer Graphics, ACM SIGGRAPH*, 14(3):262–270, 1980.

[11] G. R. Bradski. Real-time face and object tracking as a component of a perceptual user interface. In *IEEE Workshop on Applications of Computer Vision*, pages 142–149, 1998.

[12] A. Braffort, C. Collet, and D. Teil. Anthropomorphic model for hand gesture interface. In *Proc. of the CHI '94 Conference Companion on Human Factors in Computing Systems*, April 1994.

[13] M. Brand. Shadow Puppetry. In *Proc. International Conference on Computer Vision*, 1999.

[14] L. Bretzner, I. Laptev, and T. Lindeberg. Hand gesture recognition using multi-scale colour features, hierarchical models and particle filtering, 2002.

[15] H. Bunke and T. Caelli (Eds.). *Hidden Markov Models in Vision*, vol. 15(1) of *International Journal of Pattern Recognition and Artificial Intelligence*. World Scientific Publishing Company, 2001.

[16] J. Cassell. Embodied conversational interface agents. *Communications of the ACM*, 43(4):70–78, 2000.

[17] T. Choudhury, B. Clarkson, T. Jebara, and A. Pentland. Multimodal Person Recognition using Unconstrained Audio and Video. In *Proc. of the 2nd Conference on Audio- and Video-based Biometric Person Authentication*, 1999.

[18] P. R. Cohen, M. Johnston, D. McGee, S. Oviatt, J. Pittman, I. Smith, L. Chen, and J. Clow. Quickset: Multimodal interaction for distributed applications. In *Proc. of the 5th International Multimedia Conference (Multimedia '97)*, pages 31–40. ACM Press, 1997.

[19] D. Comaniciu, V. Ramesh, and P. Meer. Real-time tracking of non-rigid objects using mean shift. In *Proc. IEEE Conference on Computer Vision and Pattern Recognition*, vol. 2, pages 142–149, 2000.

[20] T. F. Cootes, G. J. Edwards, and C. J. Taylor. Active appearance models. In *Proc. European Conference on Computer Vision*, pages 484–498, 1998.

[21] T. F. Cootes and C. J. Taylor. Active shape models: Smart snakes. In *Proc. of the British Machine Vision Conference*, pages 9–18. Springer-Verlag, 1992.

[22] W. Craelius. Bionic man: Restoring mobility. *Science*, 295(5557):1018–1021, February 2002.

[23] R. Cutler and M. Turk. View-based interpretation of real-time optical flow for gesture recognition. In *Proc. of the 3rd IEEE Conference on Face and Gesture Recognition*, pages 416–421, April 1998.

[24] T. Darrell, G. Gordon, M. Harville, and J. Woodfill. Integrated person tracking using stereo, color, and pattern detection. In *Proc. IEEE Conference on Computer Vision and Pattern Recognition*, pages 601–609, June 1998.

[25] J. G. Daugman. Complete discrete 2D Gabor transform by neural networks for image analysis and compression. *IEEE Trans. Acoustics, Speech, and Signal Processing*, 36:1169–1179, 1988.

[26] J. Davis and M. Shah. Visual gesture recognition. In *Vision, Image, and Signal Processing*, vol. 141, pages 101–106, April 1994.

[27] D. DeCarlo and D. N. Metaxas. Optical flow constraints on deformable models with applications to face tracking. *Int. Journal of Computer Vision*, 38(2):99–127, 2000.

[28] J. Deutscher, A. Blake, and I. Reid. Articulated body motion capture by annealed particle filtering. In *Proc. IEEE Conference on Computer Vision and Pattern Recognition*, vol. 2, pages 126–133, 2000.

[29] R. DeValois and K. DeValois. *Spatial Vision.* Oxford Press, 1988.

[30] A. Dix, J. Finlay, G. Abowd, and R. Beale. *Human-Computer Interaction*, 2nd ed. Prentice Hall Europe, 1998.

[31] G. A. Donato, M. S. Bartlett, J. C. Hager, P. Ekman, and T. J. Sejnowski. Classifying facial actions. *IEEE Transactions on Pattern Analysis and Machine Intelligence*, 21(10):974–989, October 1999.

[32] K. Dorfmüller-Ulhaas and D. Schmalstieg. Finger tracking for interaction in augmented environments. In *IFAR*, 2001.

[33] F. Dornaika and J. Ahlberg. Face model adaptation using robust matching and the active appearance algorithm. In *IEEE Workshop on Applications of Computer Vision*, pages 3–7, December 2002.

[34] A. Doucet, N. de Freitas, and N. J. Gordon (Eds.). *Sequential Monte Carlo Methods in Practice.* Springer-Verlag, 2001.

[35] T. Ebrahimi, J.-M. Vesin, and G. Garcia. Brain-computer interface in multimedia communication. *IEEE Signal Processing Magazine*, January 2003.

[36] D. Efron. *Gesture, Race, and Culture.* King's Crown Press, 1941.

[37] P. Ekman and W. V. Friesen. The facial action coding system: A technique for the measurement of facial movement. *Consulting Psychologists Press*, 1978.

[38] R. Feris, V. Krueger, and R. Cesar Jr. Efficient real-time face tracking in wavelet subspace. In *Workshop on Recognition, Analysis and Tracking of Faces and Gestures in Real-Time Systems*, 2001.

[39] P. M. Fitts. The information capacity of the human motor system in controlling the amplitude of movement. *Journal of Experimental Psychology*, 47:381–391, 1954.

[40] A. R. J. Franois and G. G. Medioni. A modular software architecture for real-time video processing. In *Proc. of the International Workshop on Computer Vision Systems*, July 2001.

[41] W. T. Freeman and M. Roth. Orientation histograms for hand gesture recognition. In *Proc. of the International Workshop on Automatic Face and Gesture Recognition*, pages 296–301. IEEE Computer Society, June 1995.

[42] M. Fukumoto, Y. Suenaga, and K. Mase. Finger-pointer: Pointing interface by image processing. *Computers and Graphics*, 18(5):633–642, 1994.

[43] D. M. Gavrila. The visual analysis of human movement: A survey. *Computer Vision and Image Understanding*, 73(1):82–98, 1999.

[44] J. Gemmell, L. Zitnick, T. Kang, and K. Toyama. Software-enabled gaze-aware videoconferencing. *IEEE Multimedia*, 7(4):26–35, Oct.–Dec. 2000.

[45] S. B. Gokturk, J.-Y. Bouguet, and R. Grzeszczuk. A data-driven model for monocular face tracking. In *Proc. International Conference on Computer Vision*, pages 701–708, 2001.

[46] S. Gong, S. McKenna, and A. Psarrou. *Dynamic Vision: From Images to Face Recognition*. Imperial College Press, World Scientific Publishing, May 2000.

[47] W. E. L. Grimson, C. Stauffer, R. Romano, and L. Lee. Using adaptive tracking to classify and monitor activities in a site. In *IEEE Conference on Computer Vision and Pattern Recognition*, pages 22–29, June 1998.

[48] I. Haritaoglu, D. Harwood, and L. S. Davis. W$\hat{4}$: Real-time surveillance of people and their activities. *IEEE Transactions on Pattern Analysis and Machine Intelligence*, 22(8), August 2000.

[49] E. Hjelmås and B. K. Low. Face detection: A survey. *Computer Vision and Image Understanding*, 83(3):236–274, September 2001.

[50] T. Höllerer. *User Interfaces for Mobile Augmented Reality Systems*. Ph.D. thesis, Columbia University, Department of Computer Science, 2003.

[51] P. Hong, M. Turk, and T. S. Huang. Gesture modeling and recognition using finite state machines. In *Proc. of the Fourth International Conference on Automatic Face and Gesture Recognition*, pages 410–415. IEEE Computer Society, March 2000.

[52] E. Horvitz, J. Breese, D. Heckerman, D. Hovel, and K. Rommelse. The Lumiere project: Bayesian user modeling for inferring the goals and needs of software users. In *Proc. of the 14th Conference on Uncertainty in Artificial Intelligence*, July 1998.

[53] E. Horvitz and T. Paek. Harnessing models of users' goals to mediate clarification dialog in spoken language systems. In *Proc. of the 8th International Conference on User Modeling*, July 2001.

[54] C. Hu and M. Turk. Computer Vision-Based Face Tracking. Technical report, UCSB Computer Science, 2003.

[55] M. Isard and A. Blake. Condensation—Conditional density propagation for visual tracking. *International Journal on Computer Vision*, 1998.

[56] M. Isard and A. Blake. A mixed-state CONDENSATION tracker with automatic model-switching. In *Proc. International Conference on Computer Vision*, pages 107–112, 1998.

[57] H. Ishii and B. Ullmer. Tangible bits: Towards seamless interfaces between people, bits, and atoms. In *Proc. of CHI'97*, pages 234–241, 1997.

[58] T. Jebara, B. Schiele, N. Oliver, and A. Pentland. DyPERS: Dynamic personal enhanced reality system. In *Image Understanding Workshop*, November 1998.

[59] I. T. Jolliffe. *Principal Component Analysis*. Springer-Verlag, 1986.

[60] J. Jones and L. Palmer. An evaluation of the two dimensional Gabor filter methods of simple receptive fields in cat striate cortex. *J. Neurophysiology*, 58:1233–1258, 1987.

[61] M. J. Jones and J. M. Rehg. Statistical color models with application to skin detection. *Int. Journal of Computer Vision*, 46(1):81–96, January 2002.

[62] S. J. Julier, J. K. Uhlmann, and H. F. Durrant-Whyte. A new approach for filtering nonlinear systems. In *Proc. American Control Conference*, pages 1628–1632, June 1995.

[63] R. E. Kalman. A new approach to linear filtering and prediction problems. *Transactions of the ASME Journal of Basic Engineering*, pages 34–45, 1960.

[64] Y. Kameda, M. Minoh, and K. Ikeda. Three dimensional pose estimation of an articulated object from its silhouette image. In *Proc. of Asian Conference on Computer Vision*, pages 612–615, 1993.

[65] K. Karhunen. Über lineare methoden in der wahrscheinlichkeitsrechnung. *Annales Academiae Scientiarum Fennicae*, 37:3–79, 1946.

[66] L. Karl, M. Pettey, and B. Shneiderman. Speech versus mouse commands for word processing applications: An empirical evaluation. *Int. J. Man-Machine Studies*, 39(4):667–687, 1993.

[67] M. Kass, A. Witkin, and D. Terzopoulos. Snakes: Active contour models. In *Proc. International Conference on Computer Vision*, pages 259–268, 1987.

[68] A. Kendon. Some methodological and theoretical aspects of the use of film in the study of social interaction. *Emerging Strategies in Social Psychological Research*, pages 67–91, 1979.

[69] A. Kendon. "Conducting interaction: Patterns of behavior in focused encounters." In John J. Gumperz (Ed.), *Studies in Interactional Sociolinguistics 7*, Cambridge University Press, 1990.

[70] M. Kirby and L. Sirovich. Application of the Karhunen-Loève procedure for the characterization of human faces. *IEEE Transactions on Pattern Analysis and Machine Intelligence*, 12(1):103–108, January 1990.

[71] R. Kjeldsen and J. Kender. Finding skin in color images. In *Proc. of the International Conference on Automatic Face and Gesture Recognition*, pages 312–317, October 1996.

[72] N. Kohtake, J. Rekimoto, and Y. Anzai. InfoPoint: A device that provides a uniform user interface to allow appliances to work together over a network. *Personal and Ubiquitous Computing*, 5(4):264–274, 2001.

[73] T. Kurata, T. Okuma, M. Kourogi, and K. Sakaue. The hand mouse: GMM hand-color classification and mean shift tracking. In *2nd International Workshop on Recognition, Analysis and Tracking of Faces and Gestures in Real-time Systems*, July 2001.

[74] M. La Cascia, S. Sclaroff, and V. Athitsos. Fast, reliable head tracking under varying illumination: An approach based on registration of texture-mapped 3D models. *IEEE Trans. on Pattern Analysis and Machine Intelligence*, 22(4), April 2000.

[75] M. Lades, J. Vorbrüggen, J. Buhmann, J. Lange, W. Konen, C. von der Malsburg, and R. Würtz. Distortion invariant object recognition in the dynamic link architecture. *IEEE Trans. on Pattern Analysis and Machine Intelligence*, 42(3):300–311, March 1993.

[76] I. Laptev and T. Lindeberg. Tracking of multi-state hand models using particle filtering and a hierarchy of multi-scale image features. Technical Report ISRN

KTH/NA/P-00/12-SE, Department of Numerical Analysis and Computer Science, KTH (Royal Institute of Technology), September 2000.

[77] J. Lee and T.L. Kunii. Model-based analysis of hand posture. *IEEE Computer Graphics and Applications*, 15(5):77–86, 1995.

[78] Y. Li, S. Gong, and H. Liddell. Modeling faces dynamically across views and over time. In *Proc. Intl. Conference on Computer Vision*, 2001.

[79] R.-H. Liang and M. Ouhyoung. A real-time continuous gesture recognition system for sign language. In *Proc. of the Third International Conference on Automatic Face and Gesture Recognition*, pages 558–565. IEEE Computer Society, April 1998.

[80] J. Lin, Y. Wu, and T. S. Huang. Modeling the constraints of human hand motion. In *Proc. of the 5th Annual Federated Laboratory Symposium*, 2001.

[81] M. M. Loève. *Probability Theory*. Van Nostrand, 1955.

[82] J. MacCormick and M. Isard. Partitioned sampling, articulated objects, and interface-quality hand tracking. In *Proc. European Conf. Computer Vision*, 2000.

[83] I. S. MacKenzie and S. Ware. Lag as a determinant of human performance in interactive systems. In *Proc. of the ACM Conference on Human Factors in Computing Systems, INTERCHI*, pages 488–493, 1993.

[84] M. Malciu and F. Preteux. A robust model-based approach for 3D head tracking in video sequences. In *Proc. IEEE Intl. Conference on Automatic Face and Gesture Recognition (AFGR)*, pages 169–174, 2000.

[85] K. Mase. Recognition of facial expression from optical flow. *IEICE Trans.*, 74(10):3474–3483, 1991.

[86] D. McNeill. *Hand and Mind: What Gestures Reveal about Thoughts*. University of Chicago Press, 1992.

[87] D. McNeill (Ed.). *Language and Gesture*. Cambridge University Press, 2000.

[88] Medtronics, Inc. Activa Tremor Control Therapy, 1997.

[89] A. Mitiche and P. Bouthemy. Computation and analysis of image motion: A synopsis of current problems and methods. *International Journal of Computer Vision*, 19(1):29–55, 1996.

[90] B. Moghaddam and Alex Pentland. Probabilistic visual learning for object detection. In *Proc. of the 5th International Conference on Computer Vision*, pages 786–793, June 1995.

[91] A. Mohan, C. Papageorgiou, and T. Poggio. Example-based object detection in images by components. *IEEE PAMI*, 23(4):349–361, April 2001.

[92] D. D. Morris and J. M. Rehg. Singularity analysis for articulated object tracking. In *Proc. Conference on Computer Vision and Pattern Recognition*, 1998.

[93] K. Murakami and H. Taguchi. Gesture recognition using recurrent neural networks. In *ACM CHI Conference Proc.*, pages 237–242, 1991.

[94] T. A. Mysliwiec. FingerMouse: A freehand computer pointing interface. Technical Report VISLab-94-001, Vision Interfaces and Systems Lab, The University of Illinois at Chicago, October 1994.

[95] C. Nass and Y. Moon. Machines and mindlessness: Social responses to computers. *Journal of Social Issues*, 56(1):81–103, 2000.

[96] C. Nass, J. Steuer, and E. Tauber. Computers are social actors. In *ACM CHI*, pages 72–78, 1994.

[97] M.A.L. Nicolelis. Action from thoughts. *Nature*, 409:403–407, January 2001.

[98] C. Nolker and H. Ritter. GREFIT: Visual recognition of hand postures. In *Gesture-Based Communication in HCI*, pages 61–72, 1999.

[99] M. Oren, C. Papageorgiou, P. Sinha, E. Osuna, and T. Poggio. Pedestrian detection using wavelet templates. In *Proc. IEEE Conference on Computer Vision and Pattern Recognition*, June 1997.

[100] S. L. Oviatt. Ten myths of multimodal interaction. *Communications of the ACM*, 42(11):74–81, November 1999.

[101] S. L. Oviatt, P. R. Cohen, L. Wu, J. Vergo, L. Duncan, B. Suhm, J. Bers, T. Holzman, T. Winograd, J. Landay, J. Larson, and D. Ferro. Designing the user interface for multimodal speech and gesture applications: State-of-the-art systems and research directions. *Human Computer Interaction*, 15(4):263–322, 2000.

[102] P. S. Penev and J. J. Atick. Local feature analysis: A general statistical theory for object representation. *Network: Computation in Neural Systems*, 7(3):477–500, 1996.

[103] A. Pentland, B. Moghaddam, and T. Starner. View-based and modular eigenspaces for face recognition. In *Proc. IEEE Conference on Computer Vision and Pattern Recognition*, pages 84–91, June 1994.

[104] M. L. Phillips, A. W. Young, C. Senior, C. Brammer, M. Andrews, A. J. Calder, E. T. Bullmore, D. I Perrett, D. Rowland, S. C. R. Williams, A. J. Gray, and A. S. David. A specific neural substrate for perceiving facial expressions of disgust. *Nature*, 389:495–498, 1997.

[105] R. W. Picard. *Affective Computing*. MIT Press, 1997.

[106] D. A. Pollen and S. F. Ronner. Phase relationship between adjacent simple cells in the visual cortex. *Science*, 212:1409–1411, 1981.

[107] F. K. H. Quek, T. Mysliwiec, and M. Zhao. FingerMouse: A freehand pointing interface. In *Proc. International Workshop on Automatic Face and Gesture Recognition*, pages 372–377, June 1995.

[108] F. Quek, D. McNeill, R. Bryll, C. Kirbas, H. Arslan, K.E. McCullough, N. Furuyama, and R. Ansari. Gesture, speech, and gaze cues for discourse segmentation. In *IEEE Conference on Computer Vision and Pattern Recognition*, vol. 2, pages 247–254, June 2000.

[109] B. Reeves and C. Nass. *The Media Equation: How People Treat Computers, Television, and New Media Like Real People and Places*. Cambridge University Press, September 1996.

[110] J.M. Rehg and T. Kanade. Model-based tracking of self-occluding articulated objects. In *Proc. of the 5th International Conference on Computer Vision*, pages 612–617, June 1995.

[111] J. Rekimoto. NaviCam: A magnifying glass approach to augmented reality systems. *Presence: Teleoperators and Virtual Environments*, 6(4):399–412, 1997.

[112] B.J. Rhodes. The wearable remembrance agent: A system for augmented memory. *Personal Technologies Journal (Special Issue on Wearable Computing)*, pages 218–224, 1997.

[113] Y. Sato, Y. Kobayashi, and H. Koike. Fast tracking of hands and fingertips in infrared images for augmented desk interface. In *4th IEEE International Conference on Automatic Face and Gesture Recognition*, March 2000.

[114] D. Saxe and R. Foulds. Toward robust skin identification in video images. In *2nd International Face and Gesture Recognition Conference*, September 1996.

[115] A. E. Scheflen. Communication and regulation in psychotherapy. *Psychiatry*, 26(2):126–136, 1963.

[116] B. Schiele and A. Waibel. Gaze tracking based on face-color. In *Proc. of the International Workshop on Automatic Face- and Gesture-Recognition*, pages 344–349, June 1995.

[117] R. Schmidt-Feris, J. Gemmell, K. Toyama, and V. Krüger. Hierarchical wavelet networks for facial feature localization. In *Proc. of the 5th International Conference on Automatic Face and Gesture Recognition*, May 2002.

[118] S. Sclaroff and J. Isidoro. Active blobs. In *Proc. International Conference on Computer Vision*, 1998.

[119] K.-H. Seo, W. Kim, C. Oh, and J.-J. Lee. Face detection and facial feature extraction using color snake. In *Proc. IEEE International Symposium on Industrial Electronics*, vol. 2, pages 457–462, July 2002.

[120] J. Shi and C. Tomasi. Good features to track. In *Proc. IEEE Conference on Computer Vision and Pattern Recognition*, June 1994.

[121] B. Shneiderman. A nonanthropomorphic style guide: Overcoming the Humpty Dumpty syndrome. *The Computing Teacher*, 16(7), 1989.

[122] B. Shneiderman. Beyond intelligent machines: Just do it! *IEEE Software*, 10(1):100–103, 1993.

[123] B. Shneiderman. Direct manipulation for comprehensible, predictable and controllable user interfaces. In *Proc. of International Conference on Intelligent User Interfaces*, pages 33–39, 1997.

[124] B. Shneiderman. *Designing the User Interface: Strategies for Effective Human-Computer Interaction*. 3rd ed., Addison Wesley, March 1998.

[125] B. Shneiderman. The limits of speech recognition. *Communications of the ACM*, 43(9):63–65, September 2000.

[126] B. Shneiderman, P. Maes, and J. Miller. Intelligent software agents vs. user-controlled direct manipulation: A debate, March 1997.

[127] T. E. Starner and A. Pentland. Visual recognition of American sign language using hidden Markov models. In *AFGR, Zurich*, 1995.

[128] B. Stenger, P. R. S. Mendonça, and R. Cipolla. Model-based 3D tracking of an articulated hand. In *Proc. Conference on Computer Vision and Pattern Recognition*, vol. 2, pages 310–315, December 2001.

[129] R. Stiefelhagen, J. Yang, and A. Waibel. Estimating focus of attention based on gaze and sound. In *Workshop on Perceptive User Interfaces*. ACM Digital Library, November 2001.

[130] J. Ström, T. Jebara, S. Basu, and A. Pentland. Real-time tracking and modeling of faces: An EKF-based analysis by synthesis approach. In *Proc. International Conference on Computer Vision*, 1999.

[131] D. J. Sturman. *Whole Hand Input*. Ph.D. thesis, MIT, February 1992.

[132] D. M. Taylor, S. I. Helms Tillery, and A. B. Schwartz. Direct cortical control of 3D neuroprosthetic devices. *Science*, June 2002.

[133] C. Tomasi, A. Rafii, and I. Torunoglu. Full-size projection keyboard for hand-held devices. *Communications of the ACM*, 46(7):70–75, July 2003.

[134] M. Turk and A. Pentland. Eigenfaces for recognition. *J. Cognitive Neuroscience*, 3(1):71–86, 1991.

[135] M. Turk, C. Hu, R. Feris, F. Lashkari, and A. Beall. TLA-based face tracking. In *15th International Conference on Vision Interface*, May 2002.

[136] M. Turk and G. Robertson. Perceptual user interfaces. *Communications of the ACM*, 43(3):32–34, March 2000.

[137] U.S. Department of Transportation, Federal Highway Administration. Evaluation of Automated Pedestrian Detection at Signalized Intersections, August 2001.

[138] A. van Dam. Post-wimp user interfaces. *Communications of the ACM*, 40(2): 63–67, 1997.

[139] P. Viola and M. Jones. Robust real-time object detection. *International Journal of Computer Vision*, 2002.

[140] G. Welch, G. Bishop, L. Vicci, S. Brumback, K. Keller, and D. Colucci. The HiBall tracker: High-performance wide-area tracking for virtual and augmented environments. In *Proc. of the ACM Symposium on Virtual Reality Software and Technology*, December 1999.

[141] J. Wessberg, C. R. Stambaugh, J. D. Kralik, P. D. Beck, M. Laubach, J. K. Chapin, J. Kim, S. J. Biggs, M. A. Srinivasan, and M. A. L. Nicolelis. Real-time prediction of hand trajectory by ensembles of cortical neurons in primates. *Nature*, 408(361), 2000.

[142] C. D. Wickens. The effects of control dynamics on performance. In K. Boff, K. Kaufman, and J. Thomas (Eds.), *Handbook on perception and human performance—Cognitive processes and performance*, vol. 2, chapter 39. Wiley Interscience, 1986.

[143] A. Wilson and S. Shafer. XWand: UI for intelligent spaces. In *CHI*, 2003.

[144] J. R. Wolpaw, N. Birbaumer, D. J. McFarland, G. Pfurtscheller, and T. M. Vaughan. Brain-computer interfaces for communication and control. *Clinical Neurophysiology*, 113(6):767–791, June 2002.

[145] C. Wren, A. Azarbayejani, T. Darrell, and A. Pentland. PFinder: Real-time tracking of the human body. *IEEE Transactions on Pattern Analysis and Machine Intelligence*, 19(7):780–785, July 1997.

[146] C. R. Wren and A. P. Pentland. Dynamic models of human motion. In *Proc. of the 3rd International Conference on Automatic Face and Gesture Recognition*, pages 22–27. IEEE Computer Society, April 1998.

[147] H. Wu, T. Yokoyama, D. Pramadihanto, and M. Yachida. Face and facial feature extraction from color image. In *Proc. IEEE Intl. Conference on Automatic Face and Gesture Recognition (AFGR)*, pages 345–350, October 1996.

[148] L. Wu, S. L. Oviatt, and P. R. Cohen. Multimodal integration—a statistical view. *IEEE Transactions on Multimedia*, 1(4):334–331, December 1999.

[149] W. Wu, M. J. Black, Y. Gao, E. Bienenstock, M. Serruya, A. Shaikhouni, and J. P. Donoghue. Neural decoding of cursor motion using a kalman filter. In *Neural Information Processing Systems*, December 2002.

[150] Y. Wu and T. S. Huang. Vision-based gesture recognition: A review. In A. Braffort, R. Gherbi, S. Gibet, J. Richardson, and D. Teil (Eds.), *Gesture-Based Communication in Human-Computer Interaction*, vol. 1739 of *Lecture Notes in Artificial Intelligence*. Springer Verlag, 1999.

[151] Y. Wu and T. S. Huang. View-independent recognition of hand postures. In *Proc. Conference on Computer Vision and Pattern Recognition*, vol. 2, pages 84–94, 2000.

[152] Y. Wu and T. S. Huang. Hand modeling, analysis, and recognition. *IEEE Signal Processing Magazine*, May 2001.

[153] W. Zhao, R. Chellappa, and A. Rosenfeld. Face recognition: A literature survey. Technical Report Technical Report CAR-TR948, UMD CfAR, 2000.

[154] J. Xiao, T. Kanade, and J. Cohn. Robust full-motion recovery of head by dynamic templates and re-registration techniques. In *Proc. IEEE International Conference on Automatic Face and Gesture Recognition*, May 2002.

[155] M.-H. Yang, D. J. Kriegman, and N. Ahuja. Detecting faces in images: A survey. *IEEE Transactions on Pattern Analysis and Machine Intelligence*, 24(1):34–58, 1 2002.

[156] R. Yang and Z. Zhang. Eye gaze correction with stereovision for video-teleconferencing. In *European Conference on Computer Vision*, May 2002.

[157] R. Yang and Z. Zhang. Model-based head pose tracking with stereo vision. In *Proc. IEEE International Conference on Automatic Face and Gesture Recognition*, pages 255–260, 2002.

[158] S. J. Young. HTK: Hidden Markov Model Toolkit V1.5, December 1993. Entropic Research Laboratories Inc.

[159] B. D. Zarit, B. J. Super, and F. K .H. Quek. Comparison of five color models in skin pixel classification. In *Workshop on Recognition, Analysis, and Tracking of Faces and Gestures in Real-Time Systems*, pages 58–63, September 1999.

[160] J. Zhang, Y. Yan, and M. Lades. Face recognition: Eigenface, elastic matching, and neural nets. *Proc. of the IEEE*, 85(9):1423–1435, 1997.

[161] Z. Zhang. Feature-based facial expression recognition: Sensitivity analysis and experiments with a multilayer perceptron. *International Journal of Pattern Recognition and Artificial Intelligence*, 13(6):893–911, 1999.

[162] L. Zhao and C. E. Thorpe. Stereo- and neural network-based pedestrian detection. *IEEE Tran. on Intelligent Transportation Systems*, 1(3), 2000.

[163] Y. Zhu, H. Ren, G. Xu, and X. Lin. Toward real-time human–computer interaction with continuous dynamic hand gestures. In *Proc. of the Conference on Automatic Face and Gesture Recognition*, pages 544–549, 2000.

PART III
PROGRAMMING
FOR COMPUTER VISION

One of the more overlooked areas in computer vision is its programming aspect. Because of the heavy dependence of computer vision on image and range data, the proper programming environment has to efficiently handle and manipulate such data. It also has to contain the fundamental operations such as input/output and basic operations such as image convolution and 2D blob extraction. This allows the programmer to concentrate on prototyping the algorithm currently being researched.

There are two chapters in this final section that address the issue of programming for computer vision. In Chapter 11, Bradski describes the well-known Open Source Computer Vision Library (OpenCV). OpenCV is a collection of C and C++ source code and executables that are optimized for real-time vision applications. He shows how a variety of applications such as stereo, tracking, and face detection can be easily implemented using OpenCV.

In Chapter 12, François highlights and addresses architecture-level software development issues facing researchers and practitioners in the field of computer vision. A new framework, or architectural style, called SAI, is introduced. It provides a formalism for the design, implementation, and analysis of software systems that perform distributed parallel processing of generic data streams. Architectural patterns are illustrated with a number of demonstration projects ranging from single-stream, automatic, real-time video processing to fully integrated, distributed, interactive systems mixing live video, graphics, and sound. SAI is supported by an open-source architectural middleware called MFSM.

Chapter 11

OPEN SOURCE COMPUTER VISION LIBRARY (OPENCV)

Gary Bradski

11.1 Overview

The Open Source Computer Vision Library (OpenCV for short [1, 14]) is a collection of C and C++ source code and executables that span a wide range of computer vision algorithms. The code is optimized and intended for real-time vision applications. Popular application areas supported by OpenCV algorithms in decreasing order of coverage are human–computer interaction; video security; robotics and image retrieval; and factory inspection/machine vision. Look for a full handbook on OpenCV due from Springer in 2004 [12]. OpenCV supports Windows and Linux, but the code is well behaved and has been ported to many other platforms. Tested compilers include Intel's compiler version 6.0 or higher, MSVC++ 6.0 or higher, Borland C++ 5.5 or higher, and GNU C/C++ 2.95.3 or higher.

OpenCV is distributed under a BSD-style license (see Section 11.6). This license allows for royalty-free commercial or research use with no requirement that the user's code be free or open. Intel Labs (http://www.intel.com/research/mrl/index.htm) media group (http://www.intel.com/research/mrl/

research/media.htm) undertook development of this library for the following reasons:

- – Computer vision algorithms are being studied for their implications on future computer architecture. Optimized code is needed for these studies.

- – We wanted to advance external computer vision research by creating an optimized infrastructure that would avoid the need for others to endlessly reinvent and reoptimize common algorithms.

- – We wanted to hasten the adoption of computer vision applications out in the market by lowering the algorithmic "barrier to entry" for firms seeking to add computer vision functionality to their products.

OpenCV's main information Web site is located at http://www.intel. com/research/mrl/research/opencv. This site contains pointers to the user group and the download site.

The user group is at http://groups.yahoo.com/group/OpenCV/ [4]. To join the user group, you must first register (free) for Yahoogroups (http:// groups.yahoo.com/) and then sign up for OpenCV at http://groups.yahoo. com/group/OpenCV/join. The user group is the forum for reporting bugs, discussing problems, technical issues, posting/reading calls for papers or jobs, and so on. This is an active group with thousands of members. You control at any time whether you want to read it on the web only (no emails) or get daily summaries or every posting. We do not distribute names or email addresses from the list to outside parties.

The download site is on SourceForge at http://sourceforge.net/projects/ opencvlibrary/ [6]. At this site, you can obtain the various release versions of OpenCV for Windows and Linux, documentation files, and the foil set [5] from which many of the images in this chapter were taken.

11.1.1   Installation

Under Windows

Installation[1] under Windows is straightforward. Download the windows executable installation from the SourceForge site to a directory on your local disk and run it. The script installs OpenCV, registers DirectShow filters, and does other post-installation procedures. After it is finished, you may start

---

[1]Adapted with permission from the INSTALL file that comes with the download [6].

using OpenCV; you do not need to compile OpenCV into binaries unless you want to debug files.

It is also possible to build core OpenCV binaries manually from the source-code distribution for Linux (though the executable installation includes sources as well). To build manually from the Linux source, do the following.

+ Download and unpack the OpenCV-*.tar.gz package somewhere, e.g.,
  C:\MySoft\(the root folder is referred to further as
  <opencv_root>). The tree should look like this:

```
<opencv_root>
    _dsw
    cv
        include
        src
        make
    cvaux
        ...
    ...
```

+ Add <opencv_root>\bin to the system path. Under Windows 9x/ME,
  this is done by modifying autoexec.bat. Under NT/2000/XP, it
  can be done instantly at MyComputer:--right button click-->
  Properties->Advanced->Environment Variables.

+ HighGUI requires graphic libraries by default, so remove
  HAVE_JPEG, HAVE_TIFF, and HAVE_PNG from preprocessor
  definitions and libjpeg.lib, libtiff.lib, libpng.lib, and
  zlib.lib from the linker command line. The resultant HighGUI
  will be able to read and write most of JPEGs, BMPs, uncompressed
  TIFFs, PXMs and Sun raster images; capture video from AVI
  or camera via VFW; and write AVIs via VFW.

```
Building OpenCV from sources
----------------------------


You need to have a C/C++ compiler. Below are some variants:

  === Microsoft Visual C++ (6.0 or higher) ===
    This is a preferred variant, because most of the demos
    are written for it (i.e., using MFC).

    * If you plan to build DirectShow filters, acquire and
      setup DirectX SDK as described in
```

```
<opencv_root>\docs\faq.htm or
<opencv_root>\cv\include\cvstreams.h.
```

* If you plan to build an MIL-enabled version of highgui,
  set up MIL include and library paths in Developer Studio.

* If you plan to build MATLAB wrappers, you need to have
  MATLAB C/C++ interface libraries and to set up Developer
  Studio properly. Read opencv/interfaces/matlab/readme.txt
  for details.

* If you are going to build TCL\TK demo applications
  (included with the source package only), you need
  TCL\TK and BWidgets. The easiest way to obtain both is to
  download and install ActiveTcl from http://www.
  activestate.com. After installing ActiveTCL,
    - Add <tcl_root>\bin to the system path (if installer
      didn't do it).
    - Add <tcl_root>\include and <tcl_root>\lib to Developer
      Studio search paths (tools->options->directories).

Open <opencv_root>_dsw\opencv.dsw.

Choose from menu Build->Batch Build->Build.

Wait and enjoy. If you want to debug OpenCV directshow
filters, register them using regsvr32
(e.g., regsvr32 <opencv_root>\bin\CamShiftd.ax).

=== Other compilers ===
In case of other compilers, you still can build the core
libraries (cv, cvaux, highgui), algorithmic tests, and samples
(<opencv_root>\samples\c).
The following compilers are supported by default:

=== Intel compiler 6.0 or greater ===

Run nmake /f makefile.icl in the root OpenCV folder.

Because the produced binaries should be compatible with
Visual C++, you can then use the DLLs with VisualC++
-build applications etc.

```
=== Borland C++ 5.5 (free) or greater ===
```

```
    Run make -f makefile.bcc in the root OpenCV folder.
```

```
    Before running compilation, make sure <BorlandC_root>\bin is
    in the system path and <BorlandC_root>\bin\bcc32.cfg
    contains
    -I<BorlandC_root>\bcc\include -L<BorlandC_root>\bcc\lib
    -L<BorlandC_root>\bcc\lib\psdk
    (where <BorlandC_root> denotes the root folder of Borland C++
    installation).
```

```
=== GNU C/C++ 2.95.3 or greater ===
```

```
    Run mingw32-make -f makefile.gcc in the root OpenCV folder.
```

```
    Make sure that <gcc_root>\bin is in the system path.
    To build VFW-enabled highgui, read instructions in
    <opencv_root>\otherlibs\_graphics\readme.txt.
```

How to test built OpenCV binaries in Windows

Run algorithmic tests: `<opencv_root>\bin\cvtest.exe`. This will produce `cvtest.sum` and `cvtest.lst`. `cvtest.sum` should contain all OKs. Or, run samples at `<opencv_root>\samples\c`. (Note that some of the demos need AVI or Camera, e.g., motempl.c.)

How to add support for another compiler

Look at `<opencv_root>\utils\gen_make.py`. It looks at .dsp files in the specified folders and generates makefiles for all compilers it knows. GCC handling now gets a little bit ugly because the linker cannot handle long lists of files and it wasn't known if it was possible to use temporary inline files instead.

Under Linux

There are no prebuilt binaries for the Linux version (because of different C++-incompatible versions of GCC in different distributions), so you have to build it from sources. The following has been tested on RedHat 8.0 (GCC 3.2) and SuSE 8.0 (GCC 2.95.3). To build fully functional libraries and demos in Linux, use

```
+ motif (LessTif or OpenMotif) with development files.
  configure script assumes it is at /usr/X11R6/lib &
  /usr/X11R6/include/Xm.
```

+ libpng, libjpeg and libtiff with development files.

+ libavcodec from ffmpeg 0.4.6(pre) and headers.
  Earlier version does not fit because of changed interface and
  because of the GPL license (newer version is LGPL).
  However, static linking is still prohibited for non-GPL software
  (such as OpenCV), so:
    Get CVS snapshot of ffmpeg from ffmpeg.sourceforge.net
    ./configure --enable-shared
    make
    make install
    You will then have /usr/local/lib/libavcodec.so and
                    /usr/local/include/ffmpeg/*.h.

+ For building demo applications only:
  fltk 1.1.x (1.1.1 is currently the preferred one) with
  development files.
  If you do not have it, get it from www.fltk.org. In case of
  RPM-based distribution it is possible to build fltk RPMs by
      rpmbuild -ta fltk-x.y.z-source.tar.gz (for RH 8.x) or
      rpm -ta fltk-x.y.z-source.tar.gz (for others).

+ For demo applications only:
  TCL/TK 8.3.x with development files and BWidgets >=1.3.x.
  Download BWidgets from http://sourceforge.net/projects/tcllib/

Now build OpenCV:
======

1. If your distribution uses RPM, you may build RPMs via
      rpmbuild -ta OpenCV-x.y.z.tar.gz" (for RH 8.x) or
      rpm -ta OpenCV-x.y.z.tar.gz" (for others)
   where OpenCV-x.y.z.tar.gz should be put into
   /usr/src/redhat/SOURCES/ or a similar folder.

   The command will build OpenCV-x.y.z.*.rpm (there is no
   OpenCV-devel; everything is in one package).

   Then, install it by
     rpm -i --nodeps OpenCV-x.y.z.*.rpm

nodeps is needed in this version, because it cannot find
   libavcodec.so, even if it is in the path (a weird bug
   somewhere).

2. If your distribution does not support RPM, build and install it
   in the *nix traditional way:

```
./configure --with-apps  # or simply ./configure
make
make install # as root
ldconfig # as root
```

Both 1 and 2 (post-install)

The default installation path is /usr/local/lib and
/usr/local/include/opencv, so you need to add
/usr/local/lib to /etc/ld.so.conf (and run ldconfig
afterward).

How to test OpenCV under Linux:

– Run /usr/local/bin/cvtest.

– Or, compile and run simple C examples at /usr/local/share/opencv/
  samples, for example, *g++ 'opencv-config –cxxflags' -o morphology
  morphology.c 'opencv-config –libs'*. Plain gcc won't work because of
  unresolved C++ specific symbols (located in highgui).

– Or, run */usr/local/bin/{cvlkdemo|cvcsdemo|cvenv|vmdemotk}*.

11.1.2 Organization

When you have set up OpenCV, the directory organization will be something
like the following (slightly different for Linux):

```
C:\Program Files\OpenCV -- Location
   _dsw                 -- Build files under Windows
   apps                 -- Demo application files
   bin                  -- Binaries, dlls, filters ...
   cv                   -- Source Code
   cvaux                -- "Experimental" Source Code
   docs                 -- Manuals
   filters              -- DirectShow filters
   interfaces           -- Ch interpretive C, MATLAB
```

```
lib                     -- Static link libraries
otherlibs               -- highgui interface and camera support
samples                 -- Simple code usage examples
tests                   -- Source and data for test code
utils                   -- cvinfo code and other utilities
```

apps: The *apps* directory contains the following:

```
C:\Program Files\OpenCV\apps -- Location
   CamShiftDemo    -- Track color probability distributions
   Common          -- Camera routines
   cvcsdemo        -- Color tracker application
   cvlkdemo        -- Optical flow application
   HaarFaceDetect  -- Viola-Jones Boosted face detection
   HaarTraining    -- Training code for Boosted face detection
   Hawk            -- EiC Interpretive C interface
   HMMDemo         -- Face recognition using HMMs
   StereoDemo      -- Depth detection from stereo correspondence
   Tracker3dDemo   -- 3D tracking with multiple cameras
   VMDemo          -- Interpolated view (morph) between 2 images
   vmdemotk        -- Interpolated view application
```

Of special interest here is the manual in the *Hawk* directory, which details some of the functions in HighGUI (see below) that make it easy to display an image or video in a window. One-line routines can be used to attach a menu and/or sliders to the window, which can be used to interactively control OpenCV routines operating in the window. The hidden Markov model (HMM) demo does face recognition using horizontal HMMs across a face feeding into a vertical HMM down a face to do 2D face recognition.

The StereoDemo is a console-based application that uses cvcam for video capture and HighGUI for visualization (both in `otherlibs` directory). To run this demo, you need two USB cameras compatible with DirectShow that can run simultaneously. Creative WebCam is an example of such a camera. Two different cameras types might be also work (not tested). You will need to register `ProxyTrans.ax` and `SyncFilter.ax` DirectShow filters from the `opencv\bin` folder using `regsvr32.exe` utility or `opencv\bin\RegisterAll.bat` batch command file. The demo lets you calibrate the intrinsic and extrinsic parameters for the two cameras by automatically tracking a checkerboard pattern of known size and then finds stereo correspondence between the two image streams and converts that to a disparity or depth map.

Tracker3dDemo demonstrates the 3D calibration of multiple cameras (two or more) into one coordinate frame and then the 3D tracking of multiple objects from multiple views.

docs: The `docs` directory contains the HTML-based manual (index.htm, faq.htm), the license file (see Section 11.6), and directories describing some of the applications and tutorial papers.

filters: The `filters` directory contains code to build the following computer vision filters:

```
C:\Program Files\OpenCV\filters  -- Location
    CalibFilter      -- Calibrate camera lens via checkerboard
    CamShift         -- Meanshift robust object tracker
    Condens          -- Condensation particle filter tracker
    Kalman           -- Kalman filter
    ProxyTrans       -- Automatic filter calls your code
    SyncFilter       -- For use with multiple cameras
    Tracker3dFilter  -- 3D tracking with multiple cameras
```

otherlibs: The `otherlibs` directory contains the camera connection code (cvcam) and HighGUI which can be used to put up a window with still or video images and attach menus and controls to control video processing within the window.

samples: The `samples` directory is a good place to look for simple code examples of many of the functions in OpenCV.

11.1.3 Optimizations

OpenCV has two defining characteristics: (1) It spans a very wide range of functions, and (2) the code is fairly well optimized for speed. Figure 11.1 shows what code is open, what is not, and how things relate to one another. As with Intel's other performance libraries [3], when you build your application, you link against a dynamic library file, for example, `cv.lib`. At runtime, an optimized C dynamic-link library stub, `cv.dll` is loaded. This stub tries to determine what processor it is running on, and if the processor is identified, it will look for a DLL file that is specifically optimized for that processor. For example, on a Pentium 4, the specially optimized code is `OptCVw7.dll`. If such a file is found, `cv.dll` swaps itself out for this file. If the processor type cannot be determined or the corresponding optimized file cannot be found, OpenCV runs optimized C by default.

OpenCV's C code is algorithmically optimized. The processor-specific code is further optimized with inline assembly, memory alignment, and cache

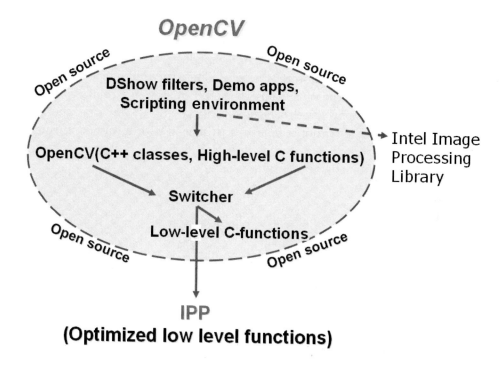

Figure 11.1. Structure of OpenCV. (*Used with permission from [5, 6].*)

considerations, and it takes advantage when possible of the internal SIMD arithmetic MMX and SSE processors. OpenCV no longer depends on Intel's Image Processing Library, but some demos still require it. Again, as long as you have the set of DLLs on your machine, you can just compile/run OpenCV applications and the switcher code inside the `cv.dll` takes care of loading the correct one and takes advantage of the performance-primitive ippCV code described below.

ippCV [2]

OpenCV comes from Intel's research labs and is fully open and free to use. Intel has many other software tools that come from other internal teams such as the Software System's Group. This other software is released commercially. Such products include the VTune code profiler, the Intel Compiler, and the multithreading preprocessor macros. Of interest here are Intel's integrated performance primitives (IPP) and specifically the 60 or so computer vision functions included under ippCV. You can get information about IPP at http://developer.intel.com/software/products/ipp/ippvm20/.

The *primitives* part of the integrated performance primitives refers to functions that operate mostly on vectors, matrices, and simple structures. These functions are re-entrant and stateless. Thus, the IPP functions tend to be easy to interface to and work well embedded in streaming-data systems.

ippCV contains many fewer functions than OpenCV, but the functions are optimized on a wider variety of platforms. The platform support includes

```
IA32 Intel
Itanium Architecture
Intel StrongARM Microarchitecture
Intel XScale Microarchitecture
```

Intel IPP supports various operating systems:

```
32-bit versions of Microsoft Windows
64-bit versions of Microsoft Windows Microsoft
Windows CE
Linux32
Linux64
ARM Linux
```

Inside ippCV, the following computer vision functions are supported:

```
Image pyramid (resample and blur)
Filters: Laplace, Sobel, Scharr, Erode, Dilate
Motion gradient
Flood fill
Canny edge detection
Snakes
Optical flow
```

Of course, there are many other non-computer vision functions supported in 2500 functions that make up IPP. IPP includes support for image processing, compression, signal processing, audio and video codecs, matrix operations, and speech recognition.

Other performance libraries

Two other libraries that serve as higher level wrappers to IPP are of interest: the signal processing library and the image processing library [3]. OpenCV was designed to work well with both IPP and these libraries. The functions covered by the signal processing and image processing libraries follow.

Signal Processing Functions:
    Data manipulation
    Windowing
    Conversion
    Arithmetic and logical operations
    Filters and transforms
    Statistics
    Audio
    Signal generation

Image Processing Functions:
    Data manipulation
    Arithmetic and logical functions
    Copy and conversions
    Geometric operations (scale, rotate, warp)
    Color conversion
    Alpha composite
    Gamma correction
    General and specific filters: FFT, DCT, wavelet transform

A final separate library to be aware of is the Math Kernel Library (MKL), which contains the full set of linear algebra and matrix manipulation routines.

11.2   Functional Groups: What's Good for What

In the Windows directory hierarchy, the OpenCV manual is located at `C:\Program Files\OpenCV\docs\index.htm`. In the manual, the functions are broken up into the following groups:

    M1.   Basic Structures and Operations
          Helper structures
          Array structures
          Arrays manipulation
          Matrix operations
          Dynamic data structures
          Sequences
          Sets
          Graphs
          Writing and reading structures

```
GetCamerasCount
GetProperty
Init
Pause
PlayAVI
Resume
SelectCamera
SetProperty
Start
Stop
```

In Section 11.2.1, we first give an overview of the functions by area in the manual. In Section 11.2.2, we give some brief ideas of what function might be good for what task. Section 11.2.3 discusses what's in the demos and sample code.

11.2.1   By area

Below, M# refers to the manual sections outlined above.

M1: Structures and matrices

The first section of the manual begins by describing structures for representing subpixel accurate points and rectangles. This is followed by array and image structures and functions for creating, releasing, copying, setting, and other data manipulations of vector or matrix data. Image and matrix logic, arithmetic, and conversion are unified and described here along with basic statistical measures such as sums, averages, standard deviations, minimums, maximums, and norms.

OpenCV contains a full set of linear algebra routines optimized for small and typical image-sized matrices. Examples of these functions are matrix multiplication, dot products, cross products, transpose, inversion, singular value decomposition (SVD), eigenimages, covariance, Mahalanobis distance, matrix log, power and exponential, Cartesian-to-polar conversion and back, and random matrices.

Dynamic structures such as linked lists, queues, and sets designed to work with images are described. There are also a full set of graph and tree structures such as that support Delaunay triangulation. This chapter ends with functions that support reading and writing of structures.

M2: Image processing

The second chapter of the manual covers a wide variety of image processing and analysis routines. It starts with a basic set of line, conic, poly, and text drawing routines, which were included to help in real-time labeling and debugging. Next, gradient, edge finding, and corner detection routines are described. OpenCV allows some useful sampling functions, such as reading pixels from an arbitrary line in an image into a vector and extracting a subpixel-accurate rectangle or quadrangle (good for rotation) from an image.

A full set of morphological operations [37] on image objects are supported along with other basic image filtering, thresholding, integral (progressive sum images), and color conversion routines. These are joined by image pyramids, connected components, and standard and gradient-directed floodfills. For rapid processing, you may find and process gray-level or binary contours of an image object.

A full range of moment-processing routines are supported, including normal, spatial, central, normalized central, and Hu moments [23]. Hough [26] and distance transforms are also present.

In computer vision, histograms of objects have been found very useful for finding, tracking, and identifying objects, including deforming and articulating objects. OpenCV provides a large number of these types of operations, such as creating, releasing, copying, setting, clearing, thresholding, and normalizing multidimensional histograms. Statistical operations on histograms are allowed, and most of the ways of comparing two histograms such as correlation, Chi-square, intersection [38, 35], and earth-mover's distance [34, 33, 32] are supported. Pairwise geometrical histograms are covered in manual section M3. In addition, you can turn histograms into probability densities and project images into these probability spaces for analysis and tracking. The chapter ends with support for most of the major methods of comparing a template to image regions, such as normalized cross-correlation absolute difference.

M3: Structural analysis

Once gray-level or binary-level image object contours are found, many operations allow you to smooth, simplify, and compare contours between objects. These contour routines allow rapid finding of polynomial approximations to objects, bounding boxes, area of objects, boundary lengths, and shape matching.

Image geometry routines allow you to fit lines, boxes, minimum enclosing circles, and ellipses to data points. This is where you'll find routines like KMeans, convex hulls, and convexity defect analysis, along with minimum area rotated rectangles. Also in this manual section is support for 2D, pairwise, geometrical histograms, as described in [24]. This chapter ends with routines that fully support Delaunay triangulation.

M4: Motion analysis and tracking

This chapter starts with support for learning the background of a visual scene in order to segment objects by background differencing. Objects segmented by this or any other method may then be tracked by converting successive segmentations over time into a motion history image (MHI) [17, 13]. Routines can take the gradient of MHIs to further find global and segmented motion regions.

The chapter then moves on to object tracking, first covering tracking of probability regions in images via the mean-shift and CamShift algorithms [11]. Tracking by energy-minimizing snakes [27] is also supported. Next, four methods of tracking by optical flow are discussed using Horn and Schunck's algorithm [22], Lucas and Kanade [30], block matching, and the recommended way, Lucas and Kanade in image pyramids [8].

This chapter concludes with two key tracking algorithms, Kalman filter and condensation tracker, based on particle filtering. There is an excellent tutorial on Kalman tracking at [41]. For condensation, see http://www.dai.ed.ac.uk/CVonline/LOCAL_COPIES/ISARD1/condensation.html.

M5: Object recognition

This chapter covers two key techniques in object recognition: eigenobjects and hidden Markov models (HMMs). There are many other recognition techniques in OpenCV, from histogram intersection [36] in M2 to boosted classifers in M7. The HMMs in this section allow HMMs to feed into other HMMs. One of the OpenCV demos uses horizontal HMMs feeding into a vertical HMM, termed *embedded HMM* (eHMM), to recognize faces.

M6: Camera calibration and 3D

OpenCV supports a full set of functions for doing intrinsic (internal camera parameters and lens distortions) and extrinsic (cameras location with respect to the outside world or other cameras) camera calibration. After calibration, functions can be called to undistort a lens or to track objects in 3D (see below). Most of these techniques were developed in [43, 45, 44, 21, 10, 9], but we added routines for finding and tracking checkerboard corners in order to help fill the matrices needed for calibration. In addition, there are routines for calculating the homography matrix, finding the fundamental matrix, and finding the epipolar lines between two images. Using calibrated, epipolar-aligned images allows us to do *view morphing*; that is, to synthesize a new object view as a weighted linear combination of the two existing views. Other techniques for stereo correspondence and multicamera tracking are supported in manual chapter M7.

This chapter further supports tracking objects in 3D. One way of doing this is to track a checkerboard object with a calibrated camera. More general ways of tracking four or more noncoplaner points on an object using weak-strong perspective iteration (POSIT algorithm) [18] are detailed.

M7: Recent experimental routines

The experimental routines would better be titled "recent routines." This chapter includes support for AdaBoost for face (or other object) detection. Routines support both boosted learning and recognition of objects. Also included are routines for finding stereo correspondence; we tried to combine the best of the faster methods to get good, dense, but fast correspondence. Note that to do stereo, you should align the cameras as parallel as possible so that the views cover as much of the same scene as possible. Monocular areas not covered by both cameras can cause shearing when the foreground object moves into the monocular area. Finally, routines supporting tracking objects using multiple (two or more) cameras are described.

M8: GUI and video I/O

This manual section covers image and video input and display from disk or cameras (for Windows or Linux) that are contained in the HighGUI library. Windows Install places source code for this in `C:\Program Files\OpenCV\ otherlibs\highgui`.

Display is covered first, with functions that allow you to put up a window and display video or images there. You can also attach sliders to the window, and they can be set to control processing parameters that you set up. Functions supporting full mouse events are also available to make interacting with images and video easy. A series of functions follow that handle reading and writing images to disk as well as common image conversions.

The next part of manual section M8 discusses functions for video I/O from either disk or camera. Writing AVIs may be done with various types of compression, including JPEG and MPEG1.

M9: Bibliography

This manual section contains a short bibliography for OpenCV, although many citations are placed directly in the function descriptions themselves.

M10: cvcam camera I/O manual

This last section of the manual is actually a submanual devoted to single or multiple camera control and capture under Windows or Linux. The library for this chapter under Windows is placed in the directory `C:\Program Files \OpenCV\otherlibs\cvcam`.

11.2.2   By task

This section provides suggestions regarding what functions might be good for a few of the popular vision tasks. Note that any of these functions may operate over multiple scales by using the image pyramid functions described in manual section M2.

Camera calibration, stereo depth maps. Camera calibration is directly supported by OpenCV, as discussed in Section 11.2.1. There are routines to track a calibration checkerboard to subpixel accuracy and to use these points to find the camera matrix and lens distortion parameters. This may then be used to mathematically undistort the lens or find the position of the camera relative to the calibration pattern. Two calibrated cameras may be put into stereo correspondence via a routine to calculate the fundamental matrix and another routine to find the epipolar lines.

From there, the "experimental section" of the manual, M7, has fast routines for computing stereo correspondence and uses the found correspondence to calculate a depth image. As stated before, it is best if the cameras are aligned as parallel as possible with as little as possible monocular area left in the scene.

Background subtraction, learning, and segmentation. There are dozens of ways that people have employed to learn a background scene. Since this is a well-used and often effective hack, I'll detail a method even if it is not well supported by OpenCV. A survey of methods may be found in [39]. The best methods are long-term and short-term adaptive. Perhaps the best current method is described in Elgammal et al.'s 2000 paper [19], in which not the pixel values themselves, but the distribution of differences, is adaptively learned using kernel estimators. We suspect that this method could be further improved by using linear-predictive adaptive filters, especially lattice filters, due to their rapid convergence rates.

Unfortunately, the above techniques are not directly supported in OpenCV. Instead, there are routines for learning mean and variance of each pixel (see "Accumulation of Background Statistics" in manual section M4). Also inside OpenCV, we could instead use the Kalman filter to track pixel values, but the normal distribution assumption behind the Kalman filter does not fit well with the typical bimodal distribution of pixel values over time. The condensation particle filter may be better at this, though at a computational cost for sampling. Within OpenCV, perhaps the best approach is just to use k-means with two or three means incrementally adapted over a window of time as each new pixel value comes in, using the same short-long scheme as employed by Elgammal et al. in [19].

Once a background model has been found, a thresholded absolute difference `cvAbsDiff` of the model with the current frame yields the candidate foreground regions. The candidate regions are indicated in a binary mask image where on = foreground candidate, off = definite background. Typically, this is a grayscale, 8-bit image so that "on" is a value of 255 and "off" is a value of zero.

Connected components. The candidate foreground region image above will be noisy, and at the image will be filled with pixel "snow." To clean it up, spurious single pixels need to be deleted by performing a morphological erode followed by a morphological dilate operation. This operation is called *morphological open* and can be done in one shot using the OpenCV function `cvMorphologyEx` with a 3×3 pixel, cross-shaped structuring element and the enumerated value `CV_SHAPE_CROSS`, and performing the open operation using `CV_MOP_OPEN` with one iteration.

Next, we want to identify and label large connected groups of pixels, deleting anything "too small." The candidate foreground image is scanned and any candidate foreground pixel (value = 255) found is used as a flood-fill seed start point to mark the entire connected region using the OpenCV `cvFloodFill` function. Each region will be marked by a different number by setting the `cvFloodFill newVal` to the new fill value. The first found region is marked with 1, then 2, and so on up to a max of 254 regions. In `cvFloodFill`, the `lo` and `up` values should be set to zero, a `CvConnectedComp` structure should be passed to the function, and `flags` should be set to 8. Once the region is filled, the area of the fill is examined (it is set in

CvConnectedComp). If the area filled is below a minimum-area-size threshold T_{size}, that area is erased by flooding it with a new value of zero, newVal
$= 0$ (regions that are too small are considered noise). If the area is greater
than T_{size}, then it is kept and the next fill value is incremented subject to it
being less than 255.

Getting rid of branch movement, camera jitter

Since moving branches and slight camera movement in the wind can cause
many spurious foreground candidate regions, we need a false detection suppression routine such as described in pages 6 to 8 of Elgammal et. al's paper
[19]. Every labeled candidate foreground pixel i has its probability recal-
culated by testing the pixel's value against each of its neighboring pixel's
probability distribution in a 5×5 region around it, N_{5x5}. The maximum
background probability calculation is assigned to that foreground pixel:

$$P_{N_{5x5}}(i_{x,y}(t)) = \max_{j \in N_{5x5}} P(i_{x,y}(t)|B_j), \tag{11.1}$$

where B_j is the background sample for the appropriate pixel j in the 5×5
neighborhood. If the probability of being background is greater than a
threshold T_{BG_1}, then that candidate foreground pixel is labeled as background. But, since this would knock out too many true positive pixels, we
also require that the whole connected region C also be found to be proba-
bilistically a part of the background:

$$P_C = \prod_{i \in C} P_{N_{5x5}}(i). \tag{11.2}$$

A former candidate foreground pixel is thus demoted to background if

$$(P_{N_{5x5}}(i) > T_{BG_1}) \wedge (P_C(i) > T_{BG_2}), \tag{11.3}$$

where T_{BG_1} and T_{BG_2} are suppression thresholds.

Stereo background subtraction. This is much like the above, except we
get a depth map from two or more cameras. The basic idea is that you
can also statistically learn the depth background. In detection mode, you
examine as foreground only those regions that are in front of your known
background. The OpenCV camera calibration and stereo correspondence
routines are of great help there (manual sections M6 and M7).

People tracking. There are innumerable ways to accomplish this task.
One approach is to fit a full physical model to a person in a scene. This

tends to be slow and is not supported in OpenCV, so it is not discussed further. For multiple cameras, we can look to the experimental section of the manual, M7, where multiple tracked areas are put into 3D correspondence. Other typical methods of tracking whole people are to sample a histogram (M2, histogram functions) or template (M2, matchtemplate) from a person, then scan through the future frames, back-projecting histogram intersections, earth-mover distances, or template-match scores to create a probability of person image. The mean-shift or CamShift algorithms (M4) can then track the recognition peaks in these images. Alternatively, people can be represented as a sequence of vertically oriented color blobs, as done in [42]. This could be accomplished in OpenCV by use of `cvKMeans2`, described in M3, to cluster colors, or by using the image statistical functions described in the array statistics section of M1, or by using the undocumented texture descriptors in the experimental section (look for "GLMC" in *cvaux.h*). A "probabilities of match" image can be made this way and tracked by mean shift as above.

Another people-tracking approach, if you can get adequate background-foreground segmentation, is to use the motion history templates and motion gradients, as described in manual section M4.

Face finding and recognition. The good way to perform face finding and face recognition is by the Viola-Jones method, as described in [40] and fully implemented in OpenCV (see section M7 in the manual and the accompanying demo in the *apps* section of the OpenCV directory).

Face recognition may be done either through eigenimages or embedded HMMs (see manual section M5), both of which have working demos in the *apps* directories.

Image retrieval. Image retrieval is typically done via some form of histogram analysis. This is fully supported via the histogram learning and comparison functions described in manual section M2.

Gesture recognition for arcade. Perhaps the best approach to gesture recognition, if you can get adequate background-foreground segmentation, is to use the motion history templates and motion gradients, as described in manual section M4. For recognition, depending on your representation, you may use the histogram functions described in M2 or Mahalanobis distances in M1, or you may do some sort of eigen trajectories using eigenobjects in M5. If hand shape is also required, you can represent the hand as a gradient histogram and use the histogram recognition techniques in M2.

The calibration, stereo, and/or 3D tracking routines described in M6 and M7 can also help segment and track motion in 3D. HMMs from section M5 can be used for recognition.

Part localization, factory applications. OpenCV was not built to support factory machine vision applications, but it does have some useful functionality. The camera calibration (manual section M6) and stereo routines (M7) can help with part segmentation. Templates can be compared with the MatchTemplate function in M2. Lines may be found with the Canny edge detector or Hough transform in M2.

There are routines for the subpixel accurate location of corners, rectangles, and quadrangles. Where to move can be aided by distance transforms. Adaptive thresholds can help segment parts, and pyramid operations can do things over multiple scales, all in M2.

Part shapes can be analyzed and recognized through the extensive collection of contour operators in M2. Motion can be analyzed and compensated for using optical-flow routines in M4. Kalman and condensation particle filters for smoothing, tracking, or predicting motion are supported as described in M4.

Flying a plane. Autonomous or semiautonomous planes are popular now for sport and military applications. A plane can be flown knowing only the horizon line. Assume that a camera has been installed such that the direction of heading is the exact center of the image and that level flight corresponds to the horizontal image scan lines being tangent to the earth's curvature. It then turns out that knowing the horizon line is enough to fly the plane. The angle of the horizon line is used for roll control, and the perpendicular distance from the line to the center of the image tells the pitch of the plane.

All we need then is a fast, robust way of finding the horizon line. A simple heurstic for finding this line was developed by Scott Ettinger [20]. Basically, we find the line through the image that minimizes the variance on both sides of the line (sky is more like sky than ground, and vice versa). This may be simply done every frame by creating an image pyramid (in manual section M2) of the aerial scene. On a much reduced scale image, we find the variance above and below the horizontal line through the center of the image using image (array) statistics (M1). We then systematically move and rotate the line until variance is minimized on both sides of the line. We can then advance to a larger or full scale to refine the line angle and location. Note that every time we move the line, we need not recalculate the variance from scratch. Rather, as points enter a side, their sum is added to that side's variance and lost points are subtracted.

OCR. The Hough transform (HoughLines) for lines in manual section M2 can be used to find the dominant orientation of text on a page. The machinery used for statistically boosted face finding described in M7 could also be used for either finding individual letters or finding and recognizing the letters. A Russian team created extremely fast letter recognition by thresholding letters, finding contours (cvFindContours, M2), simplifying the contours representing the contours as trees of nested regions and holes, and matching the trees as described in the Contour Processing Functions in M3.

The author has had quite good results from using the embedded HMM recognition techniques described in manual section M5 on text data. There are demos for both these routines (but applied to faces, not text) included in the *apps* directory of OpenCV.

11.2.3   Demos and samples

The demo applications that ship with OpenCV are discussed next.

CamShiftDemo, cvcsdemo

This is a statistically robust probability distribution mode tracker. It is based on making the tracking window of the mean-shift algorithm dynamic—this supports visual objects that can resize themselves by moving within the visual field. But this means that if your probability distributions are not compact (e.g., if they diffuse all over the image), CamShift will not work and you should switch to just mean-shift. For the demo, the distribution that CamShift tracks is just the probability of color that you selected from the video stream of images.

CamShiftDemo is the same as cvcsdemo, but uses a tcl interface.

LKDemo, cvlkdemo

This is a real-time Lucas-Kanade in the image-pyramid tracking demo. Note that Lucas-Kanade is a window-based tracker, and windows are ambiguous at object boundaries. Thus, windows on a boundary may tend to slide or stick to the the background or foreground interior. The LKDemo is the same as *cvlkdemo*; the latter just uses a tcl interface.

HaarFaceDetect, HaarTraining

This is a slight modification to the Viola-Jones AdaBoost face detector, which used Haar-type wavelets as weak feature detectors. Training code (but not the raw face database), trained parameters, and working real-time face detection are included here.

Hawk

Hawk is a window-based interactive C scripting system for working with OpenCV functions. It uses EiC interpretive C as its engine—some of High-GUI grew out of this. See other interfaces, Section 11.5, for what has replaced this earlier interface.

HMMDemo

This is a working HMM-based face detection demo complete with a sample database to train on.[2] This demo has HMMs across the face feeding into a vertical HMM down the face, which makes the final decision. You may add to the database by live camera. The HMM technique works well, except we give it uniform priors (uniformly cut up the image) to initialize training. If faces are not precisely aligned, the facial features will be blurred. This structural blurring leaves lighting as a stronger, though accidental, feature, and so this application tends to be lighting-sensitive. Putting actual eye, nose, and mouth priors into the model would probably minimize lighting dependence, but we have not tried this yet.

StereoDemo

This is a console-based stereo-depth calculation application that uses *cvcam* for video capture and HighGUI for displaying the images. See the *readme* file in ...\apps \StereoDemo for instructions on how to run this application. You will need two USB cameras that are compatible with DirectShow and that can run together. Creative WebCam is an example of such a camera. Two different cameras might be also okay.

This application allows automatic calibration of the cameras via tracking a checkerboard and then running to develop the disparity/depth image.

Tracker3dDemo

This demo uses two or more cameras calibrated together to track blobs in 3D.

---

[2]Some early members of the Russian OpenCV development team formed the training database images.

VMDemo, vmdemotk

This demo, complete with sample images in the *Media* subdirectory, uses the epipolar lines between two calibrated views to morph (interpolate) camera views anywhere between two views of an object. VMDemo and *vmdemotk* are the same except that the latter uses tcl and the former uses DirectShow filters.

Sample Code

On the Windows Install, the sample code can be found at `C:\Program Files \OpenCV\samples\c`. The same directory also contains some test images that some of the sample code operates on.

The sample codes are just simple examples of the following routines:

squares.c	– Uses contours to find colored rotated squares.
pyramid_segmentation.c	– Uses image pyramids for color segmentation.
motempl.c	– Using motion templates to track motion.
morphology.c	– Uses morphological operators.
laplace.c	– Use of Laplace operator on image.
kmeans.c	– Find clusters by K-means algorithm.
kalman.c	– Track by the Kalman filter.
fitellipse.c	– Best fit of ellipse to data.
ffilldemo.c	– Use of floodfill.
facedetect.c	– Uses AdaBoost-based face detector.
edge.c	– Uses Canny operator to edges in an image.
drawing.c	– Demos the drawing functions.
distrans.c	– Demos the distance transform function.
DemHist.c	– Demos use of several of the Histogram functions.
delaunay.c	– Performs Delaunay triangulation of an image.
convexhull.c	– Finds convex hull of a set of points.

11.3   Pictorial Tour

This section displays selected pictorial examples of some of the functions in OpenCV.

11.3.1   Functional groups

Manual section M1

This manual functional group contains static structures, image and array creation and handling, array and image arithmetic and logic, image and

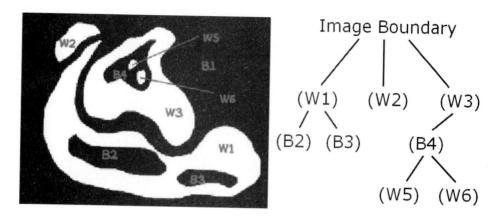

Figure 11.2. M1: Contour tree. A binary image may described as a nested series of regions and holes. (*Used with permission from [5, 6].*)

array statistics, and many dynamic structures. Trees are one structure, and as an example, a binarized image may be described in a tree form as a nested series of regions and holes as shown in Figure 11.2. Such trees may be used for letter recognition, for example.

Manual section M2

This functional group contains the basic image processing operations. For example, the Canny edge detector allows you to extract lines one pixel thick. The input is shown in Figure 11.3 and the output is shown in Figure 11.4.

Figure 11.3. M2: Canny edge detector example input. (*Used with permission from [5, 6].*)

Figure 11.4. M2: Canny edge detector example output. (*Used with permission from [5, 6].*)

Morphological operators, used to clean up and isolate parts of images, are used quite a bit in machine and computer vision. The most basic operations are dilation (growing existing clumps of pixels) and erosion (eating away at existing clumps of pixels). To do this, you use a morphological kernel that has a control point and a spatial extent, as shown in Figure 11.5.

Figure 11.5 shows morphology in 2D. Morphology can also be used in higher dimensions such as considering image brightness or color values as a

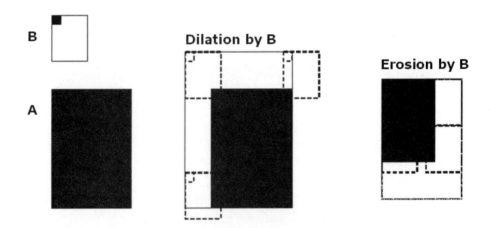

Figure 11.5. M2: Morphological kernel with control point and extent, and examples of how it grows (dilates) and shrinks (erodes) pixel groups in an image. (*Used with permission from [5, 6].*)

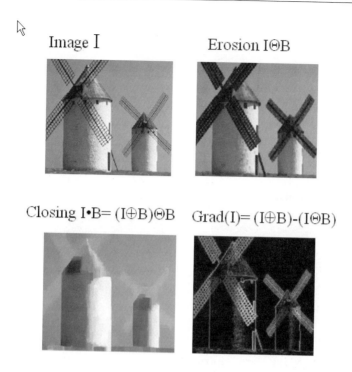

Figure 11.6. M2: Morphological examples. (*Used with permission from [5, 6].*)

surface in 3D. Morphological erode and dilate can be combined in different ways, such as to close an object by dilation followed by erode or to open, which erodes and then dilates. Gradients and bump removal or isolation can also be done, as shown in Figures 11.6 and 11.7.

The types of thresholding operations that OpenCV supports are graphically portrayed in Figure 11.8.

For computer vision, sensing at different resolutions is often necessary. OpenCV comes with an image pyramid or Laplacian pyramid function, as shown in Figure 11.9.

Floodfill is a graphics operator that is also used in computer vision for labeling regions as belonging together. OpenCV's floodfill can additionally fill upwards or downwards. Figure 11.10 shows an example of floodfilling.

In addition to the Canny edge detector, you may want to find dominant straight lines in an image even if there might be discontinuities in those lines. The Hough transform is a robust method of finding dominant straight lines in an image. In Figure 11.11, we have the raw image of a building, and Figure 11.12 shows the dominant lines found by the Hough transform.

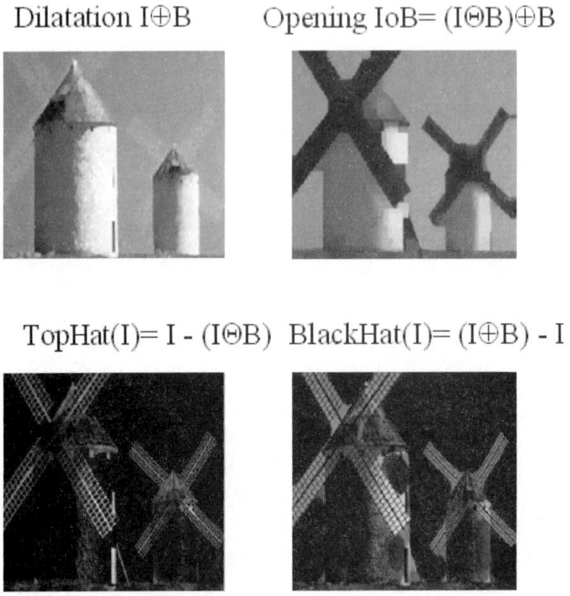

Dilatation I⊕B   Opening IoB= (I⊖B)⊕B

TopHat(I)= I - (I⊖B) BlackHat(I)= (I⊕B) - I

Figure 11.7.   M2: More morphological examples.   (*Used with permission from [5, 6].*)

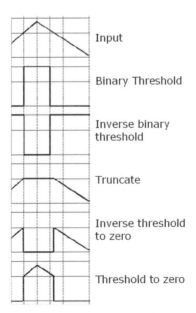

Figure 11.8.   M2: OpenCV image threshold options.   (*Used with permission from [5, 6].*)

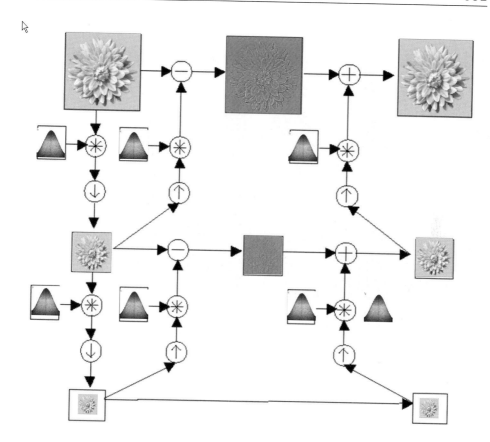

Figure 11.9. M2: Scale and Laplacian image pyramids. (*Used with permission from [5, 6].*)

The final example we show from chapter M2 is Borgefors' distance transform [7]. The distance transform calculates the approximate distance from every binary image pixel to the nearest zero pixel. This is shown in Figures 11.13 and 11.14, where the raw image is thresholded and distance transformed.

Manual section M3

The contour processing functions can be used to turn binary images into contour representations for much faster processing. The contours may be simplified and shapes recognized by matching contour trees or by Mahalanobis techniques. This is depicted in Figure 11.15 for a text recognition application.

Original image   Tolerance interval ± 5   Tolerance interval ± 6

Figure 11.10. M2: OpenCV FloodFill. (*Used with permission from [5, 6].*)

Figure 11.11. M2: Hough transform raw image. (*Used with permission from [5, 6].*)

Figure 11.12. M2: Lines found by the Hough transform for lines. (*Used with permission from [5, 6].*)

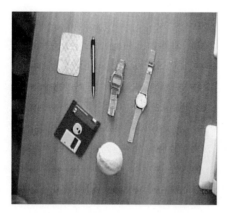

Figure 11.13.   M2: Distance transform raw image. (*Used with permission from [5, 6].*)

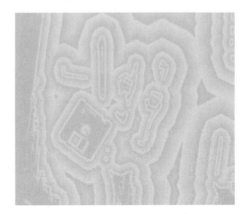

Figure 11.14. M2: Distance transform. (*Used with permission from [5, 6].*)

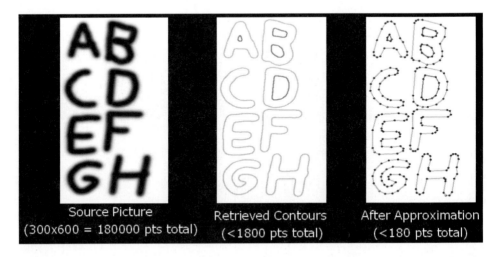

Figure 11.15. M3: Contour recognition for OCR. Contours are found, simplified, and recognized. (*Used with permission from [5, 6].*)

Manual section M4

This section supports motion analysis and object tracking. The first thing supported is background segmentation. Using running averages for means and variances, the background may be learned in the presence of moving foreground and is shown in sequence in Figure 11.16.

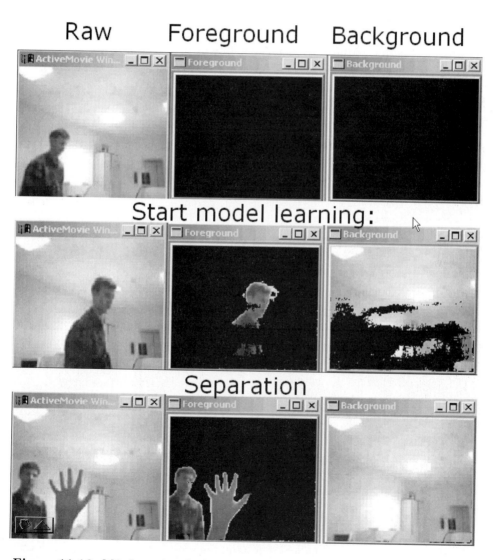

Figure 11.16. M4: Learning the background in the presence of moving foreground and segmenting the two. (*Used with permission from [5, 6].*)

Once background-foreground segmentation has been accomplished in a frame, we can use the Motion History Image (MHI) functions to group and track motions. `cvUpdateMotionHistory` creates an MHI representation by overlaying foreground segmentations one over another with a floating-point value equal to the system timestamp in milliseconds. From there, gradients (`cvCalcMotionGradient`) of the MHI can be used to find the global motion (`cvCalcGlobalOrientation`), and floodfilling can segment out local motions (`cvSegmentMotion`). Contours of the most recent foreground image may be extracted and compared to templates to recognize poses (`cvMatchShapes`). Figure 11.17 shows from left to right a downward kick, raising arms, lowering arms, and recognizing a "T" pose. The smaller circles and lines are segmented motion of limbs, the larger circle and line is global motion.

The CamShift (continuously adapting mean shift) algorithm described in [11] uses a statistically robust probability mode tracker (mean shift) algorithm to track the mode of visual probability distributions, in this case flesh, as shown in Figure 11.18.

Snakes are a classic boundary tracking algorithm based on smoothed gradient energy minimization, seen in Figure 11.19.

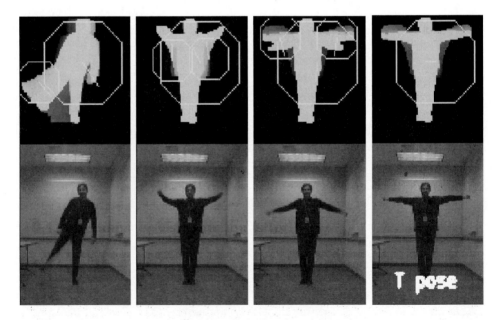

Figure 11.17. M4: The smaller circles and lines are segmented motion of limbs, the larger circle and line is global motion. The final frame uses outer-contour-based shape recognition to recognize the pose. (*Used with permission from [5, 6].*)

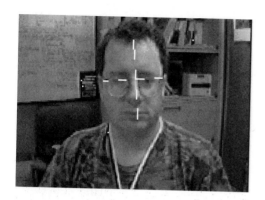 

Figure 11.18. M4: CamShift, adaptive window-based tracking of the mode of a probability distribution, in this case, probability of flesh. (*Used with permission from [5, 6].*)

Figure 11.19. M4: Snake-based tracking. (*Used with permission from [5, 6].*)

Manual section M5

There are two recognition functions in this manual section: eigenobjects and embedded HMMs. For eigenobjects, if you think of an image as a point in a huge dimensional space (one dimension per pixel), then it seems reasonable that similar objects will tend to cluster together in this space. Eigenobjects take advantage of this by creating a lower dimensional space "basis" that captures most of the variance between these objects, as depicted in Figure 11.20, with a face image basis depicted at bottom. Once a basis has been learned, we can perform face recognition by projecting a new face into the face basis and selecting the nearest existing face as being the same person with confidence proportional to the distance (Mahalanobis distance) from the new face. Figure 11.21 shows a recognition example using this basis.

Another object recognition technique is based on nested layers of HMMs [31]. Horizontal HMMs look for structure across the face and then feed their scores into an HMM that goes vertically down the face, as shown at left in Figure 11.22. This is called an embedded HMM (eHMM). At right in the figure is the default initialization for the eHMM to start its Viterbi training. Figure 11.23 shows the training and recognition process using eHMMs.

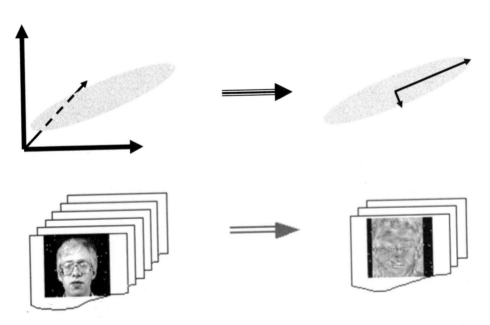

Figure 11.20. M5: Eigen idea explains most of the variance in a lower dimension.

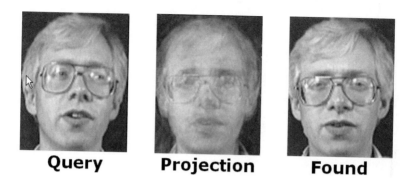

Query   Projection   Found

Figure 11.21.   M5: Eigenface recognition example.   (*Used with permission from [5, 6].*)

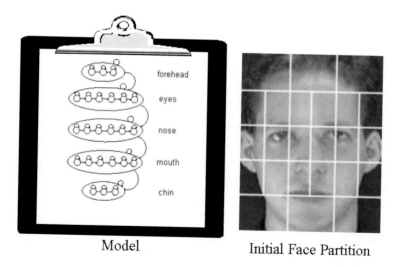

Model   Initial Face Partition

Figure 11.22.   M5: Embedded HMM for recognition.   On the left is the horizontal→vertical HMM layout; on the right is the default initialization for the HMM states. (*Used with permission from [5, 6].*)

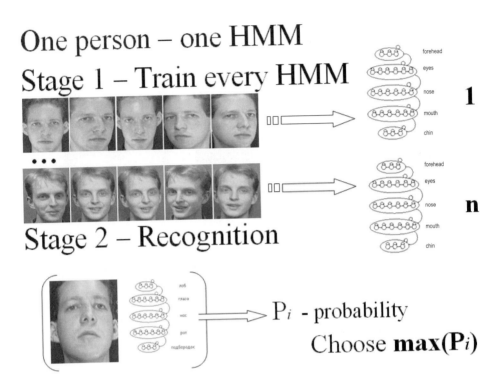

Figure 11.23. M5: Training and recognition using an embedded HMM. (*Used with permission from [5, 6].*)

Manual section M6

Functions in this section are devoted to camera calibration, image rectification, and 3D tracking. We start with a function that helps track corners in a calibration checkerboard, `cvFindChessBoardCornerGuesses`, which is shown in operation in Figure 11.24.

When a sequence of calibration points has been tracked, `cvCalibrate Camera_64d` can be used to extract camera calibration parameters. These results can then be used to undistort a lens, as shown in Figure 11.25.

After calibration, we can track a calibration checkerboard and use it to determine the 3D location of the checkerboard in each frame. This may be used for game control, as shown in Figure 11.26. OpenCV also includes support for tracking arbitrary, nonplanar objects using the POSIT (pose iteration) algorithm, which iterates between weak perspective (3D objects are 2D planes sitting at different depths) and strong perspective (objects are 3D) interpretation of points on objects. With mild constraints, POSIT

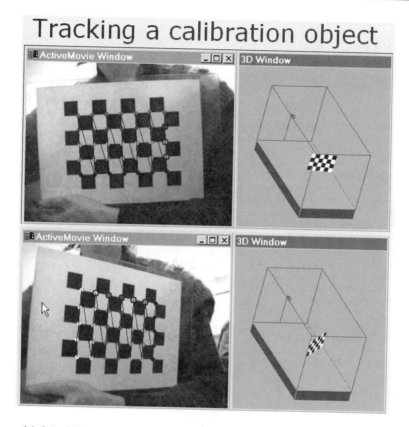

Figure 11.24. M6: Tracking a calibration checkerboard. (*Used with permission from [5, 6].*)

rapidly converges to the true 3D object pose, which may also be used to track 3D objects.

Not shown here, but in the experimental manual section M7, are functions that further use calibrated cameras for stereo vision and 3D modeling. We next pictorially describe some of the demos that ship with OpenCV.

11.3.2 Demo tour

This section shows images of some of the demos that ship with OpenCV. Not shown is our version [28] of the Viola-Jones face tracker [40], and also not shown is the two-video camera stereo demo. Shown below are screen shots of the Calibration demo and the control screen for the multiple camera 3D tracking experimental demo in Figure 11.27. Also shown are color-based tracking using CamShift, Kalman filter, and condensation in Figure 11.28.

Figure 11.25. M6: Raw image at top, undistorted image at bottom after calibration. (*Used with permission from [5, 6].*)

Figure 11.29 shows the HMM-based face recognition demo (also good for letter recognition), and finally, optical flow is shown in Figure 11.30.

11.4   Programming Examples Using C/C++

Windows setup tips:

To get programs to compile in MSVC++, there are some things you must do.

- **Release or debug mode.** MSVC++ can make either release or debug code. OpenCV ships with only release code (you can build debug

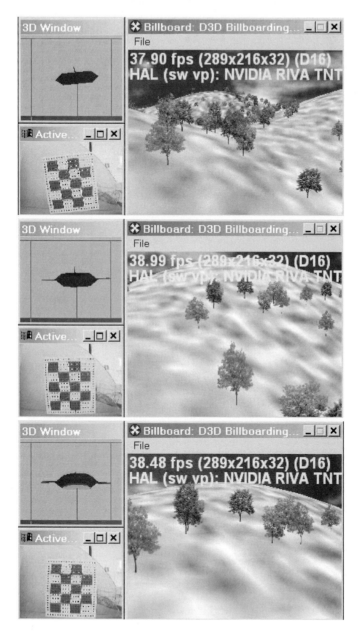

Figure 11.26. M6: Tracking a 3D object for game control. (*Used with permission from [5, 6].*)

Calibration Demo:

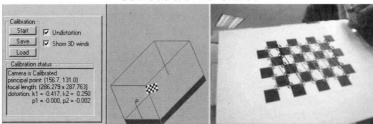

Multi-Camera 3D Tracking:

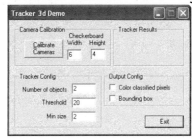

Figure 11.27. Demos: Automatic checkerboard tracking for camera calibration above and control screen for multiple-camera 3D below. (*Used with permission from [5, 6].*)

yourself). If a link error comes up requesting *cvd.lib* you are in debug mode. Go to the *Build* menu, select *Set Active Configuration*, and set it to release code.

- **Include and library file paths** must be set. Go to the *Tools* menu, select *Options*. Click on the *Directories* tab. Under *Show Directories for*, select *Include files*. Set the include path for

```
C:\Program Files\OpenCV\cv\include
C:\Program Files\OpenCV\cvaux\include
C:\Program Files\OpenCV\otherlibs\cvcam\include
C:\Program Files\OpenCV\otherlibs\highgui
```

Next, select *Library files* and set the path for `C:\Program Files\OpenCV\lib`.

- To avoid "unreferenced" function errors in link, go to *Project*, select *Settings*, and click on the *Link* tab. Under *Object/library modules* add *cv.lib cvcam.lib highgui.lib*.

CAMSHIFT-Based Tracking:

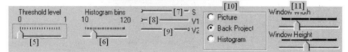

Kalman Filter-Based Tracking:

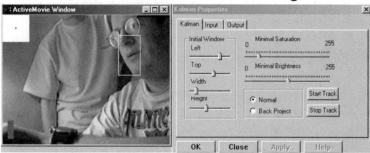

Condensation Particle Filter-Based Tracking:

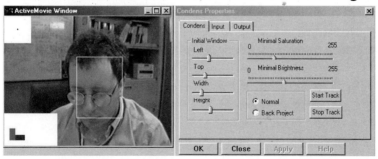

Figure 11.28. Demos: From top to bottom, CamShift tracking, Kalman filter, and condensation. (*Used with permission from [5, 6].*)

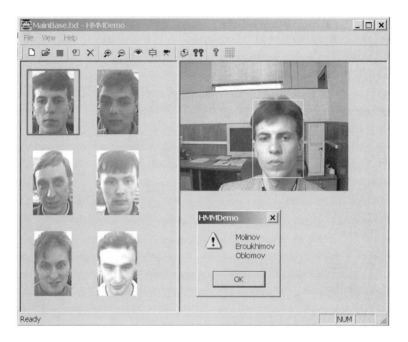

Figure 11.29. Demos: Embedded HMM-based face recognition. (*Used with permission from [5, 6].*)

Figure 11.30. Demos: Lucas-Kanade in pyramid optical flow demo. (*Used with permission from [5, 6].*)

  – Finally, make sure the Windows path has `C:\Program Files\OpenCV`
    `\bin` in it and that you've run *RegisterAll.bat* there at least once.

11.4.1   Read images from disk

The program below reads an image from disk (either from the command line
or *Mface.jpg* by default) and performs pyramid-based color region segmentation controlled by two interactive sliders. The input and output along with
slider controls are shown in Figure 11.31.

```
#ifdef _CH_
#pragma package <opencv>
#endif

#ifndef _EiC
#include "cv.h"
#include "highgui.h"
#include <math.h>
#endif
IplImage*  image[2] = { 0, 0 }, *image0 = 0, *image1 = 0;
CvSize size;
int   w0, h0,i;
int   threshold1, threshold2;
int   l,level = 4;
int sthreshold1, sthreshold2;
int   l_comp;
int block_size = 1000;
float  parameter;
double threshold;
double rezult, min_rezult;
CvFilter filter = CV_GAUSSIAN_5x5;
CvConnectedComp *cur_comp,min_comp;
CvSeq *comp;
CvMemStorage *storage;
CvPoint pt1, pt2;

void ON_SEGMENT(int a) {
    cvPyrSegmentation(image0, image1, storage, &comp,
                      level, threshold1, threshold2);
    l_comp = comp->total;
    i = 0;
    min_comp.value = 0;
    while(i<l_comp)
    {
```

```
        cur_comp = (CvConnectedComp*)cvGetSeqElem ( comp, i, 0);
        if(fabs(CV_RGB(255,0,0)- min_comp.value)>
            fabs(CV_RGB(255,0,0)- cur_comp->value))
            min_comp = *cur_comp;
        i++;
    }
    cvShowImage("Segmentation", image1);
}

int main( int argc, char** argv ) {
    char* filename = argc == 2 ? argv[1] : (char*)"Mface.jpg";
    if( (image[0] = cvLoadImage( filename, 1)) == 0 )
        return -1;
    cvNamedWindow("Source", 0);
    cvShowImage("Source", image[0]);
    cvNamedWindow("Segmentation", 0);
    storage = cvCreateMemStorage ( block_size );
    image[0]->width &= -(1<<level);
    image[0]->height &= -(1<<level);
    image0 = cvCloneImage( image[0] );
    image1 = cvCloneImage( image[0] );
    // segmentation of the color image
    l = 1;
    threshold1 =255;
    threshold2 =30;
    ON_SEGMENT(1);
    sthreshold1 = cvCreateTrackbar("Threshold1", "Segmentation",
                             &threshold1, 255, ON_SEGMENT);
    sthreshold2 = cvCreateTrackbar("Threshold2", "Segmentation",
                             &threshold2, 255, ON_SEGMENT);
    cvShowImage("Segmentation", image1);
    cvWaitKey(0);
    cvDestroyWindow("Segmentation");
    cvDestroyWindow("Source");
    cvReleaseMemStorage(&storage );
    cvReleaseImage(&image[0]);
    cvReleaseImage(&image0);
    cvReleaseImage(&image1);
    return 0;
}
#ifdef _EiC main(1,"pyramid_segmentation.c");
#endif
```

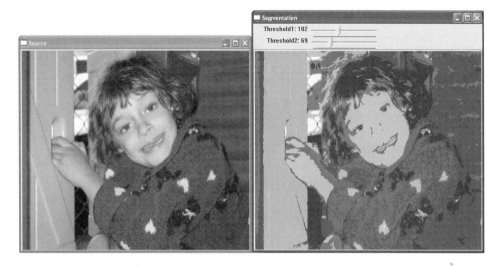

Figure 11.31. Program: Pyramid segmentation input and output. (*Used with permission from [5, 6].*)

11.4.2   Read AVIs from disk, or video from a camera

The following code example shows how to read an AVI movie from a disk and how to get images directly from the video camera.

```
#ifdef _CH_
#pragma package <opencv>
#endif

#ifndef _EiC
#include "cv.h"
#include "highgui.h"
#include
<ctype.h>
#endif

int main( int argc, char** argv ) {
    CvCapture* capture = 0;

    if( argc == 1 || (argc == 2 && strlen(argv[1]) == 1 &&
        isdigit(argv[1][0])))
        capture = cvCaptureFromCAM( argc == 2 ? argv[1][0] - '0' : 0 );
    else if( argc == 2 )
        capture = cvCaptureFromAVI( argv[1] );
    if( capture )
```

```
    {
        IplImage* gray = 0;
        IplImage* laplace = 0;
        IplImage* colorlaplace = 0;
        IplImage* planes[3] = { 0, 0, 0 };
        cvNamedWindow( "Laplacian", 0 );
        for(;;)
        {
            IplImage* frame = 0;
            int i;
            if( !cvGrabFrame( capture ))
                break;
            frame = cvRetrieveFrame( capture );
            if( !frame )
                break;
            if( !gray )
            {
                for( i = 0; i < 3; i++ )
                    planes[i] = cvCreateImage(
                            cvSize(frame->width,frame->height), 8, 1 );
                laplace = cvCreateImage(cvSize(frame->width,frame->height),
                                    IPL_DEPTH_16S, 1 );
                colorlaplace = cvCreateImage(cvSize(frame->width,
                                        frame->height), 8, 3 );
            }
            cvCvtPixToPlane( frame, planes[0], planes[1], planes[2], 0 );
            for( i = 0; i < 3; i++ )
            {
                cvLaplace( planes[i], laplace, 3 );
                cvConvertScaleAbs( laplace, planes[i], 1, 0 );
            }
            cvCvtPlaneToPix(planes[0], planes[1], planes[2], 0,
                        colorlaplace );
            colorlaplace->origin = frame->origin;
            cvShowImage("Laplacian", colorlaplace );
            if( cvWaitKey(10) >= 0 )
                break;
        }
        cvReleaseCapture( &capture );
    }
    cvDestroyWindow("Laplacian");
    return 0;
}
#ifdef _EiC
```

```
main(1,"laplace.c");
#endif
```

11.5   Other Interfaces

There are other ways to "talk" to OpenCV beyond compiled C. We detail three main methods below: Ch, MATLAB and Lush.

11.5.1   Ch

What is Ch?

Ch is a superset of C interpreter. It is designed for cross-platform scripting, 2D/3D plotting, and numerical computing. Ch was originally designed and implemented by Harry H. Cheng [16]. It has been further developed and maintained by SoftIntegration, Inc. [15, 25]

Ch supports 1999 ISO C Standard (C99), classes in C++, POSIX, X11/Motif, OpenGL, ODBC, XML, GTK+, Win32, CGI, 2D/3D graphical plotting, socket/Winsock, C LAPACK, high-level numeric functions, and shell programming.

The extensions in Ch provide the simplest possible solution for numerical computing and visualization in the C/C++ domain. Ch can interface with C/C++ binary libraries and be embedded in other application programs and hardware.

What is Ch OpenCV package?

Ch OpenCV package is Ch binding to OpenCV. It is included in the distribution of OpenCV. With Ch OpenCV, C (or C++) programs using OpenCV, C functions can readily run in Ch without compilation. Ch Standard or Professional or Student Edition and the OpenCV runtime library are needed to run Ch OpenCV.

Contents of Ch OpenCV package

Ch OpenCV package contains the following directories

1. OpenCV: Ch OpenCV package

    – OpenCV/demos: OpenCV demo programs in C readily to run in Ch

    – OpenCV/bin: OpenCV dynamic library and commands

– OpenCV/dl: dynamically loaded library

– OpenCV/include: header files

– OpenCV/lib: function files

2. Devel: Files and utilities used to build Ch OpenCV package using native C compiler and Ch SDK

– Devel/c: _chdl.c C wrappers and Makefile's

– Devel/createchf: bare OpenCV headers

– Devel/include: Ch OpenCV-specific include files

Where to get Ch and set up Ch OpenCV

Ch OpenCV can be installed in your computer following these instructions. Here we assume <CHHOME> is the directory where Ch is installed; but if, for example, Ch is installed in C:/Ch, then C:/Ch should be used instead of <CHHOME>. We also assume <OPENCV_HOME> is the directory where OpenCV is installed. If OpenCV is installed, for example, in C:/Program Files/OpenCV, then <OPENCV_HOME> should also be replaced by C:/Program Files/OpenCV.

1. If Ch has not been installed in your computer, download and install Ch from http://www.softintegration.com.

2. Move the directory OpenCV to <CHHOME>/package/opencv. <CHHOME>/package/opencv becomes the home directory of CHOPENCV_HOME for Ch OpenCV package. If you do not want to move the directory, you should add a new path to the system variable _ppath for package path by adding the following statement in the system startup file <CHHOME> /config/chrc or individual user's startup file _chrc in the user's home directory _ppath = stradd(_ppath, "<OPENCV_HOME> /interfaces/ch;"). <OPENCV_HOME> /interfaces/ch/opencv becomes the home directory CHOPENCV_HOME for Ch OpenCV package.

3. Add the system variable _ipath in the system startup file indicated in the previous step by adding the following statement: _ipath = stradd(_ipath, "<CHOPENC V_HOME> /include;"). This step is not necessary if the following code fragment is in application programs:

```
#ifdef _CH_
  #pragma package <opencv>
#endif
```

Run demo programs in directory demos:

1. Start Ch.

2. Type program name such as *delaunay.c.*

Update Ch OpenCV package for newer versions of OpenCV

The update of Ch OpenCV package for newer versions of OpenCV can be done as follows:

1. Install Ch SDK from http://www.softintegeration.com.

2. Go to ./Devel subfolder and run update_ch.ch that does the following:
   + Copies headers from <OPENCV_HOME>/cv/include and <OPENCV_HOME>/otherlibs/highgui to Devel/include subfolder
   + Generates Devel/createchf/*.h files containing bare function lists
   + Processes each function list with c2chf
   + Moves *.chf files generated by c2chf to ./OpenCV/lib subfolder and *_chdl.c files to ./Devel/c/<libname>
   + Removes automatically generated wrappers that have manually created counterparts in *_chdl.c
   + Builds .dl files using Ch SDK and places them to ./OpenCV/dl
   + Copies OpenCV DLLs to bin subfolders
   + Copies C samples from <OPENCV_HOME>/samples/c to ./OpenCV/demos

3. Copy the resultant folder ./OpenCV, which contains Ch OpenCV package, to <CHHOME>/package/OpenCV.

Programming example

Below is an example called *morphology.c*. In this example, functions cvErode() and cvDilate() are called. The prototypes of functions are as follows:

```
void cvErode(const CvArr* sourceImg, CvArr* DestImg,
            IplConvKernel*, int iterations);
void cvDilate(const CvArr* sourceImg, CvArr* DestImg,
            IplConvKernel* B, int iterations);
```

These two functions perform erosion and dilation respectively on a source image. Parameter B is a structuring element that determines the shape of a pixel neighborhood over which the minimum is taken. If it is NULL, a 33 rectangular structuring element is used. Parameter iterations is the number of times erosion is applied. Four callback functions, Opening(), Closing(), Erosion(), and Dilation(), were declared in this example. The function Opening() carries out erosion first, then dilation. The Closing() function does the opposite. Functions Erosion() and Dilation() perform only erosion and dilation respectively. The source code of program morphology.c is as follows:

```
#ifdef _CH_
#pragma package <opencv>
#endif

#include <cv.h>
#include <highgui.h>
#include <stdlib.h>

#include <stdio.h>

//declare image object
IplImage* src = 0; IplImage* image = 0; IplImage* dest = 0;

//the address of variable which receives trackbar position update
int pos = 0;

//callback function for slider, implements opening
void Opening(int id) {
    id;
    cvErode(src,image,NULL,pos);
    cvDilate(image,dest,NULL,pos);
    cvShowImage("Opening&Closing window",dest);
}
```

```
//callback function for slider, implements closing
void Closing(int id) {
    id;
    cvDilate(src,image,NULL,pos);
    cvErode(image,dest,NULL,pos);
    cvShowImage("Opening&Closing window",dest);
}


//callback function for slider, implements erosion
void Erosion(int id) {
    id;
    cvErode(src,dest,NULL,pos);
    cvShowImage("Erosion&Dilation window",dest);
}


//callback function for slider, implements dilation
void Dilation(int id) {
    id;
    cvDilate(src,dest,NULL,pos);
    cvShowImage("Erosion&Dilation window",dest);
}

int main( int argc, char** argv ) {
    char* filename = argc == 2 ? argv[1] : (char*)"baboon.jpg";
    if( (src = cvLoadImage(filename,1)) == 0 )
        return -1;

    //makes a full copy of the image including header, ROI and data
    image = cvCloneImage(src);
    dest = cvCloneImage(src);

    //create windows for output images
    cvNamedWindow("Opening&Closing window",1);
    cvNamedWindow("Erosion&Dilation window",1);

    cvShowImage("Opening&Closing window",src);
    cvShowImage("Erosion&Dilation window",src);

    cvCreateTrackbar("Open","Opening&Closing window",&pos,10,Opening);
    cvCreateTrackbar("Close","Opening&Closing window",&pos,10,Closing);
    cvCreateTrackbar("Dilate","Erosion&Dilation window",&pos,10,Dilation);
    cvCreateTrackbar("Erode","Erosion&Dilation window",&pos,10,Erosion);
```

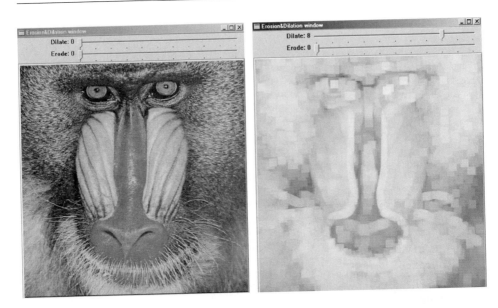

Figure 11.32. Interface CH: Raw image and after eight dilations.

```
cvWaitKey(0);
//releases header and image data
cvReleaseImage(&src);
cvReleaseImage(&image);
cvReleaseImage(&dest);
//destroys windows
cvDestroyWindow("Opening&Closing window");
cvDestroyWindow("Erosion&Dilation window");

    return 0;
}
```

Results are shown in Figure 11.32. On the left in Figure 11.32 is the original image; at right is the image after dilation. The dilation iterations was set to 8.

11.5.2 MATLAB

After installing OpenCV, you'll find the MATLAB interface at C:/Program Files/OpenCV/interfaces/matlab. These are source and executables for MATLAB wrappers for OpenCV.

To install wrappers, set path in the MATLAB to the toolbox/opencv subfolder of this folder and, optionally, to the toolbox/opencv/demos subfolder. Wrappers for OpenCV functions can be run in two ways: using cvwrap('<FunctionName>', parameters), where FunctionName is OpenCV function without cv prefix, or just by using cvFunctionName(parameters).

To build MATLAB wrappers, you need to add the <matlab_folder>/ extern/include folder to the include path (within Developer Studio) and <matlab_folder>/extern/lib /win32;   <matlab_folder>/extern/lib/win32/ microsoft/msvc60 folders to the lib path.

Contents of MATLAB OpenCV Package

As of this writing, the interface is not complete, and we welcome more mex file contributions. The interface has the following:

```
cvadaptivethreshold.m
cvapproxpoly.m
cvcalcopticalflowpyrlk.m
cvcanny.m
cvcontourarea.m
cvcontourboundingrect.m
cvcontourperimeter.m
cvdilate.m
cvdisttransform.m
cvdrawcontours.m
cverode.m
cvfindcontours.m
cvfloodfill.m
cvgoodfeaturestotrack.m
cvlaplace.m
cvmatchtemplate.m
cvminmaxloc.m
cvprecornerdetect.m
cvpyrdown.m
cvpyrup.m
cvsobel.m
cvthreshold.m
cvupdatemotionhistory.m
cvwrap.dll -- the mex wrapper
```

The demo directory contains

```
'Canny Demo',        'cannydemo'
'Contour Demo',      'contdemo'
'Flood Fill Demo',   'filldemo'
'Optical Flow Demo', 'flowdemo'
```

Figure 11.33. Interface MATLAB: Left image is the input, right image is after running Canny on it with low/high thresholds 120/200.

Programming example

MATLAB is perhaps the tersest programming language. The code below runs the Canny edge detector with (low, high) thresholds set to (120, 200) with resulting input/output shown in Figure 11.33.

```
I = imread('Clone.jpg');            %Read in an image
C = cvcanny(I(:,:,2), 120,200,3);   %Apply canny with thresholds (Low,High)
                                    %(120,200), sobel size 3
imshow(C);                          %Show the image
imwrite(C,'CloneCanny.bmp','bmp');  %Save image
```

11.5.3 Lush

For Linux and UNIX systems, Yann LeCun and Leon Bottou developed Lush [29], an object-oriented programming language with a simplified Lisp-like syntax where C may be mixed freely with simple Lisp. Lush is intended for researchers, experimenters, and engineers interested in large-scale numerical and graphic applications.

Lush is free software (under the GPL license) and runs on GNU/Linux, Solaris, and Irix. They claim a few days learning curve or your money back.

Main features are the following:

– A very clean, simple, and easy-to-learn Lisp-like syntax.

- A compiler that produces very efficient C code and relies on the C compiler to produce efficient native code (no inefficient bytecode or virtual machine).

- An easy way to interface C functions and libraries, and a powerful dynamic linker/loader for object files or libraries (*.O*, *.A*, and *.SO* files) written in other compiled languages.

- The ability to freely mix Lisp and C in a single function.

- A powerful set of vector/matrix/tensor operations.

- A huge library of over 10,000 numerical routines, including full interfaces to GSL, LAPACK, and BLAS.

- A library of image- and signal-processing routines.

- An extensive set of graphic routines, including an object-oriented GUI toolkit, an interface to OpenGL/GLU/GLUT, and a soon-to-be-available interface to the OpenInventor scene-rendering engine.

- An interface to the Simple Directmedia Layer (SDL) multimedia library, including a sprite class with pixel-accurate collision detection.

- Sound and video grabbing (using ALSA and Video4Linux).

- Several libraries for machine learning, neural net, statistical estimation, HMMs (gblearning, Torch, HTK).

- Libraries for computer vision (Intel Vision Library) and 3D scene rendering (OpenInventor, available soon).

- Bindings to the JavaVM API and to the Python C API (available soon).

Lush is geared toward research and development in signal processing, image processing, machine learning, computer vision, bio-informatics, data mining, statistics, simulation, optimization, and artificial intelligence.

11.6 Appendix A

OpenCV has a BSD-style license that basically allows royalty-free commercial and research use subject to citation of the library. In particular, you may use it in a for-profit product without the need to open your product to others. The license is as follows:

```
By downloading, copying, installing or using the software you
agree to this license. If you do not agree to this license, do not
download, install, copy or use the software.
```

```
                  Intel License Agreement
            For Open Source Computer Vision Library
```

Copyright (C) 2000, 2001, Intel Corporation, all rights reserved.
Third party copyrights are property of their respective owners.

Redistribution and use in source and binary forms, with or without
modification, are permitted provided that the following conditions
are met:

* Redistribution's of source code must retain the above copyright
  notice, this list of conditions and the following disclaimer.

* Redistribution's in binary form must reproduce the above
  copyright notice, this list of conditions and the following
  disclaimer in the documentation and/or other materials provided
  with the distribution.

* The name of Intel Corporation may not be used to endorse or
  promote products derived from this software without specific
  prior written permission.

This software is provided by the copyright holders and
contributors ''as is'' and any express or implied warranties,
including, but not limited to, the implied warranties of
merchantability and fitness for a particular purpose are
disclaimed. In no event shall the Intel Corporation or
contributors be liable for any direct, indirect, incidental,
special, exemplary, or consequential damages (including, but not
limited to, procurement of substitute goods or services; loss of
use, data, or profits; or business interruption) however caused
and on any theory of liability, whether in contract, strict
liability, or tort (including negligence or otherwise) arising in
any way out of the use of this software, even if advised of the
possibility of such damage.

11.7 Appendix B

The following is a partial "thank you" list for OpenCV:

Original OpenCV committee (vision, functionality proposals etc.):

Dr. Gary Bradski, Vadim Pisarevsky
Prof. Trevor Darrell
Prof. Irfan Essa

Prof. Jitendra Malik
Prof. Pietro Perona
Prof. Stan Sclaroff
Prof. Carlo Tomasi

People who helped to make OpenCV real (if you noticed some names
are missing, please mail to vadim.pisarevsky@intel.com)

Name	Company (time of contribution/ cooperation)	Remarks
Alekseev, Aleksey	Intel Corp.	IppCV testing
Boldyrev, Sergey	Intel Corp.	rpm spec, linux makefiles
Breen, Ed	???	EiC intepreter
Devernay, Frederic	INRIA (?)	cvcam patch (Linux version)
Don Murray	PointGrey	PointGrey SDK and patches for stereo gest. rec. code
Dubey, Premnath	???	cvcam patch (Windows vers.)
Halsall, Chris	Open Source Solutions Inc.	O-Reilly papers on OpenCV, calibration app for Linux patches for Linux version
Khamenya, Valery	BioVisioN AG	OpenCV CVS repository at SourceForge
Rodyushkin, Konstantin	Intel Corp.	OpenCV to 3dFace
Schaefer, Dirk	MD-Mathematische Dienste GmbH	Original code for Bayer->RGB pattern conv.
Veretennikov, Eugene	Intel Corp.	Testing, comp. geo. functs
Wayne W. Cheng	SoftIntegration	Ch C/C++ intepreter, OpenCV Ch toolbox (proposal + SDK)
Muehlmann, Karsten	Lehrstuhl fur Informatick V	Another stereo corresp. code (not included yet)
Bouguet, Jean-Yves	Intel	Calibration and 3D
Nefian, Ara	Intel	eHMM
Davies, Bob	Intel	Coding, test, user group
Grzeszczuk, Radek	Intel	Robust line fit

Active OpenCV forum participants (who helped us): Barbaresi,
Abramo Betti, Gabriele Cao, Ning Cawkwell, Jack Chen, Gen-Nan
Cheng, Michael Fritz, Frank Iannizotto, Giancarlo Lu, Le Kunz,
Clay Ming, Li Philomin, Vasanth Zayed, Mohamed Rocha, Jairo
Stewart, James Andr.

```
BUG reports:
------------
Small, Daniel;
Carlos Andres Rocha;
Shavit, Adi;
Zivkovic, Zoran
and other people from OpenCV forum at www.yahoogroups.com
```

Bibliography

[1] Open source computer vision library (OpenCV) main page. [Online]. Available: http://www.intel.com/research/mrl/research/opencv.

[2] [Online]. Available: http://developer.intel.com/software/products/ipp/ippvm20/.

[3] Intel performance libraries. [Online]. Available: http://developer.intel.com/software/products/perflib/.

[4] OpenCV user group. [Online]. Available: http://groups.yahoo.com/group/OpenCV/.

[5] OpenCV course foilset. [Online]. Available: http://prdownloads.sourceforge.net/opencvlibrary/CVPR01_course.zip?download.

[6] OpenCV download site. [Online]. Available: http://sourceforge.net/projects/opencvlibrary/.

[7] G. Borgefors. Distance transformations in digital images. *Computer Vision, Graphics and Image Processing*, vol. 34, pages 344–371, 1986.

[8] J.-Y. Bouguet. Pyramidal implementation of the Lucas-Kanade feature tracker. The paper is included in OpenCV distribution (algo_tracking.pdf), 2000.

[9] J.-Y. Bouguet and P. Perona. Camera calibration from points and lines in dual-space geometry. Technical Report, California Institute of Technology, 1998.

[10] J.-Y. Bouguet and P. Perona. Camera calibration from points and lines in the reciprocal space. Manuscript, 1998.

[11] G. Bradski. Computer vision face tracking as a component of a perceptual user interface. In *Proc. IEEE WACV*, included in OpenCV distribution (camshift.pdf), 1998.

[12] G. Bradski, J-Y. Bouguet, and V. Pisarevsky. Computer vision recipes with OpenCV (working title). Springer. In press, 2004.

[13] G. Bradski and J. Davis. Motion segmentation and pose recognition with motion history gradients. In *Proc. IEEE WACV*, pages 214–219, 2000.

[14] G. Bradski. The OpenCV library. *Dr. Dobb's Journal*, pages 120–125, 2000.

[15] H. H. Cheng. The Ch language environment user's guide. SoftIntegration, Inc., 2002. [Online]. Available: http://www.softintegration.com.

[16] H. H. Cheng. Scientific computing in the Ch programming language. *Scientific Programming*, 2(3):49–75, Fall 1993.

[17] J. Davis and A. Bobick. The representation and recognition of action using temporal templates. *MIT Media Lab Technical Report*, page 402, 1997.

[18] D. F. DeMenthon and L. S. Davis. Model-based object pose in 25 lines of code. In *Proc. of European Conference on Computer Vision*, pages 335–343, 1992.

[19] A. Elgammal, D. Harwood, and L. Davis. Non-parametric model for background subtraction. In *Proc. of 6th European Conference of Computer Vision*, 2000.

[20] S. M. Ettinger, M. C. Nechyba, P. G. Ifju, and M. Waszak. Vision-guided flight stability and control for micro air vehicles. In *Proc. of IEEE International Conference on Intelligent Robots and Systems*, 2002. [Online]. Available: http://www.mil.ufl.edu/~nechyba/mav/.

[21] Heikkila and Silven. A four-step camera calibration procedure with implicit image correction. In *Proc. of Conference on Computer Vision and Pattern Recognition*, 1997.

[22] B. K. P. Horn and B. G. Schunck. Determining optical flow. *Artificial Intelligence*, vol. 17, pages 185–203, 1981.

[23] M. Hu. Visual pattern recognition by moment invariants. *IRE Transactions on Information Theory*, 8(2):179–187, 1962.

[24] J. Iivarinen, M. Peura, J. Srel, and A. Visa. Comparison of combined shape descriptors for irregular objects. In *Proc. of 8th British Machine Vision Conference*, 1997. [Online]. Available: http://www.cis.hut.fi/research/IA/paper/publications/bmvc97/bmvc97.html.

[25] SoftIntegration Inc. The Ch language environment SDK user's guide, 2002. [Online]. Available: http://www.softintegration.com.

[26] J. Matas, C. Galambos, and J. Kittler. Progressive probabilistic Hough transform. In *Proc. of British Machine Vision Conference*, 1998.

[27] M. Kass, A. Witkin, and D. Terzopoulos. Snakes: Active contour models. *International Journal of Computer Vision*, pages 321–331, 1988.

[28] A. Kuranov, R. Lienhart, and V. Pisarevsky. An empirical analysis of boosting algorithms for rapid objects with an extended set of Haar-like features. Technical Report MRL-TR-July02-01, Intel Corporation, 2002.

[29] Y. LeCun and L. Bottou. Lush programing language, 2002. [Online]. Available: http://lush.sourceforge.net/index.html.

[30] B. Lucas and T. Kanade. An iterative image registration technique with an application to stereo vision. In *Proc. of 7th International Joint Conference on Artificial Intelligence*, pages 674–679, 1981.

[31] A. Nefian and M. Hayes. Hidden Markov models for face recognition. In *Proc. of IEEE International Conference on Acoustic, Speech, and Signal Processing*, vol. 5, pages 2721-2724, May 1998.

[32] Y. Rubner and C. Tomasi. Texture metrics. In *Proc. of IEEE International Conference on Systems, Man, and Cybernetics*, pages pp. 4601–4607, October 1998.

[33] Y. Rubner, C. Tomasi, and L. J. Guibas. Metrics for distributions with applications to image databases. In *Proc. of International Conference on Computer Vision*, pages 59–66, January 1998.

[34] Y. Rubner, C. Tomasi, and L. J. Guibas. The earth-mover's distance as a metric for image retrieval. *Technical Report STAN-CS-TN-98-86*, Department of Computer Science, Stanford University, September 1998.

[35] B. Schiele and J. L. Crowley. Object recognition using multidimensional receptive field histograms. In *Proc. of European Conference on Computer Vision*, vol. 1, pages 610–619, 1996.

[36] B. Schiele and J. L. Crowley. Recognition without correspondence using multidimensional receptive field histograms. *International Journal of Computer Vision*, 36(1):31–50, January 2000.

[37] J. Serra. *Image analysis and mathematical morphology*. Academic Press, 1982.

[38] M. J. Swain and D. H. Ballard. Color indexing. *International Journal of Computer Vision*, 7(1):11–32, 1991.

[39] K. Toyama, J. Krumm, B. Brumitt, and B. Meyers. Wallflower: Principles and practice of background maintenance. In *Proc. of International Conference on Computer Vision*, 1999.

[40] P. Viola and M. Jones. Robust real-time object detection. *International Journal of Computer Vision*, 2002.

[41] G. Welch and G. Bishop. An introduction to the Kalman filter. Technical Report TR95-041, University of North Carolina at Chapel Hill, 1995. [Online]. Available: http://www.cs.unc.edu/~welch/kalman/kalman_filter/kalman.html.

[42] C. Wren, A. Azarbayejani, T. Darrell, and A. Pentland. Pfinder: Real-time tracking of the human body. Technical Report 353, MIT Media Laboratory Perceptual Computing Section Technical Report, 1997. [Online]. Available: http://pfinder.www.media.mit.edu/projects/pfinder/.

[43] Z. Zhang. Parameter estimation techniques: A tutorial with application to conic fitting. *International J. of Image and Vision Computing Journal*, 15:1(59-76), 1996.

[44] Z. Zhang. A flexible new technique for camera calibration. *IEEE Transactions on Pattern Analysis and Machine Intelligence*, 22(11):1330–1334, 2000.

[45] Z. Zhang. Flexible camera calibration by viewing a plane from unknown orientations. In *Proc. International Conference on Computer Vision*, pages 666–673, September 1999.

Chapter 12

SOFTWARE ARCHITECTURE FOR COMPUTER VISION

Alexandre R. J. François

12.1 Introduction

This chapter highlights and addresses architecture-level software development issues facing researchers and practitioners in the field of computer vision. A new framework, or architectural style, called Software Architecture for Immersipresence (SAI), is introduced. It provides a formalism for the design, implementation, and analysis of software systems that perform distributed parallel processing of generic data streams. The SAI style is illustrated with a number of computer–vision related examples. A code-level tutorial provides a hands-on introduction to the development of image stream manipulation applications using Modular Flow-Scheduling Middleware (MFSM), an open source architectural middleware implementing the SAI style.

12.1.1 Motivation

The emergence of comprehensive code libraries in a research field is a sign that researchers and practitioners are seriously considering and addressing software engineering and development issues. In this sense, the introduction of the Intel OpenCV library [2] certainly represents an important milestone for Computer Vision. The motivation behind building and maintaining code libraries is to address *reusability* and *efficiency* by providing a set of standard data structures and implementations of classic algorithms. In a field like computer vision, with a rich theoretical history, implementation issues are often regarded as secondary to the pure research components outside of specialty subfields, such as real-time computer vision (see, e.g., [27]). Nevertheless, beyond reducing development and debugging time, good code design and reusability are key to such fundamental principles of scientific research as experiment reproducibility and objective comparison with the existing state of the art.

For example, these issues are apparent in the subfield of video processing/analysis, which, due to the conjunction of technical advances in various enabling hardware performance (including processors, cameras, and storage media) and high-priority application domains (e.g., visual surveillance [3]), has recently become a major area of research. One of the most spectacular side effects of this activity is the amount of test data generated, an important part of which is made public. The field has become so rich in analysis techniques that any new method must almost imperatively be compared to the state of the art in its class to be seriously considered by the community. Reference data sets, complete with ground-truth data, have been produced and compiled for this very purpose (see, e.g., [6, 7, 8, 9]). Similarly, a reliable, reusable and consistent body of code for established—and "challenger"—algorithms could certainly facilitate the performance comparison task. Some effort is made toward developing open platforms for evaluation (see, e.g., [19]). Such properties as *modularity* contribute to code reuse and to fair and consistent testing and comparison. However, building a platform generic enough to not only accommodate current methods, but also allow incorporation of other relevant algorithms, some of them not yet developed, is a major challenge. It is well known in the software industry that introducing features that were not planned for at design time in a software system is at least an extremely hard problem, and generally a recipe for disaster. Software engineering is the field of study devoted to addressing these and other issues of software development in industrial settings. The parameters and constraints of industrial software development are certainly very different from those encountered in a research environment, and thus software engineering techniques are often unadapted to the latter.

In a research environment, software is developed to demonstrate the validity and evaluate the performance of an algorithm. The main performance aspect on which algorithms are evaluated is, of course, the accuracy of their output. Another aspect of performance is measured in terms of system throughput, or algorithm execution time. This aspect is all the more relevant as the amount of data to be processed increases, leaving less and less time for storing and offline processing. The metrics used for this type of performance assessment are often partial. In particular, theoretical complexity analysis is but a prediction tool, which cannot account for the many other factors involved in a particular system's performance. Many algorithms are claimed to be "real-time." Some run at a few frames per second, but "could be implemented to run in real-time," or simply will run faster on the next generation machines (or the next, etc.). Others have been the object of careful design and specialization to allow a high processing rate on constrained equipment. The general belief that increasing computing power can make any system (hardware and software) run faster relies on the hidden or implied assumption of system *scalability*, a property that cannot and should not be taken for granted. Indeed, the performance of any given algorithm implementation is highly dependent on the overall software system in which it is operated (see, e.g., [20]). Video analysis applications commonly involve image input (from file or camera) and output (to file or display) code, the performance of which can greatly impact the perceived performance of the overall application and in some cases the performance of the individual algorithms involved in the processing. Ideally, if an algorithm or technique is relevant for a given purpose, it should be used in its best *available* implementation, on the best *available* platform, with the opportunity of upgrading either when possible.

As computer vision matures as a field and finds new applications in a variety of domains, the issue of *interoperability* becomes central to its successful integration as an enabling technology in crossdisciplinary efforts. Technology transfer from research to industry could also be facilitated by the adoption of relevant methodologies in the development of research code.

If these aspects are somewhat touched in the design of software libraries, a consistent, generic approach requires a higher level of abstraction. This is the realm of software architecture, the field of study concerned with the design, analysis, and implementation of software systems. Shaw and Garlan give the following definition in [26]:

> As the size and complexity of software systems increase, the design and specification of overall system structure become more significant issues than the choice of algorithms and data structures of computation. Structural issues include the organization of a system as a composition of components; global control structures; the protocols for communication, synchronization and data access; the assignment of functionality to design elements; the composition of design elements; physical distribution; scaling

and performance; dimensions of evolution; and selection among design alternatives. This is the *Software Architecture* level of design.

They also provide the framework for architecture description and analysis used in the remainder of this chapter. A specific architecture can be described as a set of computational *components* and a their interrelations, or *connectors*. An *architectural style* characterizes families of architectures that share some patterns of structural organization. Formally, an architectural style defines a *vocabulary* of components and connector types, and a set of *constraints* on how instances of these types can be combined to form a valid architecture.

If software architecture is a relatively young field, software architectures have been developed since the first software was designed. Classic styles have been identified and studied informally and formally, their strengths and shortcomings analyzed. A major challenge for the software architect is the choice of an appropriate style when designing a given system, as an architectural style may be ideal for some applications, while unadapted for others. The goal here is to help answer this question by providing the computer vision community with a flexible and generic architectural model. The first step toward the choice—or the design—of an architectural style is the identification and formulation of the core requirements for the target system(s). An appropriate style should support the design and implementation of software systems capable of handling images, 2D and 3D geometric models, video streams, and various data structures in a variety of algorithms and computational frameworks. These applications may be interactive and/or have real-time constraints. Going beyond pure computer vision systems, processing of other data types, such as sound and haptics data, should also be supported in a consistent manner in order to compose large-scale integrated systems, such as immersive simulations. Note that interaction has a very particular status in this context, as data originating from the user can be both an input to an interactive computer vision system and the output of a vision-based perceptual interface subsystem.

This set of requirements can be captured under the general definition of *crossdisciplinary dynamic systems, possibly involving real-time constraints, user immersion, and interaction*. A fundamental underlying computational invariant across such systems is *distributed parallel processing of generic datastreams*. As no existing architectural model could entirely and satisfactorily account for such systems, a new model was introduced.

12.1.2 Contribution

SAI (Software Architecture for Immersipresence) is a new software architecture model for designing, analyzing, and implementing applications performing distributed, asynchronous parallel processing of generic data streams. The goal of SAI is to provide a universal framework for the distributed implementation of algorithms and their easy integration into complex systems that exhibit desirable software engineering qualities such as efficiency, scalability, extensibility, reusability, and interoperability. SAI specifies a new architectural style (components, connectors, and constraints). The underlying extensible data model and hybrid (shared repository and message-passing) distributed, asynchronous parallel processing model allow natural and efficient manipulation of generic data streams, using existing libraries or native code alike. The modularity of the style facilitates distributed code development, testing, and reuse, as well as fast system design and integration, maintenance, and evolution. A graph-based notation for architectural designs allows intuitive system representation at the conceptual and logical levels, while at the same time mapping closely to processes.

MFSM (Modular Flow Scheduling Middleware) [12] is an architectural middleware implementing the SAI style. Its core element, the Fast Scheduling Framework (FSF) library, is a set of extensible classes that can be specialized to define new data structures and processes or encapsulate existing ones (e.g., from libraries). MFSM is an open source project, released under the GNU Lesser General Public License [1]. A number of software modules regroup specializations implementing specific algorithms or functionalities. They constitute a constantly growing base of open source, reusable code, maintained as part of the MFSM project. The project also includes extensive documentation, including user guide reference guide and tutorials.

12.1.3 Outline

This chapter is a hands-on introduction to the design of computer vision applications using SAI and their implementation using MFSM.

Section 12.2 is an introduction to SAI. Architectural principles for distributed parallel processing of generic data streams are first introduced, in contrast to the classic Pipes-and-Filters model. These principles are then formulated into a new architectural style. A graph-based notation for architectural designs is introduced. Architectural patterns are illustrated through a number of demonstration projects ranging from single-stream, automatic, real-time video processing to fully integrated distributed interactive systems mixing live video, graphics, and sound. A review of key architectural properties of SAI concludes the section.

Section 12.3 is an example-based, code-level tutorial on writing image-stream processing applications with MFSM. It begins with a brief overview of the MFSM project and its implementation of SAI. Design and implementation of applications for image-stream manipulation is then explained in detail using simple examples based on those used in the online MFSM user guide. The development of specialized SAI components is also addressed at the code level. In particular, the implementation of a generic image data structure (object of an open source module) and its use in conjunction with the OpenCV library in specialized processes are described step by step.

In conclusion, Section 12.4 offers a summary of the chapter and some perspectives on future directions for SAI and MFSM.

12.2   SAI: A Software Architecture Model

This section is an introduction to the SAI architectural model for distributed parallel processing of generic data streams.

The most common architectural styles for data-stream processing applications are derived from the classic Pipes-and-Filters model. It is, for example, the underlying model of the Microsoft DirectShow library, part of the DirectX suite [24]. After a brief review of the Pipes-and-Filters architectural style, of its strengths and weaknesses, a new hybrid model is introduced that addresses the identified limitations while preserving the desirable properties. This new model is formally defined as the SAI architectural style. Its component and connector types are defined, together with the corresponding constraints on instances interaction. The underlying data and processing models are explicited and analyzed. Simultaneously, graphical symbols are introduced to represent each element type. Together these symbols constitute a graph-based notation system for representing architectural designs. Architectural patterns are illustrated with a number of demonstration projects ranging from single-stream, automatic, real-time video processing to fully integrated distributed interactive systems mixing live video, graphics, and sound. Finally, relevant architectural properties of this new style are summarized.

12.2.1   Beyond pipes and filters

The Pipes-and-Filters model is an established (and popular) model for data stream manipulation and processing architectures. In particular, it is a classic model in the domains of signal processing, parallel processing, and distributed processing [10], to name but a few. This section first offers an overview of the classic Pipes-and-Filters architectural style, emphasizing its desirable properties and highlighting its main limitations. A hybrid model is then outlined, aimed at addressing the limitations while preserving the desirable properties.

The Pipes-and-Filters Style

In the Pipes-and-Filters architectural style, the components, called *filters*, have a set of inputs and a set of outputs. Each component reads (and consumes) a stream of data on its input and produces a stream of data on its output. The processing is usually defined so that the component processes its input incrementally so that it can start producing an output stream before it is finished receiving the totality of its input stream—hence the name filter. The connectors are *pipes* in which the output stream of a filter flows to the input of the next filter. Filters can be characterized by their input/output behavior: source filters produce a stream without any input; transform filters consume an input stream and produce an output stream; sink filters consume an input stream but do not produce any output. Figure 12.1 presents an overview. Filters must be strictly independent entities: they do not share state information with other filters; the only communication between filters occurs through the pipes. Furthermore, a filter's specification might include restrictions about its input and output streams, but it may not identify its upstream and downstream filters. These rules make the model highly modular. A more complete overview can be found in [26], including pointers to in-depth studies of the classic style and its variations and specializations.

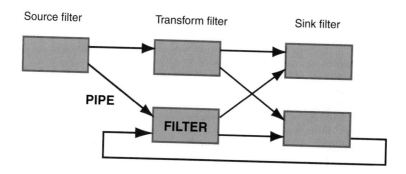

Figure 12.1. Pipes and filters.

The Pipes-and-Filters style has a number of good properties that make it an attractive and efficient model for a number of applications. It is relatively simple to describe, understand, and implement. It is also quite intuitive and allows modeling systems while preserving the flow of data. Some interesting properties result from the modularity of the model. Because of the well-defined and constrained interaction modalities between filters, complex systems described in terms of data streams flowing from filter to filter are easily understandable as a series of well-defined local transformations. The implementation details of each filter are irrelevant to the high-level understanding of the overall system as long as a logical definition of their behavior is specified (input, transformation, and output). The localization and isolation of computations facilitates system design, implementation, maintenance, and evolution. Reversely, filters can be implemented and tested separately. Furthermore, because filters are independent, the model naturally supports parallel and distributed processing.

These properties of the model provide an answer to some of the software issues highlighted in the introduction. Because of their independence, filters certainly allow reusability and interoperability. Parallelism and distributability, to a certain extent, should contribute to efficiency and scalability. It would seem that it is a perfect model for real-time, distributed parallel processing of data streams. However, a few key shortcomings and limitations make it unsuitable for designing crossdisciplinary dynamic systems, possibly involving real-time constraints, user immersion, and interaction.

The first set of limitations is related to efficiency. If the pipes are first-in-first-out buffers, as suggested by the model, the overall throughput of a Pipes-and-Filters system is imposed by the transmission rate of the slowest filter in the system. If filters' independence provides a natural design for parallel (and/or distributed) processing, the pipes impose arbitrary transmission bottlenecks that make the model nonoptimal. Figure 12.2 illustrates inefficiency limitations inherent to the distribution and transmission of data in Pipes-and-Filters models. Each filter's output data must be copied to its downstream filter(s)' input, which can lead to massive and expensive data copying. Furthermore, the model does not provide any consistent mechanism to maintain correspondences between separate but related streams (e.g. when the results of different process paths must be combined to produce a composite output). Such stream synchronization operations require data collection from different repositories, possibly throughout the whole system, raising not only search-related efficiency issues, but also dependency-driven distributed buffer maintenance issues. These can only be solved at the price

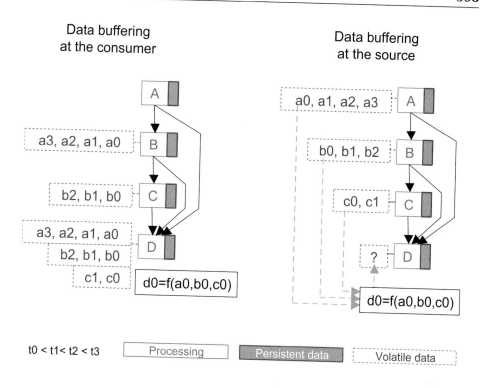

Figure 12.2. Distributed parallel processing of data streams with the Pipes-and-Filters model.

of breaking the strict filter independence rule, for example, by the introduction of a higher level system manager.

The second set of limitations is related to the simplicity of the model. As they are not part of the data streams, process parameters (state data stored in the filters) are not modeled consistently with process data (input data streams). As a result, the pure Pipes-and-Filters model is ill-suited for designing interactive systems, which can only be done at the price of increased filter complexity and deviations from some of the well-defined constraints that make for the model's attractive simplicity. For example, a common practice for interaction in Pipes-and-Filters systems is to provide access to the parameters (possibly at runtime) through control panels. This solution, however, necessitates a separate mechanism for processing and applying the interaction data. This is, for example, the solution adopted in Microsoft's DirectShow, which relies on the Microsoft Windows message-based graphical interface for parameter access. Such external solutions, however, only bypass, and do not address, the fundamental inability of the style to con-

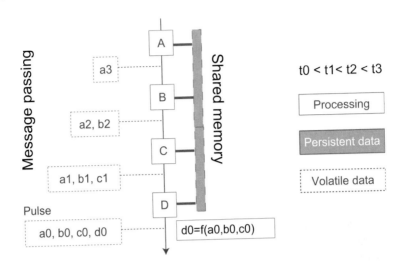

Figure 12.3. A hybrid shared repository and message passing model for distributed parallel processing of data streams.

sistently model general feedback loops—that is, subsystems in which some processes are affected by some other processes' outputs. This is a direct result of the strict communication rules between filters, and in particular, of the constraint that filters not share any state information. Yet feedback loops are a common pattern in dynamic systems, including interactive systems: interaction is indeed a processing loop in which a user is involved (as illustrated in Section 12.2.3).

A careful consideration of system requirements in the target context of applicability guided a reevaluation of the Pipes-and-Filters model and the formulation of modified constraints that preserve the positive aspects while addressing the limitations identified above. Note that other properties were also considered, but are out of the scope of this overview (e.g., runtime system evolution). These are mentioned only when relevant in the remainder of this chapter.

A Hybrid Model

A few key observations, resulting from the analysis of system requirements for real-time interactive, immersive applications allow us to formulate principles and concepts that address the shortcomings identified above. Figure 12.3 offers a synopsis.

Time.   A critical underlying concept in all user-related application domains (but by no means limited to these domains) is that of time. Whether implicitly or explicitly modeled, time relations and ordering are inherent properties of any sensory-related data stream (e.g., image streams, sound, haptics, etc.), absolutely necessary when users are involved in the system, even if not online or in real-time. Users perceive data as streams of dynamic information, that is, evolving in time. This information makes sense only if synchronization constraints are respected within each stream (temporal precedence) and across streams (precedence and simultaneity). It follows that time information is a fundamental attribute of all process data and should therefore be explicitly modeled both in data structures and in processes.

Synchronization is a fundamental operation in temporal data-stream manipulation systems. It should therefore also be an explicit element of the architectural model. A structure called *pulse* is introduced to regroup synchronous data. Data streams are thus quantized temporally (not necessarily uniformly). As opposed to the Pipes-and-Filters case, where data remains localized in the filters where it is created or used, it is grouped in pulses, which flow from processing center to processing center along streams. The processing centers do not consume their input but merely use it to produce some output that is added to the pulse. This also reduces the amount of costly data copy: in a subgraph implemented on a platform with shared memory space, only a pointer to the evolving pulse structure will be transmitted from processing center to processing center. Note that such processing centers can no longer be called filters.

Parallelism.   Figure 12.4 illustrates system latency, which is the overall computation time for an input sample, and throughput or output rate, inverse of the time elapsed between two consecutive output samples. The goal for high-quality interaction is to minimize system latency and maximize system throughput. In the sequential execution model, latency and throughput are directly proportional. In powerful computers, this usually results in the latency dictating the system throughput as well, which is arguably the worst possible case. In the Pipes-and-Filters model, filters can run in parallel. Latency and throughput are thus independent. Because of the parallelism, system latency can be reduced in most cases with careful design, while system throughput will almost always be greatly improved. The sequential behavior of the pipes, however, imposes on the whole system the throughput of the slowest filter. This constraint can actually be relaxed to yield an asynchronous parallel processing model. Instead of being held in a buffer to be processed by order of arrival, each incoming pulse is processed on arrival

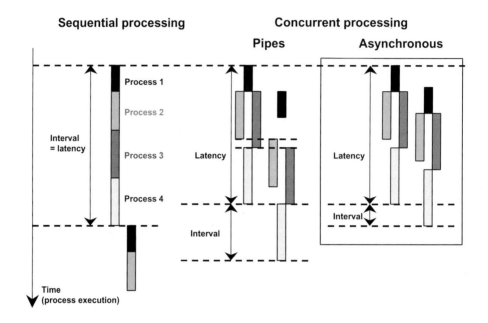

Figure 12.4. Advantage of parallelism for time-sensitive applications. Processes 2 and 3 are independent; process 4 depends on both 2 and 3. With a sequential execution model, the system latency introduced by the processing also constrains the achievable system throughput, as the maximum output rate is inversely proportional to the interval between the completion of the processes of two consecutive time samples. Parallel processing allows decorrelated latency and throughput, usually resulting in a reduction in latency and a large increase in throughput. In the Pipes-and-Filters model, the sequential behavior of the pipes imposes on the whole system the throughput of the slowest filter. In contrast, an asynchronous parallel model allows optimal throughput to be achieved.

in the processing center concurrently with any other pulse already being processed in the cell. *Achievable* throughput is now optimal. It will actually be achieved if no hardware or software resources become exhausted (e.g., computing power, memory, bus bandwidth, etc.). Of course, an asynchronous model requires explicitly implemented synchronization when necessary, but *only then*.

Data classes. The Pipes-and-Filters model explicitly separates data streams and process parameters, which is both a valid functional distinction and a source of inconsistency in the model, leading to important limitations as explained above. A reconsideration of this categorization, in the context of temporal data-stream processing, reveals two distinct data classes: *volatile* and *persistent*.

Volatile data is used, produced, and/or consumed, and remains in the system only for a limited fraction of its lifetime. For example, in a video processing application, the video frames captured and processed are typically volatile data: after they have been processed, and maybe displayed or saved, they are not kept in the system. Process parameters, on the other hand, must remain in the system for the whole duration of its activity. Note that their value can change in time. They are dynamic yet persistent data.

All data, volatile or persistent, should be encapsulated in pulses. Pulses holding volatile data flow down streams defined by connections between the processing centers in a message-passing fashion. They trigger computations and are thus called *active* pulses. In contrast, pulses holding persistent information are held in repositories, where the processing centers can access them in a concurrent shared memory access fashion. This hybrid model combining message passing and shared repository communication, combined with a unified data model, provides a universal processing framework. In particular, feedback loops can now be explicitly and consistently modeled.

From the few principles and concepts outlined above emerged a new architectural style. Because of the context of its development, the new style was baptized Software Architecture for Immersipresence.

12.2.2   The SAI architectural style

This section offers a more formal definition of the SAI architectural style. Graphical symbols are introduced to represent each element type. Together these symbols constitute a graph-based notation system for representing architectural designs. In addition, when available, the following color coding is used: green for processing, red for persistent data, blue for volatile data. Figure 12.5 presents a summary of the proposed notation.

In the remainder of this chapter, the distinction between an object type and an instance of the type will be made explicitly only when required by the context.

Components and connectors

The SAI style defines two types of components: *cells* and *sources*. Cells are processing centers. They do not store any state data related to their computations. The cells constitute an extensible set of specialized components that implement specific algorithms. Each specialized cell type is identified

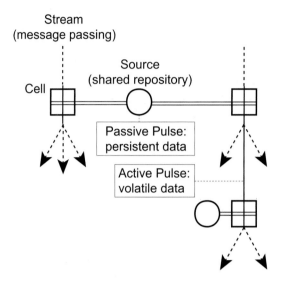

Figure 12.5. Summary of notation for SAI designs. Cells are represented as squares, sources as disks or circles. Source-cell connections are drawn as double or fat lines, while cell-cell connections are drawn as thin arrows crossing over the cells. When color is available, cells are colored in green (reserved for processing); sources, source-cell connections, and passive pulses are colored in red (persistent information); streams and active pulses are colored in blue (volatile information).

by a type name (string) and is logically defined by its input data, its parameters and its output. Cell instances are represented graphically as green squares. A cell can be active or inactive, in which case it is transparent to the system. Sources are shared repositories of persistent data. Source instances are represented as red disks or circles. Two types of connectors link cells to cells and cells to sources. Cell-to-source connectors give the cell access to the source data. Cell-to-cell connectors define data conduits for the streams. The semantics of these connectors are relaxed compared to that of pipes (which are FIFO queues): they do not convey any constraint on the time ordering of the data flowing through them.

Cell and source instances interact according to the following rules. A cell must be connected to exactly one source, which holds its persistent state data. A source can be connected to an arbitrary number of cell, all of which have concurrent shared-memory access to the data held by the source. A source may hold data relevant to one or more of its connected cells, and should hold all the relevant data for each of its connected cells (possibly with some overlap). Cell-source connectors are drawn as either double or fat red lines. They may be drawn across cells (as if cells were connected together by these links) for layout convenience. Volatile data flows in *streams*, which are defined by cell-to-cell connections. A cell can be connected to exactly one upstream cell and to an arbitrary number of downstream cells. Streams (and thus cell-cell connections) are drawn as thin blue arrows crossing over the cells.

Data model

Data, whether persistent or volatile, is held in *pulses*. A pulse is a carrier for all the synchronous data corresponding to a given timestamp in a stream. Information in a pulse is organized as a mono-rooted composition hierarchy of *node* objects. The nodes constitute an extensible set of atomic data units that implement or encapsulate specific data structures. Each specialized node type is identified by a type name (string). Node instances are identified by a name. The notation adopted to represent node instances and hierarchies of node instances makes use of nested parentheses, such as (NODE_TYPE_ID "Node name" (...) ...). This notation may be used to specify a cell's output and for logical specification of active and passive pulses.

Each source contains a *passive pulse*, which encodes the instantaneous state of the data structures held by the source. Volatile data flows in streams that are temporally quantized into *active pulses*. Pulses are represented graphically as a root (solid small disk) and a hierarchy of nodes (small circles); passive pulses may be rooted in the circle or disk representing the source.

Processing model

When an active pulse reaches a cell, it triggers a series of operations that can lead to its processing by the cell (hence the "active" qualifier). Processing in a cell may result in the augmentation of the active pulse (input data) and/or update of the passive pulse (process parameters). The processing of active pulses is carried in parallel as they are received by the cell. Since a cell process can only read the existing data in an active pulse, and never modify it (except for adding new nodes), concurrent read access does not require any special precautions. In the case of passive pulses, however, appropriate locking (e.g., through critical sections) must be implemented to avoid inconsistencies in concurrent shared-memory read/write access.

Dynamic data binding

Passive pulses may hold persistent data relevant to several cells. Therefore, before a cell can be activated, the passive pulse must be searched for the relevant persistent data. As data is accumulated in active pulses flowing down the streams through cells, it is also necessary for a cell to search each active pulse for its input data. If the data is not found, or if the cell is not active, the pulse is transmitted, as is, to the connected downstream cells. If the input data is found, then the cell process is triggered. When the processing is complete, then the pulse, which now also contains the output data, is passed downstream.

Searching a pulse for relevant data, called *filtering*, is an example of runtime data binding. The target data is characterized by its structure: node instances (type and name) and their relationships. The structure is specified as a *filter* or a composition hierarchy of filters. Note that the term filter is used here in its "sieving" sense. Figure 12.6 illustrates this concept. A filter is an object that specifies a node type, a node name or name pattern, and eventual subfilters corresponding to subnodes. The filter composition hierarchy is isomorphic to its target node structure. The filtering operation takes as input a pulse and a filter, and, when successful, returns a *handle* or hierarchy of handles isomorphic to the filter structure. Each handle is essentially a pointer to the node instance target of the corresponding filter.

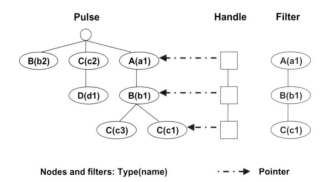

Figure 12.6. Pulse filtering. Each cell is associated with its required volatile and persistent data structures in the form of substructures called active and passive filters (respectively). Pulses are searched for these structures in an operation called filtering, which results in the creation of handles that can be used during processing for direct access to relevant nodes.

When relevant, optional names inherited from the filters allow identification of individual handles with respect to their original filters.

The notation adopted for specifying filters and hierarchies of filters is nested square brackets. Each filter specifies a node type, a node instance name or name pattern (with wildcard characters), an optional handle name, and an eventual list of subfilters, such as [NODE_TYPE_ID "Node name" *handle_id* [...] ...]. Optional filters are indicated by a star: [NODE_TYPE_ID "Node name" *handle_id*]*.

When several targets in a pulse match a filter name pattern, all corresponding handles are created. This allows the design of processes whose input (parameters or stream data) number is not fixed. If the root of the active filter specifies a pattern, the process method is invoked for each handle generated by the filtering (sequentially, in the same thread). If the root of the passive filter specifies a pattern, only one passive handle will be generated (pointing to the first encountered node satisfying the pattern).

Architectural design specification

A particular system architecture is specified at the conceptual level by a set of source and cell instances and their interconnections. Specialized cells may be accompanied by a description of the task they implement. Source and cell instances may be given names for easy reference. In some cases, important data nodes and outputs may be specified schematically to emphasize some design aspects. Section 12.2.3 contains several example conceptual graphs for various systems.

A logical-level description of a design requires specification, for each cell, of its active and passive filters and its output structure, and for each source, the structure of its passive pulse. Table 12.1 summarizes the notations for logical-level cell definition. Filters and nodes are described using the nested square brackets and nested parentheses notations introduced above. By convention, in the cell output specification, (x) represents the pulse's root, (.) represents the node corresponding to the root of the active filter, and (..) represents its parent node.

12.2.3   Example designs

Architectural patterns are now illustrated with demonstration projects ranging from single-stream, automatic, real-time video processing to fully integrated, distributed interactive systems mixing live video, graphics, and sound. The projects are tentatively presented in order of increasing complex-

Table 12.1. Notations for logical cell definition.

ClassName (ParentClass)	CELL_TYPE_ID
Active filter	[NODE_TYPE_ID "Node name" *handle_id* [...] ...]
Passive filter	[NODE_TYPE_ID "Node name" *handle_id* [...] ...]
Output	(NODE_TYPE_ID *"default output base name*–more if needed" (...) ...)

ity, interactivity, and crossdisciplinary integration. Each project is briefly introduced and its software architecture described and analyzed. Key architectural patterns, of general interest, are highlighted. These include feedback loops, real-time incremental processing along the time dimension, interaction loops, real-time distributed processing, and mixing and synchronization of multiple independent data streams.

Real-time video segmentation and tracking

The development of real-time video analysis applications was actually a steering concern during the development of the SAI style. The system presented here was used as a testbed for the implementation, testing, and refinement of some fundamental SAI concepts [18].

The video analysis tasks performed are low-level segmentation and blob tracking. The segmentation is performed by change detection using an adaptive statistical color background model (the camera is assumed stationary). A review of background maintenance algorithms can be found in [28]. Blob tracking is performed using a new multiresolution algorithm whose description is out of the scope of this overview.

Figure 12.7 shows the conceptual-level architecture of the software system. It is built around a single stream going through a number of cells. The graph can be decomposed into four functional subgraphs: capture, segmentation, tracking, visualization. The stream originates in the video input cell, which produces, at a given rate, pulses containing an image coming either from a live capture device or a video file. This cell and its source constitute the capture unit of the application. The stream then goes through the seg-

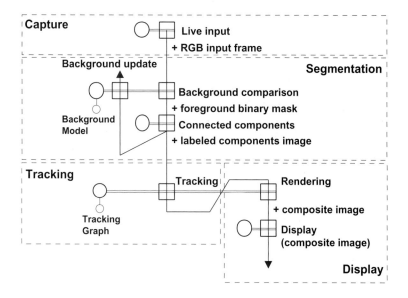

Figure 12.7. Conceptual graph for real-time color background model-based segmentation and multiresolution graph tracking application.

mentation unit, which is analyzed below. Coming out of the segmentation, the stream goes through a tracking cell. The resulting visualization unit is composed of a rendering and a display subunits. Rendering of a persistent structure (here, the tracking graph) is illustrated in a more general context in an example in the section "Live Video in Animated 3D Graphics." The display cell simply puts on the screen its input images, in this case the composite frames produced by the renderer.

A very interesting aspect of this system, and certainly the most innovative when it was introduced, is the asynchronous, parallel implementation of the segmentation algorithm, which contains a feedback loop. The corresponding conceptual graph is also a flow graph of the algorithm. Each input frame is compared with a statistical background model. For each pixel, a decision is made whether it is an observation of the background or of an occluding element. The output is a foreground binary mask. This comparison is performed by the background comparison cell. Each pulse, after going through this cell, contains the input image and a binary foreground image, which is used by the connected components cell to produce a labeled components image, added to the pulse. The input image and the labeled components image are used by the background update cell to update the distributions in the background model. Since the comparison and the update cells both use the persistent background model, they are both connected to a common source that holds the background model structure. This path forms a feedback loop in which the result of the processing of each frame is used to update the adaptive background model. Because of the asynchrony of the processing, by the time the background model is updated with the result of the processing of a given frame, many other frames might have been compared to the model. In this particular context, the quality of the result is not affected—in fact, it is common practice in background model maintenance to perform only partial updates at each frame in order to reduce computational load—and the overall system throughput permitted by this design is always significantly larger than that achievable in a sequential design. Another type of parallelism is illustrated in the branching of the stream to follow independent parallel paths. After coming out of the connected components cell, the stream follows a path to the update cell and another path through tracking and finally visualization. While pulse-level multithreading principally improves throughput, stream-level parallelism has a major impact on latency. In this case, the result of the processing should be used as soon as possible for visualization and for update, in no arbitrarily imposed order. As long as computing resources (in a general sense) are available, and assuming fair scheduling, the model allows minimal latency to be achieved.

Figure 12.8 shows two nonconsecutive frames from one of the PETS 2002 [7] test sequences, with tracked blobs and their trajectories over a few frames. Figure 12.9 presents three consecutive output frames obtained from the processing of professional racquetball videos. The ball and the players are detected and tracked in real-time. In both cases, the original colors have been altered to highlight the blobs or the trajectories in the color to grayscale conversion required for printing.

Quantitative performance metrics are discussed in Section 12.2.4. In the case of live video processing, the throughput of the system impacts the quality of the results: a higher throughput allows the background model to adapt to changing conditions more smoothly.

The modular architecture described here can be used to test different algorithms for each unit (e.g., segmentation algorithms) in an otherwise strictly identical setting. It also constitutes a foundation platform to which higher levels of processing can be added incrementally. The SAI style and its underlying data and processing models not only provide the necessary architectural modularity for a test platform, but also the universal modeling power to account for any algorithm, whether existing or yet to be formu-

Figure 12.8. Object segmentation and tracking results in a PETS02 test sequence. The examples shown are two sample output composite frames taken at different times in the original video. The original colors have been altered to highlight the segmented blobs (in white) in the color to grayscale conversion required for printing.

lated. In conjunction, the same platform also ensures that the best possible performance is achievable (provided correct architectural design and careful implementation of all the elements involved).

Real-time video painting

The Video Painting project was developed as a demonstration of the real-time video stream processing ability provided by the SAI architectural style. It also provided valuable insight on the design and implementation of algorithms performing incremental processing along the time dimension (i.e., between different samples in a same stream).

Figure 12.9. Segmentation and tracking of the players and the ball in a professional racquetball video. The examples shown are three consecutive output composite frames. The original colors have been altered to highlight the trajectories (in white) in the color to grayscale conversion required for printing.

The technical core of the application is a feature-based, multiresolution scheme to estimate frame-to-frame projective or affine transforms [21]. These transforms are used to warp the frames to a common space to form a mosaic.

The mosaic image itself is a persistent structure that evolves as more images are processed—hence the name Video Painting.

In a traditional sequential system, the transform between each pair of consecutive frames would be computed, and each frame would be added to the mosaic by combination of all the transforms between a reference frame and the current frame. Figure 12.10 shows the application graph for the SAI design. A live input unit creates the single stream with frames captured from a camera. A simple cell computes the image pyramid associated with each input frame and necessary for the multiresolution frame-to-frame transform estimation performed by the next downstream cell. Comparing two samples of the same stream requires making one persistent, which becomes the reference. Transforms are thus computed between the current frame and a reference frame. For the algorithm to work properly, though, the compared frames cannot be too far apart. The reference frame must therefore be updated constantly. Because of the asynchrony of the processing, the reference frame is not necessarily the frame "right before" the current frame in the stream. The simplistic handling of frames relationships in the sequential model is no longer sufficient. Accurate timestamps allow a more general approach to the problem. The transforms computed between two frames are no

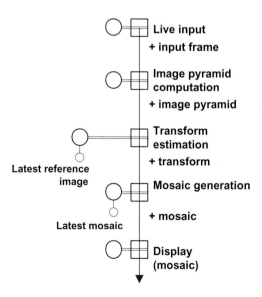

Figure 12.10.   Conceptual graph for Video Painting application (real-time mosaicking).

longer implicitly assumed to relate two consecutive frames, separated from a fixed time interval, but between two frames of arbitrary timestamps. For efficiency reasons, the mosaic is also computed incrementally. A persistent image containing the latest available version of the mosaic is maintained. Each new frame is warped and pasted on a copy of the mosaic by computing the accumulated transform from the frame to the mosaic reference frame. To that effect, with the mosaic is also stored the accumulated transform from the timestamp of the reference mosaic frame to the timestamp of the latest frame added, which is the timestamp for the mosaic. When a new frame is processed, the transform from the mosaic timestamp to the frame is computed by linear interpolation using the transform computed between the frame and the reference frame. This is possible only because time is an explicit component of the model. The transform is then composed with the accumulated transform to produce a new accumulated transform, used to warp the current frame to mosaic space. After pasting of the warped frame, the updated mosaic image is copied into the persistent mosaic image to become the latest reference. The accumulated transform is also updated. Note that a locking mechanism (e.g., critical section) is necessary to ensure consistency between the persistent mosaic and the accumulated transform as they are accessed by different threads. In this design, the updated mosaic is also added to the active pulse so that the dynamic stream of successive mosaics can be viewed with an image display cell. Figure 12.11 shows two example mosaics painted in real-time from a live video stream. The horizontal black lines are image-warping artifacts.

This application is an example where system latency and throughput directly impact the quality of the results. The transform estimation process degrades with the dissimilarity between the frames. A lower latency allows faster reference turnaround and thus smaller frame-to-frame dissimilarity. Higher throughput allows building of a finer mosaic. Embedding established algorithms into a more general parallel processing framework allows achievement of high-quality output in real-time. Furthermore, explicit handling of time relationships in the design might give some insight on real-time implementations of more complex mosaic building techniques (e.g., involving global constraints to correct for registration error accumulation).

Handheld mirror simulation

The Virtual Mirror project [16, 15] started as a feasibility study for the development and construction of a handheld mirror simulation device.

Input frame size

Figure 12.11. Video Painting results.

The perception of the world reflected through a mirror depends on the viewer's position with respect to the mirror and the 3D geometry of the world. In order to simulate a real mirror on a computer screen, images of the observed world, consistent with the viewer's position, must be synthesized and displayed in realtime. This is of course geometrically impossible, but with a few simplifying assumptions (e.g., planar world), the image transform required was made simple yet convincing enough to consider building an actual device. The current prototype system (see Figure 12.12) is comprised of an LCD screen manipulated by the user, a single camera fixed on the screen, and a tracking device. The continuous input video stream and tracker data is used to synthesize, in realtime, a continuous video stream displayed on the LCD screen. The synthesized video stream is a close approximation of what the user would see on the screen surface if it were a real mirror.

Figure 12.13 shows the corresponding conceptual application graph. The single-stream application does not involve any complicated structure. The stream originates in a live video input cell. A tracker input cell adds position and orientation data to the stream. A mirror transform cell uses the synchronous image and tracking data to synthesize the mirror simulation image, which is then presented to the user by an image display cell.

The two major difficulties in this application are (1) synchronizing the various input streams (video and tracking data) to compute a consistent result, and (2) displaying the result with a low enough latency and a high enough throughput to produce a convincing simulation. The use of pulses in the SAI model makes synchronization a natural operation, involving no

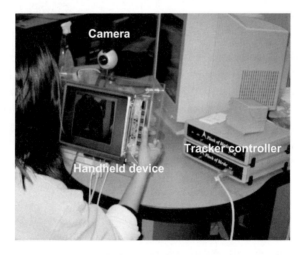

Figure 12.12. The Virtual Mirror system (handheld mirror simulation).

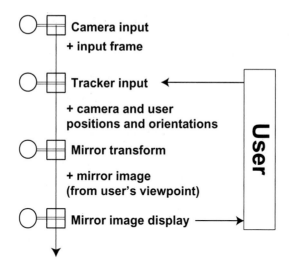

Figure 12.13. The Virtual Mirror application graph.

superfluous delays or computations. The asynchronous parallel processing model allows high frame rates and low latency.

The essential purpose of this system is interaction. Interaction can be seen as a particular data-stream loop feeding the user with a perceptual representation of the internal model (experience), collecting the user's reaction through various sensory devices, and modifying the state of the internal model accordingly (influence). From the system's point of view, these data streams are volatile, and the processes involved in producing and processing them are of the same nature as those carrying procedural internal evolution tasks. Here, the internal model is the mirror, implicit in the computations carried by the mirror transform cell. The user experiences the system through the image displayed on the handheld screen and influences it by moving her head or the screen.

Note that the way live video (frames) and corresponding tracker data are synchronized in the application as described in the conceptual graph is based on the assumption that the delay between the frame capture in the camera input cell (push mechanism acting as pulse trigger) and the capture of tracker data in the tracker input cell (pull mechanism) is small enough that no inconsistency can be perceived. This approach happens to work in this case, even when the system is running on a modest platform, but certainly does not generalize to synchronization in more sophisticated settings. This illustrates the flexibility of the SAI style, which allows us to devise general structures as well as simplified or ad hoc design when allowed by the application.

This project also illustrates the power of SAI for system integration. The modularity of the model ensures that components developed independently (but consistently with style rules) will function together seamlessly. This allows code reuse (e.g., the image node and the live video capture and image display cells already existed) and distributed development and testing of new node and cell types. The whole project was completed in only a few months with very limited resources.

The mirror simulation system can be used as a research platform. For example, it is a testbed for developing and testing video analysis techniques that could be used to replace the magnetic trackers, including face detection and tracking and gaze tracking. The device itself constitutes a new generic interface for applications involving rich, first-person interaction. A straightforward application is the Daguerréotype simulation described in [22]. Beyond technical and commercial applications of such a device, the use of video analysis and graphics techniques will allow artists to explore and interfere with what has always been a private, solitary act, a person looking into a mirror.

IMSC Communicator

The IMSC Communicator is an experimental extensible platform for remote collaborative data sharing. The system presented here is an early embodiment of a general research platform for remote interaction. From a computer vision point of view, the architectural patterns highlighted here directly apply to the design of distributed processing of multiple video streams.

Popular architectures for communication applications include client/server and Peer-to-Peer. Different elements of the overall system are considered separate applications. Although these models can either be encapsulated or implemented in the SAI style, a communication application designed in the SAI style can also be considered as a single, distributed application graph in which some cell-to-cell connections are replaced with network links. From this point of view, specific network architectures would be implemented in the SAI style as part of the overall distributed application.

The core pattern of the Communicator is a sequence of cells introducing network communication between two independent subgraphs. Figure 12.14 shows an example sequence comprising encoding, compression, and networking (send) on the emitting side; and networking (receive), decompression, and

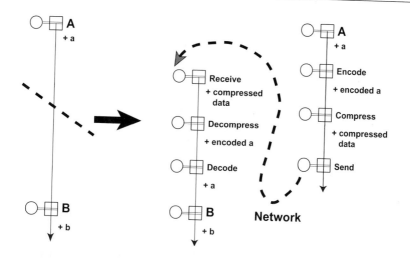

Figure 12.14. Simple generic networking for distributed applications.

decoding on the receiving side. The encoding cell flattens a node structure into a linear buffer so that the structure can later be regenerated. This encoding can be specific for various data types, as is the case for the example described here. A general encoding scheme could, for example, be based on XML. The output of an encoding cell is a character string (binary or text). The compression cell takes as input the encoded character string and produces a corresponding compressed buffer. The compression scheme used can be input data-dependent or generic, in which case it should be lossless. For the first experiments, a simple compression cell and matching decompression cell were developed that encapsulate the open source LZO library [25] for real-time lossless compression/decompression. Note that the compression step is optional. The networking cells are responsible for packetizing and sending incoming character strings on one side and receiving the packets and restoring the string on the other side. Different modalities and protocols can be implemented and tested. The first networking cells were implemented using Windows Sockets, using either TCP/IP or UDP. The decompression cell regenerates the original character string from the compressed buffer. The decoding cell regenerates the node structure into a pulse from the encoded character string.

Once a generic platform is available for developing and testing data transfer modalities, support for various specific data types can be added. The very first test system supported video only, using existing live video capture and image display cells. For the next demonstration, a new live capture cell for both image and sound was developed using Microsoft DirectShow [24]. Another new cell was developed for synchronized rendering of sound and video, also using DirectShow.

Figure 12.15 shows the conceptual graph for an early embodiment of the two-way communicator, with example screen shots. A background replacement unit, based on the segmentation by change detection (presented in Section 12.2.3), was added to the capture side to illustrate how the modularity of the architecture allows plug-and-play subgraphs to be developed independently. Having different processing of video and sound also demon-

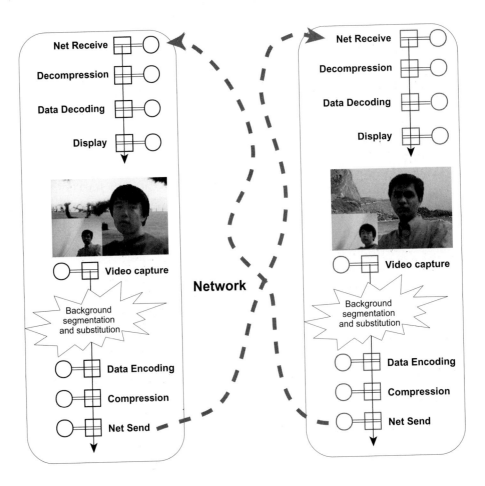

Figure 12.15. Conceptual graph for an early version of the IMSC Communicator, with support for video (synchronized image and sound).

strates the advantages of adopting an asynchronous model and of performing synchronization only when necessary, in this case in the display cell.

Live video in animated 3D graphics

The example presented in this section is a real-time, interactive application requiring manipulation of heterogenous data streams [14]. A video stream (captured live or read from a file) is used as surface texture on an animated 3D model, rendered in realtime. The (virtual) camera used for rendering can be manipulated in realtime (e.g., with a gamepad or through a head tracking device), making the system interactive. This system was developed to illustrate the design of a scene graph-based graphics toolkit and the advantages provided by the SAI style for manipulating independently and synchronizing different data streams in a complex interactive (possibly immersive) setting.

From a computer vision point of view, this example illustrates how the SAI style allows manipulation of video streams and geometric models in a consistent framework. The same architectural patterns generalize to manipulating generic data streams and persistent models.

Figure 12.16 shows the system's conceptual graph. The graph is composed of four functional subgraphs corresponding to four independent streams, organized around a central source that holds the 3D-model representation in the form of a scene graph and various process parameters for

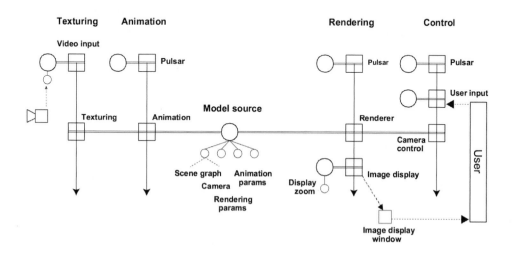

Figure 12.16. Conceptual graph for integrating real-time rendering of an animated 3D model with live video mapping and interactive camera control.

the different connected cells. The rendering stream generates images of the model. The control stream updates the (virtual) camera position based on user input. Together, the rendering and control units (including the shared source) form an interaction loop with the user. The animation stream drives the dynamic evolution of the 3D model. The texturing stream places images captured from a video source (camera or file) in the texture node in the scene graph. Interaction and modeling are now analyzed in more detail.

Interaction.   As observed above, interaction is a particular data stream loop feeding the user with a perceptual representation of the internal model (experience), collecting the user's reaction through various sensory devices, and modifying the state of the internal model accordingly (influence). From the system's point of view, these data streams are volatile, and the processes involved in producing and processing them are of the same nature as those carrying procedural internal evolution tasks, and are thus modeled consistently in the SAI style.

Any such interaction subsystem, an example of which is the user loop on the right half of Figure 12.16, will involve instances of cells belonging to a few typical functional classes: inputs, effectors, renders, displays. Input cells collect user input and generate the corresponding active pulses on streams. These components encapsulate such devices as mouse, keyboard, and 3D tracker. Effector cells use the input data to modify the state of the internal model (possibly including process parameters). In this example, the effector is the camera control cell. In some cases, the data produced by the input device requires some processing in order to convert it into data that can be used by the effector. This is the case, for example, with vision-based perceptual interfaces in which the input device is a camera, which produces a video stream that must be analyzed to extract the actual user input information. Such processes can be implemented efficiently in the SAI style (see, e.g., [18]), and be integrated consistently in an interactive application graph. Rendering and display elements produce and present a perceptual view of the internal model to the user, closing the loop. Display cells encapsulate output devices, such as screen image display, stereo image displays, and sound output. Note that in some cases, such as for haptics, input and display functions are handled by the same device.

For interaction to feel natural, latency (or lag: the delay between input action and output response) must be kept below a threshold above which the user will perceive an inconsistency between her actions and the resulting system evolution. This threshold is dependent on the medium and on the application, but studies show that latency has a negative effect on human performance [23]. For example, human performance tests in virtual environments with head-coupled display suggest a latency threshold of the order of 200ms above which performance becomes worse in terms of response time than with static viewing [11]. The same study also suggests that latency has a larger impact on performance than on frame update rate. System throughput (including higher data bandwidth) is more related to the degree of immersiveness and could influence *perceived* latency. In any case, it is quite intuitive and usually accepted that interactive systems can always benefit from lower latency and higher throughput. A careful design, an appropriate architecture, and an efficient implementation are therefore critical in the development of interactive applications.

Renderers are an example of simultaneous manipulation of volatile and persistent data. A rendering cell produces perceptual, instantaneous snapshots (volatile) of the environment (persistent) captured by a virtual device such as microphone or camera, which is itself part of the environment model. The intrinsic parallelism and synchronization mechanism of the SAI style are particularly well suited for well-defined, consistent, and efficient handling of such tasks. The rendering stream is generated by an instance of the *Pulsar* cell type, a fundamental, specialized component that produces empty pulses on the stream at a given rate. In this example, the rendering cell adds to the pulse an image of the 3D model, synthesized using the OpenGL library. Persistent data used by the rendering, apart from the scene graph, include a virtual camera and rendering parameters such as the lighting model used, and so on. An image display cell puts the rendered frames on the screen.

User-input streams can follow either a pull or push model. In the pull approach, which is used in this example, a Pulsar triggers a regular sampling of some input device encapsulated in a user-input cell and corresponding data structures. The state of the device is then used in a control cell to affect the state of the model, in this case the position and parameters of the virtual camera. In a push approach, more suitable for event-based interfaces, the input device triggers the creation of pulses with state-change-induced messages, which are then interpreted by the control cell.

Rendering and control are two completely independent streams, which could operate separately. For example, the same application could function perfectly without the control stream, although it would no longer be interactive. Decoupling unrelated aspects of the processing has some deep implications. For example, in this system, the responsiveness of the user-control subsystem is not directly constrained by the performance of the rendering subsystem. In general, in an immersive system, if the user closes her eyes,

moves her head, and then reopens her eyes, she should be seeing a rendering
of what she expects to be facing, even if the system was not able to render
all the intermediate frames at the desired rate. Furthermore, the system can
be seamlessly extended to include "spectators" that would experience the
world through the user's "eyes" but without the control, or other users that
share the same world but have separate interaction loops involving different
rendering, display, and input modalities. All such loops would then have the
same status with respect to the shared world, ensuring consistent read and
write access to the persistent data.

Modeling.   The world model itself can evolve dynamically, independently of
any user interaction. Simulating a dynamic environment requires the use of
models that describe its state at a given time and the way it evolves in time.
These models are clearly persistent (and dynamic) data in the system. Implementing the necessary SAI elements requires careful discrimination between
purely descriptive data, processes, and process parameters, and analysis of
how these model elements interact in the simulation. The design described
in [14] for 3D modeling is directly inspired from VRML (Virtual Reality
Modeling Language) [5] and the more recent X3D [4]. The main descriptive
structure of the model is a scene graph, which constitutes a natural specialization of the FSF data model. Figure 12.17 illustrates node types involved
in the scene graph model. Descriptive nodes, such as geometry nodes and the
Shape, Appearance, Material, and Texture nodes, are straightforward adaptations of their VRML counterparts. The VRML Transform node, however,
is an example of semantic collapse, combining a partial state description and
its alteration. A transform should actually be considered as an action applied to a coordinate system. It can occur according to various modalities,
each with specific parameter types. Consequently, in our model, geometric
grouping is handled with coordinate system nodes, which are purely descriptive. In order to provide a complete state description in a dynamic model,
the coordinate system node attributes include not only origin position and
basis vectors, but also their first and second derivatives. Various transforms
are implemented as cells, with their specific parameter nodes, that are not
part of the scene graph (although they are stored in the same source). Scene
graph elements can be independently organized into semantic asset graphs
by using the shared instance node, instances of which can replace individual node instances in the scene graph, as illustrated for a material node in
Figure 12.17.

Animation (i.e., scene evolution in time), is handled by cells implementing specific processes whose parameter nodes are not part of the scene graph

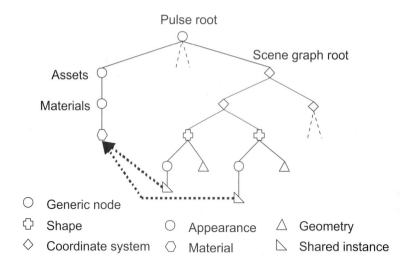

Figure 12.17. Scene graph model in the SAI style, as proposed in [14].

itself. This ensures that both persistent data, such as the scene graph, and volatile data, such as instantaneous description of graph components evolution, are handled consistently. This aspect is critical when simultaneous independent processes are in play, as it is often the case in complex simulations. Process parameters can also evolve in time, either as a result of direct feedback or through independent processes.

The analytical approach followed for modeling, by separating descriptive data, processes, and process parameters supported by a unified and consistent framework for their description and interaction, allows seamless integration of 3D graphics, video streams, audio streams, and other media types, as well as interaction streams. All these independent streams are in effect synchronized by the well defined and consistent use of shared data, in this case the scene graph and the virtual camera.

Figure 12.18 shows a rendered frame of a world composed of four spheres and four textured rectangles, rotating in opposite directions. Two fixed lights complete the model. A video stream captured live is used as texture for the rectangles. Although this setting is not immersive (in the interest of the picture), the same system can be easily made immersive by modifying the scene. The user may be placed in the center of the world, itself modeled by a cylindrical polygonal surface on which a video stream, possibly prerecorded from a panoramic camera system, is texture-mapped in realtime. The user,

Figure 12.18. A rendered frame of the (animated) 3D model with texture mapping from a live video stream.

maybe using a head-mounted display, only sees the part of the environment she is facing and may control her viewing direction using a head tracker.

This application illustrates the flexibility and efficiency of the SAI architectural style. In particular, thanks to the modularity of the framework, core patterns (e.g., rendering, control, animation) can be effortlessly mixed and matched in a plug-and-play fashion to create systems with unlimited variations in the specific modalities (e.g., mouse or tracker input, screen or head mounted display).

12.2.4   Architectural properties

By design, the SAI style preserves many of the desirable properties identified in the Pipes-and-Filters model. It allows intuitive design, emphasizing the flow of data in the system. The graphical notation for conceptual level representations give a high-level picture that can be refined as needed, down to implementation level, while remaining consistent throughout. The high modularity of the model allows distributed development and testing of particular elements and easy maintenance and evolution of existing systems. The model also naturally supports distributed and parallel processing.

Unlike the Pipes-and-Filters style, the SAI style provides unified data and processing models for generic data streams. It supports optimal (theoretical) system latency and throughput thanks to an asynchronous parallel processing model. It provides a framework for consistent representation and efficient implementation of key processing patterns such as feedback loops and incremental processing along the time dimension. The SAI style has several other important architectural properties that are out of the scope of this overview. These include natural support for dynamic system evolution, runtime reconfigurability, self monitoring, and so on.

A critical architectural property that must be considered here is *performance overhead*. Some aspects of the SAI data and processing models, such as filtering, involve nontrivial computations and could make the theory impractical. The existence of fairly complex systems designed in the SAI style and implemented with MFSM show that, at least for these examples, it is not the case.

A closer look at how system performance is evaluated and reported is necessary at this point. It is not uncommon to provide performance results in the following form: "the system runs at n frames per second on a Processor X system at z $\{K, M, G\}$Hz." For example, the segmentation and tracking system presented in the last section runs at 20 frames per second on 320×240 pixels frames. The machine used is a dual processor Pentium 4 @ 1.7 GHz with 1GB memory. Note that the system does not only perform segmentation and tracking: it also reads the frames from a file on disk or from a camera, produces a composite rendering of the analysis results, and displays the resulting images on the screen. In order to make the results reproducible, the hardware description should include the hard drive and/or camera and interface, the graphics card, and how these elements interact with each other (e.g., the motherboard specifications), and so on. In practice, such a detailed description of the hardware setting is tedious to put together and probably useless, since reproducing the exact same setting would be difficult and even undesirable. Another inaccuracy in the previous report is that it implies that 20 frames per second is the best achievable rate in the described conditions (which is incidentally not the case). Reporting best-case performance is not very useful, as a system that takes up all the computing power to produce a relatively low-level output, meaningful only in the context of a larger system, is not of much use beyond proof of concept. If the algorithm is going to be used, an evaluation of its computational requirements as part of a system is necessary.

Performance descriptions of the type described above provide only a partial and fuzzy data point that alone does not allow, for example, prediction of how the system will perform on another (hopefully) more powerful hardware platform, or how it will scale up or down in different conditions (frame rate, image resolutions, etc.). Theoretical tools such as algorithmic complexity analysis can certainly help in the prediction, if the properties of the other elements involved in the overall system is understood. Hardware components can certainly be sources of bottlenecks; for example, disk speed (for input or output) and camera rate can be limiting factors. On the software side, it is also necessary to understand not only the contribution of each algorithm, but also that of the environment in which they are implemented.

Figure 12.19 shows new experimental scalability tests performed on applications designed in the SAI style. The results are in accordance with those reported in [18]. The figure plots processing load, in percent, against processing rate, in frames per seconds, for an application performing video

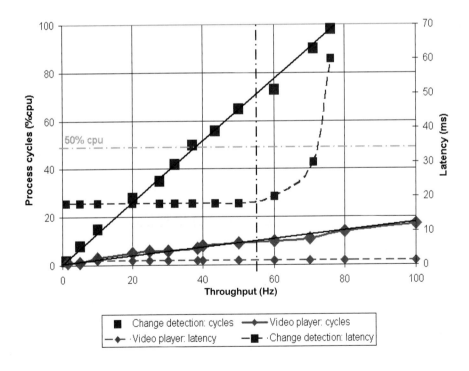

Figure 12.19. Scalability with respect to throughput.

capture, segmentation by change detection, and result visualization, and the same plot for the same application with the segmentation turned off (the image displayed is the input). Both applications *scale linearly* with respect to throughput. All conditions being equal, the difference in slope between the two curves characterizes the processing load imposed by the segmentation part. Note that these tests were performed on a dual-processor machine, so that the processor load is an average of the load of both processors. Because of the parallelism (in this case multithreading), the overall load is balanced between the two processors (by the operating system in this case). As a result, the system is completely oblivious to the 50% CPU load barrier. The figure also plots system latency, in ms, against processing rate, for both applications. The latency remains constant as long as system resources are available. In a sequential system, the latency would be directly proportional to the throughput and in fact would dictate the maximum achievable throughput. When a bottleneck is encountered (or some resources are exhausted), latency increases and system performance degrades. The segmentation application performance (in terms of latency) starts degrading around 55 frames per second, although the system can still achieve rates above 70 frames per second in these conditions.

These plots suggest that (1) as long as computing resources are available, the overhead introduced by the SAI processing model remains constant; and (2) the contribution of the different processes are combined linearly. In particular, the model does not introduce any nonlinear complexity in the system. These properties are corroborated by empirical results and experience in developing and operating other systems designed in the SAI style. Theoretical complexity analysis and overhead bounding are out of the scope of this overview.

12.3   MFSM: An Architectural Middleware

This section is an example-based, code-level tutorial on the use of the MFSM architectural middleware for implementing image-stream manipulation applications designed in the SAI style. It begins with a brief overview of the MFSM project and its implementation of SAI. Design and implementation of applications for image-stream manipulation is then explained in detail using simple examples based on those used in the online MFSM user guide. A first example illustrates application setup and cleanup, and application graph elements instantiation and connection. The development of specialized nodes is explained with a study of the implementation of a generic image data structure (object of an open source module). The use of this node type in a specialized process is explained with a study of the implementation of a custom cell used in the first example. Finally, a more complex example illustrates shared memory access for incremental computations along the time dimension.

This section constitute an image-oriented, self-contained complement to the online user guide [12].

Disclaimer. The code appearing in this section was developed by the author and is part of open source modules and examples released under the GNU Lesser General Public License [1]. They are available for download on the MFSM Web site [12]. This code is provided here in the hope that it will be useful, but *without any warranty*—without even the implied warranty of *merchantability or fitness for a particular purpose*. Complete license information is provided in the downloadable packages.

12.3.1   MFSM overview

MFSM [12] is an architectural middleware implementing the core elements of the SAI style. MFSM is an open source project, released under the GNU Lesser General Public License [1]. The goal of MFSM is to support and promote the design, analysis, and implementation of applications in the SAI style. This goal is reflected in the different facets of the project.

- The FSF library is an extensible set of implementation-level classes representing the key elements of SAI. They can be specialized to define new data structures and processes or encapsulate existing ones (e.g., from operating system services and third-party libraries).

- A number of software modules regroup specializations implementing specific algorithms or functionalities. They constitute a constantly growing base of open source, reusable code, maintained as part of the MFSM project. Related functional modules may be grouped into libraries.

- An evolving set of example applications illustrates the use of FSF and specific modules.

- An evolving set of documents provides reference and example material. These include a user guide, a reference guide and various tutorials.

Figure 12.20 shows the overall system architecture suggested by MFSM. The middleware layer provides an abstraction level between low-level services

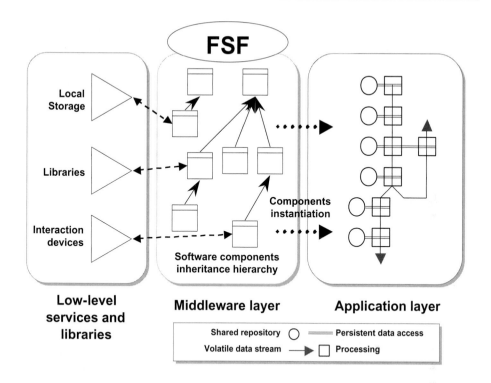

Figure 12.20. Overall system architecture suggested by MFSM.

and applications, in the form of SAI software elements. At the core of this layer is the Flow Scheduling Framework (FSF) [13, 17], an extensible set of foundation classes that implement SAI style elements. The generic extensible data model allows encapsulation of existing data formats and standards, as well as low-level service protocols and APIs, and makes them available in a system where they can interoperate. The hybrid shared repository and message-passing communication and parallel processing model supports control and asynchronous, concurrent processing and synchronization of data streams. The application layer can host a data-stream processing software system, specified and implemented as instances of SAI components and their relationships.

In its current implementation, the FSF library contains a set of C++ classes implementing SAI elements: the source, the nodes and pulses, the cells, the filters, the handles. It also contains classes for two implementation-related object types: the factories and the System. A brief overview of each object type is proposed below. An online reference guide [12] provides detailed interface description and implementation notes for all the classes defined in the FSF library.

Source.   The source class is fairly simple. It holds a single passive pulse and keeps a list of pointers to connected cells. Note that connectors are not implemented as objects, but simply lists of pointers in the components. This results from the simplicity of the role played by connectors in the SAI style. In particular, they do not perform any computation. Sources also perform garbage collecting on their passive pulse when notified of a structural change by a connected cell.

Nodes and pulses.   The FSF library contains a base node class that implements connectivity elements for building hierarchical structures. Specialized node classes, encapsulating or implementing specific data structures, are derived from the base class. In FSF, from the base node class are derived two specific classes for active and passive pulse respectively. These serve as root nodes for active and passive pulse, which are node hierarchies. Another node specialization is provided in FSF that implements nodes holding atomic type values (integer, floating point, boolean, string, etc.). These are called TypeNodes, and are implemented as template instances of a template class, itself derived from a non-template TypeNodeBase class (to allow undetermined pointers to a TypeNode object).

Cells.   The FSF library contains a base cell class that implements cell-cell and cell-source connectivity mechanisms (lists of pointers). It also implements all the computation associated with the parallel processing of active pulses. In MFSM, in each cell, each active pulse is processed in a separate thread. Each passive filtering operation is also carried in a separate thread. As a result, some precautions must be taken for concurrent data access. Thanks to inheritance and virtual functions, when implementing a specialized cell derived from the base cell, only the specific computation must be coded, in an overloaded process, virtual-member function. All the events leading to the call to this process function, including filtering, will be automatically performed. A cell can be active or inactive. For efficiency purposes, passive filtering occurs only when the cell is activated (activation fails if passive filtering fails) and when explicitly notified by the source after a cell notified the source of a structural change in the passive pulse.

Filters.   Two sets of classes are implemented for active and passive filters respectively. For each, a template class derived from a non-template base class provides support for instantiating filters corresponding to any node type. The non-template base class allows pointers to undetermined filters (e.g., in cells). The filter objects also implement the filtering functionality. Active filtering instantiates and returns a list of active handles and passive filtering instantiates and returns a list of passive handles.

Handles.   Two sets of classes are implemented for active and passive handles respectively. Active handles simply provide a pointer to a node in an active pulse. An optional handle ID, inherited during the filtering operation, allows identification of the node with respect to the filter. Passive handles play the additional role of locks for node deletion and garbage collecting (not required in active pulses). When a cell removes a node, it does not destroy it, but marks it as deleted so it is not used in subsequent filtering operations. The cell must notify the source, which in turn notifies all connected cells to refilter the passive pulse, and initiates a garbage-collecting process, which physically deletes the node when is it free of handles from ongoing process.

Factories.   Class factories are used to instantiate objects at runtime, whose type is not known at compile time. Since cells and nodes constitute extensible sets of specialized object types derived from respective base classes, Cell and Node factories are necessary for instantiating cells and nodes at runtime and modifying application graphs dynamically. Node factories also allow instantiation of filters for the corresponding node type.

System.   By convention, any application using FSF must have a single instance of the System type, a pointer to which is available as a global variable, defined in the FSF library. This object provides the reference clock and holds the lists of node and cell factories available for building application graphs. In order to use a module in an application, its node and cell factories must be declared and registered with the system instance.

12.3.2   A first image manipulation example

The online MFSM user guide [12] describes a simple program (called ex1) that generates a stream at a given frame rate and displays the timestamp of each pulse in both ms and h/m/s formats. The example in this section performs a similar task, but instead of being written to the standard output, each formatted timestamp string is written into a newly created image (using Intel's OpenCV [2]). This first example is called eximg1.

Application design and specifications

Before building a design and after identifying the system requirements, it is useful to know what building elements are available. This simple application involves a (single) stream, so the first concern is the origin of the stream. FSF contains a fundamental cell type that generates empty pulses at a given frame rate. It is called Pulsar, and Table 12.2 shows its specifications. It does not have any input. It has one parameter, of integral type, whose value is the time delay between two consecutive output pulses.

Table 12.2. `fsf::CPulsar` logical cell definition.

fsf::CPulsar (fsf::CCell)	FSF_PULSAR
Active filter	(no input)
Passive filter	[FSF_INT32_NODE "Pulse delay"]
Output	(FSF_ACTIVE_PULSE "*Root*")

In this section, it is assumed that a cell is available that is capable of looking up the timestamp of each incoming active pulse, creating an image, and printing the formatted timestamp in the image. The actual implementation of this cell is addressed later, in the Section 12.3.3. Table 12.3 shows the cell specifications.

Table 12.3. `myspace::MyCellImg` logical cell definition.

myspace::CMyCellImg (fsf::CCell)	MY_CELL_IMG
Active filter	[FSF_ACTIVE_PULSE "Root"]
Passive filter	[FSF_PASSIVE_PULSE "Root"]
Output	(IMAGE_IMAGE "*Image*")

The open source Display Module defines and implements a cell that can display an image into a window. Table 12.4 shows the ImageDisplay cell specifications.

Given these elements, it is now straightforward to formulate a design for `eximg1` from one instance of Pulsar, one instance of MyCellImg, and one instance of ImageDisplay. Figure 12.21 shows the resulting conceptual application graph.

Table 12.4. `display::ImageDisplay` logical cell definition.

display::CImageDisplay (fsf::CCell)	IMAGE_DISPLAY
Active filter	[IMAGE_IMAGE "Image"]
Passive filter	[FSF_FLOAT32_NODE "Display zoom"]
Output	(no output)

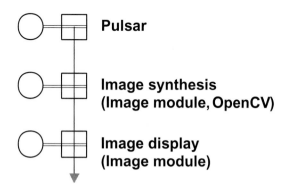

Pulsar

Image synthesis
(Image module, OpenCV)

Image display
(Image module)

Figure 12.21. Conceptual level graph for the first image manipulation example, `eximg1`.

The next few sections explain in detail how to code the simple application in C++: setting up the system, building and running the application graph, and cleaning up the objects allocated.

Getting started

The first mandatory step in using the FSF library is to allocate the unique system object that provides the reference clock and holds the node and cell factories available. The nodes and cells defined in each module used must be registered. Factories can be registered in any order. They are not necessary for precoded application graphs, although they are used by the scripting module functions (see below). Factories are necessary for dynamic, runtime application graph building and/or modification, which is out of the scope of this introduction. Each module is required to declare and implement a function called `RegisterFactories` for registering factories in the system for all nodes and cells implemented in the module.

```
// create the system
fsf::AllocateSystem();

// register system factories
fsf::RegisterFactories();

// register module factories
image::RegisterFactories();
display::RegisterFactories();

// register myspace factories
myspace::RegisterFactories();
```

Building the graph

A scripting module (namespace `scripting`) provides shortcut functions to instantiate sources and cells, instantiate nodes and place them in source pulses, and connect cells to other cells and to sources. The convention followed in the parameter order for stream connections is to plug a downstream cell into an upstream cell. The name "scripting" comes from the fact that the functions provided by this module are coding equivalents of user actions in an interactive system. In particular, the scripting module uses aspects of the MFSM implementation that are related to dynamic system evolution, such as class factories. Note that the scripting module itself does not implement any node or cell class and thus does not register any factory (there is no `scripting::RegisterFactories`).

The code for building an application graph instantiates and connects all the elements of the conceptual graph. In this simple example, the graph can be divided into three functional parts: the pulsing unit built around the Pulsar instance, the image synthesis unit built around the MyCellImg instance, and the display unit built around the ImageDisplay instance. Each subgraph in this case corresponds to one source and one cell (minimal computing units).

Each minimal unit, consisting of one cell and one source whose pulse contains the cell's parameters, can be coded following these steps:

– Instantiate the source.

– Instantiate the parameter node(s). Each node is placed in the source's passive pulse hierarchy. Optional steps for each node include setting its name and data member initial values.

– Instantiate the cell. Optional steps include setting the output base name, the active and passive filters. The cell is then connected to its source and to the cell directly upstream on the active stream, if any.

These principles can be used as guidelines and adapted to code any graph. The following code builds the graph for the example, first the pulsar unit, then the image synthesis unit, and finally the image display unit. Successful instantiation of all graph elements is checked, as failure to register the appropriate factories will result in the failure to instantiate a given cell or node.

```
// build graph
bool bSuccess=true;

//////////////////
// Pulsar unit //
//////////////////

// create the source
fsf::CSource *pPSource=new fsf::CSource;
bSuccess &= (pPSource!=NULL);

// parameter node: pulse delay
fsf::Int32Node *pPulseDelay =
  static_cast<fsf::Int32Node*>(scripting::CreateNode(
    std::string("FSF_INT32_NODE"),pPSource->GetPulse()));
bSuccess &= (pPulseDelay!=NULL);
if(bSuccess)
{
  // set name
  pPulseDelay->SetName(std::string("Pulse delay"));
  // set parameter values
  long nPulseDelay=static_cast<long>((1000.0f/fPulseRate)-1);
  pPulseDelay->SetData(nPulseDelay);
}

// cell
fsf::CCell *pPcell =
  static_cast<fsf::CCell*>(scripting::CreateCell(
    std::string("FSF_PULSAR")));
```

```
bSuccess &= (pPcell!=NULL);
if(bSuccess)
{
  // connect with source
  scripting::ConnectSource(pPcell,pPSource);
}

///////////////////////////
// Image synthesis unit //
///////////////////////////

// create the source
fsf::CSource *pMySource=new fsf::CSource;
bSuccess &= (pMySource!=NULL);

// cell
fsf::CCell *pMyCell =
  static_cast<fsf::CCell*>(scripting::CreateCell(
    std::string("MY_CELL_IMG")));
bSuccess &= (pMyCell!=NULL);
if(bSuccess)
{
  // connect with source
  scripting::ConnectSource(pMyCell,pMySource);

  // connect with Pcell
  scripting::ConnectUpstreamCell(pMyCell,pPcell);
}

/////////////////////
// Display unit //
/////////////////////

// create the source
fsf::CSource *pDisplaySource=new fsf::CSource;
bSuccess &= (pDisplaySource!=NULL);

// parameter node: display zoom
fsf::Float32Node *pDisplayZoom=static_cast<fsf::Float32Node*>(
```

```
scripting::CreateNode(
   std::string("FSF_FLOAT32_NODE"),pDisplaySource->GetPulse()));
bSuccess &= (pDisplayZoom!=NULL);
if(bSuccess)
{
  // set name
  pDisplayZoom->SetName(std::string("Display zoom"));
  // set parameter values
  pDisplayZoom->SetData(1.0f);
}


// cell
fsf::CCell *pDisplayCell =
  static_cast<fsf::CCell*>(scripting::CreateCell(
    std::string("IMAGE_DISPLAY")));
bSuccess &= (pDisplayCell!=NULL);
if(bSuccess)
{
  // connect with source
  scripting::ConnectSource(pDisplayCell,pDisplaySource);

  // connect with Pcell
  scripting::ConnectUpstreamCell(pDisplayCell,pMyCell);
}


// Check everything went OK ..
if(bSuccess==false)
{
  cout << "Some elements in the graph could not be instantiated."
       << endl;
  return (-1);
}
```


Running the graph

Once the graph is completed, the cells must be activated. The Pulsar instance is the origin of the active stream and starts generating empty pulses as soon as it is activated. The MyCellImg instance, once activated, will process incoming pulses in parallel, asynchronously. The ImageDisplay instance will render the images produced by the MyCellImg instance on the screen. The cells can be started in any order.

```
// run the cells
pMyCell->On();
pPcell->On();
pDisplayCell->On();
```

Although this aspect is not evident in this simple example, cells can be turned on and off at any time during the execution of the application, elements (sources and cells) can be connected and disconnected at any time, and new ones can be created and existing ones destroyed at any time.

Because the Display Module relies on the Microsoft Windows operating system to display images on the screen, the console application must explicitly provide for a mechanism to dispatch and process messages for the display window. The following code serves this purpose. At the same time, the GetAsyncKeyState function is used to check whether the Q key is pressed to exit the loop.

```
// message loop for windows
// + check for 'Q' key (VK_Q id 0x51) pressed
MSG msg;
while(((GetAsyncKeyState(0x51)&0x1)==0)
      && ::GetMessage( &msg, NULL, 0, 0))
{
  ::TranslateMessage( &msg );
  ::DispatchMessage( &msg );
}
```

This is the only MS Windows specific code in the application.

Clean-up

The following code stops the cells, disconnects and destroys the different elements instantiated when building the graph, and finally deallocates the unique global FSF system instance. The scripting module provides high-level functions to disconnect cells and source.

```
// Stop the cells
pPcell->Off();
pMyCell->Off();
pDisplayCell->Off();
```

```
// Clean up
scripting::DisconnectSource(pPcell);
scripting::DisconnectStream(pPcell);
delete pPcell;
delete pPSource;

scripting::DisconnectSource(pMyCell);
scripting::DisconnectStream(pMyCell);
delete pMyCell;
delete pMySource;

scripting::DisconnectSource(pDisplayCell);
scripting::DisconnectStream(pDisplayCell);
delete pDisplayCell;
delete pDisplaySource;

// Deallocate system
fsf::FreeSystem();
```

Running the program

The example implementation allows specification of the pulse rate on the command line (the default rate is 15 Hz). Figure 12.22 shows a screen shot of the display window.

12.3.3   Custom elements

If one of the goals of MFSM is to allow rapid design and development of applications from existing modules, one of its main strengths is to allow easy

Figure 12.22. Screen shot of the display window for eximg1.

specification and implementation of custom elements that will interoperate seamlessly with existing or third-party components.

The example developed in the previous section, `eximg1`, makes use of an instance of a cell type called `myspace::MyCellImg`, whose task is to look up the timestamp of each incoming active pulse, format it, and print the resulting string into a newly created image. In this section, the design and implementation of specialized SAI elements (nodes and cells) is illustrated on the customized elements of `eximg1`. First, the generic image node `image::CImage` implemented in the open source Image Module is described. Then, the design and implementation of the `myspace::MyCellImg` cell type, which makes use of the `image::CImage`, are detailed step by step.

A custom node type: `image::CImage`

One solution when designing the image node was to encapsulate an existing image structure. Unfortunately, each image processing library comes with its own image structure. Committing to a given library might prevent access to other libraries and prove restrictive in the long term. The image node defined in the Image Module provides a minimum representation to ensure its compatibility with existing image structures (in particular, that used in OpenCV). However, the image node does contain any field specific to particular image formats to ensure the widest compatibility. When needed, more specific image nodes may be derived from this base image node for leveraging specific library features. Because of inheritance properties, these specialized image nodes will be usable with all processes defined for the base image node.

Any node type specialization must be derived from the base node `fsf::CNode` or a derived node. A node type is characterized by an identification string (used to link it to its factory). A complete node type description includes a list of data members and member functions, and a short description of its semantics.

The image node is derived from the character buffer node defined in the FSF library. An image buffer is indeed a character buffer. The smallest set of parameters needed to make the character buffer usable as an image buffer are the image width and height, the number of channels, and the pixel depth. Since some libraries require data-line alignment for optimal performance, the actual aligned width (width step) must also be stored. A utility-protected member function is used to compute the aligned width.

```
class CImage : public fsf::CCharBuffer
{
protected:
  int m_nNbChannels; // Number of channels
  int m_nDepth; // Pixel depth IMAGE_DEPTH_*

  int m_nWidth; // Image width
  int m_nHeight; // Image height
  int m_nWidthStep; // Aligned width (in bytes)

  // utility protected member function
  int ComputeWidthStep(bool bNoAlign=false);
```

Any custom node class must implement a number of constructors: the default constructor, all the constructors defined in the the base node class (these must define default values for the local data members), additional constructors for specifying local data members' initial values, and the copy constructor. When necessary, the virtual destructor must also be overloaded.

```
public:
  // Default constructor
  CImage() : CCharBuffer(),
    m_nNbChannels(3), m_nDepth(IMAGE_DEPTH_8U),
    m_nWidth(0), m_nHeight(0), m_nWidthStep(0) {}

  // Constructors with default values for local data members
  CImage(fsf::CNode *pParent, DWORD dwTime=0)
    : CCharBuffer(pParent,dwTime),
      m_nNbChannels(3), m_nDepth(IMAGE_DEPTH_8U),
      m_nWidth(0), m_nHeight(0), m_nWidthStep(0) {}

  CImage(const string &strName,
         fsf::CNode *pParent=NULL, DWORD dwTime=0)
    : CCharBuffer(strName,pParent,dwTime),
      m_nNbChannels(3), m_nDepth(IMAGE_DEPTH_8U),
      m_nWidth(0), m_nHeight(0), m_nWidthStep(0) {}

  // Constructors with local data members initial values input
  CImage(int nWidth, int nHeight,
         int nNbChannels=3, int nDepth=IMAGE_DEPTH_8U,
```

```
        fsf::CNode *pParent=NULL, DWORD dwTime=0)
    : CCharBuffer(pParent,dwTime),
      m_nNbChannels(nNbChannels), m_nDepth(nDepth),
      m_nWidth(nWidth), m_nHeight(nHeight), m_nWidthStep(0) {}

  CImage(const string &strName, int nWidth, int nHeight,
        int nNbChannels=3, int nDepth=IMAGE_DEPTH_8U,
        fsf::CNode *pParent=NULL, DWORD dwTime=0)
    : CCharBuffer(strName,pParent,dwTime),
      m_nNbChannels(nNbChannels), m_nDepth(nDepth),
      m_nWidth(nWidth), m_nHeight(nHeight), m_nWidthStep(0) {}

  // Copy constructor
  CImage(const CImage&);
```

No destructor overload is needed here, since the destructor for the character buffer parent class takes care of deallocating the buffer if needed.

The custom node class must also overload a number of virtual functions that characterize the node:

- operator=: assignment operator.

- Clone: returns a copy of the node. This virtual function is necessary for runtime polymorphism. It allows allocation of an instance of a specialized node class without knowledge of the specific type at compile time.

- GetTypeID: returns the factory mapping key.

```
  // Assignment operator
  CImage& operator=(const CImage&);

  // Cloning: necessary for runtime polymorphism
  virtual fsf::CNode *Clone() { return new CImage(*this); }

  // Factory mapping key
  virtual void GetTypeID(string &str)
  { str.assign("IMAGE_IMAGE"); }
```

A set of member functions provides basic access to local data members (set and get operations). A memory allocation function and high-level

parameter and image data (buffer content) copy functions complete the set
of tools offered by the image node.

```
  void CopyParameters(const CImage&);
  void CopyImageData(const CImage&);

  // Image parameters setting
  void SetWidth(int nWidth) { m_nWidth=nWidth; }
  void SetHeight(int nHeight) { m_nHeight=nHeight; }
  void SetNbChannels(int nNbChannels) { m_nNbChannels=nNbChannels; }
  void SetPixelDepth(int nDepth) { m_nDepth=nDepth; }
  void SetWidthStep(int nWidthStep) { m_nWidthStep=nWidthStep; }

  // Image parameters access
  int Width() const { return m_nWidth; }
  int Height() const { return m_nHeight; }
  int NbChannels() const { return m_nNbChannels; }
  int PixelDepth() const { return m_nDepth; }
  int WidthStep() const { return m_nWidthStep; }

  // Memory allocation
  void Allocate(bool bNoAlign=false);
};
```

When an image node instance is created, its parameters must be set.
Constructors provide default values; set functions allow the values to be
explicitly changed. The corresponding buffer must then be allocated by a
call to the **Allocate** function. The image node instance can then be used
for processing.

Any module must define a **RegisterFactories** function that registers
its node and cell factories with the system. Following is the code for the
image::RegisterFactories function. Apart from the image node **image::
CImage**, the module also implements a number of cells that provide access
to its various data members. Their description can be found in the module
documentation, available online [12]. Since an example of cell factory registration is provided in the next section, the code for cell factory registration
has been omitted below.

```
void image::RegisterFactories()
{
  using namespace fsf;
  using namespace image;

  if(g_pSystem==NULL) return;

  // Node factories
  g_pSystem->RegisterNodeFactory(std::string("IMAGE_IMAGE"),
    new CNodeFactory<CImage>(std::string("Image node"),
    strAlex("Alexandre R.J. Francois")));

  // Cell factories
  ...
}
```

A custom cell type: `myspace::MyCellImg`

The MyCellImg cell type was introduced in Section 12.3.2, when it was used in the design of **eximg1**. Table 12.3 presents its logical definition.

Any cell type specialization must be derived from the base cell `fsf::CCell` or a derived cell. A cell type is characterized by an identification string (used to link it to its factory). A complete cell type description includes the active and passive filters, the process output, a list of data members and member functions, and a short description of the process.

Any custom cell must implement the default constructor and overload a number of virtual functions that characterize the cell:

- GetTypeID: returns the factory mapping key.

- Process: the Process function is the only one requiring significant coding, as it is the place to specialize the behavior of the cell.

- When the function is called, the binding has already succeeded, and it is executed in a separate thread.

- For the cell to be useful, the process function must be described carefully. In particular, the way the input is processed and any output generated should be carefully documented.

The following code is the declaration for the corresponding class `myspace::CMyCellImg`, derived from `fsf::CCell`.

```
class CMyCellImg : public fsf::CCell
{
public:
  CMyCellImg();

  // factory mapping key
  virtual void GetTypeID(std::string &str)
  { str.assign("MY_CELL_IMG"); }

  // specialized processing function
  virtual void Process(fsf::CPassiveHandle *pPassiveHandle,
    fsf::CActiveHandle *pActiveHandle,
    fsf::CActivePulse *pActivePulse);
};
```

The constructor sets the default output name base and instantiates both passive and active filters from the corresponding template classes.

```
CMyCellImg::CMyCellImg()
: CCell()
{
  // default output name
  m_strOutputName.assign("Image");
  // set the filters
  m_pPassiveFilter =
    new fsf::CPassiveFilter<fsf::CPassivePulse>(
      std::string("Root"));
  m_pActiveFilter =
    new fsf::CActiveFilter<fsf::CActivePulse>(
      std::string("Root"));
}
```

The specific behavior of a cell type is encoded in its overloaded process function. When the process function is executed, filtering of passive and active streams has succeeded. The active and passive handles are thus bound to the nodes satisfying the filters. When the filters are complex (i.e., hierarchies of filters), the passive and active handles point to their respective roots.

```
void CMyCellImg::Process(fsf::CPassiveHandle *pPassiveHandle,
                         fsf::CActiveHandle *pActiveHandle,
 fsf::CActivePulse *pActivePulse)
{
```

First, a pointer to the target node in the active pulse (in this case, the root) is retrieved from the active handle. In this simple example, the process does not use any persistent data: the passive filter is defined such that the passive handle points to the root of the passive pulse. There is no need to get an explicit pointer to this node.

```
  fsf::CNode *pNode =
    static_cast<fsf::CNode*>(pActiveHandle->GetNode());
```

The node timestamp is retrieved using the `fsf::CNode::GetTime` function. In this implementation, the timestamp is expressed in milliseconds. The equivalent time in hour/minute/second format is computed as follows:

```
  long h=(pNode->GetTime()/3600000);
  long m=((pNode->GetTime()-h*3600000)/60000);
  long s=(pNode->GetTime()-h*3600000-m*60000)/1000;
```

The data to be output has been computed and must now be placed into an image. The output image, of type `image::CImage`, is created by a call to one of the constructors. The constructor sets the parent pointer in the newly created node instance but does not place the node in the parent's list of subnodes. This is done after all computations on the node have been completed. This ensures that eventual concurrent filtering processes not take into account nodes that are not ready for use outside of the context in which they were created. Note that the default output base name defined in `fsf::CCell` is a string object and requires locking (using the associated CriticalSection object) to avoid problems during concurrent access. After a call to the image node constructor, the image buffer is allocated and initialized with the value 0.

```
  // Create image
  m_csOutputName.Lock();
  image::CImage *pImage =
    new image::CImage(m_strOutputName,320,240,
      3,image::IMAGE_DEPTH_8U,
```

```
    pNode,fsf::g_pSystem->GetTime());
m_csOutputName.Unlock();

// Allocate buffer
pImage->Allocate();
// Fill in with 0
memset(pImage->Data(),0,pImage->Size());
```

Once the image node instance is ready, it must be made available to the OpenCV functions, through an appropriate header. The code below creates an OpenCV image header corresponding to the image node parameters and links the header to the actual image buffer.

```
// Use OpenCV

CvSize size;
size.width=pImage->Width();
size.height=pImage->Height();

// Create IPL image header
IplImage *pIplImage =
    cvCreateImageHeader(size,IPL_DEPTH_8U,pImage->NbChannels());
// Link image data
pIplImage->imageData=pImage->Data();
pIplImage->origin=IPL_ORIGIN_BL;
```

The various results are placed into formatted character strings, which are printed in the image using the OpenCV `cvPutText` function.

```
CvFont font;
cvInitFont(&font,CV_FONT_VECTOR0,0.8,0.8,0.0,2.0);

char str[255];
sprintf(str,"Pulse time stamp:");
cvPutText(pIplImage,str,cvPoint(15,200),&font,
          CV_RGB(255,255,255));

sprintf(str,"%d ms",pNode->GetTime());
cvPutText(pIplImage,str,cvPoint(15,150),&font,
          CV_RGB(255,255,255));
```

```
sprintf(str,"%d h %2d min %2d s",h,m,s);
cvPutText(pIplImage,str,cvPoint(15,100),&font,
          CV_RGB(255,255,255));
```

When done with using OpenCV, the image header can be deleted. Note that only the header is deleted, as the image node (including its image buffer) is the product of the process function.

```
cvReleaseImageHeader(&pIplImage);
```

Finally, all computations on the image node being completed, the node can be registered in its parent's list of subnodes.

```
pNode->AddComponent(pImage);
}
```

At this point, the newly created image node is part of the active pulse. When the process function returns, a number of operations are then carried by the cell, resulting in the transmission of the augmented active pulse to all downstream cells.

In order for the cell to be usable in the system, a corresponding cell factory must be registered. Following is the code for the `myspace::Register Factories` function.

```
void myspace::RegisterFactories()
{
  using namespace fsf;
  using namespace myspace;

  g_pSystem->RegisterCellFactory(std::string("MY_CELL_IMG"),
    new CCellFactory<CMyCellImg>(
      std::string("My cell: image"),
      std::string("Alexandre R.J. Francois")));
}
```

12.3.4 A shared memory access example

This section describes a slightly more complex example, `eximg2`, an image version of the user guide example `ex2`, which computes and displays the pulse frequency on the stream. The application graph (see Figure 12.23)

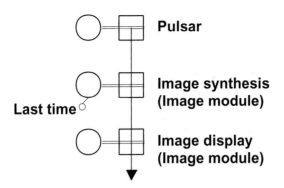

Figure 12.23. Conceptual graph for slightly more complex image manipulation example.

is very similar to that of `eximg1`. They only differ by their image synthesis unit, which is now built around an instance of a cell type called `myspace::MyCellImg2`.

Computing a pulse frequency requires computing the time delay between two consecutive pulses. Some data must therefore be shared between the threads processing each pulse: the timestamp of the last pulse processed is saved in a node on the passive pulse. It serves as the reference time from which to compute the time delay when the next pulse is processed. The following code instantiates and connects the elements of the image synthesis unit for `eximg2`. Thanks to the modularity of SAI, this image synthesis unit directly replaces that of the previous example without any modification to the rest of the application graph.

```
///////////////////////////
// Image synthesis unit //
///////////////////////////

// create the source
fsf::CSource *pMySource=new fsf::CSource;
bSuccess &= (pMySource!=NULL);

// last time stamp
fsf::Int32Node *pLastTime =
  static_cast<fsf::Int32Node*>(scripting::CreateNode(
    std::string("FSF_INT32_NODE"),pMySource->GetPulse()));
```

```
bSuccess &= (pLastTime!=NULL);
if(bSuccess)
{
  // set name
  pLastTime->SetName(std::string("Last time"));
}

// cell
fsf::CCell *pMyCell =
  static_cast<fsf::CCell*>(scripting::CreateCell(
    std::string("MY_CELL_IMG2")));
bSuccess &= (pMyCell!=NULL);
if(bSuccess)
{
  // connect with source
  scripting::ConnectSource(pMyCell,pMySource);

  // connect with Pcell
  scripting::ConnectUpstreamCell(pMyCell,pPcell);
}
```

The `myspace::MyCellImg2` cell type is in appearance quite similar to `myspace::MyCellImg`. Table 12.5 shows its logical definition.

Table 12.5. `myspace::MyCellImg2` logical cell definition.

myspace::CMyCellImg2 (fsf::CCell)	MY_CELL_IMG2
Active filter	[FSF_ACTIVE_PULSE "Root"]
Passive filter	[FSF_INT32_NODE "Last time"]
Output	(IMAGE_IMAGE *Image*)

The difference in the passive filter is reflected in the constructor code, as follows.

```
CMyCellImg2::CMyCellImg2()
: CCell()
{
  // default output name
```

```
  m_strOutputName.assign("Image");
  // set the filters
  m_pPassiveFilter =
    new fsf::CPassiveFilter<fsf::Int32Node>(
      std::string("Last time"));
  m_pActiveFilter =
    new fsf::CActiveFilter<fsf::CActivePulse>(
      std::string("Root"));
}
```

The overloaded process function encodes the specific behavior of the `myspace::MyCellImg2` cell type.

```
void CMyCellImg2::Process(fsf::CPassiveHandle *pPassiveHandle,
                          fsf::CActiveHandle *pActiveHandle,
                          fsf::CActivePulse *pActivePulse)
{
```

First, a pointer to the target node in the passive pulse is retreived from the passive handle. In this case, it is a node of type fsf::Int32Node, which contains the value of the last timestamp read. Initialization of this value is handled in the reset case described below.

```
  fsf::Int32Node *pLastTime =
    static_cast<fsf::Int32Node*>(pPassiveHandle->GetNode());
```

A pointer to the target node in the active pulse, in this case the root of the pulse, is also retrieved from the active handle.

```
  fsf::CNode *pNode =
    static_cast<fsf::CNode*>(pActiveHandle->GetNode());
```

The first time this process function is executed, the passive node containing the last timestamp value must be set to the value of the timestamp of the current active pulse, and no significant time difference can be produced. This section is executed each time the `m_bReset` flag, data member of the `fsf::CCell` class, is set to *true*.

```
  if(m_bReset)
  {
    pLastTime->SetData(pNode->GetTime());
    m_bReset=false;
  }
```

When the reset flag is not set, the regular processing is carried as follows.

```
else
{
```

First and foremost, the value of the last timestamp is collected from the passive node. The value in the node is updated immediately with the current timestamp value so that it is available as soon as possible for eventual parallel processes. In a strict parallel implementation, this read/write sequence should be placed in a critical section. In this simple example, the possible side effect is neglected. It can be made apparent by increasing the requested throughput until it becomes too high for the system on which the application is running. The throughput computed by this cell then becomes an integer fraction of the actual throughput because the reading and the updating sequence of the timestamp values is no longer in effect atomic.

```
DWORD dwLastTime=pLastTime->GetData();
pLastTime->SetData(static_cast<long>(pNode->GetTime()));
```

The remainder of the function code is quite similar to that of `CMyCell Img::Process`. The output image is created, and its data buffer is allocated and set to 0.

```
m_csOutputName.Lock();
image::CImage *pImage =
  new image::CImage(m_strOutputName,320,240,
                    3,image::IMAGE_DEPTH_8U,
                    pNode,fsf::g_pSystem->GetTime());
m_csOutputName.Unlock();

pImage->Allocate();
memset(pImage->Data(),0,pImage->Size());
```

A corresponding OpenCV header is created and linked to the image data buffer.

```
// Use OpenCV

CvSize size;
size.width=pImage->Width();
```

```
size.height=pImage->Height();

// Create IPL image header
IplImage *pIplImage =
  cvCreateImageHeader(size,IPL_DEPTH_8U,pImage->NbChannels());
// Link image data
pIplImage->imageData=pImage->Data();
pIplImage->origin=IPL_ORIGIN_BL;
```

Computed values are placed in a string, which is then written in the image buffer using the OpenCV cvPutText function.

```
CvFont font;
cvInitFont(&font,CV_FONT_VECTOR0,0.8,0.8,0.0,2.0);

char str[255];
sprintf(str,"Pulse delay:");
cvPutText(pIplImage,str,cvPoint(15,200),&font,
          CV_RGB(255,255,255));

sprintf(str,"%d ms",pNode->GetTime()-dwLastTime);
cvPutText(pIplImage,str,cvPoint(15,150),&font,
          CV_RGB(255,255,255));

sprintf(str,"%.2f Hz",1000.0f/(pNode->GetTime()-dwLastTime));
cvPutText(pIplImage,str,cvPoint(15,100),&font,
          CV_RGB(255,255,255));
```

When all OpenCV-related operations are completed, the image header can be deleted.

```
cvReleaseImageHeader(&pIplImage);
```

Finally, the output image node is added to its parent's list of subnodes.

```
    pNode->AddComponent(pImage);
  }
}
```

The example implementation allows specification of the pulse rate on the command line (the default rate is 15 Hz). Figure 12.24 show a screen shot of the image display window.

Figure 12.24. Screen shot of the image display window for `eximg2`.

12.4   Conclusion

12.4.1   Summary

The first part of this chapter introduced SAI, a software architecture model for designing, analyzing, and implementing applications performing distributed, asynchronous, parallel processing of generic data streams. The goal of SAI is to provide a universal framework for the distributed implementation of algorithms and their easy integration into complex systems that exhibit desirable software engineering qualities such as efficiency, scalability, extensibility, reusability, and interoperability.

SAI specifies a new architectural style (components, connectors, and constraints). The underlying extensible data model and hybrid (shared repository and message passing) distributed, asynchronous, parallel processing model allow natural and efficient manipulation of generic data streams, using existing libraries or native code alike. The modularity of the style facilitates distributed code development, testing, and reuse, as well as fast system design and integration, maintenance, and evolution. A graph-based notation for architectural designs allows intuitive system representation at the conceptual and logical levels, while at the same time mapping closely to processes.

Architectural patterns were illustrated through a number of computer vision-related demonstration projects ranging from single-stream, automatic, real-time video processing to fully integrated, distributed interactive systems mixing live video, graphics, sound, and so on. By design, the SAI style preserves desirable properties identified in the classic Pipes-and-Filters model. These include modularity and natural support for parallelism. Unlike the Pipes-and-Filters model, the SAI style allows optimal (theoretical) system latency and throughput to be achieved, and provides a unified framework for consistent representation and efficient implementation of fundamental processing patterns such as feedback loops and incremental processing along the time dimension.

The second part of the chapter was a code-level tutorial on MFSM, an architectural middleware implementing the SAI style. Its core element, the FSF library, is a set of extensible classes that can be specialized to define new data structures and processes or encapsulate existing ones (e.g., from libraries). MFSM is an open source project, released under the GNU Lesser General Public License. A number of software modules regroup specializations implementing specific algorithms or functionalities. They constitute a constantly growing base of open source, reusable code maintained as part of the MFSM project. The project is also thoroughly documented and comprises a user guide, a reference guide, and various tutorials.

Simple example applications illustrated the design and implementation of image-stream-manipulation applications using the SAI style and the MFSM middleware. In particular, the implementation of a generic image node (object of an open source module) and its use in conjunction with the OpenCV library in specialized cells were described step by step.

12.4.2   Perspectives

The SAI architectural style exhibit properties that make it relevant to research, educational, and even industrial projects.

Thanks to its modularity, it can accommodate today's requirements while preparing for tomorrow's applications. Using the SAI style for research can not only save time by avoiding the need to redevelop existing modules, it can also reduce the technology transfer time once the research has matured.

SAI also allows distributed development of functional modules and their seamless integration into complex systems. The modularity of the design also allows gradual development, facilitating continuing validation and naturally supporting regular delivery of incremental system prototypes. A number of crossdisciplinary research projects, in addition to those described in Section 12.2.3, are already leveraging these properties. They are producing real-time, interactive systems spanning a range of research domains.

Using the SAI style for developments in research projects can also reduce the technology transfer time once the technology has matured. From an industry point of view, the SAI style allows fast prototyping for proof-of-concept demonstrations.

For education, SAI allows classroom projects to be efficiently related to the realities of the research laboratory and of industrial software development. A project-oriented Integrated Media Systems class based on SAI and its applications in the distributed development model, was taught at USC in the Fall 2002 semester at the advanced graduate level. Rather than having small teams of students develop a same, simple, complete project, the small teams first implemented complementary parts of an overall, much more complex project (distributed soccer game). After six weeks of distributed development, sanctioned by careful incremental cross-testing, all the pieces were put together to produce a playable system. The course was an outstanding success, and regular graduate and undergraduate courses based on this model are in the planning.

Beyond those highlighted in this chapter, the SAI style has several important desirable architectural properties that make it a promising framework for many applications in various fields. These properties include natural support for dynamic system evolution, runtime reconfigurability, self-monitoring, and so on. Application of SAI in various contexts is currently being explored (e.g., interactive immersive systems). Short-term technical developments for SAI include development of a graphical user interface for system architecture design. Architecture validation and monitoring and analysis tools will be gradually integrated.

Finally, from a theoretical point of view, because of the explicit distinction between volatile and persistent data, SAI is a unified computational model that bridges the conceptual and practical gap between signal (data stream) processing on the one hand and the computation of higher level data (persistent, dynamic structures) on the other hand. As such, it might prove to be a powerful tool not only for implementing but also for modeling and reasoning about problems spanning both aspects. An example is computer vision.

12.5 Acknowledgments

The following (then) graduate students were involved in some of the projects presented in Section 12.2.3: Dr. Elaine Kang (Video Painting, Virtual Mirror), Maurya Shah (IMSC Communicator–networking and compression), Kevin Zhu (IMSC Communicator–video and sound capture and synchronized rendering).

This work has been funded in part by the Integrated Media Systems Center, a National Science Foundation Engineering Research Center, Cooperative Agreement No. EEC-9529152. Any opinions, findings, and conclusions or recommendations expressed in this material are those of the author(s) and do not necessarily reflect those of the National Science Foundation.

This work was supported in part by the Advanced Research and Development Activity of the U.S. Government under contract No. MDA-908-00-C-0036.

Bibliography

[1] Gnu Lesser General Public License [Online]. Available: http://www.gnu.org/copyleft/lesser.html.

[2] Intel Open Source Computer Vision Library [Online]. Available: http://www.intel.com/research/mrl/research/opencv/.

[3] Visual surveillance resources [Online]. Available: http://visualsurveillance.org.

[4] X3D: Extensible 3D international draft standards, ISO/IEC FCD 19775:200x [Online]. Available: http://www.web3d.org/technicalinfo/specifications/ISO_IEC_19775/index.

[5] VRML 97: The Virtual Reality Modeling Language, ISO/IEC 14772:1997, [Online]. Available: http://www.web3d.org/technicalinfo/specifications/ISO_IEC_14772-All/index.html.

[6] *PETS01: Second IEEE International Workshop on Performance Evaluations of Tracking and Surveillance*, December 2001 [Online]. Available: http://pets2001.visualsurveillance.org.

[7] *PETS02: Third IEEE International Workshop on Performance Evaluations of Tracking and Surveillance*, June 2002 [Online]. Available: http://pets2002.visualsurveillance.org.

[8] *PETS-ICVS: Fourth IEEE International Workshop on Performance Evaluations of Tracking and Surveillance*, March 2003 [Online]. Available: http://petsicvs.visualsurveillance.org.

[9] *PETS-VS: Joint IEEE International Workshop on Visual Surveillance and Performance Evaluations of Tracking and Surveillance*, October 2003 [Online]. Available: http://vspets.visualsurveillance.org.

[10] G. R. Andrews. *Foundations of multithreaded, parallel and distributed programming*. Addison Wesley, 2000.

[11] K. W. Arthur, K. S. Booth, and C. Ware. Evaluating 3D task performance for fish tank virtual worlds. *ACM Transactions on Information Systems*, 11(3): 239–265, 1993.

[12] A. R. J. François. Modular flow scheduling middleware [Online]. Available: http://mfsm.sourceForge.net.

[13] A. R. J. François. *Semantic, Interactive Manipulations of Visual Data*. PhD thesis, Dept. of Computer Science, University of Southern California, 2000.

[14] A. R. J. François. Components for immersion. In *Proc. of the IEEE International Conference on Multimedia and Expo*, August 2002.

[15] A. R. J. François and E. Kang. A handheld mirror simulation. In *Proc. of the IEEE International Conference on Multimedia and Expo*, vol. II, pages 745–748, July 2003.

[16] A. R. J. François, E. Kang, and U. Malesci. A handheld virtual mirror. In *ACM SIGGRAPH Conference Abstracts and Applications proceedings*, page 140, July 2002.

[17] A. R. J. François and G. G. Medioni. A modular middleware flow scheduling framework. In *Proc. of ACM Multimedia 2000*, pages 371–374, November 2000.

[18] A. R. J. François and G. G. Medioni. A modular software architecture for real-time video processing. In *Proc. IEEE International Workshop on Computer Vision Systems*, pages 35–49, July 2001.

[19] C. Jaynes, S. Webb, R. M. Steele, and Xiong Q. An open development environment for evaluation of video surveillance systems. In *PETS02*, pages 32–39, June 2002.

[20] M. B. Jones and J. Regehr. The problems you're having may not be the problems you think you're having: results from a latency study of Windows NT. In *Proc. of the 7th Workshop on Hot Topics in Operating Systems*, 1999.

[21] E. Y. Kang, I. Cohen, and G. G. Medioni. A graph-based global registration for 2D mosaics. In *ICPR00*, vol. I, pages 257–260, 2000.

[22] M. Lazzari, A. François, M. L. McLaughlin, J. Jaskowiak, W. L. Wong, M. Akbarian, W. Peng, and W. Zhu. Using haptics and a "virtual mirror" to exhibit museum objects with reflective surfaces. In *Proc. of the 11th International Conference on Advanced Robotics*, July 2003.

[23] I. S. McKenzie and C. Ware. Lag as a determinant of human performance in interactive systems. In *Proc. of the ACM Conference on Human Factors in Computing Systems—INTERCHI*, pages 488–493, 1993.

[24] Microsoft. DirectX [Online]. Available: http://www.microsoft.com/directx.

[25] M. F. X. J. Oberhumer. LZO [Online]. Available: http://www.oberhumer.com/opensource/lzo.

[26] M. Shaw and D. Garlan. *Software Architecture—Perspectives on an Emerging Discipline*. Prentice Hall, 1996.

[27] D. Terzopoulos and C. M. Brown (Eds.). *Real-Time Computer Vision*. Cambridge University Press, 1995.

[28] K. Toyama, J. Krumm, B. Brumitt, and B. Meyers. Wallflower: Principles and practice of background maintenance. In *Proc. of the International Conference on Computer Vision*, pages 255–261, 1999.

INDEX

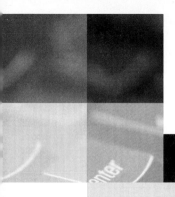

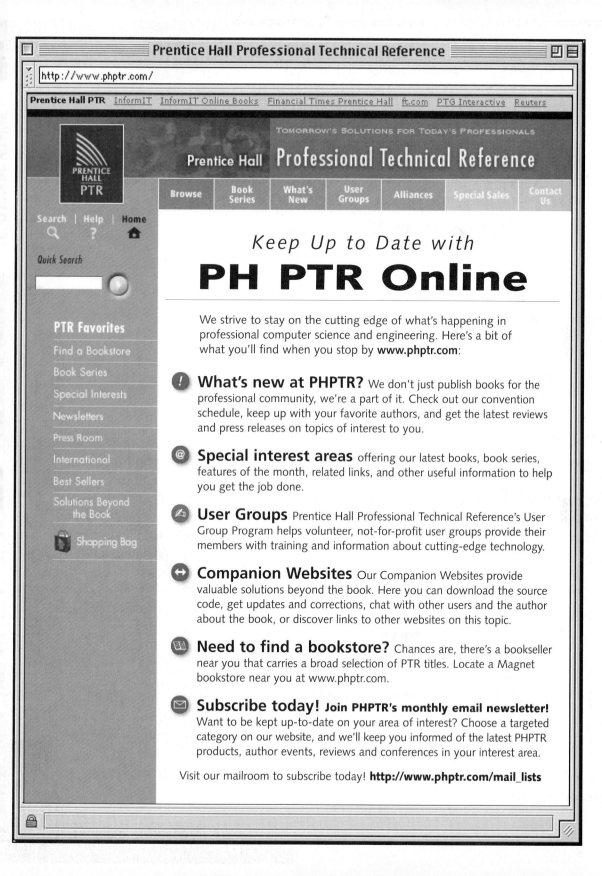

DVD Warranty

Note: The included DVDs are for the PC Only.

Pearson Technology Group warrants the enclosed DVDs to be free of defects in materials and faulty workmanship under normal use for a period of ninety days after purchase (when purchased new). If a defect is discovered in the DVDs during this warranty period, a replacement DVDs can be obtained at no charge by sending the defective DVDs, postage prepaid, with proof of purchase to:

Disc Exchange
Addison-Wesley Professional/Prentice Hall PTR
Pearson Technology Group
75 Arlington Street, Suite 300
Boston, MA 02116
Email: AWPro@aw.com

Pearson Technology Group makes no warranty or representation, either expressed or implied, with respect to this software, its quality, performance, merchantability, or fitness for a particular purpose. In no event will Pearson Technology Group, its distributors, or dealers be liable for direct, indirect, special, incidental, or consequential damages arising out of the use or inability to use the software. The exclusion of implied warranties is not permitted in some states. Therefore, the above exclusion may not apply to you. This warranty provides you with specific legal rights. There may be other rights that you may have that vary from state to state. The contents of this DVDs are intended for personal use only.